Shengwei Mei
Xuemin Zhang
Ming Cao

Power Grid Complexity

Shengwei Mei
Xuemin Zhang
Ming Cao

Power Grid Complexity

With 218 figures

TSINGHUA UNIVERSITY PRESS

Springer

Authors
Shengwei Mei
Department of Electrical Engineerging
Tsinghua University
Beijing 100084, China
E-mail: meishengwei@tsinghua.edu.cn

Xuemin Zhang
Department of Electrical Engineerging
Tsinghua University
Beijing 100084, China
E-mail: zhangxuemin@tsinghua.edu.cn

Ming Cao
Faculty of Mathematics and Natural Sciences
University of Groningen
9747 AG Groningen, the Netherlands
E-mail: m.cao@rug.nl

ISBN 978-7-302-23280-3
Tsinghua University Press, Beijing

ISBN 978-3-642-16210-7 e-ISBN 978-3-642-16211-4
Springer Heidelberg Dordrecht London New York

Library of Congress Control Number: 2010935358

Printed on acid-free paper

Springer is part of Springer Science+Business Media (www.springer.com)

Preface

Power grids are the man-made systems that have the most extensive coverage and the most complex structures in the world. As a pillar of the national economy, the safe operation of power grids is closely related to people's lives in a country. However, in recent years, the frequent large-scale blackouts around the world have shown the vulnerability of interconnected power systems. Scientists and engineers have worked together to identify the reasons for major large-scale blackouts, yet in many cases convincing analytical results are still missing, and electrical accidents occur continuously.

In 1996, after two large-scale blackouts took place in succession in the power grids of the western US, the North American Electric Reliability Corporation (NERC) set up a new group working on the dynamics of interconnected power systems, and in particular studying topics of the system stability, voltage, reactive power, control and protection. The goal was to establish a set of standard procedures and principles to ensure proper planning and reliable operation of interconnected power systems. However, seven years later, on August 14, 2003, there occurred a blackout in US and Canada that shocked the world. Compared to the two blackouts in 1996, the losses of the 8·14 blackout in US and Canada were much higher. The reason was rooted in the lack of an effective systematic study on the characteristics of power grid dynamics and the mechanism of large-scale blackouts. Moreover, there was a lack of theoretical tools to explain a number of complex phenomena, such as the evolution dynamics of power grids and the development mechanism of cascading failures. These are the difficulties in the power grid security analysis, and are also the fundamental research topics in the study of power grids worldwide. These issues must be solved to raise the level of security and reliability of power grids.

Due to the complexity of the structures and operation modes of power grids, researchers have not obtained sufficient knowledge about the nature of blackouts in large-scale interconnected power systems which are caused by cascading failures. The existing theories and computational and experimental methods are limited in the following two aspects. First, since all the departments of power systems function separately managing the systems' planning, construction and operation, the problems about the long-term evolution of the systems are divided into sub-problems under unnecessary simplification and unreasonable assumptions. Second, the current $N-1$ and $N-2$ calibration standards assume that cascading failures are small probability events, and the impact of these incidents on system blackouts can also be ignored because of the small probabilities. However, it is evident now that these overlooked cascading failures can trigger blackouts of various scales. Statistical data have shown that the probability distribution of blackouts does not satisfy the normal distribution but the power-law distribution. In other words, the risk of large-scale blackouts cannot be ignored. These facts indicate that it is necessary to establish appropriate models to study blackouts, and thus guide the planning and operation of power systems based on these models.

In recent years, researchers have applied complex system theory and the related tools in computer science, ecology, sociology, economics and other related disciplines to the study of complex power grids. While encouraging progress has been made in this direction, the existing research can only provide preliminary results on the self-organized criticality (SOC) in large-scale power grids from a statistical point of view. It is necessary to develop a systematic theoretical framework to discuss the complex characteristics of power systems, the stability of interconnected power grids, and the general methodologies of SOC and complex networks and systems.

The aim of this book is to study the occurrence and development of cascading failures in large-scale interconnected power grids using the theories and methods of SOC and complex networks. We study the spatial and temporal evolution dynamics of power grids, discuss the risk assessment indices for power grid catastrophes, and propose control strategies to prevent disastrous events.

This book consists of 14 chapters. A brief description of each chapter is given below.

Chapter 1 gives an overview of power system blackouts and SOC using tools from complexity science. It gives a brief introduction about the research progress in the field of SOC in power systems. It also provides the background information and describes the main structure of the book.

Chapter 2 is devoted to the basic concepts and theories in mathematics and statistical physics which are related to SOC. It further discusses the power-law distribution of complex systems and gives a preliminary mathematical explanation. In the end, the basic concepts of complex networks are introduced.

Chapter 3 introduces the basic concepts of organization and self-organization and then discusses the general principles of SOC based on cybernetics. In addition, the mathematical model disclosing the power system evolution mechanisms is constructed, and the SOC phenomena in large-scale power grids are discussed. We carry out further analysis using statistical data of the blackouts in the Chinese power systems. Finally, two categories of risk evaluation indices, VaR and CVaR, are discussed.

Chapter 4 describes the slow dynamics of power grids. A small-world power grids evolution model is constructed using the studied evolution mechanism.

Chapter 5 analyzes the characteristics of power flows and synchronization in small-world power grids. Then the structure vulnerability is studied.

Chapter 6 proposes a new decomposition and coordination strategy for static power girds using community structure theory.

Chapter 7 constructs a vulnerability evaluation method for transmission lines in static power grids using vulnerability theory. It then proposes an integrated vulnerability index and the corresponding evaluation method to describe more comprehensively the influence of the line fault on the reliability of static power grids.

Chapter 8 studies the dynamic performances of power systems. It introduces the simplification and equivalence method for dynamic power systems. It also discusses the synchronization problem for complex power grids.

In Chapter 9, blackout models based on the DC power flows are introduced. We classify them into two categories of the OPA and the improved OPA models. Furthermore, SOC of power systems is studied at both the macroscopic and microscopic levels.

Chapter 10 constructs a blackout model based on the AC optimal power flows. This model is composed of inner and outer loops. The inner loop simulates the fast dynamics of power systems whose criticality can be analyzed through the AC optimal power flows. The outer loop simulates the slow dynamics of power systems at a macroscopic level. It reveals the system level SOC in power grids together with its evolution mechanism.

Chapter 11 discusses a blackout model that takes into account the characteristics of reactive power and voltage. The emphasis here is on the voltage stability and voltage collapse problems. SOC in power grids is examined through the angle of reactive power and voltage.

In Chapter 12, based on optimal power flows with transient stability constraints, a blackout model is constructed, which covers the characteristics of power system dynamics on the transient, fast and slow time scales. The model is closer to the physical processes for the evolution of cascading failures, and it can also incorporate preventive and emergency control strategies in occasions of failure. Thus, it becomes possible to simulate the impact of preferred control strategies on blackouts.

Chapter 13 explains the application of SOC theory in the generation expansion planning and power network planning while Chapter 14 shows the application of SOC theory in modeling cascading failures and demonstrates the corresponding emergency management platform for complex power grids.

This book is both “comprehensive” and “innovative”. It is comprehensive in the sense that it makes use of tools from probability theory, stochastic processes, statistical physics and other fundamental theories systematically to review and summarize the general framework of SOC in complex systems concerning security and stability. The framework is then applied to a series of problems for the security and stability of large-scale interconnected power systems. It is innovative since it not only endeavors to present precisely the theoretical concepts, principles and methods with the corresponding explanations and proofs, but also seeks to introduce the latest results in the frontier of this active research field, especially those theoretical and experimental results obtained by the research team of the authors. In addition, the authors have tried to combine fundamental theories with engineering practice, and to make the texts concise and easy to understand. When discussing those mathematical concepts, such as the tools from probability theory and control theory, we have not confined ourselves to strict mathematical development. Instead, we keep the discussion at a level that is appropriate for readers who are mainly from the engineering background and have knowledge of engineering mathematics. Researchers from the field of complex systems may also find this book interesting for examples in industrial applications.

This book is not just a survey on SOC theory of complex systems and complex network theory. It summarizes the latest important progresses in the research of power grid security and stability. It can be used to develop new ideas for constructing general complexity theory for complex power grids, and provide useful guidance for the control and management of large-scale power systems. In short, we expect that this book will have impact as an important reference for academic research and engineering practice.

Several colleagues have helped us improve the quality of this book. We owe special thanks to Prof. Qiang Lu for his inspiration and constructive comments. He set the outline of the book and several results covered in this book have been obtained under his guidance. We also wish to express our sincere gratitude to Prof. Xiaoxin Zhou, Prof. Zhenxiang Han, Prof. Felix Wu and Prof. Yixin Ni, whose research results have been applied to the work in this book. The graduate students, Fei He, Xiangping Ni, Yingying Wang, Xiaofeng Weng, Mingfu Lu, Deming Xia, Gang Wang and Shengyu Wu, have carried out related researches and it was because of their strong support that the book has been completed. Finally, the first and the second authors would like to thank the National Natural Science Foundation of China (NSFC) for providing financial support under Grants 50525721 and 50595411 and the Special Fund of the National Basic Research Program of China for providing financial assistance under Grant 2004CB217902, and the third author would like to thank the support from the Dutch Organization for Scientific Research (NWO) and the Dutch Technology Foundation (STW).

Shengwei Mei and Xuemin Zhang, Beijing
Ming Cao, Groningen
October 2010

Contents

Chapter 1 Introduction

This chapter gives an overview about power system blackouts and self-organized criticality (SOC) using tools from complexity science. It also discusses briefly, based on the existing blackout models, how to apply SOC to reveal the mechanism for cascading failures in power systems. We summarize the related background information and the organization of the book in the end.

1.1 Large-Scale Power Grids as Complex Systems

Complexity science is a new discipline that aims to understand and analyze various complex dynamics and behaviors in real world. Although still young, it has already attracted considerable attention from researchers in different fields. It is rooted in classical disciplines of natural and social sciences, and focuses on the study of complexity in nature and society[1]. Modern power systems, also referred to as large-scale power grids, belong to a typical class of complex systems, whose complexity characteristics can be summarized as follows.

(1) Modern power systems are large-scale systems

Generally speaking, a modern power system consists of three subsystems. The first is the primary system, where the energy conversion, transmission, distribution and consumption take place. The second is the automatic control system, also called the secondary system, which is responsible for the secure, stable and economic operation of the power system. And the third is the power trading system. Taking the Northeast Power Grid of China in 2007 as an example. Its 500 kV and 220 kV primary system has more than 500 transformers, 700 transmission lines and 1000 generators, among which more than 300 are hydroelectric and thermal generators, and the rest are mainly wind-power generators. It is difficult to obtain the complete statistics for all the components in the whole system when taking the final 220 V distribution systems into account. The secondary system usually has more components than the primary system since in addition to the communication equipments, a large number of sensors are installed to monitor various states, such as voltage, current, and temperature, of the devices in the primary system.

In recent years, as the interconnection of regional power systems tries to optimize the use of hydroelectric, thermal and nuclear powers together with other energy resources, the power systems in China have been more and more coupled together. This trend of interconnection will be further strengthened after the completion of the projects that transmit power from the west to the east and exchange supplies between the south and the north. According to statistics, in 2004, there were 4,721 transmission lines at the 220 kV level with a total length of 168,511 km, 162 transmission lines at the 330 kV level with a total length of 16,152 km and 509 transmission lines at the 500 kV level with a total length of 63,561 km[2].

(2) Components in modern power systems are complex

Not only is the number of components in modern power systems extremely large, but the components are of different kinds as well. Generators and motors are the most common components that are used to realize the conversion between electrical and mechanical energies and their dynamics are governed by the electromagnetic induction law and Newton's second law. There are other equipments, such as coal mills, boilers, pressure pipes and reactors, that help to convert chemical, hydraulic and nuclear energies to mechanical energies. Since the 1990s, with the rapid development in electronic technologies, high-voltage direct current transmission (HVDC), thyristor controlled series compensation (TCSC) and static VAR compensator (SVC) have been utilized widely, and this makes the electric components in power systems more complicated. Different from traditional power equipments, their main functionality is to switch flexibly the power electronic devices to achieve the control of power transmission.

In particular, loads are the most complicated class of components in power systems. The modeling of dynamic loads has not been fully solved yet. Researchers usually classify loads into constant impedance loads, temperature controlled loads, compressor loads and other simplified models. In addition, various protections and controls are also important classes of complex components in power systems. For example, for a generator, there are usually excitation controllers, speed controllers, low-frequency protectors, under- and over-excitation limiters. There are also centralized control systems, such as automatic generation control (AGC) and automatic voltage control (AVC), located in dispatch centers to coordinate and control the overall performance of the whole system.

(3) Power supply and consumption need to be balanced instantaneously

Electricity is generated at power plants and then transmitted through transmission lines, transformers and switches to end-users. Electricity is transmitted with the speed of light, and so far it is yet impossible to achieve industrial-scale and high-capacity storage of electrical energy. Hence, one critical feature of power systems is that the production (generation) and consumption are completed almost simultaneously and thus the power supply and demand have to be balanced instantaneously. If the balance is broken, the system will lose its stability, which will lead to power outages of different scales.

(4) Various random factors affect the power grids' secure operation

Climate changes and human activities all force power systems to be constantly exposed to various random perturbations. For example, lightning or untrimmed trees may cause short-circuit faults on transmission lines; snow and earthquakes may destroy electricity poles and substations; in hot weather more people use air conditioners; major sport and political events may change the type of power loads. Such random factors increase the complexity in operating and managing power systems. In addition, as electricity markets are deregulated, energy management systems are more complicated. In combination with the random effects just described, the planning, operation and management of power grids are facing demanding challenges[3].

From the above four aspects, it is clear that modern power systems are large-scale complex systems. So naturally they become one subject in the complexity study. Although the modern power systems have complex dynamic characteristics, their robustness and reliability are still relatively high due to the implementation of various protections and controls. However, serious losses were caused in the few power failures in the past years. The regional interconnection brings both great social and economic benefits and growing risk for failures[4], especially cascading failures and blackouts. We list the large-scale blackouts since the 1960s in Table 1.1[5–16].

In order to prevent cascading failures and catastrophes, we have to carry out in-depth study of the catastrophic events in power systems. The traditional approach is to investigate the structural and operational states in the faulty grid to look for the causes of the accident. Analysis on the related natural environment, equipment defects, human operating errors, and relay mal-operations is carried out to identify the fault-specific strategies to improve the systems' reliability[17–21]. Such an approach has helped to realize reliable operations in the past. However, the frequent blackouts around the world since 2003 motivate us to reflect on the shortcomings in the traditional analysis and computational methods.

Table 1.1 List of the world's major power outages

Time	Location	Loss
November 4, 2006	Western Europe	8 countries in the western Europe were affected; the German industrial city Krohne was hit most; 15 regions (including Paris) in France suffered from a sudden power outage; the western European power grid was split; the load loss was 17 GW; nearly 5 million people were affected; and after 1.5 h the service was restored.
June 22, 2005	Switzerland	The national railway network was out of work and the power outage lasted for 4 h.

(Continued)

Time	Location	Loss
June 5, 2005	Moscow, Russia	The south, southwest and south-eastern parts of the city were out of power; nearly half of its urban area were paralyzed; 2 million people were affected; and the total loss was one billion US dollars.
January 18, 2005	Switzerland	A sudden power outage took place in its western region; Lausanne and Geneva were out of power; and the outage lasted for 1 h.
September 28, 2003	Italy	A 6400 MW power shortage eventually led to the collapse of the system frequency; the whole country was in blackout exception for the island of Sardinia; 57 million people were affected; the loss added up to 180 GWh; the service was restored after 20 h.
September 23, 2003	Sweden and Denmark	The lost power was 1800 MW; 5 million people were affected; it took 6.5 h to restore service.
September 1, 2003	Sydney, Australia	At least 50 buildings in the central business district were out of power and traffic jam took place in the neighborhood.
September 1, 2003	Malaysia	Five states in the north suffered from the power outage that lasted for about 4 h.
August 28, 2003	London, UK	The power loss was 724 MW; 500,000 passengers in the subway trains were stranded; the service was restored after 2 h.
August 14, 2003	East America and Canada	The power loss was 61.8 GW. More than 50 million people were affected, and the failure lasted up to 29 h. The loss was 30 billion US dollars.
January 21, 2001	Brazil	The power loss was 23,766 MW and the restoration time was 4 h.
July 29, 1999	Taiwan, China	The whole island was out of power and the loss amounted to 1 billion US dollars.
August 3, 1996	Malaysia	The whole country was in blackout and the total power loss was 5700 MW.
August 10, 1996	West America and Mexico	The system split into 4 islands, 7,500,000 people were affected, and the total power loss was 30,392 MW.
July 2, 1996	West America	The system split into 5 islands, 2,250,000 people in 15 states were affected, and the total power loss was 11,850 MW.
December 14, 1994	West America	The system split into 4 islands and 2 million people in 14 states were affected.
July 2, 1987	Tokyo, Japan	The failure lasted for 21 min, the subway system was out of order; the total load loss was 8168 MW, and more than 28 million users were affected.
January 12, 1987	West France	The power loss was 1500 MW.
December 19, 1978	France	Voltage collapsed. The total lost load was 29 GW, which was 75% of the total load. The outage lasted for 8.5 h and caused a loss of up to 300 million US dollars.
July 13, 1977	New York, US	The blackout lasted for 25 h, causing arson and looting, and the Wall Street area was out of order.
November 9, 1965	Northeast American	The blackout lasted for up to 13 h and 30 million people were affected.

In order to describe the power system dynamics, people usually start with individual electric components or sub-systems to build the system-level mathematical models in the form of high dimensional algebraic-differential equations. Such equations are then solved digitally through simulation. Although this approach is capable of generating trajectories of a system's dynamics, it is less effective for in-depth analysis of the mechanism of cascading failures and blackouts. In fact, because of the randomness, high-dimensional state-space and real-time events in power grids, the complicated dynamics in modern power systems cannot be modeled precisely using existing mathematical methods, and no analytical solution can be obtained with arbitrary precision for a given period of time. In the years to come, it will still be a great challenge to construct precise models to prevent blackouts and minimize the risk of failures. Therefore, it is in urgent need to develop new theories and tools for power network security and this motivates us to look into complexity science.

Although modern power grids can be seen as the most complex industrial systems, there are other natural and engineering systems that are of similar complexity. We can certainly learn from how physicists and mathematicians study complex systems. Physicists have been working on formulating general rules that govern complex systems that cannot be described simply by those classical results, such as Newton's Law, Maxwell's equations, relativity theory and quantum mechanics. It is widely believed that some unifying general rules can be utilized to explain various natural phenomena that are related to things varying from the universe to a quark. Obviously, we cannot start to examine all complex phenomena just by studying particles and the related physical laws, and we prefer instead to grasp the overall macroscopic system-level features. Self-organized criticality theory (SOC) is the complexity theory that is proposed to break new ground to look at complex phenomena[22–25]. The theory has attracted great attention and different SOC behaviors have been identified in natural, social and engineering systems[22].

Generally speaking, SOC implies scale invariance. It indicates the power-law distribution behind various complex phenomena, thus extreme events cannot be ignored and disasters of large and small scales are all possible to take place[22]. However, there has not been a generally accepted definition of SOC and people are still looking for necessary conditions for SOC. For this reason, new ideas and results are constantly being reported in this active field and researchers are busy with constructing new mathematical models and exploring their properties. Some generally accepted conclusions in this field are listed as follows. SOC phenomena take place in systems consisting of interacting sub-systems. Such systems are subject to both internal influences and external driving forces; the self-organized process requires that the impact of the external factors is less critical compared to the internal forces. The self-organized process drives the system towards its critical state, and such changes are similar to the phase transitions that take place when the system is at its equilibrium. Not all systems develop towards their critical states, especially when the external forces are weak. The external forces act on a slower time-scale than internal forces. The existence of thresholds is necessary but not sufficient for the occurrence of SOC. The system has several critical states, which reflect the system's local stability level. The evaluation of the distance between the system's current state and its critical state has been utilized to assess the stability margin for a given system.

As to power systems, one can take power grids as a class of typical complex systems, and then it is natural to use SOC theory to study the mechanism for disastrous events in power systems. We want to calculate secure operation margins and effectively prevent blackouts. Traditionally, the study on power grid disasters focuses on the start, development and occurrence of failures, and the collapse of the system is attributed to some isolated and specific causes, such as lightning, hail, device failure, overloaded lines, and abnormal operation conditions. However, such efforts are less effective for revealing the mechanism of catastrophes in power systems, and thus have not provided effective preventive control strategies to reduce the risk of blackouts. In fact, too much attention has been paid to individual factors and more effort needs to be made to study the collective dynamics in the whole systems. In other words, disastrous events emerge as a result of the combined influences from a variety of factors, and they are related directly to the overall system's properties and have to be analyzed following an integral approach.

In the past few years, when applying SOC theory to study power system blackouts, people have found that the statistical distributions of various blackouts all show the power-law tails[26−31]. This does not agree with the conclusions in the traditional power system study, where it is assumed that the blackout distributions are Gaussian when taking the blackout scale as the random variable[4]. From probability theory we know that, when taking Gaussian and power-law distributions as the limit distributions of a certain event series, Gaussian distributions usually imply that the events are independent of one another, and power-law distributions imply that the events are likely to be relevant. In other words, failures in power systems are not isolated accidents, and we should consider the overall dynamic characteristics governing the system for the prediction and prevention of disasters.

It is just one example of the effectiveness of SOC theory in revealing the power-law distribution from the massive data about power system blackouts. Furthermore, the view point has been transformed from the microscopic, event-specific and deterministic perspective to the macroscopic, general and stochastic one. This is, without exaggeration, a huge step in the study of power grid complexity. As a class of typical complex systems, power systems have the general characteristics of external driving forces, internal influences, interacting subsystems and critical thresholds. Since SOC theory emphasizes first the macroscopic examination of the overall system, we need to study the inherent mechanism for the changes of the system characteristics and the transition to SOC by investigating the influences from grid planning, maintenance and scheduling. The goal is to reveal the power-law relationships. In addition, we also carry out study on the microscopic level to consider the effects of generator dynamics, load dynamics and optimal power flows. To do so, we introduce later in this book a number of mathematical models to simulate power system cascading failures, analyze the macroscopic and microscopic SOC behaviors, and discuss indices to assess power grids' security levels.

Important as it is to understand fundamental physical and mathematical rules for complex power systems, we also want to modify and influence the systems so that they operate in a desirable fashion. So using SOC theory, it is hoped that new theories and methodologies can be developed for the secure operation of large-scale power systems. SOC theory provides us with a new set of tools that scales with the system of interest to identify and analyze dynamic characteristics both globally and locally. One distinguishing feature of the new framework is that the SOC methods are not constrained by the availability of detailed descriptions of the components. The focus is how complex behaviors emerge in power systems under the influence of Faraday's law, Maxwell equations, Ohm's law, and other basic physical principles and rules.

1.2 SOC Theory and Blackouts

The SOC study on complex systems was started in 1987, when researchers including Per Bak at the Brookhaven National Laboratory, Chao Tang at the University of California at Santa Barbara and Kurt Wiesenfeld at the Georgia Institute of Technology looked into a class of dissipative continuous systems consisting of thousands of components with short-range interactions. Such systems may evolve spontaneously into some critical steady state, called the SOC state, when certain conditions are satisfied[22]. In the phase space, a SOC state corresponds to a finite-dimensional attractor, which is neither stable nor unstable, but critically stable. As pointed out by Per Bak and others, when a system is in a critically stable state and is perturbed by disturbances, it evolves spontaneously to another critically stable state. Therefore, the system is always critically stable and does not arrive at an equilibrium, which implies that the system does not converge to a stable attractor. Generally speaking, the SOC state is a special type of steady state that is different from being unstable or stable. For the evolutions in complex systems, the completely stable or unstable states can hardly exist. If a system is completely stable, it then stays still and does not evolve any more. If on the other hand, a system is completely unstable, it is in transient states and thus its current state cannot be maintained in a dynamic process. Hence, only critically stable systems can both evolve and also maintain certain stability, and they are the systems that are most widely seen in the real world.

The first prototype in SOC theory is the sand pile model[22]. As shown in Fig. 1.1, the height of a sand pile increases when grains of sands are dropping randomly in a given area. When the slope of the pile at a certain location is above a threshold, the grains at that location slide down, and this in turn increases the slopes of the neighboring regions with more sand, which may further trigger an avalanche. If this does happen, some sand leaves the pile and falls off the boundaries of the given area. In this model, the amount of sand that is added to the pile equals in the statistical sense the amount of sand falling out of the pile and the slope is maintained at a critical value, which is called a critically stable state. The evolution of the sand pile model is determined by two simple rules. The first is that the difference in the slopes of neighboring locations must be upper bounded because of gravity and the second is that the sand at the locations where the slope is above the threshold determined by the first rule must slide down under the gravity force. It has been shown that in this model the avalanche scales satisfy the power-law tail distribution[22].

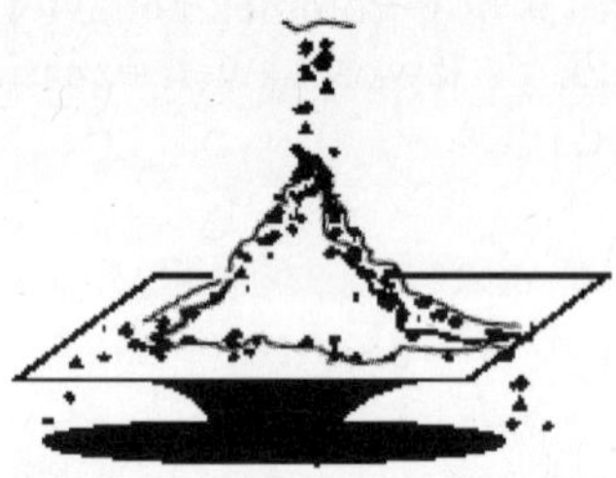

Figure 1.1 Sand pile model[22]

In [32], an analogy is made between the evolution of power systems to their critical states and the formation of sand piles. The relationships are listed in Table 1.2.

Table 1.2 Analogy between power systems and sand piles[32]

	Power system	Sand pile
System state	Network power flow distribution	Gradient profile
External driving force	Load increase	Addition of sand
Internal relaxing force	Blackout	Avalanche
Cascading failure	Line outage	Sand slide

As shown in Table 1.2, the state of a power system is the power flow distribution while that of a sand pile is the slope angle distribution; the cascading failures take forms in line outage for power systems and in sand slide for sand piles; the external driving force for power systems is the continuous load growth, and correspondingly that for sand piles is the continuous addition of sands; the internal release in power systems happens in blackouts, while that in sand piles takes place in sand collapse. When the load increases up to a certain level and the power system's internal and external forces satisfy certain conditions, the system may evolve into a critical state and blackouts of different scales may take place. Similarly, when the slope at some location in a sand pile is above a certain threshold, sand slide may occur, which may lead to avalanches of various scales. For a power system, it enters a critical state right before blackouts take place.

Furthermore, when a power system is in its critical state, any small perturbation may trigger unexpected events, which may result in cascading failures and disastrous transitions. This process is accompanied by the conversion and release of a large amount of energy. Hence, systems in critical states often show fractal structures (self-similarity) in space and flicker noise ($1/f$ noise[25]) in time. Its temporal scale invariance and long-range correlation present clear mathematical characteristics of SOC[23].

1.3 Validation of SOC Using Blackout Data

Recently Dobson, Carreras and their colleagues have analyzed the blackout data in US from 1984 to 1999 using tools from complex system theory. It has been shown that the relationship between the scales and frequencies of blackouts satisfies the power law distribution[32]. Let Q denote the blackout scale and $N(Q)$ the frequency of the blackouts of scale Q. Then

$$\ln N(Q) = a - b \ln Q \tag{1.1}$$

where a and b are constants. They have used three quantities to measure blackout scales, which are the amount of power loss in MW, the number of the affected customers and the restoration time in minutes. From the blackout data in North America from 1996 to 2002[26], they presented the statistics of the relationship between the blackout scale P_{loss} and the number of failures N in the following log-log plot in Fig. 1.2.

From Fig. 1.2, it is clear that the blackout distribution curve has the power-law

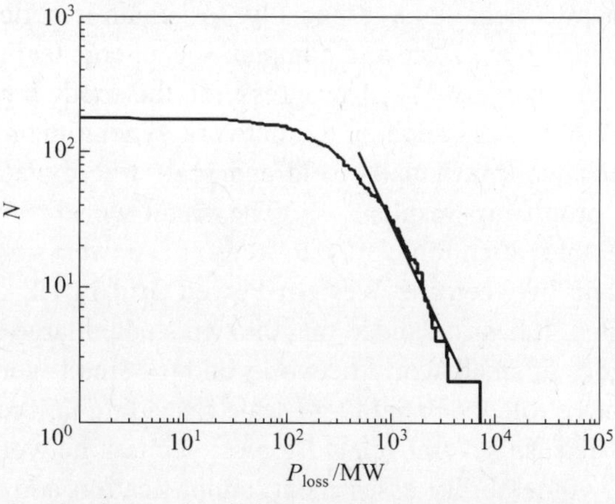

Figure 1.2 North American blackout distribution from 1996 to 2002[26]

tail, which shows the SOC characteristics. Since some blackout data were not complete and accurate enough and thus not fully reliable, Dobson et al designed an artificial power system to simulate blackouts. The obtained blackout time-series data agree with the power-law relationship[27,28].

1.4 Power Grid Complexity

Most of the existing work on SOC in power systems is based on macroscopic statistics, and lacks the analysis of the inherent evolution mechanisms, not to mention the identification of the dominant factors during the self-organized process in power grids. For this reason, researchers have tried to study the topological features of power grids in order to reveal the SOC mechanism in power systems. Great efforts have been made to analyze the SOC characteristics in interconnected power grids using complex network theory[33–35].

Complex network theory is concerned with the common features of different complex networks, such as communication networks, power grids, biological systems and social networks, and the general methods that can be used to analyze them. One of the central topics is the relationship between a network's structure and its performance. Some work develops the detailed model for subsystems first, while some others pay more attention to the overall network structures. The general procedure for the analysis of complex networks can be divided into three steps. First statistical characteristics are identified using empirical methods, and proper evaluation indices are defined. Second, a detailed network model is constructed to understand the relationship between the statistical characteristics and the performance of the network. In the end, people analyze and predict the network dynamics and then apply appropriate control strategies.

Systematic analysis has been carried out for the study of network structural features and dynamic behaviors, especially for chain reactions and accident propagations in typical large-scale and complex social, engineering and ecological networks[36,37]. There has also been progress for the modeling and analysis of collective network behaviors such as the network synchronization[38,39]. Among others, the research work on small-world and scale-free dynamic networks has shown the most promising results[40–44]. The small-world network model has attracted considerable attention for the study of power systems. In [45], a comparison is made between the Western Power Grid in US and the Northern Power Grid in China. It has concluded that the two studied large-scale power grids fall into the category of small-world networks and the small-world characteristics have played an active role to spread large-scale cascading failures in power grids. In this book, we discuss several related topics, such as network decomposition and coordination, vulnerability assessment, simplification and equivalence, and synchronization control. These results can be applied to power grids, and may serve as the first few steps towards the improvement of the structural stability and operational security of power grids.

In this book, we use the words complex networks and complex systems interchangeably since most, if not all, complex systems can be described by interconnected networks. We focus on the topological features of power grids on fault propagation, namely how structural vulnerability assists the spreading of faults. Although a number of complex network models have been investigated for the analysis of power grids, most of them only consider static topologies and simplified physical dynamics, and are thus less useful for the application in practice. Hence, there is hardly any model that is effective for the identification of structural vulnerability in real power systems. On the other hand, blackouts associated with cascading failures are typical SOC behaviors. So to assess the system safety by looking into the SOC characteristics in power system blackouts can be a promising approach, and may provide insight into the strategies and procedures to effectively prevent disasters. In addition, it is possible to construct models to describe the catastrophic evolutions in power systems and such models can be used to simulate both the slow dynamic evolution, such as network upgrade and construction, and the fast dynamics, such as unexpected faults.

1.5 Cascading Failures

To reveal the mechanism for the cascading failures in power systems, researchers have proposed different fault models by analyzing both the macroscopic topological features and microscopic component characteristics. These models can be classified into three categories. The static models describe the macroscopic topologies of power grids. They are used to evaluate, from the topological point of view, a network's vulnerability with respect to various attacks and its robustness to cascading failures[46–55]. The typical examples include the betweenness centrality model[46–48], Motter-Lai model[49–51] and effective efficiency model[52]. The second class models the component cascading failures[56–61], which includes the CASCADE model[56–58] for load transfers and the branching process model[59–61]. These models do not consider specific topologies of power grids, but instead cascading failures are consequences of component failures caused by overloading. The third class takes into account power grid dynamics[62–73], which include the OPA model based on the DC optimal power flows[32,62–66], the hidden fault model based on approximate DC power flows and hidden faults[69], and the Manchester model based on AC power flows and load shedding[70–72]. These three classes of models represent the main research progresses in recent years in power grid security study.

1.5.1 Cascading Failure Models Based on Network Topologies

(1) Betweenness centrality model

The betweenness centrality model[46–48], also called the Holme-Kim model, is mainly concerned with the overloading effect during network evolutions. It assumes that the exchange of information or energy between two nodes always takes place through the shortest path between them. Hence, the betweenness centrality is chosen to assess the loads and capacities of the nodes and lines in the network. Let $C_B(v)$ denote the betweenness centrality of node v and $C_B(e)$ denote that of edge e respectively. Then the betweenness centrality is defined by

$$C_B(v) = \sum_{(w,w')} \frac{\delta_{ww'}(v)}{\delta_{ww'}} \tag{1.2a}$$

$$C_B(e) = \sum_{v \in V} \sum_{w \in V \setminus \{v\}} \frac{\delta_{vw}(e)}{\delta_{vw}} \tag{1.2b}$$

where $\delta_{ww'}$ is the number of the shortest paths between w and w', $\delta_{ww'}(v)$ is the number of the shortest paths between w and w' that contain v, $\delta_{vw}(e)$ is the number of the shortest paths between v and w that contain e, and δ_{vw} is the total number of shortest paths between v and w.

From the Eqs. (1.2) it is clear that the betweenness centrality is used to describe the largest number of shortest paths that a node or an edge can support. So it shows how sensitive a node or an edge is with respect to load faults. In particular, when $C_B(e) > C_B^{\max}$, where $C_B^{\max}$ is the maximum value of C_B, we know that edge e is over loaded and thus will be removed. C_B is then recalculated. Repeating this process, cascading failures can thus be formed after a period of time. Similarly, if node v is overloaded, all of its adjacent lines are deleted while node v itself is still maintained and can be reconnected to the network in the subsequent process. Three characteristics are used to evaluate the system's performances, which include the number of lines, the reciprocal of the length of the shortest path and the scale of the largest connected component. From simulation, it can be shown that in order to avoid being overloaded, the capacities of nodes and lines must be increased as the scale of the network grows.

This model has its limitations when applied to real power systems because of the following reasons. First, this model is built upon the scale-free network model, and some power grids are not necessarily scale-free networks. Second, the model assumes that the network can grow in an arbitrary fashion, which does not agree with the practice in power grid construction. Last, the maximum capacities of the nodes are assumed to be the same, which is inconisitent with the real situation in power grids.

(2) Motter-Lai model

Similar to the Holme-Kim betweenness centrality model, the Motter-Lai model[49–51] also utilizes the total number of the shortest paths containing a given node as the load at that node. It is assumed that the capacity C_i of node i grows linearly with respect to its initial load L_i; in other words, $C_i = (1+\alpha)L_i$, where the constant α is called the tolerance parameter of the network and describes the speed for the growth of the capacity of node i, which can also be used to describe its capacity for handling load and thus attenuating disturbances. There are three facts about the Motter-Lai model that distinguish itself from the Holme-Kim model. The Motter-Lai model assumes that nodes have different capacities, faulted nodes are deleted from the network permanently, and the duration of a cascading failure is much shorter than that of the network construction and consequently it is not necessary to consider network growth. Although the two models simulate cascading failures following similar steps, in the Motter-Lai model the loss caused by a cascading failure is quantified in terms of the relative scale $G = N'/N$, where N and N' are the numbers of the nodes in the largest component before and after the fault respectively.

In [50], cascading failures are simulated for the western US power grid with 4941 buses using the Motter-Lai model. The results are shown in Fig. 1.3, which presents the relationship between the relative scale G and the tolerance parameter α for the cascading failures caused by removing buses from the grid. Each curve is generated by averaging over the cascading failures under five maximum degree- or load-based attacks and 50 random attacks.

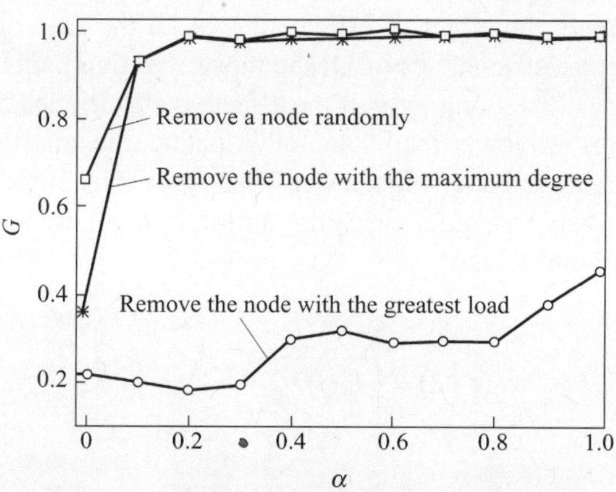

Figure 1.3 Relationship between G and α [50]

Homogenous networks, in which nodes have similar degrees, are more robust to cascading failures caused by attacks[49]. In [51], it was concluded that the degree distribution of the western US power grid satisfies the exponential law and the grid is relatively homogeneous; however, its load distribution is more asymmetric than the semi-random networks with the same link distribution. Therefore, the grid is inherently robust with respect to random removals of buses, and it is still subject to cascading failures triggered by intentional node attacks. In fact, the attack targeting at the node with the maximum load can lead to the loss of more than half of the largest connected components.

(3) Effective efficiency model

In [52], the notion of effective efficiency is introduced where a model is constructed to describe the power grid as a weighted graph associated with an $N \times N$ matrix $[e_{ij}]$, whose elements are either zero or one. When there is no connection between nodes i and j, $e_{ij} = 0$; otherwise, $e_{ij} = 1$. Here, e_{ij} is used to represent the load transmission capacity along the line between i and j, and a larger value of e_{ij} implies a stronger transmission capability. The effective efficiency of an arbitrary path between any pair of nodes is the sum of the effective efficiencies of all the lines in the path. The path that is associated with the maximum effective efficiency is called the most effective path. We further define the load $L_i(t)$ of node i to be the number of the most effective paths passing through node i at time t. We consider the capacity of node i to be $C_i = \alpha L_i(0)$, where α is the tolerance parameter, which implies that the capacity C_i is proportional to the initial load $L_i(0)$. In this model, it is assumed that the transmission between any pair of nodes always takes place through their most effective path.

In computation, the removing of the nodes that are caused by failures may change the most effective paths, and consequently the load distribution. When some nodes are overloaded, the effective efficiencies of all the adjacent lines decrease, and thus the effective efficiencies of all the most effective paths containing these nodes also decrease correspondingly. If the effective efficiencies of those originally most effective paths are lower than some other paths, transmissions redistribute to other paths with higher effective efficiencies. This process leads to load redistribution and in the end is likely to trigger cascading failures. The iteration at time t in the effective efficiency model is

$$e_{ij}(t+1) = \begin{cases} \dfrac{e_{ij}(t)}{L_i(t)/C_i}, & L_i(t) > C \\ e_{ij}(t), & L_i(t) \leqslant C \end{cases} \tag{1.3}$$

Three features of the effective efficiency model are as follows. First, when the overall network is not connected, the notion of effective efficiency can still be used to assess the network's performances. In this case, we take the effective efficiency of two nodes that are in different connected components to be zero. Second, the overloaded nodes are not removed from the network, and after the load levels of such nodes return to their normal values, the nodes are reconnected back to the network. Third, the model is based on dynamical simulations, in which a fault at one component has both direct impact on the network behavior, and also indirect impact by causing other components to be overloaded, in extreme cases even leading to cascading failures. This overcomes the limitations in traditional simulation methods where the impact of a fault is evaluated by studying the static situation when the corresponding component is removed from the network.

In the effective efficiency model, the network efficiency $E(G)$ is defined by

$$E(G) = \frac{1}{N(N-1)} \sum_{i \neq j \in G} \eta_{ij} \tag{1.4}$$

where N is the number of nodes, and η_{ij} is the efficiency of the most efficient path between nodes i and j. Using the network efficiency as a measure of network performances, simulations on cascading failures are carried out for the western US grid with 4941 buses and 6592 lines in [53]. Similar results have been obtained as those using the Motter-Lai model.

In [53], a more detailed analysis is performed for the whole North American grid with 14,009 buses and 19,657 lines. The set of buses consists of the generator bus set G_G with N_G elements, the transmission bus set G_T with N_T elements and the distribution bus set G_D with N_D elements. The average effective efficiency is then defined by

$$\bar{E} = \frac{1}{N_G N_D} \sum_{i \in G_G, j \in G_D} \eta_{ij} \tag{1.5}$$

With this definition, the initial network effective efficiency is 0.041,33. When removing buses randomly, the critical tolerance parameter is around 1.18. However, when removing buses with maximum loads, the critical tolerance parameter is then greater than 1.42. The simulation results about the relationship between the network overall effective efficiency and the tolerance parameter is about the same as that for the western US power grid.

Simulations have shown that with different tolerance parameters, the buses in the North America grid can be divided into three groups, according to their different influences on the effective efficiency when being removed from the grid. The results are shown in Fig. 1.4. The removal of the first group of buses, denoted by pluses, only changes the effective efficiency slightly, and about 60% of all the buses belong to this category. The busses belonging to the second group are denoted by diamonds, squares and dots. When they are removed from the grid, the effective efficiency loss D is

$$D = D_0(1 - x^{\beta} / (K^{\beta} + x^{\beta})) \tag{1.6}$$

where $x = 1 - \alpha$, $D_0 = 0.25$ is the greatest loss, K is from 1 to 1.2, and $\beta \approx 2$. The third group of buses is denoted by triangles. When $\alpha < 1.4$, their curves are similar to those of the second group; however, when $\alpha = 1.4$, the loss decreases dramatically to a small value.

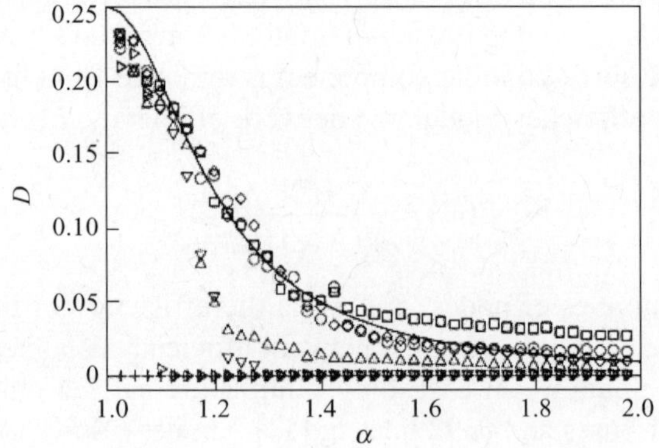

Figure 1.4 Relationship between the effective efficiency loss D and the tolerance parameter α [53]

The following conclusions are drawn in [52]. (a) The initial disturbance in a cascading failure needs to be greater than 1% of the initial load. (b) The removal of 0.33% buses with the highest loads may cause the loss of 40% of the effective efficiency. (c) Even if only one bus with the highest load and largest degree is removed, the effective efficiency may still decrease by 25%. The conclusions in [35] are that the vulnerability of the transmission grid should be paid sufficient attention to, and there are preventive strategies such as building more transmission lines and power stations to distribute the load from those heavily loaded buses and control the spread of cascading failures. Distributed generation may be an even more environment friendly strategy.

It needs to be pointed out that the three models just described have not been powerful enough to reveal the dynamical mechanisms for the evolution of power girds, especially for their structural stability and SOC analysis. The main reason is that the power grid has been idealized and the models are built on static statistics on node degrees, shortest paths and other quantities. These models cannot accurately reflect the actual dynamics in power systems, such as the power flow distribution and load redistribution.

1.5.2 Component Cascading Failure Dependent Model

(1) CASCADE model

The basic ideas of the CASCADE model[56–58] are as follows. The system under consideration consists of n identical components. They are all initially in their normal states with independent initial loads $L_1, L_2, \cdots, L_n$, which are randomly chosen in the range $[L_{\min}, L_{\max}]$. A random disturbance D is then exerted on all the loads. If the load of some component is above its threshold value, then a fault takes place at that component, whose load is then redistributed to other components that are working under normal conditions. This process may cause cascading failures.

In [58], the influence of operation parameters, such as the average load level L, has been studied for the evolution of cascading failures and system collapse. It is found that when the load levels of all components are low, the faults are more or less independent and the blackout distribution satisfies the exponential distribution. So the system is less likely to experience large-scale failures. It is also found that when the load levels increase to some critical values, the fault distribution exhibits the power-law tail characteristics and when the load level keeps increasing, large-scale blackouts may happen.

Although the CASCADE model can qualitatively simulate and analyze blackouts, it still has some limitations. It assumes that all the components and their interactions are identical and ignores the network topology when computing load redistribution. Neither does it consider the changes in power generation. Compared to real power systems, the CASCADE model is idealized and only considers the load redistribution caused by faulted components. Since the load redistribution is only roughly approximated, the computation results cannot be applied directly to real systems.

(2) Branching process model

In order to simulate accurately cascading failures in real power systems, branching processes are introduced in [59–61], and correspondingly the model constructed is called the branching process model. The system considered consists of an infinite number of components and the faulted component in the previous stage may still be faulted in the current stage according to a certain probability. The process finishes when no fault occurs. After analyzing the statistical characteristics of the total number of faulted components, it is found that when the average fault probability, also considered to be the fault propagation speed, reaches a certain threshold value, the fault probability distribution satisfies the power-law distribution with the coefficient being -1.5. When the system is at its critical and supercritical state, it runs with great risk[60].

When applying the branching process model into power systems, we need to consider two problems. First, we need to relax the assumption that the system has infinite elements. We also need to relate the system's load level to the fault propagation speed to study the criticality. The branching process model is in fact an approximation of the CASCADE model, and in [59, 61] it has been shown through numerical analysis that the binomial form of the CASCADE model in its limit is the branching process model, and the branching process model when simulated through the Poisson distribution can approximate the CASCADE model.

It is further shown in [61] that there is a linear relationship between the fault propagation speed and the system's load level.

In short, the branching process model is actually an approximated CASCADE model. It takes the average failure probability as the cascading failure propagation speed. Although the branching process model has clear physical meanings, it is too simplified to reflect closely the dynamics in real power systems and thus has only been applied to a limited number of systems.

1.5.3 Power Grid Dynamics and Blackout Models

The causes for cascading failures in interconnected power grids are usually extremely complicated. From power system dynamics, the blackout mechanism can be roughly described as follows. Initially, each component in the grid operates with a certain initial load. When one or more components are out of service, their loads are redistributed to other components, which may be overloaded and lead to another round of load redistribution. This may further develop into cascaded overloading faults, and eventually form large-scale blackouts. Power grid cascading failures are continuous dynamical processes, involving generators, transmission lines, transformers, buses, breakers, switches, and other components. Usually, the failures at buses, transmission lines and breakers, the maloperation at relay protections and the sudden heavy load redistributions are the main reasons for cascading failures. Once such events take place, relay protections, when working properly, act promptly to cut off the faulted lines. However, such actions may force other lines to be overloaded or other bus voltages to be greatly apart from their rated values. So more protections are triggered, which may in the end lead to cascading failures or even the collapse of the whole system. The difficulties in simulating and analyzing cascading failures are mainly the variety of triggering events, the complexity of the cascading process and the characteristics of power system dynamics. In the cascading failure analysis methods described before, only static or simplified dynamic processes are considered and as a result, it is hard to trace the power flow dynamics. In this context, researchers have proposed several cascading failure models that consider power grid dynamics by taking into account both macroscopic topological features and microscopic physical component properties.

(1) OPA model

The ORNL-PSerc-Alska (OPA) model is a blackout model proposed by the researchers at the Oak Ridge National Laboratory (ORNL), the Power System Engineering Research Center of the University of Wisconsin (PSerc), and the University of Alaska[32], and the name of the model comes from the combination of the first letters of the names of the three research institutes. This model takes power systems as dynamical systems similar to sand piles that have both fast and slow dynamics. The slow dynamics corresponds to adding sand grains to the pile, and the fast dynamics corresponds to sand slide or collapse.

In fact, the core of the OPA model is the DC power flow model that involves transmission lines, loads, generators and many other components. It studies the global dynamics of the blackouts in a power system by inspecting load changes. The model is based on the simplified DC power flow mode. Its slow dynamics simulate the long-term developments in power systems that are determined by the pair of competing forces of load growth and transmission capacity improvement, which may drive the system spontaneously to its critical states. Its fast dynamics simulate daily operations. Line outage is the main form of the faults that may trigger cascading failures. The computation starts from a solution to the power flow problem. Then a random fault is generated. If there is a line out of service, the power flow is recalculated by solving the DC optimal power flow problem. This process is repeated until there is no faulted line any more.

Since the OPA model is based on DC optimal power flows, its computation is mainly to solve a linear programming problem, in which the variables to be optimized are the generators' outputs and the loads to be shed, and the objective function is

$$C = \sum P_i(t) - W \sum P_j(t) \tag{1.7}$$

where $\sum P_i(t)$ denotes the total output of all the generators at time t, $\sum P_j(t)$ is the total load at time t, and W is the cost for shedding loads. The OPA model requires that the generators and transmission lines should operate within their capacities in addition to the power balancing. Cascading failures can thus be simulated by cutting off with a certain probability those heavily loaded lines operating near their limits. Simulations in [32] have shown that as the load grows, the time that customers are fully serviced decreases while the average line load and the number of faulted lines increase significantly. It is shown in [62] that the average line load is independent of how the system evolves to its equilibrium, and with some load growth rate, the system's blackout distribution satisfies the power law when the load increases to some critical level. In [63, 64], it is reported that the OPA model has two critical points, one corresponding to the load shedding when generators reach their limits and the other corresponding to the load shedding because of line outages. The blackout distribution satisfies the power law and the risk for large-scale blackouts increases dramatically when the load level is critical.

The OPA model can be further improved in several aspects. The number of buses considered in the model is still small and as a result the small-scale system under study is quite different from actual power grids. It assumes that the dynamics of the same type of components are identical and control strategies are realized by adjusting a few parameters in the model. The relationships between the parameters of the model and those of a practical system are not so clear and cannot be used directly to study the SOC characteristics for power grid planning, operation, automation and control. Most importantly, the OPA model is developed upon the DC power flow, and in theory it works only for high-voltage low-load power grids and cannot describe the reactive power and voltage characteristics. This is undesirable since cascading failures and blackouts usually take place in overloaded networks.

(2) Hidden failure model

A hidden failure is a fault in a relay protection equipment that has not been detected under normal working conditions and might be exposed in abnormal situations and cause the malfunction of the relay system. For example, a relay may mistakenly cut off a component after a switching action[67]. Although hidden failures are events with small probability, their impact can be catastrophic[68]. Hence, a hidden failure model is developed in [67] to simulate cascading failures and blackouts. The process starts from a random initial line outage. If the power flow through a line adjacent to the faulted line is above a pre-set threshold value, that line is tripped as well. Otherwise, the hidden failure mechanism is used to decide whether the line should be tripped. Power flows are recalculated after each tripping action until the end of the development of the cascading failure. We describe this process in more detail as follows:

Step 1 Randomly select a transmission line and trip it to generate the initial fault.

Step 2 Recalculate the DC power flow.

Step 3 Check for overloaded lines and trip them when they are found.

Step 4 When there is no overloaded line in the previous step, trip the line that is adjacent to the one that is tripped last time. If there are more than one such lines, we trip the one with the highest tripping probability.

Step 5 Check the connectivity of the power grid. If it breaks into several islands, then we do the calculations within each island.

Step 6 Shed load when necessary to maintain the system's stability and keep the power flows along the lines being within the corresponding transmission capacities. This requires solving a linear programming problem to determine where and how to shed load.

Step 7 When there is no lines to be tripped any more, record the total load loss and the sequence of the line outages, and then quit the simulation. Otherwise, go back to Step 2.

The hidden failure model is used in [68] to analyze power system cascading failures. The simulation results have shown that when the load level is critical, the blackout distribution has the power-law characteristics. It needs to be pointed out that, if the OPA model only considers fast dynamics, it becomes essentially the same as the hidden failure model. So the hidden failure model is only applicable to simulate high-voltage low-load power grids and this model cannot be used to describe the long-term evolution of power grids.

(3) Manchester model

The Manchester model is a cascading failure model based on AC power flows[70–72], which distinguishes itself from previous cascading failure models that are based on DC power flows. Since solving for DC power flows only requires solutions to linear programming problems, it is usually not necessary to consider the existence and convergence of solutions. However, the AC power flows are related to nonlinear programming problems that are constrained by nonlinear constraints, so the solution may no longer exist. Therefore, in the Manchester model, shedding load is used to guarantee the convergence of the AC power flows, which is an important innovation of the model. The calculation flow chart is shown in Fig.1.5.

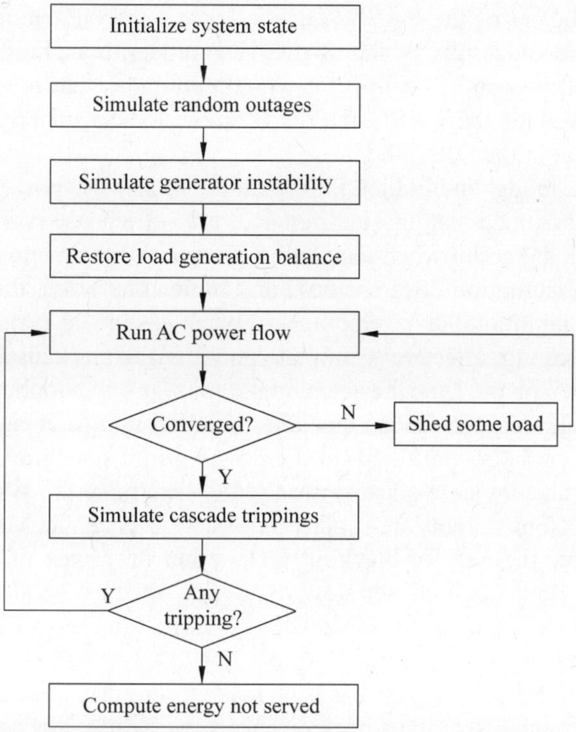

Figure 1.5 Manchester cascading model for failures in power systems

The main drawback of the Manchester model is that it only considers fast dynamics, but ignores slow dynamics, such as the improvement of generation and transmission capacities and load growth.

In summary, the OPA model, hidden failure model and Manchester model all take into account power system dynamics. The OPA model is built upon the sand pile model in SOC theory and cascading failures are simulated through the computation of simplified DC power flows that involves both fast and slow dynamics. The hidden failure model is constructed by the computation of approximate DC power flows and the hidden failure mechanism. It can effectively simulate cascading failures. The Manchester model is based on AC power flows, it utilizes load shedding and power flow redistribution to simulate cascading failures.

Simulations of the above models have shown that when some operation parameters, such as the system load level and the failure propagation speed, reach a certain threshold, the blackout distributions have the power-law tails. In [56], the differences have been presented for the power-law distributions of the three models at their critical points. The relationships between the related parameters, especially the critical parameters, are further investigated in [59, 74–76]. More specifically, the relationship between the average failure propagation speed and the load level is shown to be roughly linear[59]. The relationship between the operation parameters of the CASCADE model and the OPA model is shown in [75]. The simulation results of the CASCADE model, branching process model and OPA model are compared in [76], which concludes that the average failure propagation speed of the CASCADE is to some extent linearly related to the system load level in the OPA model.

Both the OPA model and hidden failure model utilize DC power flows, so they have implicitly assumed that phase differences are sufficiently small, which is true when the load is light. But when simulating blackouts, the system is often heavily loaded, so the assumption does not hold any more. Therefore, these two models cannot describe faithfully the development of blackouts in real systems. In addition, the two models cannot effectively simulate power blackouts caused by inadequate reactive powers. For the Manchester model, although the adoption of AC power flow calculations can overcome the above shortcomings, it assumes that the outputs of the generators are fixed. In the power grid dispatching, optimal power flows that minimize the network loss are used to determine the system's operating point, and generators' outputs are usually adjustable. Hence, the Manchester model may overestimate the risk for blackouts. The other drawback of the Manchester model is that it does not consider slow dynamics, such as the slow load growth, and it is thus less capable to analyze the system's long-term evolution from a macroscopic view.

Although considerable progress has been made to simulate and analyze cascading failures, all the models that have been discussed so far use the basic idea that line outage may lead to massive load transfer and thus more lines may be overloaded when simulating cascading failures. On the other hand, statistics of most of the large-scale blackouts find that a considerable number of blackouts were triggered by large disturbances, such as three-phase short-circuit permanent fault. In other words, transient instability plays a key role in the development of blackout since the actions of shedding generations and loads and tripping off lines may in the end result in cascading failures. Hence, transient stability constraints should be considered in the models for blackouts. In addition, the blackout models using SOC theory have been mainly deployed for qualitative analysis, and the critical variables or critical features obtained through such models cannot be directly applied to the assessment of power grid security, and cannot be utilized directly to design effective prevention and control strategies.

Hence, a desirable blackout model should consider the system transient dynamics and power flows, while avoid the demanding computation for the solutions to high dimensional differential-algebraic equations. In this book, we aim to develop system evolution models that consider jointly different time scales from transient dynamics to power flow dynamics and long-term network construction and development. We want to obtain macroscopic statistical critical features, identify the system's weak spots and design control strategies to prevent blackouts.

1.6 Book Overview

Generally speaking, the power systems' dynamics in blackouts may be rooted in the unstable structures when the systems are at their SOC states. However, to identify the characteristics during disastrous transitions is one of the most difficult problems in nonlinear sciences. Although cascading failures and blackouts in power grids can be generally taken as SOC phenomena in complex power systems, there is not yet a SOC theoretical framework that consists of the nonlinear modeling, simulation platforms and the identification of indices for security assessment. In this book, we first give an introduction of the basic concepts and models in SOC theory from the viewpoint of power system study. Then we further explore models that describe the evolution of power grids in different time scales. We reveal and analyze critical features during disastrous transitions, introduce risk assessment methods, and propose prevention and control strategies.

The main purpose is to develop power grid complexity theory that bridges the gap between SOC and complex network theories and the technologies in the secure operation of power systems. We choose to use general mathematical tools and basic power system analysis methods in order to make the book readable to most of the interested readers and it can be conveniently referenced in real practice. We try to present the materials in a rigorous and logical way, and readers with only basic knowledge about probability theory and other mathematical tools can still follow the main concepts, algorithms, methodologies and results without difficulty. We keep the need to solve practical engineering problems in mind, paying special attention to the security assessment problem in large-scale power grids.

The thread of the book is SOC theory and complex network theory, which are both used to quantitatively analyze dynamics in complex power grids from macroscopic and microscopic views[77,78]. The required mathematical tools include probability theory, stochastic processes, fluctuation, power-law distributions, and self-organized criticality. The complex network characterizations are used to study the evolution mechanism of power grids, and consequently we are able to establish the small-world power grid evolution model and study the associated power flows, synchronization characteristics and vulnerability tests. We also discuss the decomposition and coordination of static power grids, and more importantly the simplification and equivalence of dynamic power grids and their synchronization control methods. Various blackout models are constructed taking into consideration the transient dynamics, power flows and power grid planning.

The complexity analysis and security assessment for power grids are challenging research topics that appeal to researchers from several disciplines, such as complexity science, dynamical systems and power systems. Significant progress has been made in this area in the recent years and this book may not be comprehensive enough to cover all the aspects. But we hope to help readers to better understand the foundations and frontiers of SOC and complex network theories and their applications in the study of complex power grids.

References

[1] Auyang S Y. Foundations of Complex-System Theories: in Economics, Evolutionary Biology, and Statistical Physics. New York: Cambridge University Press, 1998

[2] Zhou Y L, Wang Y L, Zhao M Y. Statistics of operation situation of protective relayings and automation devices of power systems in China in 2004. Power System Technology, 2005, 29(16): 42 – 48 (in Chinese)

[3] Lu Q, Mei S W. Vital basic research on modern power systems geared to the 21st Century. Progress in Natural Science, 2000, 10(8): 585 – 593

[4] Li W. Risk Assessment of Power Systems: Models, Methods, and Applications. New York: Wiley, 2005

[5] Bhatt N B. August 14, 2003 US-Canada blackout. IEEE PES General Meeting, Denver, CO, 2004

[6] Venkatasubramanian V, Li Y. Analysis of 1996 Western American electric blackouts. Bulk Power System Dynamics and Control-Ⅵ, Aug. 2004: 22 – 27

[7] Anderson G, Donalek P, Farmer R. Causes of the 2003 major grid blackouts in north America and Europe, and recommended means to improve system dynamic performance. IEEE Transactions on Power Systems, 2005, 20(4): 1922 – 1928

[8] Union for the coordination of electricity transmission (UCTE). UCTE releases detailed interim report on the disturbances of 4 Nov. http://www.ucte.org. Accessed Nov. 30, 2006

[9] Bialek J W. Blackouts in the US/Canada and continental Europe in 2003: Is liberalization to blame? 2005 IEEE Power Tech, Jun. 2005

[10] Corwin J L, Miles W T. Impact assessment of the 1977 NewYork City blackout. US Department of Energy, Washington DC, 1978

[11] Sforna M, Delfanti M. Overview of the events and causes of the 2003 Italian blackout. Power Systems Conference and Exposition, 2006: 301 – 308

[12] Ohno T, Imai S. The 1987 Tokyo blackout. IEEE PES Power System Conference and Exposition, 2006: 314 – 318

[13] Makarov Y V, Reshetov V I, Stroev V A, et al. Blackout in North American and Europe: analysis and generalization. 2005 IEEE Power Tech, Jun. 2005

[14] Corsi S, Sabelli C. General blackout in Italy Sunday September 28, 2003, h03:28:00. IEEE Power Engineering Society General Meeting, Jun. 2004, 2: 1691 – 1702

[15] Li C Y, Sun Y Z, Chen X Y. Recommendation to improve power system security: Lessons learned from the Europe blackout on November 4, 2007. 42nd UPEC, 529 – 533

[16] Kiesling L. The North American blackouts and electricity policy: alternatives to transmission construction. Economic Affairs, 2004, 24: 53 – 57

[17] Liu C C, Vittal V, et al. The strategic power infrastructure defense system. IEEE Control Systems Magazine, 2000, 13(8): 40 – 52

[18] Perez L G, Flechsig A J, Venkatasubramanlan V. Modeling the protective system for power system dynamic analysis. IEEE Transactions on Power Systems, 1994, 9(4): 1963 – 1973

[19] Kundur P. Power System Stability and Control. New York: McGraw-Hill, 1994

[20] Taylor C W. Power System Voltage Stability. New York: McGraw-Hill, 1994

[21] IEEE System Dynamic Performance Subcommittee of the Power System Engineering Committee of the PES. Voltage Stability of Power System: Concepts, Analytical Tools, and Industry Experience. Technical Report 90 TH 0358-2PWRS, IEEE, 1990

[22] Bak P. How Nature Works: the Science of Self-Organized Criticality. New York: Copernicus Press, 1996

[23] Bak P, Chen K. Self-organized criticality. Scientific American, 1991, 264(1): 26 – 33

[24] Bak P. Self-Organized Criticality. Cambridge: Cambridge University Press, 1998

[25] Bak P, Tang C, Wiesenfeld K. Self-organized criticality: an explanation of $1/f$ noise. Physical Review Letters, 1987(59): 381 – 384

[26] North American electric reliability corporation. http://www.nerc.com/~dawg/database.html

[27] Carreras B A, Newman D E, Dobson I, et al. Initial evidence for self-organized criticality in electric power system blackouts. Proceedings of Hawaii International Conference on System Science. Maui, Hawaii, USA, Jan. 2000

[28] Carreras B A, Newman D E, Dobson I, et al. Evidence for self-organized criticality in electric power system blackouts. Proceedings of the 34th Annual Hawaii International Conference on System Sciences, Jan. 2001: 705 – 709

[29] Yu Q, Guo J B. Statistics and self-organized criticality characters of blackouts in China electric power systems. Automation of Electric Power Systems, 2006, 30(2): 16 – 21 (in Chinese)

[30] Yu Q, Guo J B. Study on self-organized criticality of power system blackouts in China. Power System Technology, 2006, 30(6): 1 – 5 (in Chinese)

[31] Weng X, Hong Y, Xue A, et al. Failure analysis on China power grid based on power law. Journal of Control Theory and Application, 2006, 4(3): 1 – 4

[32] Dobson I, Carreras B A, Lynch V E, et al. An initial model for complex dynamics in electric power system blackouts. Proceedings of the 34th Hawaii International Conference on System Sciences, Maui, Hawaii, Jan. 2001: 710 – 718

[33] Mei S W, Xue A C. Power network analysis based on the complex approaches. Proceedings of the 23rd Chinese Control Conference, Wuxi, Aug. 2004: 1233 – 1237
[34] Mei S W, Xue A C, Zhang X M. On power system blackout modeling and analysis based on self-organized criticality. Science in China (Series E), 2008, 51(2): 209 – 219
[35] Sun K, Han Z X. Analysis and comparison on several kinds of model of cascading failure in power system. IEEE/PES Transmission and Distribution Conference & Exhibition: Asia and Pacific, Dalian, China, 2005: 1 – 7
[36] Simkin M V, Roychowdhury V P. Stochastic modeling of citation slips. Scientometrics, 2005, 62(3): 367 – 384
[37] Boykin P O, Roychowdhury V P. Leveraging social networks to fight spam. Computer, 2005, 38(4): 61 – 68
[38] Sarshar N, Roychowdhury V P. Multiple power-law structures in heterogeneous complex networks. Physical Review E, 2005, 72(2): 026144
[39] Sarshar N, Roychowdhury V P. Scale-free and stable structures in complex ad hoc networks. Physical Review E, 2004, 69(2): 026101
[40] Albert R, Barabasi A L. Statistical mechanics of complex networks. Review of Modern Physics, 2002, 74: 47 – 97
[41] Erdos P, Renyi A. On the evolution of random graphs. Publications of the Mathematical Institue of the Hungarian Academy of Sciences, 1960, 5: 17 – 60
[42] Watts D J, Strogatz S H. Collective dynamics of “small-world” networks. Nature, 1998, 393(6684): 440 – 442
[43] Barabasi A L, Albert R. Emergence of scaling in random networks. Science, 1999, 286: 509 – 512
[44] Milgram S. The small-world problem. Psychology Today, 1967(2): 60 – 67
[45] Meng Z W, Lu Z X, Song J Y. Comparison analysis of the small-world topological model of Chinese and American power grids. Automation of Electric Power Systems, 2004, 28(15): 21 – 29 (in Chinese)
[46] Holme P. Edge overload breakdown in evolving networks. Physical Review E, 2002, 66(2): 036119
[47] Holme P, Kim B J. Vertex overload breakdown in evolving networks. Physical Review E, 2002, 65(1): 066109
[48] Holme P, Kim B J, Yoon C N, et al. Attack vulnerability of complex networks. Physical Review E, 2002, 65(1): 056109
[49] Lai Y C, Motter A E, Nishikawa T. Attacks and cascades in complex networks. Lecture Notes in Physics, 2004, 650: 299 – 310
[50] Motter A E, Lai Y C. Cascade-based attacks on complex networks. Physical Review E, 2002, 66(2): 065102
[51] Zhao L, Park K, Lai Y C. Attack vulnerability of scale-free networks due to cascading breakdown. Physical Review E, 2004, 70(3): 035101
[52] Cructiti P, Latora V, Marchiori M. Model for cascading failures in complex networks. Physical Review E, 2004, 69(4): 045101
[53] Sun K. Complex networks theory: a new method of research in power grid. IEEE/PES

Transmission and Distribution Conference & Exhibition: Asia and Pacific, Dalian, China, 2005: 1 – 6

[54] Albert R, Albert I, Nakarado G L. Structual vulnerability of the North American power grid. Physical Review E, 2004, 69(2): 025103

[55] Latora V, Marchiori M. Vulnerability and protection of intrastucture networks. Physical Review E, 2004, 71(1): 015103

[56] Dobson I, Chen J, Thorp J S, et al. Examining criticality of blackouts in power system models with cascading events. Proceedings of the 35th Annual Hawaii International Conference on System Sciences, Jan. 2002: 7 – 10

[57] Dobson I, Carreras B A, Newman D E. A probabilistic loading-dependent model of cascading failure and possible implications for blackouts. Proceedings of the 36th Annual Hawaii International Conference on System Sciences, Jan. 2003: 65 – 74

[58] Dobson I, Carreras B A, Newman D E. A loading dependent model of probabilistic cascading failure. Probability in the Engineering and Informational Sciences, 2005, 19(1): 15 – 32

[59] Dobson I, Carreras B A, Newman D E. A branching process approximation to cascading load-dependent system failure. The 37th Hawaii International Conference on System Sciences, Hawaii, Jan. 2004

[60] Dobson I, Carreras B A, Newman D E. A criticality approach to monitoring cascading failure risk and failure propagation in transmission systems. Electricity Transmission in Deregulated Markets, Conference at Carnegie-Mellon University, Pittsburgh PA, Dec. 2004

[61] Dobson I, Carreras B A, Newman D E. Branching process models for the exponentially increasing portions of cascading failure blackouts. The 38th Hawaii International Conference on System Sciences, Hawaii, Jan. 2005

[62] Carreras B A, Lynch V E, Dobson I, et al. Modeling blackout dynamics in power transmission networks with simple structure. Proceedings of the 34th Annual Hawaii International Conference on System Sciences, Jan. 2001: 719 – 727

[63] Carsreras B A, Lynch V E, Newman D E, et al. Dynamics, criticality and self-organization in a model for blackouts in power transmission systems. Proceedings of the 35th Annual Hawaii International Conference on System Sciences, Jan. 2002

[64] Carreras B A, Lynch V E, Dobson I, et al. Complex dynamics of blackouts in power transmission systems. Chaos, 2004,14(3): 643 – 652

[65] Carreras B A, Lynch V E, Dobson I, et al. Critical points and transitions in an electric power transmission model for cascading failure blackouts. Chaos, 2002,12(4): 985 – 994

[66] Mei S W, He F, Zhang X M, et al. An improved OPA model and blackout risk assessment. IEEE Transactions on Power Systems, 2009, 24(2): 814 – 823

[67] Chen J, Thorp J S, Dobson I. Cascading dynamics and mitigation assessment in power system disturbances via a hidden failure model. International Journal Electrical Power and Energy Systems, 2005, 27(4): 318 – 326

[68] Tamaronglak S. Analysis of power system disturbances due to relay hidden failures. Virginia Polytechnic and State University, Blacksburg, Virginia, 1994

[69] Phadeke A G, Thorp J S. Expose hidden failures to prevent cascading outages. IEEE Computer Application in Power, 1996, 9(3): 20 – 23

[70] Nedic D P, Dobson I, Kirschend S, et al. Criticality in a cascading failure blackout model. Proceedings of the 15th Power Systems Computation Conference, Liege, Belgium, Aug. 2005

[71] Rios M A, Kirschen D S, Jayaweera D, et al. Value of security: modeling time dependent phenomena and weather conditions. IEEE Transactions on Power Systems, 2002, 17(3): 543 – 548

[72] Kirschen D S, Jayaweera D, Nedic D P, et al. A probabilistic indicator of system stress. IEEE Transactions on Power Systems, 2004, 19(3): 1650 – 1657

[73] Mei S W, Xue A C, Zhang X M. On power system blackout modeling and analysis based on self-organized criticality. Science in China, Series E: Technology sciences, 2008, 51(2): 209 – 219

[74] Dobson I, Carreras B A, Lynch V E, et al. Complex systems analysis of series of blackouts: cascading failure, criticality, and self-organization. Bulk Power System, Dynamics and Control-Ⅵ, Cortina d'Ampezzo, Italy, Aug. 2004

[75] Carreras B A, Lynch V E, Newman D E, et al. Dynamical and probabilistic approaches to the study of blackout vulnerability of the power transmission grid. Proceedings of the 37th Hawaii International Conference on System Sciences, Hawaii, Jan. 2004

[76] Dobson I, Carreras B A, Lynch V E, et al. Estimating failure propagation in models of cascading blackouts. The 8th International Conference on Probability Methods Applied to Power Systems, Ames Iowa, Sep. 2004

[77] Mei S W, Ni Y X, Wang G, Wu S Y. A Study of self-organized criticality of power system under cascading failures based on AC-OPF with voltage stability margin. IEEE Transactions on Power Systems, 2008, 23(4): 1719 – 1726

[78] Mei S W, Wu S Y, Zhang X M, et al. Power system blackout model with transient constraints and its criticality. European Transactions on Electrical Power 2011, 21(1): 59 – 69

Chapter 2 SOC and Complex Networks

The main thread of this chapter is the power law distribution, which is the most common mathematical description of SOC. We first introduce some important concepts related to SOC, such as the scale-free distribution, fluctuation and long-range correlation. We then discuss the mathematical description of the mechanism of phase transitions and extreme events in complex systems. We also review some key definitions in complex networks, such as the node degree distribution, average path length, clustering coefficient and betweenness centrality. Special attention is paid to the small-world networks and scale-free networks, which are the most popular models used for complex networks. Finally, topics of community structures, central node identification, structural vulnerability and network synchronization are covered in order to investigate the relationship between a network's topology and its functionalities.

As one of the theories developing at the frontier of scientific studies, the theory of complex systems, though growing with great momentum, has not been defined and thus investigated based on a generally accepted framework. Most of the adopted notions and concepts are rooted in probability theory, stochastic processes and statistical physics, and consequently the main material of this book may not be so easy to follow at least for some of the readers. In this chapter, we review the basics that are critical to the understanding of the main results about SOC and complex networks. The goal is to point out a few key issues to understand "why a complex system is complex"[1].

2.1 Random Variables

2.1.1 Probability

Various phenomena in nature and human society can be roughly divided into deterministic and random events. Those that are guaranteed to (or not to) happen belong to deterministic events. Those that exhibit different outcomes when tested repeatedly under the same conditions and thus cannot be predicted a priori are said to be random.

A possible outcome of a test (random experiment) is called a basic event. The set of all the basic events is called the sample space, denoted by Ω.

Correspondingly, a point in the sample space, i.e. a basic event, is also called a sample, denoted by ω. A subset of interest in the sample space is called a random event, or simply an event, denoted by A. In a test, if the outcome basic event ω is contained in A, i.e., $\omega \in A$, then we say that event A happens. For a random experiment, although the outcome of each single test is hard to predict, the statistical characteristics of the outcomes after a large number of tests obey certain rules, which can be delineated mathematically using tools from probability theory[2]. The most basic and important notion in probability theory is probability, which is used to describe the likelihood of the occurrence of an event.

Definition 2.1 If the random event A takes place n_A times in a test of n trials, then the ratio

$$f_n(A) = \frac{n_A}{n} \tag{2.1}$$

is defined to be the frequency of event A in the test.

Definition 2.2 The probability $P(A)$ of event A is the limit of its frequency as the number of trials in the test approaches infinity, namely

$$P(A) = \lim_{n\to\infty} f_n(A) = \lim_{n\to\infty} \frac{n_A}{n} \tag{2.2}$$

The above definition of probability is mathematically concise and physically intuitive, but has been less powerful in pushing the study of random phenomena forward. It is Kolmogorov from the former Soviet Union who defined the probability theoretic framework based on set theory and measure theory, and proposed the axiomatization definition of probability[3]. We review his definitions below.

Definition 2.3 Let F be the union of some subsets of a non-empty set Ω satisfying

(1) $\Omega \in F$;

(2) If $A \in F$, then $\overline{A} \in F$;

(3) If $A_i \in F$, $i = 1, 2, \cdots$, then $\bigcup_{i=1}^{\infty} A_i \in F$.

Then we say F is a σ-algebra (domain), and (σ, F) is a measurable space.

Definition 2.4 Consider a measurable space (σ, F). Let $P(A)$ be a real function of event A defined on F such that

(1) $P(A) \geqslant 0, \forall A \in F$

(2) $P(\Omega) = 1$

(3) For mutually exclusive and countable events $A_i \, (i = 1, 2, \cdots) \in F$, it follows

$$P\left(\bigcup_{i=1}^{\infty} A_i\right) = \sum_{i=1}^{\infty} P(A_i) \tag{2.3}$$

Then $P(A)$ is the probability of event A. We call the triple (Ω, F, P), consisting of the sample space Ω, the events' σ-algebra and the probability P defined on F, the probability space of the test. Here the countability used in condition (3) refers to the existence of a one-to-one mapping between the set of interest and the set of natural numbers.

Definition 2.4 is sometimes referred to as the axiomatic definition of probability, for which Definition 2.2 can be taken as a special case.

Example 2.1 Tossing unbiased coins. Let $\omega = 1, 0$ denote the head and tail of an unbiased coin respectively. Then $\Omega = \{0,1\}$ is the corresponding sample space, and $F = \{\{0,1\},\{1\},\{0\},\varnothing\}$ is a σ-algebra. We define the following real-valued function:

$$P(\{1\}) = p,\ P(\{0\}) = q = 1 - p,\ P(\{0,1\}) = 1,\ P(\phi) = 0$$

Then (Ω, F, P) is the probability space, where P is the probability defined on F.

When $A_i (i = 1, 2, \cdots, n) \in F$ are mutually exclusive, we have

$$P\left(\bigcup_{i=1}^{n} A_i\right) = \sum_{i=1}^{n} P(A_i) \tag{2.4}$$

and this is called the additive property of probability.

Now we define independent events.

Definition 2.5 We say the events $A_i, i = 1, 2, \cdots, n$ are mutually independent if for $\forall k = 2, 3, \cdots, n,$ it follows that

$$P\left(\bigcap_{j=1}^{k} A_{i_j}\right) = \prod_{j=1}^{k} P(A_{i_j}),\ \forall 1 \leqslant j \leqslant k \tag{2.5}$$

where $i_1, i_2, \cdots, i_k$ are any natural numbers satisfying $1 \leqslant i_1 < i_2 < \cdots < i_k \leqslant n.$

2.1.2 Random Variables and Their Distributions

Definition 2.6 Consider the probability space (Ω, F, P) and a real function $X(\zeta)$ defined on Ω. If for any x it holds that

$$\{\zeta \mid X(\zeta) \leqslant x\} \subseteq F \tag{2.6}$$

then $X(\zeta)$ is a random variable in (Ω, F, P).

Random variables can be classified into two categories: (1) discrete random variables, whose possible values are countable; and (2) non-discrete random variables, which include mainly continuous random variables and are the most important in practice.

Example 2.2[4] Rolling two dice. Let ω denote the numbers showing on two dice, and then $\Omega = \{(1,1),(1,2),\cdots,(6,6)\}$ is the sample space. Let X denote the sum of the numbers of the two dice, and then $X(\omega)$ takes eleven different possible values $(2,3,\cdots,12)$. So X is a random variable in Ω.

In practice, to describe a random phenomenon clearly, we usually need to use several random variables. For example, the position of a plane in the sky is described by three random variables (X,Y,Z). So n random variables $X_1, X_2, \cdots, X_n$ can be exploited jointly as an n-dimensional random variable (random vector) written as $(X_1, X_2, \cdots, X_n)$.

The statistical characteristics of random variables can be described by their distribution functions.

Definition 2.7 Let X be a random variable in (Ω, F, P). Then the real function

$$F(x) = P\{X \leqslant x\}, \quad x \in \mathbb{R} \tag{2.7}$$

is called the distribution function (or distribution for simplicity) of X and we also say X satisfies the distribution F.

From the above definition it is clear that the probability for the random variable X to be in the interval $(x_1, x_2]$ is

$$P\{x_1 < X \leqslant x_2\} = P\{X \leqslant x_2\} - P\{X \leqslant x_1\} = F(x_2) - F(x_1) \tag{2.8}$$

As a special case,

$$P\{X > x_1\} = 1 - P\{X \leqslant x_1\} = 1 - F(x_1) \tag{2.9}$$

In addition, since the distribution of a discrete random variable is in the form $P(X = x_k) = p_k$, the distribution function of X can be written as

$$F(x) = P\{X \leqslant x\} = \sum_{x_k \leqslant x} P\{X = x_k\} = \sum_{x_k \leqslant x} p_k \tag{2.10}$$

by using the countable and additive properties of probability. The joint distribution function of an n-dimensional random variable $(X_1, X_2, \cdots, X_n)$ is

$$F_X(x_1, x_2, \cdots, x_n) = P\{X_1 \leqslant x_1, X_2 \leqslant x_2, \cdots, X_n \leqslant x_n\} \tag{2.11}$$

Example 2.3 Tossing an unbiased coin for n times. Let $\omega = 1, 0$ denote heads and tails respectively. Let the random variable X_i be the value of ω_i on the ith toss. Then

$$\begin{aligned} F_X(x_1, x_2, \cdots, x_n) &= P_n\{X_1 \leqslant x_1, X_2 \leqslant x_2, \cdots, X_n \leqslant x_n\} \\ &= P\{X_1 \leqslant x_1\} P\{X_2 \leqslant x_2\} \cdots P\{X_n \leqslant x_n\} \end{aligned}$$

is the joint distribution of the tossing of the coin for n times, where P has been defined as the probability function in Example 2.1.

In particular, for independent tests, if there are only two outcomes with probabilities p and q in each test, then $p+q=1$. This model was first considered by Bernoulli when studying the relationship between frequency and probability, so it is called the Bernoulli model or Bernoulli test[3].

2.1.3 Continuous Random Variables

Definition 2.8 Let X be a random variable on (Ω, F, P), and $F(x)$ is its distribution function. If for any x, there exists a nonnegative and integrable function $f(x)$ such that

$$F(x) = \int_{-\infty}^{x} f(t)\mathrm{d}t \tag{2.12}$$

then X is called a continuous random variable, $F(x)$ a continuous distribution function, and $f(x)$ the probability density function (or density function for short) for X.

A continuous random variable X is uncountable and at the same time the sum of the probabilities that it takes all possible values is always one. Therefore, the probability that a continuous random variable takes its value at a specific real number is zero, and it has to be described by a probability distribution function $F(x)$. In particular, if the derivative of the $F(x)$ exists, then

$$f(x) = \frac{\mathrm{d}F(x)}{\mathrm{d}x} = \lim_{\Delta x \to 0} \frac{F(x+\Delta x) - F(x)}{\Delta x} = \lim_{\Delta x \to 0} \frac{P\{x < X \leqslant x + \Delta x\}}{\Delta x} \tag{2.13}$$

For a differentiable joint distribution function of an n-dimensional random variable $(X_1, X_2, \cdots, X_n)$, we have

$$f_X(x_1, x_2, \cdots, x_n) = \frac{\partial^n F_X(x_1, x_2, \cdots, x_n)}{\partial x_1 \partial x_2 \cdots \partial x_n} \tag{2.14}$$

which is called the joint probability density function of the n-dimensional random variable. The joint distribution function of the $m (m < n)$ components of the n-dimensional random variable is called the m-dimensional marginal distribution function, which can be obtained from the corresponding n-dimensional joint distribution function $F_X(x_1, x_2, \cdots, x_n)$ by

$$F_X(x_1, x_2, \cdots, x_m) = F_X(x_1, x_2, \cdots, x_m, \infty, \cdots, \infty) \tag{2.15}$$

Furthermore, consider the random variables X and Y that satisfy a non-monotonic relationship $Y = g(X)$, i.e. the inverse map $X = h(Y)$ is not unique in the sense that a specific Y may correspond to two values of X such that $x_1 = h_1(y)$ and $x_2 = h_2(y)$. If X takes its values in the interval $(x_1, x_1 + \mathrm{d}x_1)$ or $(x_2, x_2 + \mathrm{d}x_2)$, then Y takes its values in $(y, y + \mathrm{d}y)$. So

$$f_Y(y)\mathrm{d}y = f_X(x_1)\mathrm{d}x_1 + f_X(x_2)\mathrm{d}x_2 \tag{2.16}$$

Substituting x_1, x_2 by $h_1(y), h_2(y)$ respectively, and in view of the fact that probability cannot be negative, we know $\mathrm{d}x_1/\mathrm{d}y, \mathrm{d}x_2/\mathrm{d}y$ should be in the form of their absolute values. Then

$$f_Y(y) = |h_1'(y)| f_X[h_1(y)] + |h_2'(y)| f_X[h_2(y)] \tag{2.17}$$

The above calculations can be generalized to the case when a specific Y corresponds to several values of X, namely

$$f_Y(y) = |h_1'(y)| f_X[h_2(y)] + |h_2'(y)| f_X[h_2(y)] + \cdots \tag{2.18}$$

Now we introduce a few widely used continuous random variables.

Definition 2.9 (Normal (Gaussian) distribution) If the probability density function of a continuous random variable satisfies

$$f(x) = \frac{1}{\sqrt{2\pi}\sigma} \mathrm{e}^{-\frac{(x-m)^2}{2\sigma^2}}, \quad -\infty < x < \infty \tag{2.19}$$

where $\sigma > 0$ and m is a constant, then X is said to follow a normal distribution with parameters m and σ, denoted by $X \sim N(m, \sigma^2)$.

In the real world, a large number of random phenomena follow the normal distribution, such as the heights of adult males in a given region and the errors in the measurements of a certain mechanical part.

Definition 2.10 (Uniform distribution) If a continuous random variable X takes its value evenly likely in the interval (a, b), i.e., its probability density function is

$$f(x) = \begin{cases} 1/(b-a), & x \in (a,b) \\ 0, & x \notin (a,b) \end{cases} \tag{2.20}$$

then X is said to follow the uniform distribution, and its corresponding distribution function is

$$F(x) = \begin{cases} 0, & x \leqslant a \\ (x-a)/(b-a), & x \in (a,b) \\ 1, & x \geqslant b \end{cases} \tag{2.21}$$

Definition 2.11 (Power-law distribution) If the probability density function of a continuous random variable X is

$$f(x)=\begin{cases}-x^{\alpha}(\alpha+1), & x\geqslant 1\\ 0, & x<1\end{cases} \tag{2.22}$$

where $\alpha<-1$, then we say the random variable X follows the power-law distribution with parameter α, and its distribution function is

$$F(x)=\begin{cases}1-x^{\alpha+1}, & x\geqslant 1\\ 0, & x<1\end{cases} \tag{2.23}$$

Definition 2.12 (Exponential distribution) If the probability density function of a continuous random variable X is

$$f(x)=\begin{cases}\lambda \mathrm{e}^{-\lambda x}, & x>0\\ 0, & x\leqslant 0\end{cases} \tag{2.24}$$

where $\lambda>0$, then we say X follows the exponential distribution with the parameter λ and its distribution function is

$$F(x)=\begin{cases}1-\mathrm{e}^{-\lambda x}, & x>0\\ 0, & x\leqslant 0\end{cases} \tag{2.25}$$

2.1.4 Characteristics of Random Variables

Once the distribution function of a random variable X is known, all its probabilistic characterizations can be obtained accordingly. However, the probability distributions are usually difficult to obtain in practical engineering systems, and usually only certain characteristics, not every detail about the distribution, are sufficient for the analysis purposes. So we focus on discussing several important characteristics of random variables next.

Definition 2.13 For a discrete random variable X with its probabilities $p_k(k=1,2,\cdots)$, if $\sum_{k=1}^{\infty}|x_k|p_k<\infty$, then its mathematical expectation (or statistical mean) is defined by

$$E[X]=\sum_{k=1}^{\infty}x_k p_k \tag{2.26}$$

which is also simply called expectation or mean.

Definition 2.14 For a continuous random variable X with its probability density $f_X(x)$, if $\int_{-\infty}^{\infty} |x| f_X(x)\mathrm{d}x < \infty$, then its mathematical expectation (or statistical mean) is defined by

$$E[X] = \int_{-\infty}^{\infty} x f_X(x)\mathrm{d}x \tag{2.27}$$

which is also simply called expectation or mean.

The expectation of a random variable X is always a scalar, usually denoted by m or m_X. The expectation of the function $g(X)$ of a random variable X can be calculated as follows.

(1) For a discrete random variable X with its probabilities $p_k (k = 1, 2, \cdots)$, if $\sum_{k=1}^{\infty} |g(x_k)| p_k < \infty$, then the expectation of $g(X)$ is

$$E[g(X)] = \sum_{k=1}^{\infty} g(x_k) p_k \tag{2.28}$$

(2) For a continuous random variable X with its probability density $f_X(x)$, if $\int_{-\infty}^{\infty} |g(x)| f_X(x)\mathrm{d}x < \infty$, then the expectation of $g(X)$ is

$$E[g(x)] = \int_{-\infty}^{\infty} g(x) f_X(x)\mathrm{d}x \tag{2.29}$$

Definition 2.15 The expectation of the kth power of a random variable X is

$$E[X^k] = \int_{-\infty}^{\infty} x^k f_X(x)\mathrm{d}x, \quad |E[X^k]| < \infty \tag{2.30}$$

which is called X's k^{th} moment about zero.

Specifically,

when $k = 0, E(X^0) = 1$;

when $k = 1, E(X^1)$ is the expectation of X;

when $k = 2, E(X^2)$ is the second moment of X, which is also called its mean square value.

Definition 2.16 The expectation of the kth power of the deviation $X - m_X$ for the random variable X with its expectation m_X is calculated by

$$E[(X - m_X)^k] = \int_{-\infty}^{\infty} (x - m_X)^k f_X(x)\mathrm{d}x, |E[(X - m_X)^k]| < \infty \tag{2.31}$$

which is called the kth central moment of X.

Specifically,

when $k = 0, E[(X - m_X)^0] = 1$;

when $k = 1, E[(X - m_X)^1] = E[X] - m_X = 0$;

when $k=2, E[(X-m_X)^2]$ is the second central moment, which is the most important and most commonly used moment among all the central moments. It is usually called the variance of *X*, denoted by $D(X)$, σ_X^2 or $\langle X^2 \rangle$. It shows the variation of the random variable with respect to its expectation.

Definition 2.17 Consider the 2-dimensional random variable (*X*,*Y*) with the joint probability density function $f_{XY}(x,y)$. The expectation

$$E[X^n Y^k] = \int_{-\infty}^{\infty}\int_{-\infty}^{\infty} x^n y^k f_{XY}(x,y)\mathrm{d}x\mathrm{d}y, \quad |E[X^n Y^k]| < \infty \tag{2.32}$$

is called the $(n+k)^{\text{th}}$ joint moment about zero. In particular, $E(XY)$ is the most commonly used joint moment, which shows the correlation between the random variables *X* and *Y*, and it is usually denoted by R_{XY}, namely

$$E[XY] = \int_{-\infty}^{\infty}\int_{-\infty}^{\infty} xy f_{XY}(x,y)\mathrm{d}x\mathrm{d}y = R_{XY} \tag{2.33}$$

Definition 2.18 For a 2-dimensional random variable (X,Y) with the joint probability density function $f_{XY}(x,y)$, the expectation

$$E[(X-m_X)^n (Y-m_Y)^k] = \int_{-\infty}^{\infty}\int_{-\infty}^{\infty} (x-m_X)^n (y-m_Y)^k f_{XY}(x,y)\mathrm{d}x\mathrm{d}y \tag{2.34}$$

is called the $(n+k)^{\text{th}}$ joint central moment. Here, it is required that

$$|E[(X-m_X)^n (Y-m_Y)^k]| < \infty$$

Note that $E[(X-m_X)(Y-m_Y)]$ is the most important joint central moment, which shows the correlation between the random variables *X* and *Y* and is called the covariance of *X* and *Y*, denoted by C_{XY} or $Cov(X,Y)$, i.e.

$$\begin{aligned} E[(X-m_X)(Y-m_Y)] &= \int_{-\infty}^{\infty}\int_{-\infty}^{\infty} (x-m_X)(y-m_Y) f_{XY}(x,y)\mathrm{d}x\mathrm{d}y \\ &= C_{XY} \end{aligned} \tag{2.35}$$

If $C_{XY}=0$, we say that *X* and *Y* are uncorrelated. The covariance can also be calculated by

$$\begin{aligned} C_{XY} &= E[(X-m_X)(Y-m_Y)] \\ &= E[XY] - m_X m_Y \\ &= R_{XY} - m_X m_Y \end{aligned} \tag{2.36}$$

Definition 2.19 For two random variables *X* and *Y*, if $\sigma_X \neq 0$, and $\sigma_Y \neq 0$, then we call

$$\rho_{XY} = \frac{C_{XY}}{\sigma_X \sigma_Y} \tag{2.37}$$

the correlation coefficient of X and Y.

The correlation coefficient describes the linear correlation between the two random variables, which has the following properties:

(1) $-1 \leqslant \rho_{XY} \leqslant 1$;

(2) If X and Y have linear relationship with probability 1, namely

$$P\{Y = aX + b\} = 1 \text{ for some constants } a \text{ and } b,$$

then $|\rho_{XY}| = 1$.

(3) If X and Y are linearly uncorrelated, then $|\rho_{XY}| = 0$, or equivalently

$$C_{XY} = 0 \tag{2.38}$$

which is again equivalent to

$$D[X \pm Y] = D[X] + D[Y] \tag{2.39}$$

or

$$R_{XY} = E[X]E[Y] \tag{2.40}$$

Definition 2.20 For any random variable X, we call

$$f(t) = E[\mathrm{e}^{\mathrm{i}tX}] = E[\cos tX] + \mathrm{i}E[\sin tX] \tag{2.41}$$

the characteristic function of X.

Since $|\mathrm{e}^{\mathrm{i}tX}| \leqslant 1$ or equivalently $\cos(tX) \leqslant 1$ and $\sin(tX) \leqslant 1$, the characteristic function always exists for any random variable. Obviously, the characteristic function has the following properties.

(1) The characteristic function $f(t)$ is a complex function of real variables;

(2) If the characteristic function of the random variable X is $f(t)$, then the characteristic function of $a + bX$ is $\mathrm{e}^{\mathrm{i}ta} f(bt)$;

(3) $f(0) = 1$.

In particular, the characteristic function of the standard normal distribution $X \sim N(0,1)$ is $f(t) = \mathrm{e}^{-t^2/2}$, and the characteristic function of the general normal distribution $X \sim N(\mu, \sigma^2)$ is $f(t) = \mathrm{e}^{\mathrm{i}\mu t - \sigma^2 t^2/2}$.

2.2 Stochastic Processes

Stochastic processes are usually considered to be the probability theory involving dynamics. The probabilistic methods discussed before mainly deal with static situations, and study the rules for one or more random variables in the probability space (Ω, F, P). In real-world problems, we also want to know how random phenomena evolve with time, so we need to introduce time-varying random variables. Stochastic processes are concerned with time-varying random variables of up to infinite dimensions. Roughly speaking, stochastic processes can be taken as the evolution of time-varying random variables[5].

2.2.1 Characteristics of Stochastic Processes

Definition 2.21 Let (Ω, F, P) be the probability space. If $X(t,\varsigma)$ is a random variable in (Ω, F, P) for any $t \in [0, \infty)$, then $\{X(t,\varsigma); t \geqslant 0\}$ is called a stochastic process in (Ω, F, P), denoted by $X(t)$. In particular, for any fixed $\varsigma, \{X(t,\varsigma), t \geqslant 0\}$ is a real-valued function defined on $t \in [0, \infty)$, which is called a trajectory of the stochastic process.

Now we give an example.

Example 2.4 Tossing a coin for an infinite number of times. Let $\omega = 1,0$ denote heads and tails respectively. Then the set of all possible outcomes is

$$\Omega = \{\omega = (\omega_1, \omega_2, \cdots, \omega_n, \cdots) : \omega_n = 0,1; n = 1,2,\cdots\}$$

Let

$$X_n(\omega) = \omega_n$$

Then the random sequence $\{X_n, n \geqslant 1\}$ is a stochastic process, where the set of X_n's values $S = \{0,1\}$ is the state space. For any fixed $\omega \in \Omega, \{X_n, n \geqslant 1\}$ is an infinite sequence of 0 or 1, which is a trajectory of $\{X_n, n \geqslant 1\}$.

In view of the above example, we consider a particle that moves at every step either to its left or right according to the outcome of tossing a coin. It starts from the zero position, and moves to its right if the outcome is a head and to the left otherwise. We denote the particle's position after n steps by S_n. Then S_n can be determined by the Bernoulli distribution. More specifically, we first introduce the independent random variables Z_n with the probabilities

$$P\{Z_n = 1\} = p, \quad P\{Z_n = -1\} = q = 1 - p$$

Then

$$S_{n+1} = S_n + Z_n = S_0 + \sum_{k=1}^{n} Z_k$$

Since Z_n and $2X_n - 1$ have the same distribution, we have

$$S_{n+1} = S_n + Z_n = S_0 + \sum_{k=1}^{n} (2X_k - 1)$$

In fact, we have just described a special example of stochastic processes, which is called a random walk.

Example 2.5 Let (Ω, F, P) be a probability space and $\{M_n, n \geqslant 1\}$ be independent identically distributed random variables, satisfying

$$P\{M_n = 1\} = p, \ P\{M_n = 0\} = 1 - p = q, \quad n = 1, 2, \cdots$$

Let

$$X_n = X_0 + \sum_{i=1}^{n} (2M_i - 1), \quad n = 1, 2, \cdots$$

where X_0 is a random variable taking its values in integers, and usually set to be 0. The random variable sequence $\{M_n, n \geqslant 1\}$ is called a Bernoulli sequence, and the random variable sequence $X = \{X_n, n \geqslant 1\}$ is called a one-dimensional simple random walk. In particular, when $p = 0.5$, it is called the one-dimensional simple symmetric random walk.

It is clear from the above definition that the stochastic process $X(t, \zeta)$ is a collection of time series $\{X(t, \zeta_1), X(t, \zeta_2), \cdots, X(t, \zeta_m), \cdots\}$ corresponding to all possible outcomes $\zeta \in \Omega$. For a fixed $t_1 \in T, X(t_1)$ is a one-dimensional random variable.

Definition 2.22 Consider a stochastic process $X(t)$ in the probability space (Ω, F, P). For any fixed $t \in T$, we call

$$F_X(x;t) = P\{X(t) \leqslant x\}, \qquad t \in T \tag{2.42}$$

the one-dimensional distribution of the stochastic process $X(t)$. In addition, if the partial derivative of $F_X(x;t)$ exists, then

$$f_X(x;t) = \frac{\partial F_X(x;t)}{\partial x} \tag{2.43}$$

is called the one-dimensional probability density function of the stochastic process $X(t)$.

For two fixed time instants t_1, t_2, the states $X(t_1), X(t_2)$ constitute the two dimensional variable $[X(t_1), X(t_2)]$ of the stochastic process $X(t)$, and the distribution is

$$F_X(x_1, x_2; t_1, t_2) = P\{X(t_1) \leqslant x_1, X(t_2) \leqslant x_2\} \tag{2.44}$$

When the second order joint partial derivative of $F_X(x_1, x_2; t_1, t_2)$ with respect to x_1, x_2 exists, we call

$$f_X(x_1, x_2; t_1, t_2) = \frac{\partial^2 F_X(x_1, x_2; t_1, t_2)}{\partial x_1 \partial x_2} \tag{2.45}$$

the two-dimensional probability density.

Similar to random variables, we define several characteristics for stochastic processes.

Definition 2.23 The expectation of a stochastic process $X(t)$ is

$$E[X(t)] = \int_{-\infty}^{\infty} x f_X(x;t)\mathrm{d}x = m_X(t) \tag{2.46}$$

Definition 2.24 The mean square of a stochastic process $X(t)$ is

$$E[X^2(t)] = \int_{-\infty}^{\infty} x^2 f_X(x;t)\mathrm{d}x = \phi_X^2(t) \tag{2.47}$$

Definition 2.25 The variance of a stochastic process $X(t)$ is

$$\begin{aligned} D[X(t)] &= E\{[X(t) - m_X(t)]^2\} \\ &= \int_{-\infty}^{\infty} [x - m_X(t)]^2 f_X(x;t)\mathrm{d}x \\ &= \sigma_X^2(t) \end{aligned} \tag{2.48}$$

Definition 2.26 The autocorrelation function of a stochastic process $X(t)$ is

$$\begin{aligned} R_X(t_1,t_2) &= E[X(t_1)X(t_2)] \\ &= \int_{-\infty}^{\infty}\int_{-\infty}^{\infty} x_1 x_2 f_X(x_1,x_2;t_1,t_2)\mathrm{d}x_1\mathrm{d}x_2 \end{aligned} \tag{2.49}$$

The autocorrelation function is one of the most important characteristics of stochastic processes.

Definition 2.27 The autocovariance of a stochastic process $X(t)$ is

$$\begin{aligned} C_X(t_1,t_2) &= E\{[(X(t_1) - m_X(t_1))(X(t_2) - m_X(t_2))]\} \\ &= \int_{-\infty}^{\infty}\int_{-\infty}^{\infty} [x_1 - m_X(t_1)][x_2 - m_X(t_2)] f_X(x_1,x_2;t_1,t_2)\mathrm{d}x_1\mathrm{d}x_2 \end{aligned} \tag{2.50}$$

When $t_1 = t_2 = t$, the autocovariance of $X(t)$ is just the variance, namely

$$\begin{aligned} C_X(t,t) &= E\{[X(t) - m_X(t)]^2\} \\ &= D[X(t)] \\ &= \phi_X^2(t) - m_X^2(t) \\ &= \sigma_X^2(t) \end{aligned} \tag{2.51}$$

Definition 2.28 The autocorrelation coefficient of $X(t)$ is

$$\rho_X(t_1,t_2) = \frac{C_X(t_1,t_2)}{\sigma_X(t_1)\sigma_X(t_2)} \tag{2.52}$$

where $\sigma_X(t_1) \neq 0$, and $\sigma_X(t_2) \neq 0$.

It needs to be pointed out that in Definitions 2.23 – 2.27, we require that the generalized integrations must converge.

2.2.2 Stationary Stochastic Processes

There are two classes of stochastic processes. One is called non-stationary stochastic processes, whose probability characteristics change with time, and the other is called stationary stochastic processes, whose probability characteristics remain the same as time evolves. In this book, we mainly focus on the latter.

Definition 2.29 For a stochastic process, if its n-dimensional distribution is time invariant, namely

$$f_X(x_1,x_2,\cdots,x_n;t_1+\Delta t,t_2+\Delta t,\cdots,t_n+\Delta t)=f_X(x_1,x_2,\cdots,x_n;t_1,t_2,\cdots,t_n) \quad (2.53)$$

then we say it is a strictly stationary stochastic process.

The probability characteristics of a strictly stationary stochastic process are not affected when shifted in time. Even when time approaches infinity, its probability characteristics should remain the same. In other words, a strictly stationary stochastic process does not have a starting or ending time. So such stochastic processes do not exist in real life. However, for analysis purposes, if the probability characteristics change only slightly during the time period of interest, then the stochastic process can be seen as strictly stationary.

For a strictly stationary stochastic process, its one-dimensional and two-dimensional probability densities and characteristics have the following properties:

Proposition 2.1 The one-dimensional probability density function and characteristics of a strictly stationary stochastic process are independent of time t.

In fact, according to the definition of strictly stationary stochastic processes, when $\Delta t=-t_1$, we have

$$f_X(x_1;t_1)=f_X(x_1;t_1+\Delta t)=f_X(x_1;0)=f_X(x_1) \quad (2.54)$$

Thus, the expectation, mean variance and variance are all independent of time, which are usually denoted by m_X,ϕ_X^2 and σ_X^2 respectively.

Proposition 2.2 The two-dimensional probability density function and characteristics of strictly stationary stochastic processes are only dependent on the time difference $\tau=t_2-t_1$, and independent of the starting time t_1.

The two-dimensional probability density of a strictly stationary stochastic process is

$$\begin{aligned} f_X(x_1,x_2;t_1,t_2) &= f_X(x_1,x_2;t_1+\Delta t,t_2+\Delta t) \\ &= f_X(x_1,x_2;0,t_2-t_1) \\ &= f_X(x_1,x_2;\tau) \end{aligned} \quad (2.55)$$

where $\Delta t=-t_1$.

Therefore, the autocorrelation function is the function of the single variable τ, namely

$$R_X(t_1,t_2)=R_X(\tau) \quad (2.56)$$

Similarly, the autocovariance function is

$$C_X(t_1,t_2) = C_X(\tau) = R_X(\tau) - m_X^2 \tag{2.57}$$

When $t_2 = t_1 = t, \tau = 0$, we have

$$C_X(0) = \sigma_X^2 = R_X(0) - m_X^2 \tag{2.58}$$

It is clear from the definition of a strictly stationary stochastic process that its finite dimensional distributions are invariant when shifted in time. This is a strong condition and difficult to verify. So in practice, people usually discuss stationary processes using second moments and the most commonly used stationary process is called wide-sense stationary stochastic processes.

Definition 2.30 For a stochastic process $X(t)$, if its one-dimensional and two-dimensional distributions are independent of the starting time, namely

$$f_X(x_1,x_2;t_1+\Delta t,t_2+\Delta t) = f_X(x_1,x_2;t_1,t_2) \tag{2.59}$$

and in addition, if its expectation is constant, its autocorrelation function depends only on the time difference, and its mean square is finite, namely

$$\begin{cases} E[X(t)] = m_X \\ R_X[t_1,t_2] = E[X(t_1)X(t_2)] = R_X(\tau) \\ E[X^2(t)] < \infty \end{cases} \tag{2.60}$$

then $X(t)$ is said to be a wide-sense stationary stochastic process (or generalized stationary stochastic process).

Remark 2.1 In this book, if not specified otherwise, we take stationary stochastic processes as wide-sense stationary stochastic processes.

Now we give a simple example for wide-sense stationary stochastic processes.

Example 2.6 Consider the sinusoidal series with random phases

$$X(n) = A\sin(2\pi fnT_s + \Phi)$$

where A, f, T_s are constants, and Φ follows the uniform distribution on $[0,2\pi]$. Obviously, for each Φ in $[0,2\pi]$, the corresponding sample $X(n)$ is a sinusoid signal. Its expectation is

$$E(X(n)) = \int_0^{2\pi} A\sin(2\pi fnT_s + \varphi)\frac{1}{2\pi}\mathrm{d}\varphi = 0$$

and its autocorrelation is

$$R_X(n_1,n_2) = \frac{A^2}{2}\cos[2\pi f(n_2-n_1)T_s]$$

Furthermore, since $m_X = 0$, its autocovariance is

$$C_X(m) = R_X(m) = \frac{A^2}{2}\cos(2\pi f m T_s)$$

where $m = n_2 - n_1$. Hence, the above random-phase sinusoidal series is a wide-sense stationary stochastic process.

The autocorrelation function of a stationary stochastic process has the following properties:

(1) The value of the autocorrelation function at $\tau = 0$ is the nonnegative mean square of the stochastic process, namely

$$R_X(0) = E[X^2(t)] = \phi_X^2 \geqslant 0 \tag{2.61}$$

(2) The autocorrelation function and autocovariance function are even functions with respect to τ, namely

$$R_X(-\tau) = R_X(\tau) \tag{2.62a}$$

$$C_X(-\tau) = C_X(\tau) \tag{2.62b}$$

(3) The autocorrelation function and autocovariance function reach their maximum at $\tau = 0$, namely

$$R_X(0) \geqslant |R_X(\tau)| \tag{2.63a}$$

$$C_X(0) = \sigma_X^2 \geqslant |C_X(\tau)| \tag{2.63b}$$

(4) If $X(t)$ is a periodic stationary process, satisfying $X(t+T) = X(t)$, then its autocorrelation function is also periodic with the same period as that of $X(t)$, namely

$$R_X(t+T) = R_X(t) \tag{2.64}$$

(5) If the stationary process has no periodic components, then

$$\lim_{|\tau|\to\infty} R_X(\tau) = R_X(\infty) = m_X^2 \tag{2.65}$$

(6) If the mean of the stationary process is m_X, then the mean of its autocorrelation function is m_X^2, namely

$$R_X(\tau) = C_X(\tau) + m_X^2 \tag{2.66}$$

In order to describe solely the linear correlation of the states of a stationary process at two different time instants, we need to normalize the autocorrelation function. We define the autocorrelation coefficient of a stationary process $X(t)$ to be

$$\rho_X(\tau)=\frac{C_X(\tau)}{\sigma_X^2}=\frac{R_X(\tau)-m_X^2}{\sigma_X^2} \tag{2.67}$$

Then $\rho_X(\tau)=0$ implies that the two signals are linearly unrelated, and $|\rho_X(\tau)|=1$ implies the strongest correlation.

In addition, for a fixed τ_0, if $X(t)$ and $X(t+\tau)$ are unrelated for all $\tau>\tau_0$, then τ_0 is called the correlation time. We give three different definitions for the correlation time which are used in different engineering applications.

Definition 2.31 The correlation time τ_0 is defined to be the length of the time interval across which the autocorrelation coefficient decreases from 1 to 0.05, i.e.

$$|\rho_X(\tau_0)|=0.05 \tag{2.68}$$

Definition 2.32 The correlation time τ_0 for a stationary process without high frequency components is defined to be

$$\tau_0=\int_0^\infty \rho_X(\tau)\mathrm{d}\tau \tag{2.69}$$

Definition 2.33 For a stationary process with high frequency components satisfying $\rho_X(\tau)=a_X(\tau)\cos\omega_0\tau$, the correlation time τ_0 is defined to be the integration of the envelop $a_X(\tau)$ of the autocorrelation function, namely

$$\tau_0=\int_0^\infty a_X(\tau)\mathrm{d}\tau \tag{2.70}$$

2.2.3 Ergodic Processes

To characterize completely all the statistical characteristics of a stochastic process, we need, at least in theory, to have the n-dimensional probability density or the n-dimensional distribution function or all the sample trajectories, which are usually difficult to realize in practice because of the required large number, which may tend to infinity, of tests or observations. To deal with this difficulty, people have been looking for conditions under which the observation of a single sample function during a finite time interval can be used as the sufficient statistics for the whole process.

The Russian mathematician Khinchin proved that[5] under certain conditions, there exists a class of stationary stochastic processes having the property that the time averages of any of its sample functions converge to the corresponding averages of this process in the statistical sense. The stochastic processes with this property are called ergodic stochastic processes (or ergodic processes for short). We define it mathematically as follows:

Definition 2.34 For a stationary stochastic processs *X*(*t*), if its time average over a sufficiently long period of time converges to its statistical average with probability one, namely

$$P\{\lim_{n\to\infty} \text{various time averages} = \text{the corresponding statistical averages}\} = 1 \quad (2.71)$$

then we say $X(t)$ is strictly ergodic, and $X(t)$ is called a strict-sense ergodic process.

Definition 2.35 Consider a stationary stochastic process $X(t)$. If

$$\overline{X(t)} = E[X(t)] = m_X \quad (2.72)$$

exists with probability one, then we say the mean of $X(t)$ is ergodic.

If

$$\overline{X(t)X(t+\tau)} = E[X(t)X(t+\tau)] = R_X(\tau) \quad (2.73)$$

exists with probability 1, the autocorrelation function of $X(t)$ is said to be ergodic.

If the mean and autocorrelation function of $X(t)$ are both ergodic, then we say $X(t)$ is a wide-sense ergodic process.

The ergodicity of a stochastic process has important practical meanings. For a general stochastic process, its time average is a random variable while for an ergodic process, it tends to a deterministic value. Said differently, the time average of the different sample functions of an ergodic process can be considered to be the same. Thus, the time average (the statistical average of the whole process) of an ergodic process can be calculated by the time average of a random sample function, that is

$$E[X(t)] = \lim_{T\to\infty} \frac{1}{2T} \int_{-T}^{T} x(t)\mathrm{d}t \quad (2.74)$$

$$R_X(\tau) = \lim_{T\to\infty} \frac{1}{2T} \int_{-T}^{T} x(t)x(t+\tau)\mathrm{d}t \quad (2.75)$$

The above calculations are convenient to use in solving various practical engineering problems, and this is exactly the significance for introducing the concept of ergodicity.

Ergodic processes are always stationary, but the converse is generally not true. Now we give three theorems to determine ergodicity[5].

Theorem 2.1 The mean of a stationary stochastic process is ergodic if and only if

$$\lim_{T\to\infty} \frac{1}{T} \int_{0}^{2T} \left(1 - \frac{\tau}{2T}\right)[R_X(\tau) - m_X^2]\mathrm{d}\tau = 0 \quad (2.76)$$

Theorem 2.2 The autocorrelation function of a stationary stochastic process is ergodic if and only if

$$\lim_{T\to\infty}\frac{1}{T}\int_0^{2T}\left(1-\frac{\tau}{2T}\right)[R_\Phi(\tau)-E^2[\Phi(t)]]\mathrm{d}\tau=0 \tag{2.77}$$

where

$$\Phi(t)=X(t+\tau_1)X(t)$$
$$R_\Phi(\tau)=E[\Phi(t+\tau)\Phi(t)]$$
$$E\left[\Phi(t)\right]=E[X(t+\tau_1)X(t)]=R_X(\tau_1)$$

Theorem 2.3 A stationary Gaussian process is ergodic if and only if

$$\int_0^\infty |R_X(\tau)|\,\mathrm{d}\tau<\infty$$

holds when the mean is zero, and the autocorrelation function is continuous.

In practice, it is difficult to prove whether a stationary stochastic process satisfies the conditions described above. In fact, since all the samples of the same stochastic process are affected by the same random factors, different sample functions have the same probability distribution. So most stationary stochastic processes can be approximately considered to be ergodic. In practice, ergodicity is usually taken as a conjecture, and then checked by experiments.

2.2.4 Estimate of Time Domain Characteristics of Random Signals

Experiments are necessary when studying the statistical characteristics of random signals. Various random signals arise in practice, such as recorded earthquake waves, and measured human brain waves and so on. They are actually realizations (samples) of random signals. It is almost impossible to obtain in theory the statistical characteristics, such as the expectation, variance, autocorrelation function and power spectrum density, of these random signals. Only by using experiments can we estimate the statistical characteristics.

The statistical experimental study is based on the assumption that the random signals of interest are ergodic. Under this assumption, the statistical characteristics of the random signals can be estimated using samples.

(1) Estimated expectation

$$\hat{m}_X=\frac{1}{N}\sum_{i=0}^{N-1}x_i \tag{2.78}$$

When the sample data are independent, the estimate of (2.78) is uniformly unbiased, and thus valid. But if the contrary is true, the consistency decreases, and the estimated variance is greater than that in the independent data case, but less than $1/N$ of the variance of the signal.

(2) Estimated variance

$$\hat{\sigma}_X^2 = \frac{1}{N}\sum_{i=0}^{N-1}(x_i - \hat{m}_X)^2 \tag{2.79}$$

Equations (2.78) and (2.79) are both maximum likelihood estimations under the assumption that $\{X(n)\}$ satisifies the Gaussian distribution. When the density function of $\{X(n)\}$ is unknown or does not follow the Gaussian distribution, Eqs. (2.78) and (2.79) can still be used, in which case the expectation estimator is uniformly unbiased, and the variance estimator is asymptotically uniformly unbiased. However, here when the number of samples is finite, the estimations are not guaranteed to be optimal since they are no longer the maximum likelihood estimations.

(3) Estimated autocorrelation function

$$\hat{R}_X(m) = \frac{1}{N}\cdot\sum_{i=0}^{N-|m|-1} x_i x_{i+|m|}, \qquad m = 0,1,\cdots,N-1 \tag{2.80}$$

When the $X(t)$ is an independent Gaussian process, $\hat{R}_X(0)$ is a maximum likelihood estimation. When the X_i of $\{X(n)\}$ are independent but not Gaussian distribution, $\hat{R}_X(0)$ is an asymptotically uniformly unbiased estimation. When X_i are dependent, we still use the same estimator although it is not optimal and even not asymptotically uniformly unbiased any more.

In addition, the number of summands in $\hat{R}_X(m)$ is usually less than N. When m is large, the number of summands is very small. When $m \geqslant N$, $x_i x_{i+|m|} = 0$. In this case, we assume

$$\hat{R}_X(m) = 0, \qquad m \geqslant N \tag{2.81}$$

(4) Estimated probability density

The estimation of the multi-dimensional probability density is complicated, and here we only introduce a simple estimation method for one-dimensional probability density, which is called the histogram method. This method also assumes that the distribution function is ergodic.

For a set of sample data $x_0, x_1, \cdots, x_{N-1}$ of a stochastic process $X(t)$, the histogram of $X(t)$ can be obtained by following the steps below.

Step 1 Identify the minimum

$$a = \min\{x_0, x_1, \cdots, x_{N-1}\}$$

and the maximum

$$b = \max\{x_0, x_1, \cdots, x_{N-1}\}$$

Step 2 Divide the intervals.

Divide the interval $[a,b)$ into K non-overlapping subintervals by choosing the points $a_0 < a_1 < \cdots < a_k < \cdots < a_K$ where $a_0 \leqslant a$ and $a_K > b$. Then the subintervals are $[a_{k-1}, a_k)$ with $k = 1, 2, \cdots, K$.

Step 3 Calculate the empirical frequency of the data.

Count the number N_k of points that fall into the interval $[a_{k-1}, a_k)$, namely the number of x_n that satisfies

$$a_{k-1} \leqslant x_n < a_k, \qquad n = 0, 1, 2, \cdots, N-1$$

Then the empirical frequency is defined by

$$f_k = \frac{N_k}{N}, \qquad k = 1, 2, \cdots, K \tag{2.82}$$

Obviously, $0 \leqslant f_k \leqslant 1$, and $\sum_{k=1}^{K} f_k = 1$. When N is sufficiently large, f_k can be approximately taken as the probability for the random variable X to take values in $[a_{k+1}, a_k)$.

Step 4 Calculate the probability density estimation in $[a_{k-1}, a_k)$

$$\hat{f}(x) = \frac{f_k}{a_k - a_{k-1}}, \qquad k = 1, 2, \cdots, K; x \in [a_{k-1}, a_k) \tag{2.83}$$

2.3 Frequency Domain Characteristics — Power and Spectrum

2.3.1 Power Spectral Density of Real Stochastic Processes

Consider a deterministic signal $s(t)$ which is an aperiodic real function of time t. The condition for the existence of its Fourier transform is that $s(t)$ satisfies the Dirichlet condition for all t, that is $\int_{-\infty}^{\infty} |s(t)| \, \mathrm{d}t < \infty$ is absolutely integrable, which can be equivalently stated as $\int_{-\infty}^{\infty} |s(t)|^2 \, \mathrm{d}t < \infty$ (the total energy of $s(t)$ is finite).

Suppose $s(t)$ satisfies the above condition, then we have the following Fourier transforms.

(1) Fourier forward transform from the time domain to the frequency domain

$$S(\omega) = \int_{-\infty}^{\infty} s(t) \mathrm{e}^{-\mathrm{j}\omega t} \mathrm{d}t \tag{2.84}$$

$S(\omega)$ is also called the spectrum of $s(t)$.

(2) Fourier inverse transform from the frequency domain to the time domain

$$s(t) = \frac{1}{2\pi}\int_{-\infty}^{\infty} S(\omega)\mathrm{e}^{\mathrm{j}\omega t}\mathrm{d}\omega \tag{2.85}$$

Furthermore, for a real signal $s(t)$, the Parseval equality is

$$\int_{-\infty}^{\infty}[s(t)]^2\,\mathrm{d}t = \frac{1}{2\pi}\int_{-\infty}^{\infty}|S(\omega)|^2\,\mathrm{d}\omega \tag{2.86}$$

where the left-hand side describes the total energy of $s(t)$ for all time t, and the right-hand side is the integration of $|S(\omega)|^2$ over all the frequencies. So $|S(\omega)|^2$ is also called the power spectral density of $s(t)$.

Since a sample function of a stochastic process evolves as t goes to infinity, and its total energy can be infinite, the Dirichlet condition may not be satisfied. So the frequency spectrum of the sample function may not exist. On the other hand, since the average power is always finite, or said differently the average power of the sample function always satisfies the Dirichlet condition, the power spectrum always exists.

If one wants to apply the Fourier transform to any real stochastic process, some modifications have to be used and truncating is the simplest choice.

Definition 2.36 For the sample function $X(t,\zeta_k) = x_k(t)$ of a real stochastic process $X(t)$, its truncation function is

$$x_{kT}(t) = \begin{cases} x_k(t), & |t| \leqslant T \\ 0, & |t| > T \end{cases} \tag{2.87}$$

where $T>0$ is a positive number.

If T is finite, the truncation function $x_{kT}(t)$ satisfies the absolutely integrable condition, so its Fourier transform exists, namely

$$\chi_{kT}(\omega) = \int_{-\infty}^{\infty} x_{kT}(t)\mathrm{e}^{-\mathrm{j}\omega t}\mathrm{d}t = \int_{-T}^{T} x_k(t)\mathrm{e}^{-\mathrm{j}\omega t}\mathrm{d}t \tag{2.88}$$

$$x_{kT}(t) = \frac{1}{2\pi}\int_{-\infty}^{\infty}\chi_{kT}(\omega)\mathrm{e}^{\mathrm{j}\omega t}\mathrm{d}\omega \tag{2.89}$$

It can be seen from the above equations that $\chi_{kT}(\omega) = \chi_T(\omega,\zeta_k)$ is the frequency spectrum function of $x_{kT}(t)$. In addition, from the Parseval equation, we have

$$\int_{-\infty}^{\infty} x_{kT}^2(t)\mathrm{d}t = \int_{-T}^{T} x_k^2(t)\mathrm{d}t = \frac{1}{2\pi}\int_{-\infty}^{\infty}\left|\chi_{kT}(\omega)\right|^2\mathrm{d}\omega \tag{2.90}$$

The frequency spectrum of the truncation function of a stochastic process has clear physical meanings. For example, if $X(t)$ denotes voltage with noise, then

$\int_{-T}^{T} x_k^2(t)\mathrm{d}t$ is the total energy consumed by a 1 Ω resistor during the time interval $(-T,T)$ under the sample noise ζ_k. The time average limit of the total energy is

$$P_k = \lim_{T\to\infty}\frac{1}{2T}\int_{-T}^{T} x_k^2(t)\mathrm{d}t = \lim_{T\to\infty}\frac{1}{4\pi T}\int_{-\infty}^{\infty}\left|\chi_{kT}(\omega)\right|^2 \mathrm{d}\omega \tag{2.91}$$

In fact, P_k is the time average power of the sample function $x_k(t)$.

For different ζ_k, the sample function $x_k(t)$ are different, and so are the P_k. Corresponding to experimental results $\zeta \in \Omega$, the set $\{P_k\} = P_\Delta(\zeta)$ of the average power P_k of all the sample functions is a random variable, namely

$$\begin{aligned} P_\Delta(\zeta) &= \lim_{T\to\infty}\frac{1}{2T}\int_{-T}^{T} X^2(t,\zeta)\mathrm{d}t \\ &= \lim_{T\to\infty}\frac{1}{4\pi T}\int_{-\infty}^{\infty}\left|X_T(\omega,\zeta)\right|^2 \mathrm{d}\omega \end{aligned} \tag{2.92}$$

where $X(t,\zeta) = X(t)$ is a stochastic process, and $X_T(\omega,\zeta) = X_T(\omega)$ is the frequency spectrum of its truncation function.

Taking the statistical average of $P_\Delta(\zeta)$, we have

$$\begin{aligned} P &= E[P_\Delta(\zeta)] \\ &= \lim_{T\to\infty}\frac{1}{2T}\int_{-T}^{T} E[X^2(t)]\mathrm{d}t \\ &= \lim_{T\to\infty}\frac{1}{4\pi T}\int_{-\infty}^{\infty} E[|X_T(\omega)|^2]\mathrm{d}\omega \end{aligned} \tag{2.93}$$

We call the P determined by the above equation the mean power of the stochastic process $X(t)$.

We denote the integration on the right-hand side by

$$G_X(\omega) = \lim_{T\to\infty}\frac{1}{2T}E[|X_T(\omega)|^2] \tag{2.94}$$

Then the integration of the $G_X(\omega)$ over the whole frequency domain is the mean power of $X(t)$. So $G_X(\omega)$ describes the distribution of the mean power of $X(t)$ at different frequencies, and thus we say $G_X(\omega)$ is the power spectral density of $X(t)$.

The power spectral density is an important statistical characteristic of stochastic processes in the frequency domain, and it has the following properties:

(1) It is non-negative

$$G_X(\omega) \geqslant 0 \tag{2.95}$$

(2) It is real

$$G_X^*(\omega) = G_X(\omega) \tag{2.96}$$

(3) It is symmetric, namely an even function of ω

$$G_X(\omega) = G_X(-\omega) \tag{2.97}$$

(4) It is integrable

$$\int_{-\infty}^{\infty} G_X(\omega)\mathrm{d}\omega < \infty \tag{2.98}$$

When we write $G_X(\omega)$ into the following form of a rational function

$$G_X(\omega) = G_0 \frac{\omega^{2m} + a_{2m-2}\omega^{2m-2} + \cdots + a_0}{\omega^{2n} + b_{2n-2}\omega^{2n-2} + \cdots + b_0} \tag{2.99}$$

then from properties (1) and (4), we know that:

(1) $G_0 > 0$;

(2) The denominator has no real roots (no poles on the real axis), and $n > m$.

The deterministic signal $s(t)$ and its spectrum $S(\omega)$ are a pair of Fourier transforms between the time domain and frequency domain. Similarly, one can show that the autocorrelation function of a stationary stochastic process and its power spectral density are also a pair of Fourier transforms. So if $R_X(\tau)$ is absolutely integrable, then

$$G_X(\omega) = \int_{-\infty}^{\infty} R_X(\tau)\mathrm{e}^{-\mathrm{j}\omega\tau}\mathrm{d}\tau \tag{2.100}$$

and

$$R_X(\tau) = \frac{1}{2\pi}\int_{-\infty}^{\infty} G_X(\omega)\mathrm{e}^{\mathrm{j}\omega\tau}\mathrm{d}\omega \tag{2.101}$$

These relationships are described in the well-known Wiener-Khinchin theorem[6].

2.3.2 Cross-Power Spectral Density of Real Stochastic Processes

For two given stochastic processes $X(t)$ and $Y(t)$, their *cross-power spectral density* is

$$G_{XY}(\omega) = \lim_{T\to\infty} \frac{1}{2T} E[X_T^*(\omega)Y_T(\omega)] \tag{2.102}$$

and the cross average power is

$$P_{XY} = \frac{1}{2\pi}\int_{-\infty}^{\infty} G_{XY}(\omega)\mathrm{d}\omega \tag{2.103}$$

Similarly, the cross-power spectral density of $Y(t)$ and $X(t)$ is defined by

$$G_{YX}(\omega)=\lim_{T\to\infty}\frac{1}{2T}E[Y_T^*(\omega)X_T(\omega)] \tag{2.104}$$

The relationship between the above two cross-power spectral densities is

$$G_{XY}(\omega)=G_{YX}^*(\omega) \tag{2.105}$$

If $X(t),Y(t)$ are joint stationary, we have

$$G_{XY}(\omega)=\int_{-\infty}^{\infty}R_{XY}(\tau)\mathrm{e}^{-\mathrm{j}\omega\tau}\mathrm{d}\tau \tag{2.106}$$

$$R_{XY}(\tau)=\frac{1}{2\pi}\int_{-\infty}^{\infty}G_{XY}(\omega)\mathrm{e}^{\mathrm{j}\omega\tau}\mathrm{d}\omega \tag{2.107}$$

So for two jointly stationary real stochastic processes, their cross-power spectral density and cross-correlation function are a pair of Fourier transforms.

The cross-power spectral density (cross spectral density for short) of two stochastic processes has different properties compared with those of a single stochastic process. It is no longer a nonnegative real even function with respect to the frequency ω. Now we give several properties of the cross-power spectral density.

(1) The cross-power spectral density is not even, namely

$$G_{XY}(\omega)=G_{YX}^*(\omega)=G_{YX}(-\omega) \tag{2.108}$$

(2) The real part of the cross-power spectral density is even with respect to ω, namely

$$\mathrm{Re}[G_{XY}(\omega)]=\mathrm{Re}[G_{XY}(-\omega)] \tag{2.109}$$

$$\mathrm{Re}[G_{YX}(\omega)]=\mathrm{Re}[G_{YX}(-\omega)] \tag{2.110}$$

(3) The imaginary part of the cross-power spectral density is odd with respect to ω, namely

$$\mathrm{Im}[G_{XY}(\omega)]=-\mathrm{Im}[G_{XY}(-\omega)] \tag{2.111}$$

$$\mathrm{Im}[G_{YX}(\omega)]=-\mathrm{Im}[G_{YX}(-\omega)] \tag{2.112}$$

(4) If $X(t),Y(t)$ are orthogonal, then

$$G_{XY}(\omega)=G_{YX}(\omega)=0 \tag{2.113}$$

(5) If $X(t),Y(t)$ are uncorrelated with constant mean values m_X,m_Y, then

$$R_{XY}(t,t+\tau)=m_Xm_Y \tag{2.114}$$

$$G_{XY}(\omega)=G_{YX}(\omega)=2\pi m_Xm_Y\delta(\omega) \tag{2.115}$$

(6) The time average of the cross-correlation function and the cross-power spectral density are a pair of Fourier transforms, namely

$$\overline{R_{XY}(t,t+\tau)} \rightleftharpoons G_{XY}(\omega) \tag{2.116}$$

$$\overline{R_{YX}(t,t+\tau)} \rightleftharpoons G_{YX}(\omega) \tag{2.117}$$

where $\overline{(\cdot)} = \lim_{T\to\infty}\frac{1}{2T}\int_{-T}^{T}(\cdot)\mathrm{d}t$ denotes the time average and $\rightleftharpoons$ means that its left- and right-hand sides are a pair of Fourier transforms.

2.4 Brownian Motion and Noise

Brownian motion is a stochastic process with continuous parameters and continuous state space. It is the simplest, and yet most important stochastic process and many other stochastic processes can be seen as its generalizations. At the same time, Brownian motion is also a basic model used in the complexity study.

2.4.1 Brownian Motion

The study on Brownian motion started in England[4]. In 1827, the British botanist Robert Brown found that the pollen particles suspended in liquid move randomly. In the late 19th century, this phenomenon was explained by the hitting of a large number of molecules. Later in 1905, Albert Einstein gave the first theoretical explanation for Brownian motion, and described systemically its mechanism. Nowadays, Brownian motion has been used as a model in different fields of natural and social sciences, and become a popular model in complex systems science.

Roughly speaking, Brownian motion can be described by the displacement $\{X(s): 0 \leqslant s \leqslant t\}$ of a particle following a random walk during the finite time interval $[0,t]$.

Definition 2.37 A stochastic process is said to be a Brownian motion if it satisfies the following three conditions:

(1) Independent increment: the increments $X(t_1)-X(s_1), X(t_2)-X(s_2), \cdots, X(t_n)-X(s_n)$ of non-overlapping intervals $(s_1,t_1], (s_2,t_2], \cdots, (s_n,t_n]$ are independent;

(2) $X(t+h)-X(t)$ satisfies the Gaussian distribution $N(0,\lambda t)$ with mean 0 and variance λt;

(3) For any fixed $\varsigma, X(t,\varsigma)$ is a continuous function of t.

In addition, when $\lambda = 1$, it is called a standard Brownian motion.

From the definition of the correlation function, we know Brownian motion has the following properties.

(1) For the correlation function,

$$C_X(s,t) = Cov(X(s), X(t)) = \min(\lambda s, \lambda t) \tag{2.118}$$

In particular,

$$\langle X^2(t)\rangle = Cov(X(t), X(t)) = \lambda t \tag{2.119}$$

$$E(|X(t) - X(s)|^2) = \lambda |t - s| \tag{2.120}$$

(2) Shift invariance: if $\{X(t), t \geqslant 0\}$ is a Brownian motion, then $\{X(t) - X(a), t \geqslant 0\}$ with a being a constant is also a Brownian motion.

(3) Scale invariance: if $\{X(t), t \geqslant 0\}$ is a Brownian motion, $\left\{\frac{X(at)}{\sqrt{a}}, t \geqslant 0\right\}$ with a being a constant is also a Brownian motion.

Remark 2.2 The variance of a Brownian motion increases with time, so it is a non-stationary stochastic process.

Now we introduce several characteristics of Brownian motion.

(1) The mean displacement variance of a Brownian motion $X(t)$ is linearly proportional to t, namely

$$\langle X^2(t)\rangle \propto t \tag{2.121a}$$

where $\propto$ denotes linear relationships.

(2) The autocorrelation coefficient of a Brownian motion is

$$\rho_X(\tau) = \frac{\langle X(t) \cdot X(t+\tau)\rangle}{\langle X^2(t)\rangle} = \mathrm{e}^{-\frac{\tau}{T}} \tag{2.121b}$$

where T is called the characteristic time.

(3) The power spectrum of a Brownian motion is

$$S(\omega) = \int_0^\infty \rho_X(\tau) \cdot \mathrm{e}^{-\mathrm{j}\omega\tau} \mathrm{d}\tau \propto \omega^{-2} \tag{2.121c}$$

Next, we introduce the fractal dimensional Brownian motion.

Definition 2.38 A continuous stochastic process is called a *fractal dimensional Brownian motion* if it satisfies

(1) $P(X(0) = 0) = 1$

(2) For any time instants s and t, $X(s) - X(t)$ satisfies the Gaussian distribution $N(0, |s-t|^{2\alpha})$ where α is a given parameter.

In particular, when $\alpha = 1/2$, it becomes the standard Brownian motion.

The main difference between a Brownian motion and a fractal dimensional Brownian motion is that the increments of the former are independent while those of the latter are not. If taking time zero as the reference, the correlation coefficient

$C(t)$ of the past increment $\{X(0)-X(-t)\}$ and the future increment $\{X(t)-X(0)\}$ is

$$C(t)=t^{2\alpha-1}-1 \tag{2.122}$$

In view of the physical meaning of the correlation coefficient $C(t)$, the fractal dimensional Brownian motion has the following three forms:

(1) When $\alpha=1/2$, the correlation coefficient is zero, and the series is independent;

(2) When $0<\alpha<1/2$, the correlation coefficient is negative, and the series is negatively correlated;

(3) When $1/2<\alpha<1$, the correlation coefficient is positive, and the series is positively correlated.

So the parameter α can be used to represent the series' correlation degree.

The squared displacement of a fractal dimensional Brownian motion satisfies

$$\langle x^2(t)\rangle \propto t^{2\alpha} \tag{2.123}$$

Next, we investigate the power spectrum of fractal dimensional Brownian motion using the dimensional analysis method. Suppose $S(\omega)\propto\omega^{-\beta}$ where β is the power spectrum index. Since $t^{2\alpha}$ is measured in the unit of power, and so is $\omega\cdot S(\omega)$, it can be concluded that

$$\omega\cdot S(\omega)=\omega\cdot\omega^{-\beta}\propto t^{2\alpha}\propto\omega^{-2\alpha} \tag{2.124}$$

So

$$\beta=2\alpha+1 \tag{2.125}$$

Since in general $0<\alpha<1$, we know $1<\beta<3$.

2.4.2 Noise

We introduce three classes of typical random noises which include white noise, brown noise and $1/f$ noise[7].

(1) White noise

A stationary process $X(t)$ is called a white noise process (or white noise for short) if its mean is zero, its power spectral density distributes uniformly along the whole frequency axis $(-\infty,\infty)$, and satisfies

$$S_X(\omega)=\frac{1}{2}N_0 \tag{2.126}$$

where N_0 is a positive real number.

According to the Wiener-Khinchin theorem, the autocorrelation function of the white noise $X(t)$ is

$$R_X(\tau) = \frac{1}{2} N_0 \delta(\tau) \tag{2.127}$$

which is a δ function with the intensity being the power spectral density. The autocorrelation coefficient is

$$\rho_X(\tau) = \frac{R_X(\tau)}{R_X(0)} = \begin{cases} 1, & \tau = 0 \\ 0, & \tau \neq 0 \end{cases} \tag{2.128}$$

which implies that the signals at any two time instants are independent. Since the power spectral density of white noise is constant, and its autocorrelation function is impulsive, it is convenient to be used as a noise in the calculation with other signals. Furthermore, since the power spectrum of white noise is infinitely wide, its average power is infinite. Although all physical stochastic processes have finite average powers, in practice as long as the bandwidth of a noise is sufficiently wider than the studied signal and its spectral density is approximately even, then it can be treated as white noise without causing much error.

(2) Brown noise

For the frequency spectral density function of a random noise signal $x(t)$

$$\hat{X}(\omega) = \int_{-\infty}^{\infty} x(t) \mathrm{e}^{-\mathrm{j}\omega t} \mathrm{d}t \tag{2.129}$$

we have

$$x(t) = \frac{1}{2\pi} \int_{-\infty}^{\infty} \hat{X}(\omega) \mathrm{e}^{\mathrm{j}\omega t} \mathrm{d}\omega \tag{2.130}$$

From the Wiener-Khinchin theorem, we know that the power spectrum $S(\omega)$ is the square of the norm of its frequency spectral density $\hat{X}(\omega)$, namely

$$S(\omega) = |\hat{X}(\omega)|^2 \propto \omega^{-\beta} \tag{2.131}$$

where $\beta > 0$ is the power spectrum index.

If $x(t)$ is a Brownian motion, then $\beta = 2$, and the spectrum in Eq. (2.131) is the spectrum of the Brownian motion noise. So Brownian motion is also called *Brown noise*.

The white and brown noises are closely related and we briefly discuss their relationship as follows.

Differentiating both sides of Eq. (2.130), we have

$$\frac{\mathrm{d}x(t)}{\mathrm{d}t} = \frac{1}{2\pi} \int_{-\infty}^{\infty} (\mathrm{j}\omega) \hat{X}(\omega) \mathrm{e}^{\mathrm{j}\omega t} \mathrm{d}\omega \tag{2.132}$$

Comparing Eq. (2.132) with Eq. (2.129), it is clear that $\hat{X}(\omega)$ changes to $(\mathrm{j}\omega)\hat{X}(\omega)$, and so the power spectrum of $\hat{x}(t)$ is

$$G(\omega)|_{\dot{x}(t)}=|(\mathrm{j}\omega)\hat{X}(\omega)|^2\propto\omega^{-\beta+2}=\omega^{-(\beta-2)} \tag{2.133}$$

which implies that the power spectrum index changes from β to $\beta-2$.

For a Brownian motion, the power spectrum index of its differentiated signal changes from $\beta=2$ to $\beta=0$, i.e. $G(\omega)\propto\omega^0$, which is the power spectrum of white noise. Therefore, white noise can be obtained by differentiating brown noise and correspondingly brown noise can be obtained by integrating white noise.

(3) $1/f$ noise

The noise that falls between white noise and brown noise with $0<\beta<2$ is called $1/f$ *noise*, which has distinct characteristics in the time domain. Its amplitude or power is inversely proportional to its frequency. In other words, the lower the frequency is, the greater the amplitude will be. This is why it is called $1/f$ noise.

The power spectrums of these three classes of noises are shown in Fig. 2.1[7].

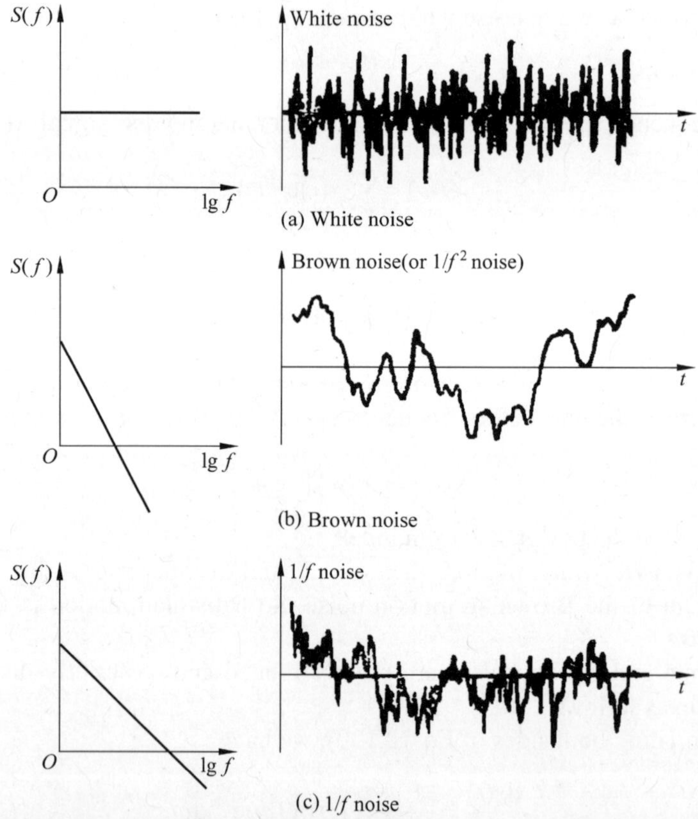

Figure 2.1 Three classes of noises and their power spectrums

2.5 Central Limit Theorem

The normal distribution is one of the most important distributions in probability theory. It has been proven that random variables in various engineering, natural and social problems on both macroscopic and microscopic scales satisfy the normal distribution. The central limit theorem to be discussed in this section shows that a random variable when affected by a large number of independent finite-variance small random factors follows the normal distribution in general.

2.5.1 General Form of the Central Limit Theorem

Definition 2.39 Suppose the expectation and variance of a random variable series $\{X_n\}$ exist, and let

$$\varsigma_n = \left(\sum_{j=1}^{n} X_j\right)^* = \frac{\sum_{j=1}^{n} X_j - E\left[\sum_{j=1}^{n} X_j\right]}{\sqrt{D\left[\sum_{j=1}^{n} X_j\right]}} \tag{2.134}$$

where the superscript * denotes the normalization of the random variable, which implies that the expectation and variance of ς_n are 0 and 1 respectively. We say $\{X_n\}$ satisfies the central limit theorem if it converges to the normal distribution, denoted by $\varsigma_n \xrightarrow{d} Z \sim N(0,1)$, that is

$$F_{\varsigma_n}(x) = P(\varsigma_n \leqslant x) \to \Phi(x), n \to \infty, \quad \forall x \in \mathbb{R} \tag{2.135}$$

where $\Phi(x)$ is a standard normal distribution.

We introduce several commonly used forms of the central limit theorems below[8].

Theorem 2.4 (Lindeberger-Levy) Let $\{X_n\}$ be independently identically distributed. Suppose its variance is positive and finite. Then $\{X_n\}$ satisfies the central limit theorem and ς_n converges uniformly to $N(0,1)$ for all $x \in \mathbb{R}$, namely

$$\varsigma_n = \frac{\sum_{j=1}^{n}(X_j - \mu)}{\sqrt{n}\sigma} = \frac{1}{\sqrt{n}}\sum_{j=1}^{n} X_j^* \xrightarrow{d} Z \sim N(0,1) \tag{2.136}$$

As a special case for the Linderberger-Levy theorem, we consider the De Moivre-Laplace limit theorem that applies to the Bernoulli tests. It shows that the normal distribution can be taken as the approximation of the binomial distribution.

Corollary 2.1 (De Moivre-Laplace) Let the successful number of the n-Bernoulli tests with parameter p be μ_n. Then

$$\lim_{n\to\infty} P\left(a \leqslant \frac{\mu_n - np}{\sqrt{npq}} \leqslant b\right) = \frac{1}{\sqrt{2\pi}}\int_a^b \mathrm{e}^{-t^2/2}\mathrm{d}t \tag{2.137}$$

or

$$\lim_{n\to\infty} P(k_1 \leqslant \mu_n \leqslant k_2) = \Phi\left(\frac{k_2 - np}{\sqrt{npq}}\right) - \Phi\left(\frac{k_1 - np}{\sqrt{npq}}\right) \tag{2.138}$$

and $P\left(\left|\frac{\mu_n}{n} - p\right| < \varepsilon\right) = 2\Phi\left(\varepsilon\sqrt{\frac{n}{pq}}\right) - 1$.

Theorem 2.5 (Lindeberger) Let $\{X_n\}$ be an independent random variable series satisfying the Lindeberger condition

$$\lim_{n\to\infty} \frac{1}{B_n^2}\sum_{k=1}^{n}\int_{|x-\mu_k|>\tau B_n}(x-\mu_k)^2 \mathrm{d}F_{X_k}(x) = 0 \tag{2.139}$$

Then $\{X_n\}$ satisfies the central limit theorem, where $\mu_k = E[X_k]$, $B_n^2 = D\left(\sum_{k=1}^{n} X_k\right)$.

Corollary 2.2 (Lyapunov) Let $\{X_n\}$ be an independent random variable series. If there exists $\delta > 0$ such that

$$\frac{1}{B_n^{2+\delta}}\sum_{k=1}^{n} E|X_k - \mu_k|^{2+\delta} \to 0 \tag{2.140}$$

Then Eq. (2.135) holds for all $x \in \mathbb{R}$.

Corollary 2.3 Let $\{X_n\}$ be an independent random variable series. If there exists a sequence of constant $k_n > 0$ such that

$$\max_{1\leqslant j\leqslant n} |X_j| \leqslant k_n, \quad \lim_{n\to\infty}\frac{k_n}{B_n} = 0 \tag{2.141}$$

Then Eq. (2.135) holds for all $x \in \mathbb{R}$.

Remark 2.3 Theorem 2.5 does not require the random variable series to be identically distributed.

The central limit theorem indicates that if the studied random variable is the sum of many independent random variables, and each individual random variable has only small influence on the sum, then the random variable satisfies approximately the normal distribution. This explains the wide existence of the normal distribution. In particular, if the events are independent, then the event series satisfies the normal distribution. When the central limit theorem does not hold, the events are no longer independent but dependent.

2.5.2 Donsker Invariance Principle

For stochastic processes, we have the following Donsker invariance principle which is similar to the central limit theorem.

Theorem 2.6 If $\{X_n, n > 1\}$ is an independent identically distributed random variable series, and $E(X_i) = 0, D(X_i) = \sigma^2 > 0$. Let

$$S_n = \sum_{i=1}^{n} X_i \tag{2.142}$$

and for any $t \in [0,1]$ define

$$B_n(t) = \frac{S_{[nt]}}{\sqrt{n}\sigma} \tag{2.143}$$

Then for the finite-dimensional distribution, we have

$$B_n(t) \to B(t), \quad n \to +\infty \tag{2.144}$$

where $B(t)$ is the standard normal distribution, i.e. the standard Brownian motion, and $[nt]$ is the integer part of nt.

The Donsker invariance principle is an important generalization of the central limit theorem, which, similar to the central limit theorem, implies the importance of the Brownian motion. It shows that an independent identically distributed stochastic process with finite variance converges to Brownian motion in distribution.

2.6 Scale Invariance

Scale invariance exists widely in various complex systems, and it is related closely to concepts like the power-law distribution, fractal and multi-scale systems. The power-law distribution to be discussed later is a typical scale-free distribution. The word "fractal" was coined by the French American mathematician Benoit Mandelbrot. It is used to describe some irregular, shattered geometric entities or evolving patterns whose components are similar to the whole in a certain way, and the dimensions of the entities or patterns may not be integers. Although fractal is one of the typical examples for system complexity, it also obeys implicitly some simple and regular rules, one of which is the scale-invariant property. Scale invariance is also referred to as self-similarity for physical quantities under the changes of their scale systems. The concept of scale symmetry is also used to describe the fact that some objects after being expanded or reduced in size still have the same properties as the original.

(1) Kolmogorov turbulence[7]

The speed difference $\Delta v(r)$ of two particles that are r distance away from each other in turbulence is a random signal, and its scale index is $\alpha = 1/3$, namely

$$\Delta v(r) \propto r^{\alpha}, \quad \alpha = \frac{1}{3} \tag{2.145}$$

In addition, its variance is

$$D[\Delta v(r)] = \sigma^2_{\Delta v(r)} = \langle \Delta v(r)^2 \rangle$$

and its power spectrum $S(k)$ satisfies

$$\langle (\Delta v(r))^2 \rangle \propto r^{\frac{2}{3}} \tag{2.146}$$

$$S(k) \propto k^{-\frac{5}{3}} \tag{2.147}$$

It is clear from the above calculation that

$$\langle \Delta v(\lambda r)^2 \rangle \propto \lambda^{\frac{2}{3}} \langle \Delta v(r)^2 \rangle \tag{2.148}$$

$$S(\lambda k) \propto \lambda^{-\frac{5}{3}} S(k) \tag{2.149}$$

which implies that Kolmogorov turbulence is scale invariant with $\frac{2}{3}$ and $-\frac{5}{3}$ being the scale indices.

(2) Brownian motion

As discussed previously, the particale's displacement $x(t)$ in a one-dimensional Brownian motion follows the normal distribution, namely

$$p(x,t) = \frac{1}{2\sqrt{\pi v t}} \mathrm{e}^{-\frac{x^2}{4vt}} \tag{2.150}$$

Taking x and t as scales, we perform the following scale transformation:

$$x \to \lambda^{1/2} x, \quad t \to \lambda t \tag{2.151}$$

Then

$$p(\lambda^{1/2}x,\lambda t)=\lambda^{-1/2}p(x,t) \tag{2.152}$$

It is easy to prove that the overall probability is the same under the above transformation, i.e.

$$\int_{-\infty}^{\infty}p(x,t)\mathrm{d}x=\int_{-\infty}^{\infty}p(\lambda^{1/2}x,\lambda t)\mathrm{d}(\lambda^{1/2}x)=1 \tag{2.153}$$

So Brownian motion has the scale-invariance property.

(3) Scale-Free networks

In complex network theory, degree is a simple but important concept describing node properties. It is defined to be the number of edges that are adjacent to a given node. Intuitively, a node with a greater degree might be more important than those with smaller degrees. The node degree distribution can be described by the distribution function $p(k)$, which shows the probability that a randomly chosen node is with the degree k. If the degree distribution satisfies the power law

$$p(k)\propto k^{-r} \tag{2.154a}$$

then the network is said to be scale-free.

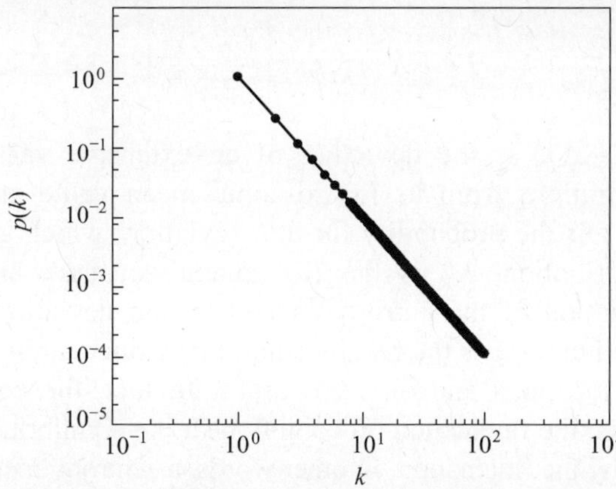

Figure 2.2 Degree distribution of a scale-free network

From Fig. 2.2 we can see that the node degrees of a scale-free network have no obvious characteristic scale. In fact, for any $\lambda>0$, we have

$$p(\lambda k)=(\lambda k)^{-r}=\lambda^{-r}p(k) \tag{2.154b}$$

which shows the self-similarity of scale-free networks.

A typical example of scale-free networks is the airline network in which most network nodes are small airports, but there also exist a few big airports connecting to smaller ones. In general, scale-free networks are not even. In other words, in a scale-free network, most nodes have relatively small degrees, and a few nodes have relatively large degrees. Here, the nodes with large degrees can be considered as the hubs of the network.

2.7 Fluctuation

Fluctuation is often used to explain self-organized criticality (SOC) theory, and it comes from thermodynamics in classical physics. It is closely related to concepts like critical states, phase transition and long-range correlation. Although the definition of critical states in SOC is not completely equivalent to phase transition in thermodynamics, they nevertheless share similarities. In this section we discuss fluctuation and its related properties.

Fluctuation refers to the spontaneous changes of some macroscopic physical quantities from its statistical mean or most probable states which are caused by microcosmic irregular motions in the systems. In classical physics, it is used to describe small deviations that a system at its equilibrium state may still experience. For a system at its equilibrium state, the probability that the fluctuation occurs follows the Einstein equation

$$P(\{\delta X\}) \propto \exp\left[\frac{(\delta^2 S)_0}{2K_B}\right] \tag{2.155}$$

where $\{\delta X\} = \{X - X_0\}$ is the deviation of an extensive variable X (such as substance concentration) from its macroscopic mean value at the equilibrium state X_0, $P(\{\delta X\})$ is the probability for this deviation, which generally satisfies the Gaussian distribution, K_B is the Boltzmann constant, and $(\delta^2 S)_0$ is the second-order deviation of the entropy caused by the deviation of X. Equation (2.155) can be used to discuss the relationship of X's fluctuation in its equilibrium state at different locations and different times. In fact, the equation gives the relationship between the fluctuation probability near the equilibrium and the change in entropy caused by the fluctuation. In other words, as entropy increases, fluctuation reduces on average and thus the system remains stable in its equilibrium region. But the situation is different when the system lies far away from its equilibrium region. Being around unstable states, fluctuation does not decrease, but is amplified. This drives the system to evolve from its unstable state to a new stable state.

From the perspective of stochastic processes, fluctuation can be classified into two categories. One is called internal fluctuation that is caused by the system itself. Its microscopic mechanism is the irregular thermal motion and the local interaction among particles within their neighborhoods. The other is caused by external noise. At a macroscopic level, a system evolves in a complicated environment, and is always under the influence of various external forces. Some external noise might get amplified under certain conditions, which may lead to changes in the system's macroscopic behavior. Around the equilibrium region, although noise may affect the system's stability, and drive the system to some extent away from its stable state, the system's stabilization capability is strong near its equilibrium region and as a result fluctuation is attenuated and the system remains stable. However, when the system state is far from the equilibrium region, external noise and internal fluctuation may reinforce each other through positive feedback, and various small disturbances and fluctuations are amplified until driving the system to lose its stability and evolve towards a new operating point.

On the other hand, when there is no phase transition and the Einstein fluctuation equation holds, the effect of fluctuation compared to the magnitude of the system's macroscopic mean value is always small, and can usually be ignored. In fact, even when the fluctuation reaches the scale of the distance between the particles in the system, the system states are still unrelated. But when approaching phase transition, the fluctuation in the densities of particles and entropy can grow extremely large, and long-range correlation may take place. In principle, the abnormal behavior of fluctuation and long-range correlation near the phase-transition point (critical point) for a system in equilibrium are both rooted in the microscopic interactions among the particles in the system.

For a system in its non-equilibrium state under the combined influence of nonlinear dynamics, ordered spatio-temporal structures may appear. Their appearance implies the existence of long-range correlation for events at different locations and times. Here, the long-range correlation is used in the sense of statistical physics, and it can also be discussed from the angle of probability theory.

When considering all factors that affect fluctuation, the system states cannot just be described by a macroscopic variable $\{X(r,t)\}$, we need to use the distribution functions of the system's various microscopic degrees of freedom as well. However, if the distribution function $\rho(\{r_i\},\{p_i\},t)$ of an n-particle system at its non-equilibrium state is used, it becomes too complicated mathematically. In addition, people are usually not too interested in the details of the system's microscopic states. For example, for a chemical reaction, people are mainly concerned with the concentration of a certain reactant or product, but not particles' momentums or position distributions. Because of this, the Prigogine school has developed a theory that is in the middle of macroscopic descriptions and statistical physics. It uses tools from probability theory to deal with the systems at non-equilibrium states. The main ideas are as follows.

First, the systems' macroscopic states are described by a set of random variables $\{X_i\}(i=1,2,\cdots,n)$. Assume if the system is in a certain global state, then all the X_i take their values between $\{X_i\}$ and $\{X_i+\delta X_i\}$; Let $P(\{X_i,t\})\{\delta X_i\}$ denote the probability that the system is in that global state, then from statistical physics it has the following relationship with the distribution function ρ

$$P(\{X_i\},t)\{\delta X_i\} = \int_{\{X_i,X_i+\delta X_i\}} \{\mathrm{d}r_j\}\{\mathrm{d}p_j\}\rho(\{r_j\},\{p_j\},t) \tag{2.156}$$

where $\delta X_i = X_i - X_i^0$ is the deviation of X_i from its reference X_i^0, which is the fluctuation of X_i.

The correlation between the values of X at the locations r and r' is the spatial correlation that can be computed by

$$Cov(\delta X(r), \delta X(r')) = G_{XX}(r,r') \tag{2.157}$$

If the fluctuations at r and r' are independent, then $G_{XX}(r,r') = 0$.

2.8 Power Law and Correlation

Two classes of power law relationships are considered in SOC theory: one is the algebraic power law relationship, and the other is the probability density function in the exponential form. The commonality lies in the linear relationship in the log-log coordinates.

2.8.1 Power Law

The algebraic power law relationship means that a variable N can be written as the exponential of the other variable s, namely

$$N(s) \propto s^{\alpha} \tag{2.158}$$

where $\alpha \in \mathbb{R}$ is a constant.

The probability power law is defined as follows.

Definition 2.40 If the probability density function of a random variable can be written as the exponential of s, that is

$$N(s) \propto s^{\alpha}, \quad \alpha < 0 \tag{2.159}$$

then s is said to satisfy the power law distribution.

If not specified otherwise, in the following we always take the power law distribution as the probability power law distribution. For the probability distribution or its distribution density function, α is usually negative. For the statistical data obtained from time series analysis, α can be positive or negative. For example, let s be the released energy in an earthquake and $N(s)$ be the number of earthquakes with the released energy being s. Taking the logarithm of both sides of Eq. (2.159), we have

$$\lg N(s) = -\tau \lg s \tag{2.160a}$$

where $\tau = -\alpha$. This shows the linear relationship between $\lg(N(s))$ and $\lg(s)$, and every segment of the linear curve has the same geometric properties of the whole. We also have

$$N(\lambda s) = \lambda^{-\tau} N(s) \tag{2.160b}$$

So the power law distribution implies self-similarity and scale invariance.

In view of the central limit theorem in Section 2.5, the physical meaning of the random variables in the power law distribution is completely different from those in the normal distribution. The normal distribution as a limit distribution implies that the events or random variable series are independent, but the power law distribution indicates scale invariance, long-range correlation and event dependence[9]. It should be pointed out that the power law distribution in practical problems usually takes place at the tails of probability density functions, which is sometimes called the power law tail distribution. We discuss it next.

2.8.2 Levy Distribution

In engineering problems, we often encounter sharp peaks and heavy (long) tails in probability distributions[7–11], which means that a random variable might take a large value with certain probabilities. For example, the annual precipitation in Beijing is 500 – 600 mm, but the precipitation of 50 – 100 mm in a single day is also quite possible. Paul Pierre Levy studied this class of distributions in depth, and obtained their mathematical properties. So this distribution is named after him[12].

From the central limit theorem we know that normal distributions are the most widely seen probability distribution in natural and engineering systems, and the physical meaning behind it is that the probability for extreme events is much smaller than that of normal events. However, long-tail power law distributions deal with different situations where extreme events or random variables of large values can still happen with certain probability that cannot be ignored. In order to further explain the difference between the normal distribution and Levy distribution, we use the following two examples. First consider the heights of adults which usually satisfy the normal distribution. The probability that an adult's height is twice of that of another is almost zero. Now assume that adults' health statuses can be quantized. Then the probability of an adult's health status is twice of that of another person is not a small probability and cannot be ignored any more.

Generally speaking, the Levy distribution is closely related to stable distributions.

Definition 2.41 Suppose random variables X_1 and X_2 are independent and identically distributed with the random variable X. If for any real numbers $a, b > 0$, there exist a positive number c and a real number d such that $aX_1 + bX_2$ and $cX + d$ are identically distributed, then we say X has a stable distribution. In particular, if $d = 0$, then X is said to have a strictly stable distribution.

Example 2.7 Consider a random variable $X \propto N(\mu, \delta^2)$ satisfying the Gaussian distribution. For any real numbers $a, b > 0$, we can always find $aX \propto N(a\mu, (a\delta)^2)$, $bX \propto N(b\mu, (b\delta)^2)$, and

$$c = \sqrt{a^2 + b^2}, \quad d = (a + b - c)\mu$$

such that $aX + bX$ and $cX + d$ are identically distributed.

The above example shows that Gaussian distributions are stable.

Definition 2.42 We say the random variable *X* follows the Levy skew alpha distribution (Levy distribution for short), if its characteristic function takes the form

$$\varphi(t) = \exp[\mathrm{i}\delta t - |\gamma t|^{\alpha} (1 - \mathrm{i}\beta \operatorname{sgn}(t) \Phi)] \tag{2.161}$$

where sng(t) is the sign function of t, and

$$\Phi = \begin{cases} \tan\left(\dfrac{\pi\alpha}{2}\right), & \alpha \neq 1 \\ -\dfrac{2}{\pi}\lg t, & \alpha = 1 \end{cases} \tag{2.162}$$

We denote $X \propto L(\alpha, \beta, \delta, \gamma)$. Here, $\alpha \in (0, 2]$ measures the widths of the peak and tail; $\beta \in [-1, 1]$ indicates the degree of symmetry, in particular, when $\beta = 0$, the distribution is symmetric; δ is the mean value; γ measures the deviation from the mean value.

Similar to the Gaussian distribution, the Levy distribution is also stable according to its characteristic function (2.161). It can be shown that the probability density function of a random variable with Levy distribution can be described by its continuous Fourier transformation

$$f(x; \alpha, \beta, \delta, \gamma) = \frac{1}{2\pi} \int_{-\infty}^{\infty} \varphi(t) \mathrm{e}^{-\mathrm{i}tx} \mathrm{d}t \tag{2.163}$$

In particular, when $\alpha < 2$, the asymptotic behavior is

$$f(x) \to \frac{\alpha c^{\alpha} (1 + \beta) \sin(\pi\alpha / 2) \Gamma(\alpha) / \pi}{|x|^{1+\alpha}} \tag{2.164}$$

where Γ is the Gamma function. When $\alpha < 1$ and $\beta = 1$ (resp. -1), the tail disappears from the left (resp. the right). So for the Levy distribution, the following long-tail probability distribution holds

$$f(x) \propto c x^{-\alpha-1}, \quad |x| \text{ is large} \tag{2.165}$$

where c is called the tail parameter, and $0 \leqslant \alpha \leqslant 2$. When α decreases from 2 to 0, the tail grows fatter and fatter. This "heavy tail" behavior leads to the infinite variance of the stable distribution when $\alpha < 2$. The mean is not well defined when $\alpha \leqslant 1$ and is δ when $\alpha > 1$. Equation (2.165) shows that when $\alpha < 2$, the asymptotic behavior of the Levy distribution follows the power law.

From the properties of the Levy distribution, we have the following results.

Proposition 2.3 If $X \propto L(\alpha, \beta, 0, \gamma)$, then $wX \propto L(\alpha, \beta, 0, w\gamma)$.

It should be pointed out that the analytical expressions of the Levy distribution and its probability density function are usually difficult to obtain except for the following three cases.

When $\alpha = 1, \beta = 0$, the probability density function is

$$f(x) = \frac{1}{\pi} \frac{\gamma}{(x-\delta)^2 + \gamma^2}, \quad x \in (\delta, \infty) \tag{2.166}$$

which satisfies the Cauchy distribution.

When $\alpha = 2, \beta = 0$, the probability density function is

$$f(x) = \frac{1}{\sqrt{2\pi}\sigma} \mathrm{e}^{-\frac{(x-\delta)^2}{2\sigma^2}}, \quad x \in (\delta, \infty) \tag{2.167}$$

which is the normal distribution.

When $\alpha = 1/2, \beta = 1$, the probability density function is

$$f(x) = \sqrt{\frac{\gamma}{2\pi}} \frac{1}{(2x-\delta)^{3/2}} \mathrm{e}^{-\frac{\gamma}{2(x-\delta)}}, \quad x \in (\delta, \infty) \tag{2.168}$$

which is a typical long-tail power-law distribution. Note that $\delta > 0$ in Eqs. (2.166) to (2.168).

Figure 2.3 shows the Levy probability density function in log-log coordinates when $\alpha = 2, \alpha = 1$ and $\alpha = 1/2$. It can be seen that when $\alpha \neq 2$, the extreme events corresponding to $x > 10$ can still happen.

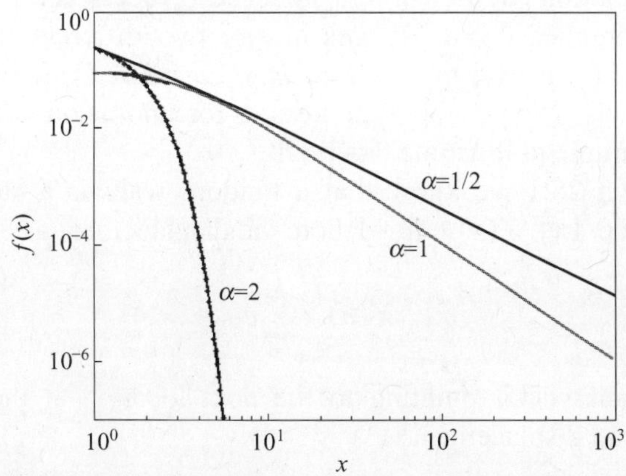

Figure 2.3 Levy distributions with different values of α[7]

For the long-tail power-law distributions in Eq. (2.165), we perform the scale transformation

$$x' = c^{1/\alpha} x \tag{2.169}$$

and then get the following self-similar relationship

$$f(c^{1/\alpha} x) = c^{-1/\alpha} f(x) \tag{2.170}$$

which implies that the random phenomena described by the Levy distribution also exhibit typical scale-invariant characteristics.

2.8.3 Levy Distribution and Long-Range Correlation

Studies have shown that the sharp-peak long-tail Levy distribution lead to diffusions that are much faster than that caused by Brownian motion. We take the Kolmogorov turbulence as an example[7]. The speed difference $\Delta v(r)$ for two particles that are r distance away is

$$\Delta v(r) \propto \varepsilon^{1/3} r^{1/3} \tag{2.171}$$

where $\varepsilon > 0$ is a positive constant. Then

$$t \propto \varepsilon^{-1/3} r^{2/3} \tag{2.172}$$

So the diffusion variance is

$$\sigma_r^2 = \langle r^2 \rangle \propto t^3 \tag{2.173}$$

which shows that the diffusion speed of the Kolmogorov turbulence is much faster than that of the Brownian motion shown in Eq. (2.121). This phenomenon is called anomalous diffusion in physics.

If the autocorrelation coefficient of a stochastic process $X(t)$ is almost always nonzero, which implies that the states at any two different times are always correlated, then $X(t)$ is said to be *long-range correlated*. Study has shown that long-range correlation is an important reason for anomalous diffusion. We use the following example to illustrate this point.

From Definition 2.21 we know that a random walk is a stochastic process defined on a lattice. Let $X(n) = X_n$ denote the displacement after n steps. Then

$$X(n) = \sum_{i=1}^{n} e_i \tag{2.174}$$

where e_i is the unit vector pointing to the next location at the ith step. Since $E(e_i) = 0$, the mean displacement is

$$E(X(n)) = 0 \tag{2.175}$$

Suppose the studied random walk is a Brownian motion, then

$$Cov(e_i, e_j) = 0 \tag{2.176}$$

which shows that the steps in the movement are uncorrelated. The mean square displacement is then

$$\langle X^2(n)\rangle = \left\langle \left(\sum_{i=1}^{n} e_i\right)^2 \right\rangle = n + 2\sum_{i>j}^{n} Cov(e_i, e_j) = n \tag{2.177}$$

namely

$$\langle X^2(n)\rangle = n \tag{2.178}$$

This explains that the random walk in Brownian motion can only cause normal diffusive processes.

On the other hand, if $X(t)$ is long-range correlated, the current step is always related to the previous and e_i and e_j are related

$$Cov(e_i, e_j) \propto \frac{A}{|i-j|^{\kappa}} \tag{2.179}$$

where $0 < \kappa < 1, A > 0$ are constant. Then

$$\langle X^2(n)\rangle = n + 2\sum_{i>j}^{n} Cov\langle e_i, e_j\rangle = n + Bn^{2-\kappa} \tag{2.180}$$

where B is a constant determined by A, κ.

The second term on the right-hand side of Eq. (2.180) is much greater than the first, so

$$(X^2(n)) \propto n^{2-\kappa} \tag{2.181}$$

which shows that the anomalous diffusion is a result of the long-range correlation of the random walk. In comparison, the current step in Brownian motion is uncorrelated to the previous, and so the mean displacement variance is positively proportional to time.

In fact, long-tail Levy distribution indicates that the probability of extreme events cannot be ignored, so the system may have large fluctuations with certain probabilities, which causes the states at different times to be correlated, and in the end leads to the anomalous diffusion. We provide more details as follows.

Step 1 The relationship between the displacement $X(t)$ and the speed $v(t)$ is determined by

$$X(t) = \int_0^t v(t')\mathrm{d}t' \tag{2.182}$$

Step 2 We look for the relationship between the distance variance and the autocorrelation coefficients.

Differentiating $\langle X^2(t)\rangle$, we have

$$\frac{\mathrm{d}\langle X^2(t)\rangle}{\mathrm{d}t} = \left\langle 2X(t)\frac{\mathrm{d}X(t)}{\mathrm{d}t}\right\rangle = \langle 2X(t)v\rangle = 2\int_0^t \langle v(t)v(t')\rangle \mathrm{d}t' \tag{2.183}$$

Set $t' - t = \tau$, then

$$\frac{\mathrm{d}\langle X^2(t)\rangle}{\mathrm{d}t} = 2\langle v^2\rangle \int_0^t \rho(\tau)\mathrm{d}\tau$$

where $\rho(\tau) = \dfrac{\langle v(t)v(t+\tau)\rangle}{\langle v^2(t)\rangle}$ is the autocorrelation coefficient of the stochastic process $v(t)$. Thus

$$\langle X^2(t)\rangle = 2\int_0^t \langle v^2\rangle \int_0^t \rho(\tau)\mathrm{d}\tau\mathrm{d}t \tag{2.184}$$

Step 3 Define the long-range correlation function

$$R(\tau) \propto \tau^{-\beta}, \beta \in (0,1) \tag{2.185}$$

Step 4 We get the diffusion variance for $X(t)$

$$\langle X^2(t)\rangle \propto t^{2-\beta} \tag{2.186}$$

which implies that Levy distribution results in anomalous diffusion, compared to the diffusion variance $\langle X^2(t)\rangle \propto t$ corresponding to Brownian motion.

Note that an important property of the Levy power-law distribution is that its variance $\delta_X^2 = \langle X^2(t)\rangle$ is usually divergent (e.g., the Cauchy distribution when $\alpha = 1$). This implies that the corresponding mean value of fluctuation is infinite, and so large fluctuation may happen with high probability.

In summary, if the random variable of the system states satisfies the long-tail power-law distribution, the random variable series is long-range correlated, and thus the scaling index and power spectrum index are greater than those of the standard Brownian motion. Therefore, the diffusion process is much more active, and the system shows the abnormal dynamics. On the other hand, in case of the power-law distribution, the system is scale invariant both in time and space. The temporal scale invariance implies that the waiting time of the system's transition from one state to another is scale invariant, and thus the duration of extreme events can be much longer than those expected in the general Gaussian distribution. The spatial scale invariance implies that the system's state transition distance is scale invariant, and then large fluctuation can lead to fast diffusion, and finally result in abnormal dynamics. All these conclusions are built upon the ideas developed in thermodynamics and other physical processes discussed in Section 2.7. However, for a specific stochastic process, what is usually conveniently available is a time series which is a sample or trajectory of a stochastic process. Then to determine the correlation of the corresponding stochastic process needs to rely on two analytical methods that are widely used in engineering practice.

2.8.4 Scaled Windowed Variance (SWV) Method

It is difficult to determine directly whether the dynamics of a system satisfy the power law, but we can tell whether the system has non-Gaussian distributions and some other complicated properties by studying the autocorrelation of its events on a long time scale. If the autocorrelation function of the time series decays asymptotically following the power law, the series is considered to be correlated on the long time scale. However, when the time step is chosen to be large, noises sometimes dominate the signal, and thus the correlation is difficult to be computed. Researchers have proposed different methods to address this challenge. One early popular approach is the Rescaled Range-Statistics (*R*/*S*) method, proposed by Mandelbrot and Wallis using ideas from hydrology analysis. Another method to be discussed in detail first is the scaled windowed variance (SWV) method.

In 1997, Cannon et al. summarized the SWV method[13], and we give a brief review as follows:

Consider the time series

$$X = \{X_t : t = 1, 2, \cdots, n\} \tag{2.187}$$

Construct the corresponding Brown motion series

$$Y = \{Y_t : t = 1, 2, \cdots, n\} \tag{2.188}$$

where $Y_t = \sum_{i=1}^{t} X_i$. Let $m = 1, 2, \cdots, n$, and we construct a new series from Y

$$Y^{(m)} = \left\{Y_u^{(m)} : u = 1, 2, \cdots, \frac{n}{m}\right\} \tag{2.189}$$

where $Y_u^{(m)} = \{Y_{um-m+1}, \cdots, Y_{um}\}$. Then compute the standard deviation series

$$\sigma_m = \frac{\sum_{u=1}^{n/m} \sigma_m^{(u)}}{n/m} \tag{2.190}$$

where $\sigma_m^{(u)}$ is the standard deviation of $Y_u^{(m)}$.

It is assumed in the SWV method that if the original time series X has the power law characteristics, so does the standard deviation series[13], namely

$$\sigma_m \propto m^H \tag{2.191}$$

where H is the Hurst exponent.

Taking logarithm on both sides of the equation, we have

$$\lg \sigma_m = H \lg m \tag{2.192}$$

Then we can examine the power law characteristic of X in logarithmic coordinates according to the linearity of σ_m.

In addition, from Hurst exponent theory[14], the H exponent of the time series X can be used to tell its correlation. The details are listed as follows:

(1) When $1 > H > 0.5$, X is positively correlated on long time scales;

(2) When $0.5 > H > 0$, X is negatively correlated on long time scales;

(3) When $H = 1.0$, X is deterministic;

(4) When $H = 0.5$, X is uncorrelated.

The above criteria show that the deviation of the H exponent from 0.5 reflects the ratio of random factors over deterministic factors.

2.8.5 *R*/*S* (Rescaled Range-Statistics) Method

Based on the early work by the British hydrologist Harold Edwin Hurst[15], Mandelbrothe and Wallis proposed the *R*/*S* analysis method[16] and we give a brief review as follows.

Consider a time series

$$X = \{X_t : t = 1, 2, \cdots, n\} \tag{2.193}$$

For any positive integer τ, define the mean value series of X to be

$$\langle X \rangle_\tau = \frac{1}{\tau} \sum_{t=1}^{\tau} X(t) \tag{2.194}$$

Then the average variance of X is

$$S(\tau) = \left(\frac{1}{\tau} \sum_{t=1}^{\tau} \{X(t) - \langle X \rangle_\tau\}^2 \right)^{\frac{1}{2}} \tag{2.195}$$

Further define the accumulated deviation of X as

$$X(t, \tau) = \sum_{n=1}^{t} \{X(n) - \langle X \rangle_\tau\}, \quad 1 \leqslant t \leqslant \tau \tag{2.196}$$

and the limit range of X as

$$R(\tau) = \max_{1 \leqslant t \leqslant \tau} X(t, \tau) - \min_{1 \leqslant t \leqslant \tau} X(t, \tau), \quad \tau = 1, 2, \cdots \tag{2.197}$$

When studying the Nile reservoir storage, river flow, mud sediments and annual rings of trees, Hurst found that for the time series generated from these natural phenomena, the ratios of the limit ranges over the variances (*R*/*S*) usually satisfy

$$R/S \propto (c\tau)^H \tag{2.198}$$

Taking logarithm on both sides, we have

$$\lg(R/S) \propto H\lg\tau + H\lg c \tag{2.199}$$

which implies that the *R*/*S* series has the power law characteristics.

2.8.6 *R*/*S* Method and Levy Power-Law Distribution

Stable Levy distributions serve as the main theoretical basis for *R*/*S* analysis. We provide more details as follows.

Consider a random variable ξ_i which satisfies the Levy distribution

$$\xi_i \propto L(\alpha, \beta, 0, \gamma) \tag{2.200}$$

Since the sum of the random variables satisfying stable distributions satisfies a stable distribution, the mean value $X_n = \frac{1}{n}\sum_{i=1}^{n}\xi_i$ after *n* time steps satisfies

$$nX_n \propto L(\alpha, \beta, 0, n^{1/\alpha}\gamma) \tag{2.201}$$

or equivalently

$$X_n \propto L(\alpha, \beta, 0, n^{1/\alpha-1}\gamma)$$

Furthermore, from Proposition 2.3 it is clear that

$$\frac{nX_n}{n^{1/\alpha}} \propto L(\alpha, \beta, 0, \gamma) \tag{2.202}$$

Let

$$\xi_n = \frac{X_n}{n^{1/\alpha-1}} \tag{2.203}$$

Then

$$\xi_n \propto L(\alpha, \beta, 0, \gamma) \tag{2.204}$$

Hence, we know that ξ_n and ξ_i are random variables with identical distributions. In addition,

$$\frac{X_n}{\xi_n} = n^{1/\alpha - 1} \tag{2.205}$$

or equivalently

$$\ln\left(\frac{X_n}{\xi_n}\right) = \left(\frac{1}{\alpha} - 1\right)\ln n \tag{2.206}$$

which implies that for a random variable with the Levy power-law distribution, statistical data can be constructed to show the power law characteristics.

2.8.7 Generalized Central Limit Theorem

The central limit theorem shows that as the number of finite-variance random variables increases, the distribution of their sum approaches the normal distribution. Note that normal distributions are stable. In fact, stable distributions play an important role in the central limit theorem, which is validated by the following generalized central limit theorem.

Theorem 2.7[11] (generalized central limit theorem) A nontrivial random variable Z satisfies a stable distribution if and only if there exist independently and identically distributed random variables $X_1, X_2, \cdots, X_n$ and constants $a_n > 0, b_n \in \mathbb{R}$ such that

$$a_n(X_1 + X_2 + \cdots + X_n) - b_n \xrightarrow{d} Z \tag{2.207}$$

Compared to the central limit theorem, the generalized version states that when the assumption about the finite variance does not hold, one can only conclude that the limit distribution is stable. In particular, the stable limit distribution may have a power-law tail. The generalized central limit theorem shows that as the number of the random variables with the decreasing power-law tail distribution $1/|x|^{\alpha+1}$ increases, the distribution of their sum approaches a stable distribution $f(x;\alpha,0,c,0)$ where c is a constant.

2.9 Statistical Characteristics of Network Topologies

The study of complex networks is concerned with the statistical properties of a large number of nodes with their interconnections. Network topologies usually determine various network properties and functionalities. To better understand the behavior of complex networks one needs to study more in detail the networks' statistical properties. In the following, we cover four main statistical indices.

2.9.1 Degree

For an undirected network G, we use $V = \{v\}$ and $E = \{e\}$ to denote the set of nodes and the set of edges respectively. The degree d_v of a node v is the number of edges that are adjacent to node v, i.e.

$$d_v = \sum_{e \in E} \delta_e^v \tag{2.208}$$

where

$$\delta_e^v = \begin{cases} 1, & \text{edge } e \text{ is adjacent to node } v \\ 0, & \text{otherwise} \end{cases} \tag{2.209}$$

In a weighted network, the degree d_v of node v is the sum of the weights of all the edges that are adjacent to v. The mean of all nodes' degrees is called the average degree $\overline{d}$ of the network. The degree distribution is an important topological characterization of the network. It is usually described by the probability $p(d)$ which is the probability for any given node to be with degree d [17].

2.9.2 Average Path Length

For an un-weighted network, the distance between two nodes is defined to be the number of edges in the shortest path connecting these two nodes. In a weighted network, the distance is then defined to be the sum of the weights of all the edges in the shortest path. The diameter of a network is defined to be the maximum distance between any pair of nodes in the network.

Thus, the average path length L of a network is the average distance of all possible pairs of nodes in the network:

$$L = \frac{1}{N(N-1)} \sum_{i \neq j} w_{ij} \tag{2.210}$$

where N is the number of nodes in the network, and w_{ij} is the distance between nodes i and j. L is related to the degree of separation of nodes in the network, or in some sense the size of the network.

If a complex network is under attack, it might be divided into several disconnected sub-networks, in which case the average path length becomes infinite. To deal with this situation, researchers have proposed to use the average inverse geometric distance, which is defined by

$$L^{-1} = \frac{1}{N(N-1)} \sum_{i \neq j} \frac{1}{w_{ij}} \tag{2.211}$$

Note that the average inverse geometric distance is not the reciprocal of the average geometric distance. Roughly speaking, when the average inverse geometry distance is great, the network is more tightly connected.

One of the most important discoveries in the field of complex networks is that most large-scale real-world networks have short average path lengths. This fact inspired the concept of the small world effect [18], which is originated from the famous "small world" test by Milgram[19]. In fact, the popular notion of "six degrees of separation" also comes from this test [20].

2.9.3 Clustering Coefficient

The clustering coefficient measures the degree to which the nodes in a network tend to cluster together[17]. For a node i, let k_i be the number of its adjacent nodes. Then there are potentially $k_i(k_i-1)/2$ edges between these k_i nodes. Let t_i be the number of edges that exist. Then the clustering coefficient C_i for node i is defined by

$$C_i = \frac{2t_i}{k_i(k_i-1)} \tag{2.212}$$

Correspondingly, the clustering coefficient C for the whole network is defined to be the average of the clustering coefficients of all nodes.

Obviously, only when the network is completely connected, can its clustering coefficient be one. In general, the clustering coefficient is always less than one. Note that in most of the real-world large-scale networks, nodes tend to get clustered. Although their clustering coefficients are much less than one, their sizes can be much larger than those of random networks.

2.9.4 Betweenness

We need to distinguish node betweenness and edge betweenness [21]. The vertex betweenness is the ratio of the number of the shortest paths containing node i over the number of all the shortest paths. Edge betweenness can be defined similarly. Betweenness describes the influence of a node or edge on the whole network. For example, in a social network, the distribution of betweenness shows the social position of the people involved, and can be used to identify crucial persons in the network[17].

2.10 Network Models

In this section, we introduce two typical complex network models and explain the mechanisms for their evolutions. However, we want to clarify that there is still not a single method that is capable to produce a complex network that has exactly the same statistical properties as those of any given real-world network.

2.10.1 Small-World Networks

The small-world network model was first proposed by Watts and Strogatz in 1998[22]. It is a breakthrough in the study of complex networks. The networks constructed by Watts and Strogatz are obtained by operations on regular networks, and fall in between regular networks and random networks. To be more specific, for a node in a regular network, we choose one of its adjacent edges with probability p, and then rewire it to connect to a randomly chosen node. Hence, regular networks and random networks are special cases of such networks corresponding to p being 0 and 1 respectively.

Due to the rewiring process, the degrees of nodes in a small-world network are not the same, and the degree distribution peaks around the average degree and diminishes quickly away from the peak[17]. Compared with regular networks and random networks, small-world networks have small average path lengths and large clustering coefficients (see Table 2.1). This phenomenon is called the small-world effect[22]. The properties of small-world networks imply that a few random links can change the topological features of a network significantly[17].

Table 2.1 Comparison of three classes of networks

Statistical characteristics	Regular network	Small-world network	Random network
Clustering coefficient	Large	Large	Small
Average path length	Large	Small	Small

Empirical research shows that, real-world networks' average path lengths L are close to those of random networks denoted by L_{rand}, and their clustering coefficients C are much larger than those of random networks denoted by C_{rand}. We provide more details in Table 2.2, where $\bar{d}$ is the average degree of the network. Table 2.2 implies that real-world networks are neither entire regular nor random, but more close to small-world networks. The network of human contact is a typical small-world network, and its structural characteristics assist the spread of epidemics. This is because virus can easily spread out from one node to its neighboring nodes, and in this way even a remote friend of a given infected person can still get affected. More relevant to the topic of this book, studies have shown that typical power grids are usually small-world networks[23].

Table 2.2 Small-world phenomena in real-world networks[18]

Network	Number of nodes	$\bar{d}$	L	L_{rand}	C	C_{rand}
WWW (in terms of websites)	153,127	35.21	3.1	3.35	0.1078	0.000,23
Internet (in terms of domains)	3,015 − 6,209	3.52 − 4.11	3.7 − 3.76	6.36 − 6.18	0.18 − 0.3	0.001
Movie star network	225,226	61	3.65	2.99	0.79	0.000,27
MEDLINE Coauthorship network	1,520,251	18.1	4.6	4.91	0.066	0.000,011
Math coauthorship network	70,975	3.9	9.5	8.2	0.59	0.000,054
E.coli reflection figure network	315	28.3	2.62	1.98	0.59	0.09
Silwood Park food network	154	4.75	3.40	3.23	0.15	0.03
Power grid	4941	2.67	18.7	12.4	0.08	0.005

2.10.2 Scale-Free Networks

Although the small-world model can better describe some features of the complex networks in the real world, theoretical analysis has shown that its node distribution does not agree with the power law $P(d) = d^{-\alpha}$ that exhibits in most of the large-scale real-world networks, such as communication network, social network, biological network, etc[24]. As to random networks and regular networks, their node distributions are narrow and only a few nodes are with degrees far from the mean value, so the average degree can be seen as one feature of the node degree. The power-law distribution in log-log coordinates is a relatively slowly decreasing curve and has no peak compared to exponential distributions. Most of the nodes have low degrees and only a few nodes have a lot of adjacent edges. When varying the measurement unit (scale) by a factor λ, the trend of the function remains the same. For this reason, such networks with the power-law degree distribution are called scale-free networks, and the power-law degree distribution is referred to as the scale-free characteristic of a network[1].

Table 2.3 Comparison of the node degree distributions of three classes of networks

Type	Degree distribution $P(d)$
Regular network	Delta distribution
Random network	Poisson distribution
Scale-free network	Power-law distribution

In order to explain the evolution mechanism of scale-free networks, Barabasi and Albert proposed the well-known BA model[1,26]. They incorporated two critical notions to generate a network, which are the growth and preferential attachment. Growth means that new nodes are added sequentially to the existing network, and preferential attachment means that a newly added node is biased to be linked to the nodes in the network with higher degrees. The BA model has revealed some universal mechanism for self-organized evolution in complex systems[17] and is thus considered to be a milestone in the study of complex networks.

Improvement has been made later and new models have been proposed, such as the steady growth model[27–29], accelerated growth model[30], local-world evolution model[31], gradual aging model[32], adaptive competition model[33], etc. Note that many, although not all, real networks are scale-free networks, and the degree distributions of some of them are close to truncated exponential distributions. In fact, the scale-free characteristics enable the networks to be robust against random attacks but also make them vulnerable against hostile attacks.

Research has shown that besides the small-world effect and scale-free characteristics, real-world networks also have other important statistical features, such as mixed mode characteristics[30], degree-related characteristics[34,35], etc.

2.11 Community Structure Theory

Many practical networks have "community structures"[36,37]. In other words, the whole network consists of several sub-networks called communities. Nodes within the same community have close connections and the connections between communities are relatively sparse, as shown in Fig. 2.4.

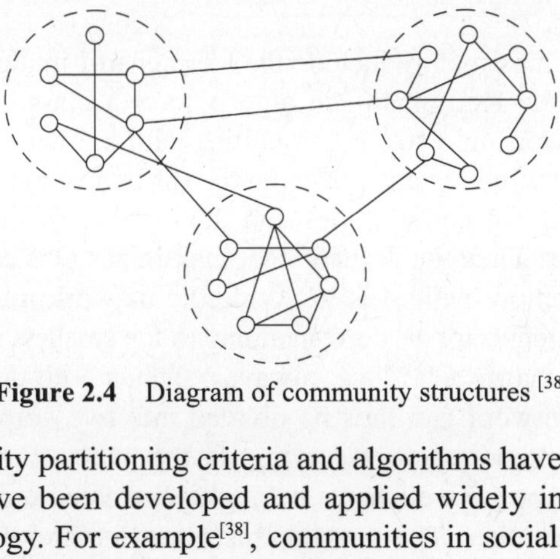

Figure 2.4 Diagram of community structures [38]

Various community partitioning criteria and algorithms have been proposed, and related theories have been developed and applied widely in biology, computer science, and sociology. For example[38], communities in social networks are social groups sharing similar interest or social background, communities in citation networks are research papers within the same research area, and communities in the world wide web are websites with related topics. To identify effectively these communities is helpful for understanding related network functionalities, and formulating rules to decompose and coordinate.

Currently, the methods used to identify community structures use mainly ideas from graph theory, computer science and sociology. According to the number of communities involved at each step, the algorithms can be divided into bisection and multi-section algorithms. The former includes the Kernighan-Lin algorithm and spectral bisection method using the Laplacian eigenvectors and the latter includes agglomerative algorithms and splitting algorithms. Typical agglomerative algorithms are the Newman fast agglomerative algorithm and clique percolation method and typical splitting algorithms are the GN algorithm, fast splitting algorithm, dissimilarity-based splitting method, and the splitting method using information centrality.

2.11.1 Laplacian Matrix

The weighted Laplacian matrix L is critical for community structure analysis. For an undirected and connected graph G consisting of N nodes, its weighted Laplacian matrix is the $N \times N$ symmetric matrix whose diagonal element L_{ii} is the sum of the weights of the edges adjacent to node i, and off-diagonal element L_{ij} is the opposite of the weight of the edge connecting nodes i and j. One can check that L is a positive semidefinite matrix, and we denote its eigenvalues by $\lambda_1 < \lambda_2 \leqslant \cdots \leqslant \lambda_n$ where $\lambda_1 = 0$.

2.11.2 Bisection Algorithms

The bisection algorithms mainly include the Kernighan-Lin algorithm and spectral bisection method. The Kernighan-Lin algorithm is a class of greedy algorithm that divides the network into two communities with known sizes. It introduces a gain function Q first which is defined to be the difference between the number of the edges within the two communities and the number of the edges connecting the two communities. Then the division that maximizes Q is constructed.

The spectral bisection method is to divide the network into two communities by calculating the eigenvector v_2 corresponding to the smallest non-zero eigenvalue λ_2 of the Laplacian matrix L[39,40]. v_2 always contains both positive and negative elements, and the network can thus be divided into two communities according to the signs of the elements corresponding to the nodes. In other words, nodes corresponding to the positive elements belong to one community, and all the other nodes belong to the other. The spectral bisection method can only effectively divide the network into two communities, and λ_2 can be used as an indication for the effects of the division. The network is usually better divided with a smaller λ_2.

2.11.3 Multi-Section Algorithms

Multi-section algorithms mainly include the agglomerative algorithm and splitting algorithm. The basic idea of the agglomerative algorithm is as follows. An index is chosen to measure the quality for dividing the network, and initially each node itself is seen as a community. Then communities are merged to improve the division in the gradient direction of the index until the network is merged into a single community. The optimal community number and corresponding partition pattern can thus be determined according to the value of the chosen index. The splitting algorithms work the other way around. They remove in sequence the edge that is the least tightly connected to the network.

Both the agglomerative algorithm and splitting algorithm have to define how tightly an edge is connected to the network, for which edge weights, betweenness and other indices can be used. We show the relationship between the two classes of algorithms in Fig. 2.5.

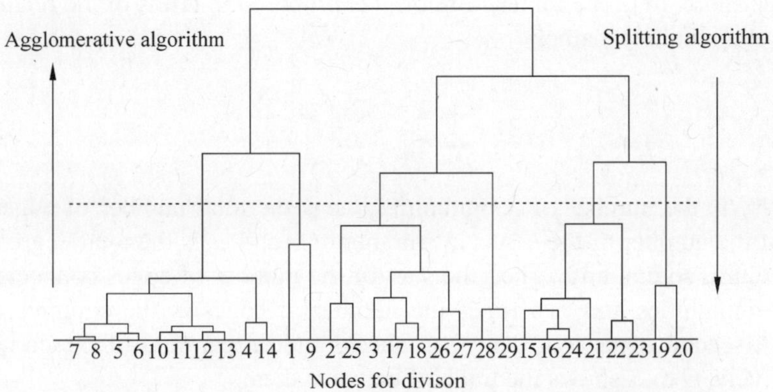

Figure 2.5 Diagram of the agglomerative algorithm and splitting algorithm

2.11.4 Evaluation Index

In order to quantitatively evaluate the quality of dividing the network and identify the optimal partition pattern, Newman et al. introduced the notion of module degree Q[40], and applied it to weighted networks[37]. The module degree of a weighted network is defined by

$$Q = \frac{1}{2m}\sum_{i=1}^{n}\sum_{\substack{j=1\\j\neq i}}^{n}\left\{\left[(-L_{ij}) - \frac{L_{ii}L_{jj}}{2m}\right]\delta(i,j)\right\} \tag{2.213}$$

where

$$L_{ii} = \sum_{j=1, j\neq i}^{n} (-L_{ij}) \tag{2.214}$$

$$m = \sum_{i=1}^{n} (L_{ii} / 2) \tag{2.215}$$

Here, L_{ij} and L_{ii} are the elements of the Laplacian matrix L, and m is the sum of weights of all the edges in the network. If nodes i and j are in the same community, then $\delta(i,j)=1$, otherwise $\delta(i,j)=0$. Q describes the relationship between the actual and expected node connectedness. If we set the L_{ii} of each node stay the same and generate a random network, then $L_{ii}L_{ij}/2m$ is the expected weight of the edge connecting nodes i and j. The difference between the connectedness in a given network and that in a random network shows the topological feature of the network. If the sum of the weights of the inner connections within a community equals the expected value, then $Q=0$. The closer Q is to 1 the more obvious the community structure is.

Furthermore, in [37] another form of Q is proposed in terms of the relationship among communities, namely

$$Q = \sum_{r=1}^{N_c} \frac{e_{rr}}{2m} - \sum_{r=1}^{N_c} \left(\frac{a_r}{2m} \right)^2 \tag{2.216}$$

where N_c is the number of communities, m is the total number of edges, e_{rr} is twice of the number of the edges in community r and a_r is the total degree of the nodes within community r, i.e., the sum of the number of edges connected with other communities and two times the number of edges in the community. The first and second items represent the cohesion and separation of the community respectively. Q thus shows the tradeoff between them.

The value of Q can be used to identify the feasible and effective way of division. It has been shown that[37] a desirable partition is achieved when Q is equal to or greater than 0.3. So by comparing the optimal Q in different networks one can evaluate whether a network is dividable. Furthermore, Q is easy to compute, it has a clear physical meaning and its peak value is usually below 2[37]. Hence, it is easy to perform network division using Q. In addition, by identifying Q's maximum value, the optimal network partition can be determined correspondingly.

2.12 Center Nodes

People usually use the centrality of a vertex to determine the relative importance of a vertex within the network[41,42]. Freeman proposed the degree centrality, closeness centrality and betweenness centrality[43,44] while Bonacich used the eigenvector centrality[41] and Wuchty defined the geometric centrality[42]. Generally speaking, a network may have a set of center nodes. Then centrality is a property determined by each vertex's topological position. The distribution of centrality reflects the level of centralization in the whole network. In this section, we discuss the concepts of the degree centrality, closeness centrality and betweenness centrality.

2.12.1 Degree Centrality

In undirected graphs, the degrees of nodes are related to the importance of the node in the network [43]. A node with a higher degree is connected to more nodes. If the network is directed, we need to distinguish two types of degree centralities according to in-degrees and out-degrees [43]. In social networks, a larger in-degree means that the node is higher in hierarchy or with better reputation since other nodes try to connect to it directly. Conversely, a large out-degree means that the node has good communication skills, can articulate its opinions well, and is thus influential.

The disadvantage of using the degree centrality is that only immediate links are considered. Some nodes may connect to local nodes by many edges, but are not tightly connected to other nodes in the network. In this case, the node with high degree may just be a local center node.

2.12.2 Closeness Centrality

Closeness centrality is concerned with the shortest distance between a node and all other nodes that are reachable from it[43]. Here, distance can be defined differently and once a definition is picked, the distance between any two nodes can be calculated. Then the distance distribution shows the centrality of the network.

2.12.3 Betweenness Centrality

The betweenness centrality counts the number of times that a node occurs in the shortest paths between any other pair of nodes in an undirected network[43]. It indicates the topological importance and capability of this node. If the node often locates in the shortest paths of other node pairs, then it is highly influential on information propagation in the network. Freeman further generalized this concept to directed networks and developed relative index to measure the centrality[43]. However, note that in some real-world networks, the communication between nodes depends on not only shortest paths, but all available paths as well.

2.13 Vulnerability

The study on vulnerability is to investigate the impact of random failures and intentional attacks. Here, random failures are usually caused by the removal of some nodes or edges randomly while intentional attacks are usually caused by attacks targeting at the most critical nodes or edges. Researchers have proposed different indices to evaluate networks' vulnerability and also designed parameter adjustment strategies to improve the robustness of the network structures against various attacks.

2.13.1 Node Vulnerability

In 2000, Albert and his colleagues published the paper, "Error and attack tolerance in complex networks"[45] in *Nature*. They investigate the vulnerability of different networks under different types of attacks. Several indices have been proposed to reveal the network vulnerability, which include the ratio of the scale of the largest connected sub-network over that of the whole network and the average shortest path length in the largest connected sub-network. Their results have shown that scale-free networks are more robust than random networks when facing random attacks while scale-free networks is more vulnerable when facing intentional attacks since the network collapses once some key nodes are removed. This fact is rooted in the key role played by those nodes with high degrees.

Other researchers have later studied the vulnerability of other complex networks. It has been found that most networks are robust when facing random attacks but vulnerable when facing targeted attacks, such as the removal of the nodes with the highest degree. The conclusion holds for the protein networks[46], food webs[47–49], E-mail networks[48] and internet[49]. In [50], the authors have considered the intentional attacks targeting at both nodes and edges that are selected according to the betweenness centrality[42–44].

2.13.2 Edge Vulnerability

There are three types of indices to evaluate the impact of removing edges on the performances of a network.

(1) Global indices

Typical indices in this category include the mean geometric distance, mean inverse geometric distance, edge betweenness centrality, and so on. They measure the shortest paths between nodes from the viewpoint of the whole network. Generally speaking a system is vulnerable if the length of the shortest path between all pair of nodes increases a lot after attacks.

(2) Local indices

One typical index in this category is the clustering coefficient, which is usually used to check the impact of losing an edge on the connectivity of its local neighborhood.

(3) Scale of the largest connected sub-network

If the network splits after a fault, the ratio of the number of nodes in the largest connected sub-network over the total number of nodes in the whole network can be used to measure the severity of the fault. The ratio is

$$S = \frac{N_i}{N_0} \tag{2.217}$$

where N_0 is the total number of nodes before the fault and N_i is the number of nodes in the largest connected sub-network after the fault takes place at the ith edge.

2.14 Synchronization

We discuss the network synchronization problem in two cases, namely state synchronization and phase synchronization[51].

2.14.1 State Synchronization

State synchronization refers to the situation when the states of all the nodes in a network become the same. Several network models have been used in this context, such as the continuous-time diffusively coupled dynamical networks, linearly diffusively coupled time-varying dynamical network, and networks with time delays. We provide more details below.

(1) Continuous-time diffusively coupled dynamical networks

The network considered consists of N nodes with identical self-dynamics. The state equations are

$$\dot{x}_i = f(x_i) - c\sum_{j=1}^{N} a_{ij} h(x_j), \quad i = 1, 2, \cdots, N \tag{2.218}$$

where x_i is the state of node i, c is the constant coupling strength, $h(\cdot): \mathbb{R} \to \mathbb{R}$ is the inner coupling function, also called the nodes' output functions, which are identical for all the nodes, and the coupling matrix $A = (a_{ij}) \in R^{N\times N}$ describes the topology of the network. It is usually assumed[52,53] that A satisfies the diffusive conditions

$$a_{ii} = -\sum_{j=1, j \neq i}^{N} a_{ij}, \quad i = 1, 2, \cdots, N \tag{2.219}$$

$$\sum_{j} a_{ij} = 0, \quad i = 1, 2, \cdots, N \tag{2.220}$$

Note that $h(x_j) = 0$ when the states of all the nodes become the same and the network is synchronized.

Different indices have been proposed to evaluate a network's capability to get synchronized[54]. One is the second smallest eigenvalue λ_2 of the coupling matrix A, and another is the ratio of the largest eigenvalue over the second smallest eigenvalue λ_N / λ_2. Some networks cannot get synchronized and it is always important to examine the self-dynamics function $f(\cdot)$ and inner coupling function $h(\cdot)$ to check whether the network can be synchronized.

Various network parameters have been examined to check their influences on the network's synchronization capabilities. For example, people have looked at the influence of the coupling strength and network scale on nearest-neighbor networks, globally coupled networks, star-shaped networks and regular networks. The influence of the edge connection probability p has been studied with respect to the *NW* small-world network[55] and *BA* scale-free networks[56]. Algorithms have been proposed to construct optimal network structures for synchronization starting from the *BA* scale-free network[57]. From the analysis on the robustness and vulnerability, it is found that the synchronization capability can be strengthened if some random connections are added.

(2) Linearly diffusively coupled time-varying dynamical networks

Consider a linearly diffusively coupled time-varying network consisting of N nodes with identical self-dynamics, whose state equations are

$$\dot{x}_i = f(x_i) - c\sum_{j=1}^{N} a_{ij}(t) h(t) x_j, \quad i = 1, 2, \cdots, N \tag{2.221}$$

where $h(t) \in \mathbb{R}$ is the inner coupling function at time t and $A(t) = (a_{ij}(t)) \in \mathbb{R}^{N \times N}$ is the coupling matrix at time t satisfying

$$a_{ii}(t) = -\sum_{j=1, j \neq i}^{N} a_{ij}(t), \quad i = 1, 2, \cdots, N \tag{2.222}$$

It has been shown that the synchronization capability is determined by the eigenvalues and eigenvectors of the inner coupling matrix $H(t)$ and coupling matrix $A(t)$ [58,59]. Using tools from algebraic graph theory, a lower bound on the coupling strength can be determined for the purpose of synchronization[60].

Following this approach, researchers have constructed the lower bounds for regular networks and small-world networks[60,61], and shown how the lower bounds depend on network parameters.

(3) Networks with time delay

Consider a network with time delay consisting of N nodes with identical self-dynamics, whose state equations are

$$\dot{x}_i = f(x_i) - c\sum_{j=1}^{N} a_{ij} h x_j(t-\tau), \quad i = 1, 2, \cdots, N \tag{2.223}$$

Synchronous criteria have been constructed[62], and the relationship among the time delay, coupling strength and the network's synchronization capability have also been discussed[63].

2.14.2 Phase Synchronization

We say two coupled nodes are synchronized in phase if the phases of the two nodes satisfy

$$| n\theta_i - m\theta_j | < s^* \tag{2.224}$$

for some integers n and m and a constant s^*. The study on phase synchronization has been carried out on symmetric networks and asymmetric networks.

(1) Symmetrically coupled networks

A popular symmetric network model is the Kuramoto oscillator model[64,65]:

$$\dot{\theta}_i = \omega_i + \sum_{j=1}^{N} \Gamma_{ij}(\theta_j - \theta_i), \quad i = 1, 2, \cdots, N \tag{2.225}$$

where θ_i is the phase of the ith oscillator and ω_i is its natural frequency. When all the couplings have the same strength, we have

$$\Gamma_{ij}(\theta_j - \theta_i) = \frac{c}{N}\sum_{j=1}^{N} \sin(\theta_j - \theta_i) \tag{2.226}$$

where $c \geqslant 0$ is the coupling strength. Then

$$\dot{\theta}_i = \omega_i + \frac{c}{N}\sum_{j=1}^{N} \sin(\theta_j - \theta_i), \quad i = 1, 2, \cdots, N \tag{2.227}$$

The critical coupling strength c^* has been constructed[64,65]. For small-world networks, the synchronization capability is influenced by the edge connection probability p and network scale N[66] and for the BA scale-free network, similar analysis has been done for the influences of the node degree d and network scale N[67].

(2) Asymmetrically coupled networks

Consider the node dynamics

$$\dot{\theta}_i = \omega_i - \frac{c}{N}\sum_{j=1}^{N} C_{ij} a_{ij} \sin(\theta_j - \theta_i), \quad i = 1, 2, \cdots, N \tag{2.228}$$

where $C_{ij} = c / d_i$ is the coupling strength from nodes i and j, c is a constant, and d_i is the degree of node i. If node i is connected to node j, then $a_{ij} = -1$; otherwise, $a_{ij} = 0$. So the node dynamics can be rewritten as

$$\dot{\theta}_i = \omega_i - \frac{c}{d_i}\sum_{j=1}^{N} a_{ij} \sin(\theta_j - \theta_i), \quad i = 1, 2, \cdots, N \tag{2.229}$$

It was pointed out in [68] that for different node degree distributions, systems (2.227) and (2.229) always have the same critical coupling strength c^*, and the same conclusion holds also for different network topologies $A = (a_{ij})_{N\times N}$.

2.15 Notes

In this chapter, we have discussed basic concepts and methods in SOC theory and complex network theory. These concepts and methods are not isolated, but tightly connected together.

One of the central topics in complex system theory is to study abnormal dynamics, which include, but not limited to, extreme events and catastrophe. The power-law distribution is an effective macroscopic description for the possible occurrence of extreme events. So we have focused on discussing the power law in this chapter. Firstly, the power law complements the central limit theorem in that the latter only describes the common normal events and dynamics in natural or man-made systems, e.g. the Gaussian distribution and Brownian motion, but is not applicable to abnormal dynamics. Second, after setting up the mathematical model of complex systems as stochastic processes, we can investigate further whether the system has anomalous diffusion by looking at long-range correlations. If anomalous diffusion does occur, then it indicates that large displacements and fluctuations are frequent, and the long-tail Levy power-law distribution dominates. Several important concepts are related, such as noise, scale invariance and fluctuation.

Complex network theory combines ideas from graph theory with observations from real-world complex networks. It focuses on analyzing the influence of network topologies on the functionalities of networks. It has been developed with great momentum in the past decade. The studied networks can be dynamic or static. We have discussed community structures, network evolution, and synchronization. These research topics provide effective tools for the control of complex networks.

In short, in this chapter we have covered quite a few different concepts and methods, and introduced several subjects in mathematics and physics. They serve as the base for better understanding the materials to be introduced in the next few chapters.

References

[1] Barabasi A L, Albert R. Emergence of scaling in random networks. Science, 1999, 286: 509 – 512

[2] Peebles P Z. Probability, Random Variables, and Random Signal Principles (3rd Edition). New York: McGraw-Hill, 1993

[3] Spanos A. Probability Theory and Statistical Inference. Cambridge, UK: Cambridge University Press, 1999

[4] Albeverio S, Blanchard P, Streit L. Stochastic Processes and Their Applications in Mathematics and Physics. Boston: Kluwer Academic Publishers, 1990

[5] Todorovic P. An Introduction to Stochastic Processes and Their Applications. New York: Springer-Verlag, 1992

[6] Tsien H S. Engineering Cybernetics. New York: McGraw-Hill, 1954

[7] Liu S D, Liang F M, Liu S S, et al. Chaos and Fractals in Natural Science. Beijing: Peking University Press, 2003 (in Chinese)

[8] Petrov V V. Sums of Independent Random Variables/Translated from Russian by A. A. Brown. Berlin: Springer-Verlag, 1975

[9] Bak P. How Nature Works: The Science of Self-Organized Criticality. New York: Copernicus Press, 1996

[10] Nicolis G, Prigogine I. Self-Organization in Nonequilibrium Systems. New York: Wiley, 1977

[11] Proakis J G. Algorithms for Statistcal Signal Processing. Beijing: Tsinghua University Press, 2003

[12] Ryberg T H. The normal inverse Gausian Levy process: simulation and approximation. Communication in Statistics: Stochastic Models, 1997, 13: 887 – 910

[13] Cannon M J, Percival D B, Caccia D C, Raymond G M, Bassingthwaighte J B. Evaluating scaled windowed variance methods for estimating the Hurst coefficient of time series. Physica A, 1997, 241: 606 – 626

[14] Wang X T , Qiu W Y, Ren F Y. Option pricing of fractional version of the Black-Scholes model with Hurst exponent *H* being in (1/3, 1/2). Chaos, Solitons & Fractals, 2001, 12 (3): 599 – 608

[15] Hurst H E. Long-term storage capacity of reservoirs. Transactions of the American Society of Civil Engineers, 1951, 116: 770 – 808

[16] Mandelbrot B B, Walis J R. Noah, Joseph, and operational hydrology. Water Resources Research, 1969, 4(5): 909 – 918
[17] Guo L, Xu X M. Complex Networks. Shanghai: Shanghai Scientific & Technological Education Publishing House, 2006 (in Chinese)
[18] Albert R, Barabasi A L. Statistical mechanics of complex networks. Review of Modern Physics, 2002, 74: 47 – 97
[19] Milgram S. The small-world problem. Psychology Today, 1967, 2: 60 – 67
[20] Guare J. Six Degrees of Separation: A Play. New York: Vintage, 1990
[21] Goh K I, Oh E, Kahng B, et al. Betweenness centrality correlation in social networks. Physical Review E, 2003, 67: 017101
[22] Watts D J, Strogatz S H. Collective dynamics of "small-world" networks. Nature, 1998, 393(6684): 440 – 442
[23] Meng Z W, Lu Z X, Song J Y. Comparison analysis of the small-world topological model of Chinese and American power grids. Automation of electric power systems, 2004, 28(15): 21 – 29 (in Chinese)
[24] Dobson I, Carreras B A, Newman D E. A probabilistic loading-dependent model of cascading failure and possible implications for blackouts. Proceedings of the 36th Annual Hawaii International Conference on System Sciences, Jan. 2003: 65 – 74
[25] Barrat A, Weight M. On the properties of small-world network models. European physical Journal, 2000, B13: 547 – 560
[26] Barabasi A L, Albert R. Mean-field theory for scale-free random networks. Physica A, 1999, 272: 173 – 187
[27] Bianconi G, Barabasi A L. Competition and multiscaling in evolving networks. Europhysics Letters, 2001, 54: 436 – 442
[28] Liu Z H, Lai Y C, Ye N, et al. Connective distribution and attack tolerance of general networks with both preferential and random attachments. Physics Letters A, 2002, 303: 337 – 344
[29] Dorogovtsev S N, Mendes J F F, Samukhin A N. Structure of growth networks with preferential linking. Physical Review Letters, 2000, 85: 4633 – 4636
[30] Shi D H, Chen Q H, Liu L M. Markov chain-based numerical method for degree distribution of growing networks. Physical Review E, 2005, 71: 036140
[31] Li X, Chen G. A local world evolving network model. Physica A, 2003, 328: 274 – 286
[32] Newman M E J. Mixing patterns in networks. Physical Review E, 2003, 67: 026126
[33] Pastor-Satorras R, Vazquez A, Vespignani A. Dynamical and Correlation Properties of the Internet. Physical Review Letters, 2001, 87: 258701
[34] Zhou T, Wang B H, Hui P M, et al. Topological properties of integer networks. Physica A, 2006, 367: 613 – 618
[35] Vazquez A, Pastor-Satorras R, Vespignani A. Large-scale topological and dynamical properties of the Internet. Physical Review E, 2002, 65: 066130
[36] Newman M E J, Girvan M. Finding and evaluating community structure in networks. Physical Review E, 2004, 69: 026113
[37] Newman M E J. Analysis of weighted networks. Physical Review E, 2004, 70: 056131
[38] Girvan M, Newman M E J. Community structure in social and biological networks. Procedings of the National Academic of Sciences, 2002, 99: 7821 – 7826

[39] Pothen A, Simon H, Liou K P. Partitioning sparse matrices with eigenvectors of graphs. SIAM Journal on Matrix Analysis and Applications, 1990, 11(3): 430 – 452

[40] Newman M E J. The structure and function of complex networks. SIAM Review, 2003, 45(2): 167 – 256

[41] Bonacich P. Power and centrality: a family of measures. The American Journal of Sociology, 1987, 92(5): 1170 – 1182

[42] Wuchty S, Stadler P F. Centers of complex networks. Journal of Theoretical Biology, 2003, 223(1): 45 – 53

[43] Freeman L C. A set of measures of centrality based upon betweenness. Sociometry, 1997, 40: 35 – 41

[44] Freeman L C. Centrality in networks: I. conceptual clarification. Social Networks, 1979, 1: 215 – 239

[45] Albert R, Jeong H, Barabasi A L. Error and attack tolerance in complex networks. Nature, 2000, 406: 378 – 382

[46] Jeong H, Mason S, Barabasi A L, et al. Lethality and centrality in protein networks. Nature, 2001, 411: 41 – 42

[47] Dunne J A, Williams R J, Martinez N D. Network structure and biodiversity loss in food webs: robustness increases with connectance. Ecology Letters, 2002, 5: 558 – 567

[48] Newman M E J, Forrest S, Balthrop J. Email networks and the spread of computer viruses. Physical Review E, 2002, 66(3): 035101

[49] Magoni D. Tearing down the internet. IEEE Journal on Selected Areas in Communications, 2003, 21(6): 949 – 960

[50] Home P, Kim B J, Yoon C N, et al. Attack vulnerability of complex Networks. Physical Review E, 2002, 65(1): 056109

[51] Wang X F, Chen G. Synchronization in complex dynamical networks, J. Systems Science and Complexity, 2003, 16: 1 – 14

[52] Pecora L M, Carroll T L. Master Stability functions for synchronized coupled systems. Physical Review Letters, 1998, 80: 2109

[53] Barahona M, Pecora L M. Synchronization in Small-World Systems. Physical Review Letters, 2004, 89: 054101

[54] Kocarev L, Amato P. Synchronization in power-law networks. Chaos, 2005, 15: 024101

[55] Wang X F, Chen G R. Synchronization in small-world dynamical networks. International Journal of Bifurcation and Chaos, 2002, 12: 187 – 192

[56] Wang X F, Chen G R. Synchronization in scale-free dynamical networks: robustness and fragility. IEEE Transactions on Circuits and Systems I, 2002, 49(1): 54 – 62

[57] Fan J, Li X, Wang X F. On synchronization in scale-free dynamical networks. Physica A, 2005, 349: 443 – 451

[58] Lv J, Chen G. A time-varying complex dynamical network model and its controlled synchronization criteria. IEEE Transactions on Automatic Control, 2005, 50: 841 – 846

[59] Lv J, Yu X, Chen G. Chaos synchronization of general complex dynamical networks. Physica A, 2004, 334: 281 – 302

[60] Belykh I V, Belykh V N, Hasler M. Connection graph stability method for synchronized coupled chaotic systems. Physica D, 2004, 195: 159 – 187

[61] Belykh I V, Belykh V N, Hasler M. Blinking model and synchronization in small-world networks with a time-varying coupling. Physica D, 2004, 195: 188 – 206
[62] Li C, Chen G. Synchronization in general complex dynamical networks with coupling delays. Physica A, 2004, 343: 236 – 278
[63] Dhamala M, Jirsa V K, Ding M. Enhancement of neural synchrony by time delay. Physical Review Letters, 2004, 92(7): 074104
[64] Kuramoto Y. Chemical Oscillations, Waves and Turbulence. Berlin: Springer-Verlag, 1984
[65] Strogatz S H. From Kuramoto to Crawford: exploring the onset of synchronization in populations of coupled oscillators. Physica D, 2000, 143: 1 – 20
[66] Hong H, Choi M Y, Kim B J. Synchronization on small-world networks. Physical Review E, 2002, 65: 026139
[67] Moreno Y, Pacheco A F. Synchronization of Kuramoto oscillators in scale-free networks. Europhysics Letters, 2004, 68: 603 – 609
[68] Li X. Uniform synchronous criticality of diversely random complex networks. Physica A, 2006, 360(2): 629 – 636

Chapter 3 Foundation of SOC in Power Systems

This chapter first introduces the notions of organization and self-organization and then discusses the general principles of SOC based on cybernetics. In addition, we construct the mathematical model describing the power system evolution mechanisms, and discuss the SOC phenomena in large-scale power grids. Further analysis is carried out using statistical data about the blackouts in the Chinese power systems. Finally, two categories of risk evaluation indices, VaR and CVaR, are discussed.

It has been reported that SOC behavior exhibits in a large number of complex systems in various forms, such as sand piles, earthquakes, and forest fires. The concept has been successfully applied to economics and biology. More and more researchers believe that SOC can be taken as a universal model for complex system dynamics, and the understanding of SOC can help to analyze characteristics of disasters and catastrophes, and then to design preventive control strategies. Although SOC theory was established quite recently, it has developed with great momentum absorbing contributions from researchers with diverse backgrounds, such as economy, physics, computer science, biology, meteorology, and engineering sciences[1−24].

SOC theory has several main subjects, which include statistical physics, probability theory, stochastic processes, nonlinear system theory and complex system theory. There is no clearly cut boundary for the domain of the study of SOC and there is even no generally accepted definition for SOC, not to mention the necessary conditions under which SOC behavior takes place. This is especially true in the power system literatures. As we know, power systems belong to a class of dissipative dynamical systems that expand geographically in space and evolve dynamically in time. They are not only time-varying, nonlinear, and open systems, but also complex systems with uncertainties and influenced by social and economic impacts. The close relationship between load growth and economic development makes the planning and construction in power systems an integral part of human societies. On the other hand, power systems may also bring constraints to the social and economic development in human societies. The large-scale, active and regular interactions between the open power systems and human activities lead to complex evolutionary phenomena in power systems, in particular self-organized events. In the process of planning and construction, power system engineers always seek optimal schemes, such as the plan with the minimum cost. Such practice dominates the self-organization in power systems and drives the system to evolve to its critical states, at which small disturbances may trigger cascading events that develop into catastrophe. When implementing recovery strategies after a blackout, optimal schemes are again preferred. This is another spontaneously organized evolution in power systems, which may again lead the system towards its critical states. Hence, power systems may develop into critical states following different paths.

Power systems have self-organization characteristics at the macroscopic and microscopic levels. For the former, a system's security margin decreases and its risk for large-scale blackouts increases with the development of economy and correspondingly the growth of load demand. Thus new power plants and transmission lines have to be built to reduce the risk of contingency. This pair of counterbalancing forces may push the system to operate in a critical state. For the latter, a system usually adopts some control strategies, such as adjusting generators' outputs and shedding loads, to maintain the stable operation and reliable power supply when facing random load perturbations, device faults or short circuit failures etc. Disturbances may deteriorate the dynamic performances of a system, and even cause instability. Control actions at the same time aim to stabilize the system. The system may thus arrive at a critical state at the microscopic level under the actions of these two forces when certain conditions are satisfied.

The research on the mechanism of power system catastrophe and the actions in response using complex system theory is a new and promising field that develops fast in recent years. SOC in power systems, as one of the frontiers in complexity science, becomes a hot topic in the study of the security of large-scale power systems. In this chapter, we summarize the main concepts and results of SOC that have proven to be useful for the analysis of power systems; and we also discuss theories and methods that can be used to reveal the SOC characteristics of power systems with their implications.

3.1 Organization and Self-Organization

The concept of "organization" is used in several fields in modern science. In particular, it is one of the most fundamental concepts in systems science to reveal the mechanisms for the start, growth, preservation, evolution and management of all entities in nature and human society [1]. In a strict sense, an organization is a collective entity that is established by people to achieve some objective collaboratively, such as political parties, labor unions, business corporations and military alliances. Nowadays, organizations are not just the cells of the society, but the foundation as well. In a broad sense, an organization is an interconnected system that consists of various interacting elements. Systems theory, control theory, information theory, dissipativity theory and synergetics are all theories to study organized systems from different angles. Here the interacting elements are not confined to people. For example, a swarm of honey bees is an organization organized around the queen with clear hierarchies. In this book, we use the concept of organization in this broader sense.

When organization is taken as a class concept, self-organization is then a sub-class concept. The counterpart of self-organization is called hetero-organization. SOC theory first appeared in the middle of the nineteenth century. Darwin's evolutionism can be viewed as the SOC theory in biology, and natural selection is one of the principals for self-organization. The social transition theory proposed by Marx tries to apply self-organization principles to social studies. Phase transition theory is related to the SOC in physics. These three are the SOC theories in different disciplines, but are not built upon the general frameworks and methodologies of SOC[1].

In the mid-twentieth century, the situation has changed fundamentally. After decades of study on principles and rules for phase transitions at equilibrium states, conditions have been ready for research on phase transitions at non-equilibrium states. The pioneering system scientists, especially Ashby and Wiener, proposed to construct the general theory for self-organization and Ashby has even finished a monograph entitled *Principles of Self-Organization*[2].

In the 1960s, there emerged several schools of scholars who aimed to reveal the general self-organization rules and the representative works include Prigogine's dissipativity structure theory[3], Haken's synergetics[4] and M. Eigen's hypercycles[5]. All these theories are rooted in the frontier of contemporary sciences and constitute the theoretical framework for SOC. Since the 1980s, the Santa Fe school has promoted actively the development of SOC theory under the joint efforts of Stuart A. Kauffman[6,7], Christopher G. Langton[8] and John Henry Holland[9].

However, up to now, there are still quite a number of problems in SOC theory that need to be solved, especially those with respect to the general principals and frameworks. On one hand, each school has developed insightful concepts, rules and methodologies, which show convincingly the promising future of SOC. On the other hand, different schools come from different backgrounds, and the corresponding theories are usually applicable only to their own subjects but not the other's. In extreme cases, viewpoints taken by different schools are even not consistent with one another[1].

In this book, an organization is said to be self-organized if there have been no specific external influences during its temporal and spatial evolution. Here, by "specific" we mean that the structures or functions are not exerted on the system through pre-specified manners[1]. In contrast to self-organization, hetero-organization is usually formed by external commands, constructive influences, and restrictive boundary conditions. Philosophically, self-organization and hetero-organization form a pair of contradictions, and they are repelling and co-existing. As pointed out by Haken[10–12], the boundary between self-organized systems and man-made systems is not strict. Self-organization is still possible to appear in a man-made system under certain scenarios.

3.2 SOC Based on Cybernetics

3.2.1 Mathematical Description of the System

A system is the integration of a collection of interacting and interdependent components that behaves and functions according to certain laws. An open, controlled, and perturbed system can be described as follows:

$$\dot{x}_i = f_i(x, u_i, w_i), \quad i = 1, 2, \cdots, n \tag{3.1}$$

where $x_i \in \mathbb{R}^{n_i}$ is the ith element of the system state $x = (x_1^{\mathrm{T}}, x_2^{\mathrm{T}}, \cdots, x_n^{\mathrm{T}})^{\mathrm{T}}$, $u_i \in \mathbb{R}^{m_i}$ is the control input, $w_i \in \mathbb{R}^{k_i}$ is exogenous disturbance and f_i describes the interactions among different components. Set $u = (u_1^{\mathrm{T}}, u_2^{\mathrm{T}}, \cdots, u_n^{\mathrm{T}})^{\mathrm{T}}$, $w = (w_1^{\mathrm{T}}, w_2^{\mathrm{T}}, \cdots, w_n^{\mathrm{T}})^{\mathrm{T}}$, and $f = (f_1^{\mathrm{T}}, f_2^{\mathrm{T}}, \cdots, f_n^{\mathrm{T}})^{\mathrm{T}}$, then the system can be rewritten as

$$\dot{x} = f(x, u, w) \tag{3.2}$$

which is the general system model that we work with.

As to multi-dimensional and large-scale systems, we assume that the ith subsystem has a state y_i whose time scale is different from that of x_i. Let $y = (y_1^{\mathrm{T}}, y_2^{\mathrm{T}}, \cdots, y_n^{\mathrm{T}})^{\mathrm{T}}$, then the multi-dimensional subsystem can be described by

$$\begin{cases} \dot{x}_i = f_i(x, y, u_i, w_i) \\ \varepsilon_i \dot{y}_i = g_i(x, y, u_i, w_i) \end{cases} \tag{3.3}$$

where $\varepsilon_i > 0 (i = 1, 2, \cdots, n)$ is a sufficiently small positive number. Furthermore, set $g = (g_1^{\mathrm{T}}, g_2^{\mathrm{T}}, \cdots, g_n^{\mathrm{T}})^{\mathrm{T}}$ and $\varepsilon = (\varepsilon_1^{\mathrm{T}}, \varepsilon_2^{\mathrm{T}}, \cdots, \varepsilon_n^{\mathrm{T}})^{\mathrm{T}}$, then the multi-dimensional large-scale system can be described by

$$\begin{cases} \dot{x} = f(x, y, u, w) \\ \varepsilon \dot{y} = g(x, y, u, w) \end{cases} \tag{3.4}$$

Generally speaking, when the x_i changes within a given range during the time to be considered, the above system can be written as a set of differential algebraic equations (DAE):

$$\begin{cases} \dot{x} = f(x, y, u, w) \\ 0 = g(x, y, u, w) \end{cases} \tag{3.5}$$

If the system's parameters change as time evolves, or to be more specific, the parameter $s_i \in \mathbb{R}^{p_i}$ of the ith subsystem is time-varying, we describe such changes using discrete difference equations. Let $s = (s_1^{\mathrm{T}}, s_2^{\mathrm{T}}, \cdots, s_n^{\mathrm{T}})^{\mathrm{T}}$, then the dynamics of the ith subsystem are described by the following differential-difference-algebraic equations (DDAE):

$$\begin{cases} \dot{x}_i = f_i(x, y, s_i, u_i, w_i) \\ 0 = g_i(x, y, s_i, u_i, w_i) \\ s_i(t_{k+1}) = \varphi_i(s(t_k), x, y, u_i, w_i) \end{cases} \tag{3.6}$$

Let $\varphi = (\varphi_1^{\mathrm{T}}, \varphi_2^{\mathrm{T}}, \cdots, \varphi_n^{\mathrm{T}})^{\mathrm{T}}$, then the system evolution equations are

$$\begin{cases} \dot{x} = f(x, y, s, u, w) \\ 0 = g(x, y, s, u, w) \\ s(t_{k+1}) = \varphi(s(t_k), x, y, u, w) \end{cases} \tag{3.7}$$

Remark 3.1 When computing the systems' dynamics, Eq. (3.7) are usually used as constraints. One needs to construct objective functions which are determined by practical considerations. Then a complete model is established for searching for u. In fact, this procedure can be taken as a time-varying optimization problem; in other words, as time evolves, the optimization problem changes as well.

Remark 3.2 Since the control u can be written as a function of the system's states and outputs, the system described by Eq. (3.7) is an open, self-organized, and multi-component dynamical system.

3.2.2 System Evolution and SOC

The evolution of a complex system is a complicated progress, which can be roughly classified into the dynamics caused by the external and internal actions respectively. The system's output is determined jointly by internal and external forces as shown in Fig. 3.1.

We discuss the relationship between the system evolution and SOC using Fig. 3.1. We provide three different angles: slow dynamics, fast dynamics and control.

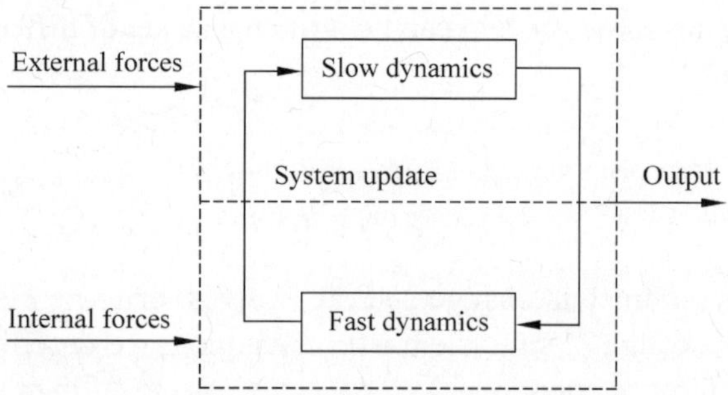

Figure 3.1 A system under external and internal forces

In the long-term dynamic processes (slow dynamics), the system is running in multi-equilibrium states and evolves to its critical state spontaneously. The mechanism is that the system continuously adjusts its state to absorb the external energy supplies and at the same time dissipates through its dissipative components. At the macroscopic level, the overall system is always in the transient state from one equilibrium to another. In the short-term dynamic processes (fast dynamics), the external forces remain constant approximately, therefore the system is an autonomous system, whose dynamic behaviors are governed by its own intrinsic properties. However, the system's behavior can become unpredictable because of the existence of random events and uncertain factors. If we study the SOC using the angle of control theory, we can distinguish internal and external variables, and when the interactions between the two satisfy a certain relationship, the system may enter a critical state. Then any disturbance, no matter large or small, may lead the system into unexpected changes, even dramatic events in the end.

In short, because of the system's complexity and uncertainty, it mainly operates in the transient process from one equilibrium to another, namely at non-equilibrium states. This drives the system to evolve into self-organized states, which further facilitate the separation and collaboration among the system's main components in different time scales. Various macroscopic spatial and temporal structures are thus formed through fluctuations, amplified by nonlinear effects and propagated via system-level fluctuations. Hence, most SOC phenomena are affected by both internal nonlinearities and external disturbances.

3.3 Evolution Models for Power Systems

The evolution of power systems involves dynamic processes on different time scales. The construction cycle of power plants and grids ranges from years to decades, the increase of electrical loads is evaluated by the unit of years and changes with the transition of seasons, and the daily load curves always show peaks and valleys. These can all be classified into the mid- or long-term dynamics, which are also called slow dynamics. In comparison, disturbances like sudden load jump and lightening activate the system's fast or transient processes, which include switch operations and their electromagnetic transients in the unit of microseconds, electromechanical transients in the unit of seconds, power flow adjustments in the unit of seconds or minutes, and transformer tap changing in the unit of minutes. So the evolution model of power systems has distinct multi-timescale features. For convenience, we divide the power system evolutions into three timescales: slow, fast and transient dynamics as shown in Fig. 3.2.

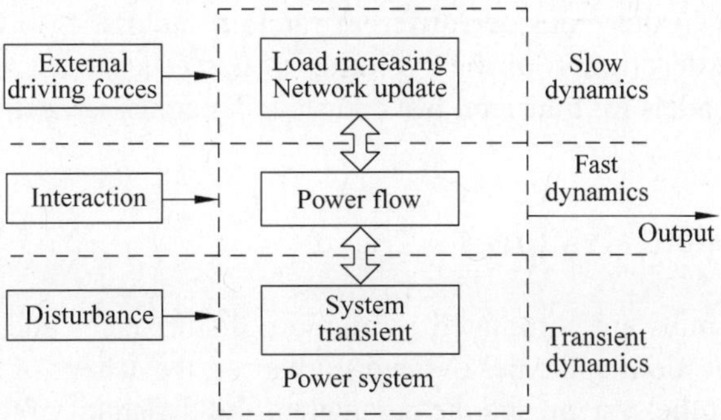

Figure 3.2 Power system evolution

In Fig. 3.2, slow dynamics include the continuous load increase and the grid upgrade, which are evolutions driven by external forces. Fast dynamics are induced by the changes in load demand and power supply and the redistribution of power flows after line faults, which are mainly determined by Ohm's law, Kirchhoff's law and other physical rules. Transient dynamics are the sequence of dynamic processes after unexpected disturbances, such as three-phase short-circuit faults, which are usually determined by Faraday's law of induction, Lenz's law, Newton's law and the control logics and strategies that are designed beforehand.

The evolution of power systems can be described by the following set of differential-difference-algebraic equations (DDAE):

$$\begin{cases} \dot{x} = f(x, y, s, u, w) \\ 0 = g(x, y, s, u, w) \\ s(t_{k+1}) = \varphi(x, y, s(t_k), u, w) \end{cases} \tag{3.8}$$

where the set of differential equations $\dot{x} = f(x, y, s, u, w)$ correspond to the transient dynamics of generators and loads, the set of algebraic equations $0 = g(x, y, s, u, w)$ are the power flow constraints, which are the fast dynamics; and the set of difference equations $s(t_{k+1}) = \varphi(x, y, s(t_k), u, w)$ describe slow dynamics including the load increase and power grid constructions. Here x, y, s, u, w are the state, output, discrete parameters, control and disturbance respectively. The state consists of the power angels and frequencies of the generators and the output includes the active and reactive powers through the lines and bus voltages the discrete parameters are determined by the upgrades of the system; the control consists of the adjustments of generation outputs, relay actions, switches, and other preventive and emergency control actions; the disturbances include short-circuit faults, incorrect operation of dispatchers and misoperation of relays.

Eq. (3.8) reveals the inherent driving mechanism for the system evolutions and takes into consideration various factors in long-term evolutions. From the viewpoint of system science, it is difficult to establish a precise model that describes all kinds of power system dynamics and then search for solutions. It is almost impossible to do so if we further consider different random factors. In order to provide a comprehensive description of the evolution of power systems, in this section we introduce the models for transient, fast and slow dynamics respectively.

3.3.1 Transient Dynamics

Transient dynamics are stimulated by random disturbances and thus are related closely to the locations and types of the disturbances, the actions of relay protections, the structures of the system, and the parameters and dynamics of the system's components. It is usually needed to evaluate the system's stability after disturbances, namely whether the system can maintain its synchronous stability, transient voltage stability and frequency stability. If the system becomes unstable, but the stability control devices respond properly, the corresponding actions such as tripping generators, fast switching off valves, and shedding loads can be implemented immediately (within tens or hundreds of milliseconds). These actions change the structures and parameters of the system and drive the system back to its stable operation point before the disturbances or to some other new stable operation point. However, when the faults are so severe that control devices are not able to stabilize the system anymore, actions like underfrequency and undervoltage load shedding start to trip generators and loads, which may cause cascading failures. In this process, if some loads are lost, then blackout happens.

Now we discuss different aspects of the transient dynamic process.

1) Disturbance

There are different types of disturbances in power systems, and here we mainly consider impulsive load disturbances, short-circuit faults and line outages. Impulsive load disturbances are caused by the sudden jump in loads, and they lead to dynamic processes in generators. Correspondingly excitation regulators and governors regulate the generators' voltages and active powers to catch up with the increasing loads. Taking into account the starting processes of induction motors, the load themselves may also experience dynamic processes.

As to the short-circuit faults, they can be classified into three-phase, two-phase, single-phase grounding, and two-phase grounding faults. Except for the three-phase short-circuit fault, all the other types of short-circuit faults cause the three-phase unbalance, and thus the symmetrical component method is usually used for these cases[25].

The line outage faults are usually caused by external forces or misoperation of protections, and they change the topological structures of power grids.

2) Relay protection

Relay protection is vital in the operation of power systems. When faults occur in the system, large electric current might be generated and thus the protections separate the faulted equipments from the system, and so the other devices may not be damaged and thus maintain the system's stability. The protection actions are usually described by discrete variables that affect the system's topologies. The actions of relay protections in transient dynamics in general have delays of tens of milliseconds.

3) Generator bus dynamics

Generally speaking, generator bus dynamics include the dynamics of the generator itself and the dynamics of the excitation system and the governor. Their relationships are shown in Fig. 3.3[26]. The dynamics of the generator can be further classified into the swing equations described by Newton's second law and the electromagnetic induction equations described by Faraday's law. The mathematical description of the electromagnetic induction is as follows.

(1) Per unit stator voltage equations

$$\begin{aligned} e_d &= p\psi_d - \psi_q\omega_r - R_a i_d \\ e_q &= p\psi_q + \psi_d\omega_r - R_a i_q \\ e_0 &= p\psi_0 - R_a i_0 \end{aligned} \tag{3.9}$$

where ψ_d, ψ_q and ψ_0 are the stator's flux linkages along *d*, *q* and 0 axes; p is the differential operator; ω_r is the rotor's angular velocity; e_d, e_q, e_0, i_d, i_q and i_0 are the stator's voltages and currents along *d*, *q* and 0 axes; and R_a is the armature resistance per phase.

(2) Per unit rotor voltage equations

$$\begin{cases} e_{fd} = p\psi_{fd} + R_{fd} i_{fd} \\ 0 = p\psi_{1d} + R_{1d} i_{1d} \\ 0 = p\psi_{1q} + R_{1q} i_{1q} \\ 0 = p\psi_{2q} + R_{2q} i_{2q} \end{cases} \tag{3.10}$$

where ψ_{fd} is the flux linkage of field winding, ψ_{1d} is the flux linkage of the

d-axis amortisseur circuit, ψ_{1q} and ψ_{2q} are the flux linkages of the q-axis amortisseur circuits q_1 and q_2, R_{fd}, R_{1d}, R_{1q}, R_{2q}, i_{fd}, i_{1d}, i_{1q} and i_{2q} are the equivalent resistances and currents of these rotor windings; and e_{fd} is the field voltage[26].

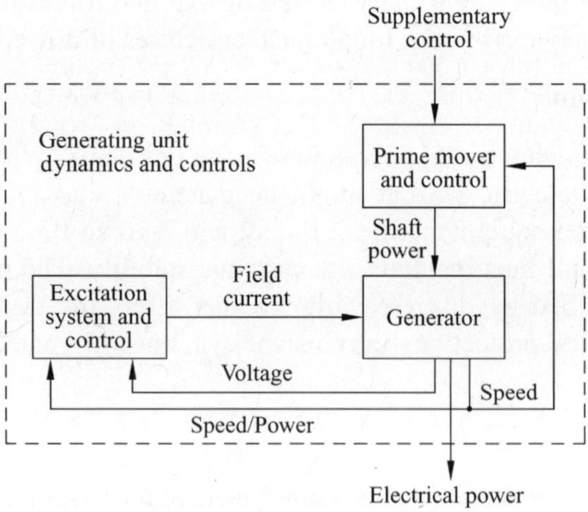

Figure 3.3 Dynamics of generator buses

(3) Per unit stator flux linkage equations

$$\begin{cases} \psi_d = -(L_{ad} + L_l)i_d + L_{ad}i_{fd} + L_{ad}i_{1d} \\ \psi_q = -(L_{aq} + L_l)i_q + L_{aq}i_{1q} + L_{aq}i_{2q} \\ \psi_0 = -L_0 i_0 \end{cases} \tag{3.11}$$

where L_{ad} and L_{aq} are the mutual inductances between stator and the d-axis(q-axis) rotor windings, and L_l is the leakage inductance.

(4) Per unit rotor flux linkage equations

$$\begin{cases} \psi_{fd} = L_{ffd}i_{fd} + L_{f1d}i_{1d} - L_{ad}i_d \\ \psi_{1d} = L_{f1d}i_{1d} + L_{11d}i_{1d} - L_{ad}i_d \\ \psi_{1q} = L_{11q}i_{1q} + L_{aq}i_{2q} - L_{aq}i_q \\ \psi_{2q} = L_{aq}i_{1q} + L_{22q}i_{2q} - L_{aq}i_q \end{cases} \tag{3.12}$$

where L_{ffd}, L_{11q} and L_{22q} are the self-inductances of the field winding, and the d-axis and q-axis amortisseur circuits respectively, and L_{f1d} is the mutual inductance between the field winding and the q-axis amortisseur winding.

(5) Per unit air-gap torque

$$T_e = \psi_d i_q - \psi_q i_d \tag{3.13}$$

(6) Swing equations

$$\dot{\delta} = \omega_0 \Delta\omega_r \tag{3.14}$$

$$\Delta\dot{\omega}_r = \frac{1}{2H}(T_{\mathrm{m}} - T_{\mathrm{e}} - K_D \Delta\omega_r) \tag{3.15}$$

where δ and $\Delta\omega_r$ are the angle and angular velocity of the rotor, T_{m} is the mechanical torque, T_{e} is the electromagnetic torque, H is the combined moment of the inertia of the generator, and K_D is the damping coefficient.

4) Excitation system

Figure 3.4 shows the functional block diagram of a typical excitation system for a large synchronous generator[26].

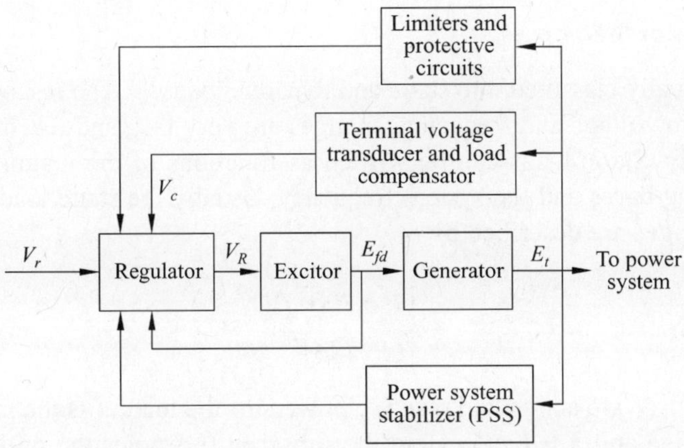

Figure 3.4 Functional block diagram of a synchronous generator's excitation system

The excitation systems have taken different forms over the years, which can be classified into three broad categories based on the utilized excitation power source: DC, AC and static excitation systems. The basic function of an excitation system is to provide direct current to the synchronous machine's field winding. In addition, the excitation system controls the generator's terminal voltage, and thus enhances the system's stability. Its protective functions ensure that the synchronous machine, excitation system, and other equipments operate within their capability limits.

5) Governor

The governing system adjusts the active power according to the commands generated by the automatic generation control (AGC). It can also maintain the system's frequency within an acceptable range by its rotational speed measurement device and feedback controller. Figure 3.5 shows the functional relationship between the basic elements associated with the governing system[26].

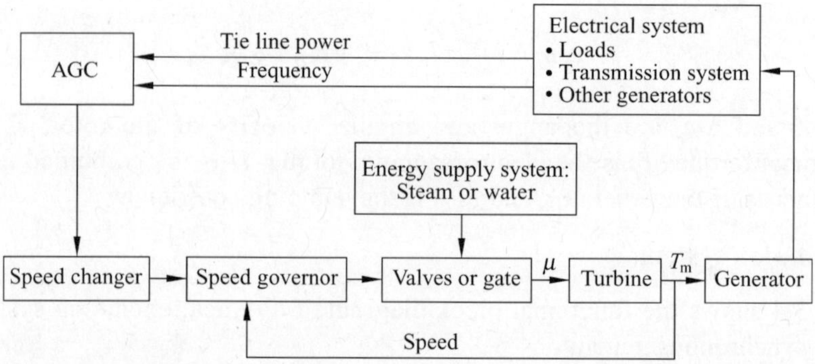

Figure 3.5 Functional block diagram of the governing system

6) Dynamics of load buses

Loads are usually classified into static and dynamic loads[26]. The responses of most static loads to voltage and frequency changes are very fast, and reach their steady states quickly. Static loads can be written as functions of the magnitudes of the corresponding buses and the system frequency. Usually the static loads depending only on voltages are described by

$$\begin{cases} P = P_0(\bar{V})^a \\ Q = Q_0(\bar{V})^b \end{cases} \tag{3.16}$$

where P and Q are active and reactive powers of the load, V is the magnitude of the bus voltage and $\bar{V} = V/V_0$. Here the subscript 0 denotes the initial values of the corresponding variables and a, b are constants.

An alternative description of static loads is the following polynomial model:

$$\begin{cases} P = P_0(p_1\bar{V}^2 + p_2\bar{V} + p_3) \\ Q = Q_0(q_1\bar{V}^2 + q_2\bar{V} + q_3) \end{cases} \tag{3.17}$$

which is commonly referred to as the ZIP model because it consists of constant impedance (Z), constant current (I), and constant power (P) loads. The coefficients p_1 to p_3 determine the proportions of each type of load, where $p_1 + p_2 + p_3 = 1$.

When further considering the frequency dependency, the loads can be modeled by

$$\begin{cases} P = P_0(\bar{V})^a(1 + K_{pf}\Delta f) \\ Q = Q_0(\bar{V})^b(1 + K_{qf}\Delta f) \end{cases} \tag{3.18}$$

or

$$\begin{cases} P = P_0(p_1\bar{V}^2 + p_2\bar{V} + p_3)(1 + K_{pf}\Delta f) \\ Q = Q_0(q_1\bar{V}^2 + q_2\bar{V} + q_3)(1 + K_{qf}\Delta f) \end{cases} \tag{3.19}$$

where $\Delta f = f - f_0$ and typically $K_{pf} \in [0, 3.0]$ and $K_{qf} \in [-2.0, 0]$.

Under various circumstances, it is usually necessary to take into consideration load dynamics. This is especially true for the study of inter-area oscillations, voltage stability, and long-term stability. Typically, 60% – 70% of the total load is of the motor type. Therefore, the characterization of the motor dynamics is usually the most significant step in the modeling of the load dynamics. In Fig. 3.6, we show the simplified equivalent circuit for motors[26].

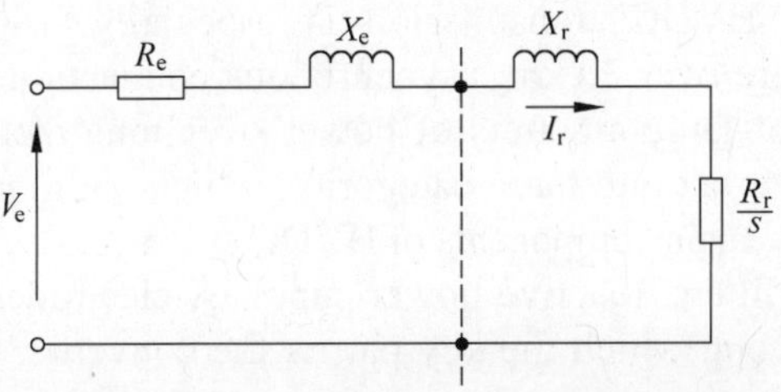

Figure 3.6 Equivalent circuit for evaluating torque-slip relationships

The rotor's acceleration is determined by

$$T_e - T_m = J\frac{d\omega_m}{dt} \tag{3.20}$$

where

$$T_e = 3\frac{p_f}{2}\left(\frac{R_r}{s\omega_s}\right)\frac{V_e^2}{(R_e + R_r/s)^2 + (X_e + X_r)^2},$$

R_e and X_e are the equivalent resistance and the leakage inductance of the stator windings respectively, R_r and X_r are the resistance and the leakage inductance of the rotor windings respectively, T_m is the mechanical load torque that varies with speed, T_e is the electromagnetic torque, V_e is Thevenin's equivalent voltage, s is the slip, ω_m is the rotor's angular velocity in the unit of radians per second, J is the polar moment of inertia of the rotor and the connected load, and ω_s is the synchronous speed.

The modeling of loads is complicated since a typical load bus that is used in stability studies corresponds to a large number of hugely different devices, such as fluorescent and incandescent lamps, refrigerators, heaters, compressors, motors, furnaces, and so on. The exact composition of load is difficult to estimate, which changes with time (hour, day, season), weather, and economic situations. Even if load composition could be known exactly, it would still be impractical to model in detail each individual load component because there are usually millions of such components that are supplied by a power system. Therefore, load representation has always been done after a considerable amount of simplification. So when studying different topics in power system analysis, different models have to be established and adopted according to the features of the problem at hand. This holds not only for load modeling, but also for the SOC and complexity study for large-scale power grids throughout this book.

7) HVDC and other FACTS

In recent years, conversion equipments have become smaller and less costly while their reliability has also been improved significantly. These developments have enabled the widespread of the high voltage direct current (HVDC) transmission technology[25,27,28]. HVDC transmission is especially useful for underwater transmissions that are over 30 km, asynchronous connection of two AC systems and transmissions of large amounts of power over long distances. HVDC links can be broadly classified into three categories, which are monopolar, bipolar and homopolar links. The main components of HVDC systems are converters, smoothing reactors, harmonic filters, reactive power supplies, electrodes, DC lines and AC circuit breakers, among which the key part is the converter. Converters perform AC/DC conversions and provide a means for controlling power flows through the HVDC link. By controlling or adjusting the operation and network parameters through large-capacity power electronic devices, flexible alternating transmission systems (FACTS) facilities optimize the system's operation and improve its transmission capabilities[25].

For an existing traditional power system without FACTS, the parameters of transmission lines are constant, and the system regulates the active and reactive powers of its key generator in its operation. Although such conventional systems can adjust the LTC taps and the series and shunt compensator to change system parameters, and open or close lines to modify network topologies, all such actions have to be carried out mechanically, and consequently the regulation speed is too slow to satisfy the requirements in transient dynamics. Compared with the generators' actions to control velocities and speeds, the system has few options to control the network efficiently. This is exactly where the FACTS can be utilized to overcome these limitations.

Nowadays, the FACTS devices include parallel static var compensator (SVC), static synchronous compensator (STATCOM), thyristor controlled series capacitor (TCSC), static synchronous series compensator(SSSC), unified power flow controller (UPFC), and thyristor controlled phase shifting transformer (TCPST). Among them, SVC, STATCOM and TCSC have been applied widely in engineering systems. There are UPFC prototypes, while the other FACTS devices have not been applied in practical projects yet.

8) Stability control devices

Power system may lose stability after disturbances. The instability may be in the forms of generators' loss of synchronization and the severe rise or fall in voltages and frequencies. Said differently, when a system loses its stability, some of its physical states cannot be maintained within an acceptable range.

When the power system is in normal conditions, even three-phase short circuit cannot cause instability directly. This is because of relay protection and stability control devices that are programmed with control strategies beforehand or enabled with online computation capabilities to deal with single or some types of double failures. After a fault happens, the stability control devices first check whether there are preset responding strategies, such as tripping off generators or loads. However, since there are faults that are unpredictable, the stability control devices are not effective for every fault.

9) Underfrequency and undervoltage load shedding

To ensure stability, power systems are installed with a large number of underfrequency and undervoltage load shedding devices that are used to cut loads according to the local voltages, frequencies and their rates of change to maintain the power balance. However, such devices can only acquire local information, and may not be able to act cooperatively over a large area.

3.3.2 Fast Dynamics

When the power system is stable after some disturbances, one can study its dynamics using the steady-state model that is usually in the form of power flow equations or their expansions. The fast dynamics discussed in this book are in the time scale of minutes or hours, which are much longer than the time scale of milliseconds and seconds in transient dynamics. During fast dynamics, economic dispatch, preventive control and long time-delay overcurrent protection take actions. Economic dispatch and preventive control regulate the active and reactive power outputs via dispatching centers to guarantee the safe and economic operations of power systems.

Now we briefly introduce power flow equations, economic dispatch, preventive control and long time-delay overcurrent protection.

1) Power flow equations

For an n-bus system, its power flow equations in polar coordinates can be written as

$$\begin{cases} P_{gi} - P_{di} - \sum_{j=1}^{n} (V_i V_j B_{ij} \sin\theta_{ij} + V_i V_j G_{ij} \cos\theta_{ij}) = 0 \\ Q_{gi} - Q_{di} - \sum_{j=1}^{n} (V_i V_j G_{ij} \sin\theta_{ij} - V_i V_j B_{ij} \cos\theta_{ij}) = 0 \end{cases} \tag{3.21}$$

where P_{gi}, Q_{gi}, P_{di} and Q_{di} are the generation active power, generation reactive power, active load and reactive load at bus i respectively, G_{ij} and B_{ij} are the conductance and susceptance from bus i to bus j, $\theta_{ij} = \theta_i - \theta_j$ is the phase-angle difference between bus i and bus j, and V_i denotes the magnitude of the voltage at bus i.

For a system in operation, its dispatching center can acquire the state information from various measurement devices, such as remote terminal unit (RTU) and phase measurement unit (PMU). The information about bus voltages, line powers, tap positions, and switch statuses are collected to monitor the system's operation conditions and network loss, and in the end to perform economic dispatches.

But it is usually difficult to obtain the exact solution to the above power flow equations. In fact, because of the large scale of the network, the measurement data of power systems are asynchronous (which can be improved by PMU) and sometimes there are malfunctioning measurement units. So state estimation programs have to be used in dispatching centers to deal with the errors or even huge mistakes in the measurement data and get approximate power flow solutions. State estimation programs can identify wrong data by making use of the redundancy in the data, and then adopt algorithms like the least square method to approximate the power flow solution. It is worth mentioning that the calculation and analysis software at the dispatching center may also have bugs which can lead to potential faults.

To calculate power flows, we usually divide the buses into three categories which are *PQ*, *PV* and slack buses. Here, *PQ* buses have constant active and reactive power, and can be used to simulate constant power loads, *PV* buses have constant active power and voltage magnitudes, and they are used to simulate generators with automatic voltage regulators (AVR) and governors (GOV); slack buses have constant voltage magnitudes and phase angles[25]. There is no slack bus in real systems and it is set for the convenience of calculation. Using the classification just described, all the unknowns can be calculated according to Eq. (3.21). The most popular algorithms include Newton-Raphson algorithm, Gauss-Sediel algorithm and *PQ* decomposition algorithm[25].

Since the set of power flow equations (Eq. (3.21)) is nonlinear in high-dimensional space, it is common to encounter divergence, which is due to two reasons. One is that the initial values are too far away from the solution while the solution exists, and the other is that the equations do not have a solution at all, which, in studying voltage stability, usually means that a saddle-node bifurcation has occurred and the system's voltages collapse. The blackout in Tokyo in 1987 is such a typical example.

In addition, constraints on generators' outputs and load buses' voltages should also be considered in practical power flow equations. Thus, *PQ* buses may sometimes become *PV* buses and *PV* buses become *PQ* buses. As to AC-DC power systems, the AC-DC alternating iterative algorithm is usually used[27], and we do not go into details here.

2) Economic dispatch

The main task of economic dispatch is to solve for optimal power flows (OPF), and then regulate generators' outputs and other controllable devices to maintain the operation state of the system at the optimum. Mathematically speaking, OPF is a nonlinear optimization problem. For a system with n buses and m lines, the objective function at time k might be to minimize the network active power loss as shown in Eq. (3.22)

$$\min F(P,Q) = \sum_{i\in G} P_{gi,k} - \sum_{j\in L} P_{dj,k} \tag{3.22}$$

where G is the generator set, L is the load set, $P_{gi,k}$ is the active power output of the generator at bus i, and $P_{dj,k}$ is the active load at bus j, and P and Q are the generators' active and reactive power outputs, which are adjustable.

In the meantime, some constraints must be satisfied, which are listed as follows:

(1) Power flow equations (the same as Eq. (3.21))

(2) Generator output constraints

$$\begin{cases} P_{gi,k}^{\min} \leqslant P_{gi,k} \leqslant P_{gi,k}^{\max}, & i \in G \quad (3.23) \\ Q_{gi,k}^{\min} \leqslant Q_{gi,k} \leqslant Q_{gi,k}^{\max}, & i \in G \quad (3.24) \end{cases}$$

where $P_{gi,k}$ and $Q_{gi,k}$ are the active and reactive power outputs of the ith generator at time k respectively, $P_{gi,k}^{\max}$ and $Q_{gi,k}^{\max}$ are the corresponding upper limits and $P_{gi,k}^{\min}$ and $Q_{gi,k}^{\min}$ are the lower limits.

(3) Bus voltage constraints

$$V_{i,k}^{\min} \leqslant V_{i,k} \leqslant V_{i,k}^{\max} \tag{3.25}$$

where $V_{i,k}$ is the magnitude of the voltage at bus i at time k, and $V_{i,k}^{\max}$ and $V_{i,k}^{\min}$ are the corresponding upper and lower limits of $V_{i,k}$.

(4) Transmission capacity constraint

$$-F_{l,k}^{\max} \leqslant F_{l,k} \leqslant F_{l,k}^{\max}, \quad l = 1,2,\cdots,m \tag{3.26}$$

where $F_{l,k}$ is the power flow of line l at time k, and $F_{l,k}^{\max}$ is the capacity of line l at time k.

3) Preventive control

The automatic security analysis system is widely used in modern power systems. Once the system identifies that its operation point is out of the stability region or predicts that the system will lose its stability under the anticipated fault, the system changes its operation point by regulating the generators' outputs and other controllable devices to drive the system back to safe states.

4) Overcurrent protection

Overcurrent protection includes instantaneous overcurrent protection and long time-delay overcurrent protection. In fast dynamics, the long time-delay overcurrent protections of transmission lines and transformers may act to trip overcurrent equipments if the current lasts for more than a preset threshold. During the 8·14 (Aug.14) North America blackout, five 345 kV transmission lines were tripped in the first hour after the first outage event with time intervals being 22, 9, 5 and 29 minutes[29].

In practice, misoperation and maloperation may happen in the overcurrent protection devices because of wrong settings and signal interference. According to the investigation report of UCTE[30] for the 11·4 (Nov. 4) Europe blackout in 2006, the dispatchers of the E.ON power grid calculated the regular power flows before the outage of the line Conneforde-Diele, and wrongly assumed that the protection setting of the Wehrendorf station in the RWE zone was 2000A, which should be 1900A instead. Because of this mistake, they thought the system had satisfied the $N-1$criterion. After this blackout, the simulation results have shown that although there was no facility overloaded, line Landesbergen-Wehrendorf had been overloaded when line Bechterdissen-Elsen was tripped.

3.3.3 Slow Dynamics

The time scale of slow dynamics is in the unit from days up to tens of years. This process is affected by load fluctuation, power plant construction and grid development.

1) Load fluctuation

The fluctuation of load first shows some regularity in short terms. For example, the load changes daily with the same characteristics of peaks and valleys. At the same time, the fluctuation is also unpredictable in long terms because of social and environmental events and changes, such as sudden weather changes and big sport events. Load fluctuation requires that the active and reactive power in the power flow equations have to be modified, namely

$$\begin{cases} P_{di,k+1} = m_{i,k+1} P_{di,k}, & i \in L \\ Q_{di,k+1} = n_{i,k+1} Q_{di,k}, & i \in L \end{cases} \tag{3.27}$$

where L is the set of load buses, $P_{di,k}$ and $Q_{di,k}$ are the active and reactive loads at bus i at time k, and $m_{i,k+1}$ and $n_{i,k+1}$ denote the growth rates of the active and reactive loads at bus i at time $k+1$ respectively.

2) Power plant construction and grid development

In order to meet the demand of increasing loads and optimize energy supply structures, new power plants, substations and transmission lines are being constructed, and outdated equipments are being replaced. So power systems evolve all the time. Generally speaking, power construction plans must be determined several years (usually 5 or 10 years) before their implementation and the following questions must be considered:

(1) In the system where are the weak areas that are prone to faults and blackouts?

(2) What is the load level in 5 or 10 years from now and how is the load distributed?

(3) Where are the energy resources that are easy to be utilized? How difficult is it to build new transmission lines with what cost?

(4) How will the newly built plants, transmission lines and equipments influence the security and reliability of the system?

(5) What is the budget for the planned new plants, transmission lines and other equipments?

(6) How soon are the development projects expected to finish?

Two reasons make it extremely difficult to come up with an optimal plan. First, since the plan is made for the system after a rather long construction period, it is hard to predict precisely the load demands in 5 or 10 years. Unexpected load growth patterns can easily lead to the shortage or surplus in power supply. Second, power systems develop under various constraints. New projects have to be planned and implemented facing issues associated with environmental concerns, huge capital expenditures, changing energy distribution and most importantly the system's stability. However, the system parameters that are used to check the security of the overall system are difficulty to determine precisely.

The topology and parameters of the system change after new plants and transmission lines are constructed and new equipments are installed. For instance, when additional generator sets are installed in a plant, the plant's generation capacity increases which implies that the maximum active and reactive power outputs of the plant increase accordingly:

$$\begin{cases} P_{gi,k+1}^{\max} = \lambda_{i,k+1} P_{gi,k}^{\max}, \ i \in G \\ Q_{gi,k+1}^{\max} = \lambda_{i,k+1} Q_{gi,k}^{\max}, \ i \in G \end{cases} \tag{3.28}$$

where $\lambda_{i,k+1} > 1$ is the increase rate of the generator capacity at time k+1, G is the set of the upgraded generators, and $P_{gi,k}^{\max}$ and $Q_{gi,k}^{\max}$ are the upper limits of the active and reactive powers of the generators at bus i at time k.

If we further consider the transient dynamics after the upgrade of the power plant, we need to modify the dynamic parameters of the generators:

$$\begin{cases} x_{d,gi,k+1} = \dfrac{x_{d,gi,k}}{\lambda_{i,k+1}}, \ i \in G \\ x'_{d,gi,k+1} = \dfrac{x'_{d,gi,k}}{\lambda_{i,k+1}}, \ i \in G \end{cases} \tag{3.29a}$$

$$\begin{cases} x_{q,gi,k+1} = \dfrac{x_{q,gi,k}}{\lambda_{i,k+1}}, \ i \in G \\ x'_{q,gi,k+1} = \dfrac{x'_{q,gi,k}}{\lambda_{i,k+1}}, \ i \in G \end{cases} \tag{3.29b}$$

$$H_{gi,k+1} = \lambda_{i,k+1} H_{gi,k}, \ i \in G \tag{3.29c}$$

where $x_{d,gi,k}$ and $x'_{d,gi,k}$ are the d-axis synchronous reactance and transient reactance of the generators at bus i at time k respectively, $x_{q,gi,k}$ and $x'_{q,gi,k}$ are the q-axis synchronous reactance and transient reactance of the generators at bus i at time k respectively and $H_{gi,k}$ is the inertia time constant of the generators at bus i at time k.

In addition, if parallel lines are built to strengthen the transmission capacity in weak areas, the parameters of transmission lines must be updated as well:

$$F^{\max}_{j,k+1} = \mu_{j,k+1} F^{\max}_{j,k} \tag{3.30}$$

where $\mu_{j,k+1}$ is the upgrade rate at time k+1, and $F^{\max}_{j,k}$ is the transmission capacity of line j at time k. Then for line j, we have

$$\begin{cases} r_{j,k+1} = \dfrac{r_{j,k}}{\mu_{j,k+1}} \\ x_{j,k+1} = \dfrac{x_{j,k}}{\mu_{j,k+1}} \\ b_{j,k+1} = \mu_{j,k+1} b_{j,k} \end{cases} \tag{3.31}$$

where $r_{j,k}$, $x_{j,k}$, and $b_{j,k}$ are the resistance, reactance and susceptance of line j at time k.

When new generation buses are added to the system, the system topology changes consequently, and thus the power flow equations and the system's differential equations have to be augmented correspondingly.

3) Control technologies and power markets

Advanced control technologies enhance power grids' security level while helping them operate at economic points. Novel control devices and strategies, such as wide area measurement system (WAMS), hybrid automatic voltage control (HAVC)[31,32], and nonlinear control methods[33,34], have influenced positively the evolution (including slow dynamics) of power systems. The upgrade and development in control technologies themselves improve further the respond speed, control precision and effective range of the controllers.

Power markets also play important roles in the system evolution. In order to maximize the profits of the power plants and various consumers, generation plans and power consumption patterns can be modified and as a result the system's evolution, especially its slow dynamics, is affected in the end.

In summary, the evolution process of power systems are extremely complicated, which include transient, fast and slow dynamics that develop in three different time scales. We summarize different factors and attributes related to the power system evolution according to their different time scales in Table 3.1.

Table 3.1 Factors influencing the power system evolution

Class	Factors	Time scale
Slow dynamics	Load variation in long terms	Year
	Construction of power plants and lines	Several Years
	Power forward market	Season~Year
	Controller upgrade	Year
Fast dynamics	Power spot market	Hour
	AVC	Minute
	Economic dispatch	Minute
	Preventive control	Minute
	Delay overcurrent protection	Minute
	AGC	5 – 8 Seconds
Transient dynamics	Governor	Second
	Excitation regulator	Second
	Generator	Second
	Motor	Second
	Temperature controlled load	Second
	Underfrequency and undervoltage load shedding	Second
	Stability control facility	Tens of milliseconds
	HVDC and FACTS	Millisecond
	Relay protection	Tens of milliseconds
	Disturbances such as short circuits	Millisecond

3.4 SOC Phenomena in Power System Blackouts

The overall behavior of power system blackouts can be caused by the unstable structures of the system in its critical states, and it is an extremely difficult problem in nonlinear science to identify the characteristics for instability. In this book, we study blackout phenomena using SOC theory and analyze the SOC phenomena in power system blackouts in order to reveal the indices for critical characteristics and the blackout mechanisms in power systems. The ultimate goal is to propose novel and effective preventive control strategies. There are two approaches to achieve this goal. The first is to construct mathematical models for power system blackouts that serve as the foundation for further analysis and calculations. This is challenging because of the complexity of power systems, but we intend to address this challenge head on and in fact the main content of this book is devoted to this approach. The second is to study the time series data of power system blackouts using the method discussed in Chapter 2 to back out the critical characters after frequency spectrum analysis. The obtained SOC characteristics can be further used to provide temporal and spatial information for the prediction of future blackouts of various scales. We discuss this approach in this section and of course, the effectiveness of this approach relies on the quantity and quality of the collected blackout data. In recent years, Dobson, Carreras and their colleagues have investigated the blackout data in US during the period of 1984 – 1999. They found that the blackout scales and frequencies satisfy the power law distribution[35]. In this section, we show similar results for the Chinese power grids.

3.4.1 Power Law Characteristics of Blackouts in the Chinese Power Grids

Our analysis is based upon the data of blackout events during 1988 – 1997 in the Chinese power systems[36]. The blackout scale is measured by the energy supply shortage E, the power loss P_{loss} and the restoration time t.

Figure 3.7 shows the distribution of supply shortage in logarithmic coordinates

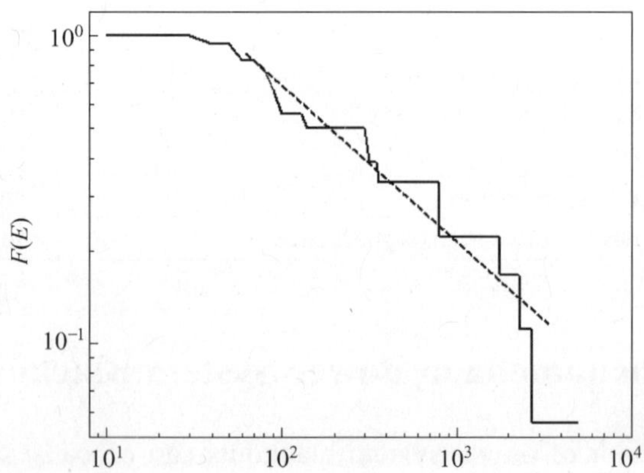

Figure 3.7 Probability vs. blackout scale evaluated by energy supply shortage

where the horizontal axis denotes E and the vertical axis denotes the corresponding probability $F(E)$.

Table 3.2 shows the slope of the fitted straight line that describes the relationship between E and $F(E)$.

Table 3.2 Data analysis of the energy supply shortage

Range of the energy supply shortage E/(MW·h)	Slope of the fitted straight line
50 – 5500	−0.7898

Remark 3.3 We always use the probability distribution of a variable x when stating power system safety analysis results using SOC theory. The exact meaning of this statement is that if x is chosen to denote the scale of some events, e.g. blackouts, then the probability distribution $F(x)$ is the probability of all the events whose scales are greater than x. For instance, in Fig. 3.7 we give the probability distribution of the energy supply shortage. We use this definition of the probability distribution throughout the book if not specified otherwise.

Figure 3.8 shows the distribution of the power loss in logarithmic coordinates, where the horizontal axis denotes the power loss P_{loss} and the vertical axis denotes the probability distribution $F(P_{\text{loss}})$.

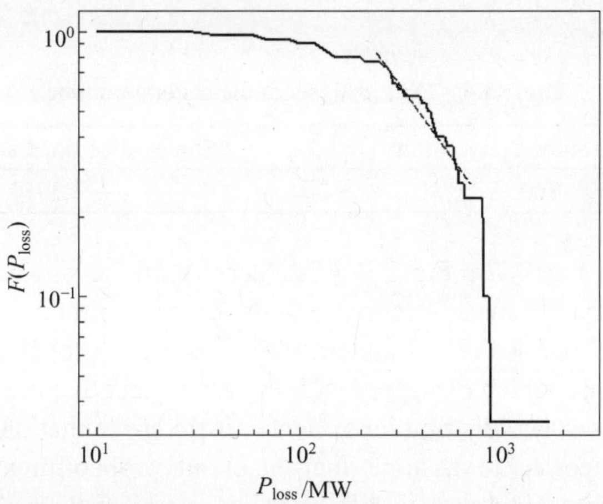

Figure 3.8 Probability vs. blackout scale evaluated by power loss

Table 3.3 shows the slope of the fitted straight line describing the relationship between P_{loss} and $F(P_{\text{loss}})$.

Figure 3.9 shows the distribution of the restoration time in logarithmic coordinates, where the horizontal axis denotes the restoration time t, and the vertical axis denotes the probability distribution $F(t)$.

Table 3.3 Data analysis of the power loss

Range of power loss P_{loss} /MW	Slope of the fitted straight line
200 – 600	–0.8641

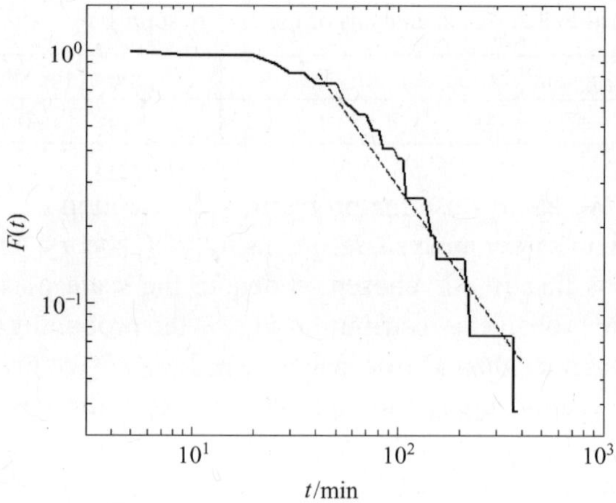

Figure 3.9 Probability vs. blackout scale measured by restoration time

Table 3.4 shows the slope of the fitted straight line describing the relationship between t and $F(t)$.

Table 3.4 Data analysis of the restoration time

Range of restoration time t/min	Slope of the fitted straight line
50 – 380	–1.3950

As shown from Fig. 3.7 to Fig. 3.9, it is clear that all the tails of the three curves can be fitted by straight lines. According to the central limit theorem in Chapter 2, when taking blackout scales as random variables, their probability distributions should be Gaussian, which disagrees with the above fact that the tails of the three curves follow power law distributions. Hence, there are limitations of the classical power system theories to assume that blackout probabilities obey Gaussian distributions. In fact, Gaussian distributions are related to the assumption that the random events are uncorrelated while power law distributions implies, on the contrary, the events are correlated. Said differently, blackouts in power systems are not independent from one another. When predicting and preventing blackouts, we should think more about the global dynamics at the system level.

We can explain the nonlinearity for the initial segments of the above curves as follows. The data used in the analysis are only the statistics of the major blackouts.

Take Fig. 3.7 as an example. The curve exhibits linearity when the energy supply shortage is above 30 MW·h ($\approx 10^{1.5}$). If more statistics are available for the faults whose scales are below 30MW·h, the curve should be linear in a bigger range. We can also explain the nonlinearity for the ending segments of the above curves as follows. The available samplings become less when the blackout scales are above a certain level and the validity of the data does not hold any more in a statistical sense.

Furthermore, when we use $f(x)$ to denote the probability density function, the expected blackout scale is

$$E(X > y) = \int_y^\infty x f(x)\mathrm{d}x = \int_y^\infty x \cdot x^{-\alpha}\mathrm{d}x = \frac{1}{2-\alpha} x^{2-\alpha}\Big|_{x=y}^{\infty} \tag{3.32}$$

When $\alpha < 2, E(x) \to \infty$, which implies that the expected blackout scale is unbounded and the whole system is likely to collapse. However, on the other hand, according to the traditional theory, the blackout probability decreases exponentially as the blackout scale increases, which implies that the expected blackout scale must be bounded and in fact approaches zero eventually. Hence, the risk for the collapse of the whole system has been underestimated in traditional theories.

3.4.2 Autocorrelation of Blackouts in the Chinese Power Grids Based on SWV Method

In this section, we analyze the blackout data of the power grids in China during 1988 – 1997 using the SWV method[36]. We also investigate the relationship between the blackout frequencies and scales in order to reveal the complex dynamic characteristics from a macroscopic perspective. We evaluate the blackout scales using the numbers of blackouts, energy shortage, power loss and restoration time. Four time series $X = \{X_t : t = 1, 2, \cdots, n\}$ are constructed, in which the intervals for the adjacent elements are one day. The value of X is taken to be the number of blackouts (Which equals 1 if blackout happens and 0 otherwise), energy shortage, power loss and restoration time respectively.

We choose four cases in this section which are discussed in detail as follows.

1) Utilizing the blackout data during 1988–1997, we calculate the Hurst exponents using the SWV method and the results are shown in Table 3.5.

Furthermore, we plot the data in log-log coordinates and the fitted curves are shown in Fig. 3.10, where the horizontal axis denotes time t and the vertical axis denotes the chosen expression for blackout scales.

Table 3.5 Hurst exponents of blackouts in China from 1988 to 1997

Time window length	Number of blackouts	Energy shortage	Power loss	Restoration time
300–3000	0.9509	0.7144	0.7455	0.8716
300–1500	0.8243	0.7339	0.7829	0.8827
300–1000	0.8976	0.8558	0.8872	0.9545
100–3000	0.9437	0.7959	0.7974	0.9103
100–2000	0.9332	0.8597	0.8573	0.9565
100–1000	0.9276	0.9731	0.9292	0.9977

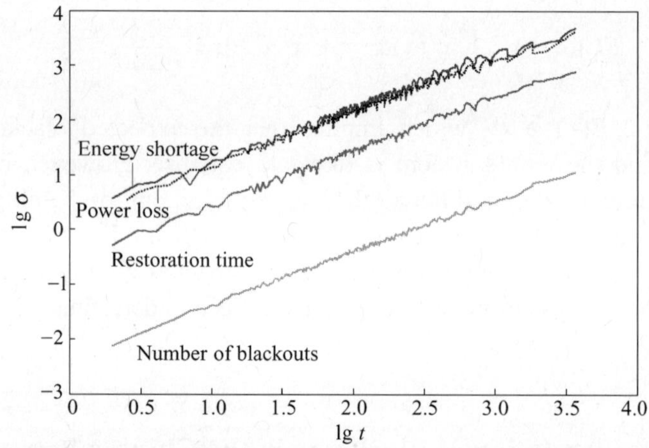

Figure 3.10 Analysis of the blackout data from 1988 to 1997 in China using the SWV method

2) Using the blackout data from 1988 to 1991, we calculate the Hurst exponents using the SWV method.

Table 3.6 Hurst exponents of blackouts in China from 1988 to 1991

Time window length	Number of blackouts	Energy shortage	Power loss	Restoration time
30–1000	0.8702	1.0205	0.9697	1.0035
30–500	0.9283	1.0714	1.0229	0.9989
100–1000	0.8386	1.0097	0.9561	1.0254
100–500	0.9245	1.1290	1.0464	1.0526
200–1000	0.7720	0.9759	0.9184	1.0441
300–950	0.7278	0.8472	0.9005	1.0028
250–1000	0.7295	0.8489	0.8855	0.9906

Similar to case 1, we plot the curves in Fig. 3.11.

3) Using the blackout data in China from 1994 to 1997, we calculate the Hurst exponents using the SWV method (see Table 3.7).

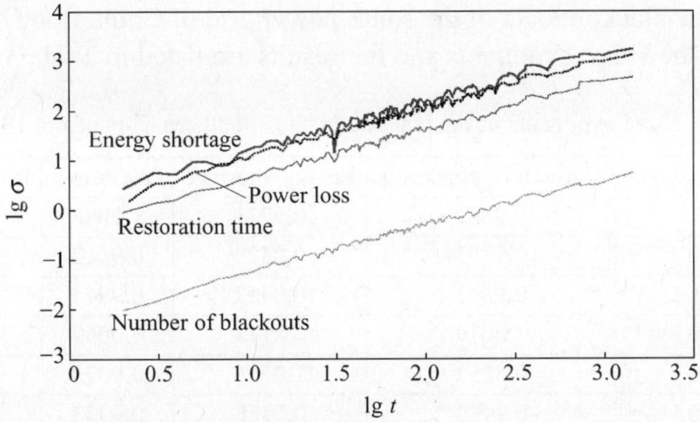

Figure 3.11 Analysis of the blackout data in China from 1988 to 1991 using the SWV method

Table 3.7 Hurst exponents of the blackout data in China from 1994 to 1997

Time window length	Number of blackouts	Energy shortage	Power loss	Restoration time
100–1000	0.8470	1.1610	0.8891	0.9903
230–800	0.6604	0.8665	0.7705	0.7159
300–800	0.5550	0.9189	0.6788	0.6405
320–800	0.5061	0.6916	0.6505	0.6008
325–810	0.4928	0.7344	0.6381	0.6360
320–820	0.5024	0.8164	0.6333	0.6941

We plot the calculation results in Fig. 3.12.

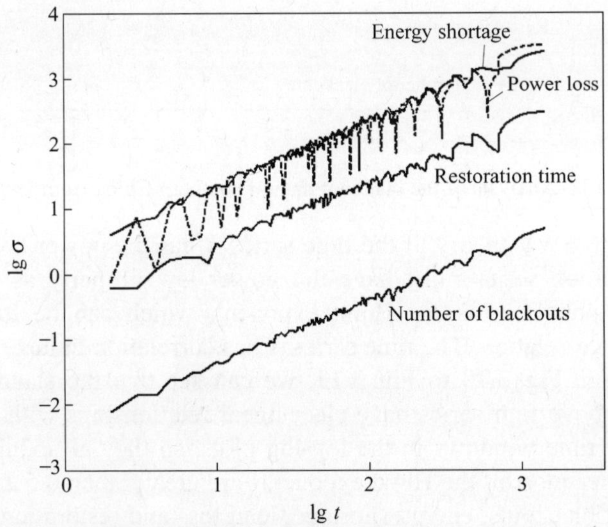

Figure 3.12 Analysis of the blackout data in China from 1994 to 1997 using the SWV method

4) Using the blackout data of the south power grid of China from 1988 to 1997, we calculate the Hurst exponents and the results are listed in Table 3.8.

Table 3.8 Hurst exponents of the blackout data in southern China from 1988 to 1997

Time window length	Number of blackouts	Energy shortage	Power loss	Restoration time
50–500	0.9831	0.9721	1.0086	0.9838
50–1500	0.9251	0.9323	0.9512	0.8961
100–500	0.9643	0.9457	0.9915	0.9504
100–1000	1.0019	0.8717	1.0059	0.9454
100–2000	0.8733	1.0775	0.9072	0.9812
200–1000	0.9982	0.7768	0.9943	0.8852
200–2000	0.8360	1.1116	0.8761	0.9756

Similar to the first three cases, we plot the curves for the computation results in Fig. 3.13.

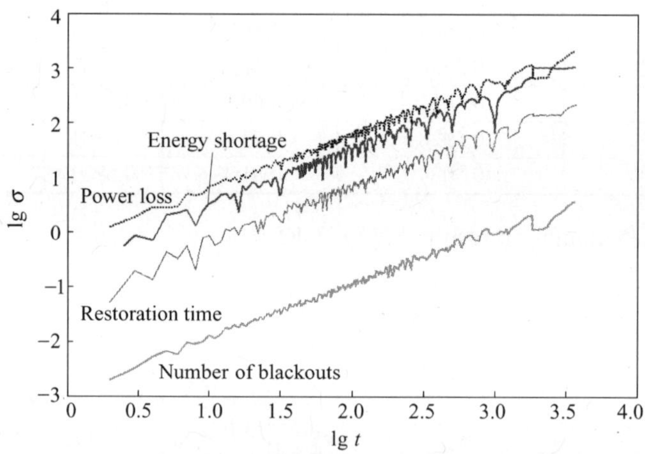

Figure 3.13 Analysis of the blackout data of southern China from 1988 to 1997

According to SWV theory, if the time series X has the power law property, its standard deviation series must have the power law property as well, and the power law exponent H is the Hurst exponent, which can be further used to determine the correlation. The time series X has correlation in large time scales if $1>H>0.5$. From Fig. 3.10 to Fig. 3.13, we can see that the standard deviation series of the above four series have clear linear relationships with respect to the lengths of the time windows in the log-log plots, so they all exhibit power law properties. Since most of the Hurst exponents are greater than 0.5 and less than 1, the number of blackouts, energy shortage, load loss and restoration time all have long-term correlations.

There are 46 blackouts in the sample space from 1988 to 1997 and about 4 blackouts on average each year. In the meanwhile, the network structure changes greatly from 1988 to 1997, so the computation result based on the collected data has its own limitation. This explains why a few of the Hurst exponents in the tables are greater than 1 or less than 0.5.

3.4.3 Autocorrelation of Blackouts in the Chinese Power Grids Based on *R*/*S* Method

In this section, we analyze the black data of the power grids in China from 1988 to 1997 using *R*/*S* method[36] and calculate the Hurst exponent for power loss and restoration time. The Hurst exponent for power loss is shown in Table 3.9.

Table 3.9 Hurst exponent for power loss in China from 1988 to 1997

Range of power loss/MW	*H*
800–3600	0.7052

We plot the computation results for power loss in Fig. 3.14.

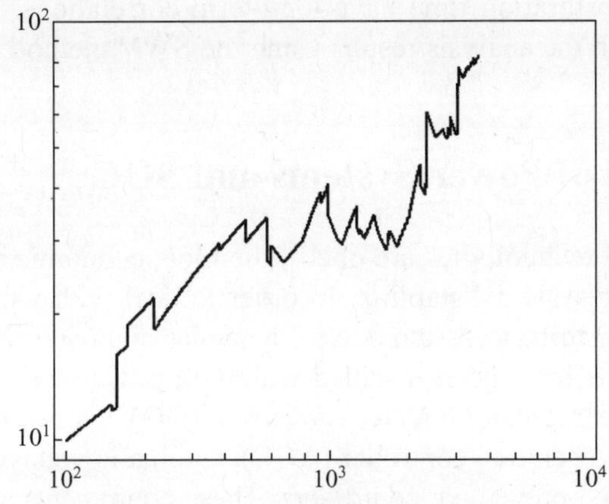

Figure 3.14 Analysis of the blackout data for power loss using *R*/*S* method

The Hurst exponent for the restoration time is shown in Table 3.10.

Table 3.10 Hurst exponent for the restoration time from 1988 to 1997

Range of restoration time /min	*H*
100 – 3000	0.5508

We plot the computation result for the restoration time in Fig. 3.15.

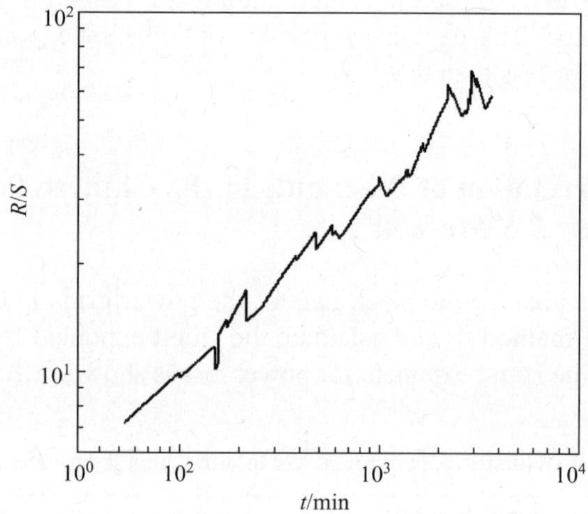

Figure 3.15 Analysis the blackout data for the restoration time using *R*/*S* method

From the above figures and tables, it is clear that the standard deviation series of the two time series have the power law properties. Since all of the Hurst exponents are greater than 0.5 and less than 1, both of the two time series of power loss and restoration time have long-term correlations. These conclusions are consistent with the analysis results using the SWV method.

3.5 Control of Power Systems and SOC

Advanced control technologies are one of the most economic and effective ways to improve power systems' stability. In order to deal with various uncertainties, such as external disturbances, unmodeled dynamics and parameter errors, different control equipments have been installed widely in power systems, which include automatic voltage regulator (AVR), governor (GOV), power system stabilizer (PSS), automatic generation controller (AGC), automatic voltage controller (AVC), relay protection and preventive controllers. These equipments stabilize the system and reduce the blackout probabilities by regulating generators' voltages, active and reactive power, and isolating quickly faulted components.

On the other hand, with the development and implementation of control technologies, modern power systems are approaching their operating limits gradually under the pressure to make full use of the existing equipment to maximize profit. This in turn motivates researchers and engineers to develop more advanced control strategies and equipments.

Hence, it is fair to say that the advances in control technologies in power systems are the dissipative forces that prevent the system from evolving into its critical state. At the same time, the increase in load demands and the actions to maximize economic efficiency are the driving forces that propel the system to evolve towards its critical states. In fact, there would be no blackouts if control actions could be perfect. However, the design and implementation of control devices are constrained by several factors. There are always simplifying assumptions at the design stage; control devices might age with time; relay protection might misoperate or maloperate sometimes; there are unpredictable uncertainties in the system. For these reasons, potential weakness in the controllers may lead to cascading failures and blackouts. So higher requirements are raised for control technologies to improve the precision, reliability and robustness of control devices to effectively prevent catastrophes. We will discuss more about this in the latter chapters.

3.6 Blackout Risk Assessment in Power Systems

There is a great deal of randomness in real power systems and it is extremely difficult to reveal the mechanism for blackouts only taking a microscopic perspective. If, on the other hand, a macroscopic perspective using SOC theory is taken to study power system blackouts, we need so many blackout data for statistical analysis that we overstretch the research resources because sufficient data have to be collected over a long period of time with the collaboration from all the related departments. In order to obtain reliable data in a short period of time, computer simulations have to be relied on to simulate the long-term development of power systems. Using the model we discussed previously covering the transient, fast and slow dynamics, we can generate simulation data to study faults that might develop into cascading failures and thus identify the system's SOC characteristics.

Traditional power system reliability theory starts from the analysis of individual electric component, and then develops the computation gradually to larger scale to estimate the system's reliability. This method has clear physical explanation, but is difficult to be applied to large systems for online computations because of the daunting computational complexity. In this context, SOC theory provides new ideas to describe at the system level the reliability of power systems. To describe the reliability quantitatively, some indices must be defined first. Statistically, the tails of the fault distribution functions indicate the distribution characteristics for blackouts. In the log-log plots of the fault distribution curves in the previous sections, we have shown that the tails show linearity. However, if we apply directly the linear relationship in risk analysis, the results are not satisfactory because of the discrete nature of the statistical data and the difference between blackouts on different scales. The latter makes it extremely hard to identify precisely the critical point using only linear fitting.

Thus, in this section we introduce the notions of VaR and CVaR[37–39] into the assessment of the risk for power system blackouts. Using these two indices, one can evaluate the blackout risk quantitatively and consequently the risks in different systems or the same system operating at different operating points can be compared quantitatively. The criticality characteristics can be clearly revealed at the same time.

3.6.1 VaR

To evaluate investment risks, the financial firm JPMorgan first introduced the VaR risk assessment method when developing its risk metrics system[38]. Now the VaR has become the main stream model in financial analysis worldwide. The VaR is the abbreviation for "value at risk" that is developed using tools from probability statistics. It refers to the potential largest loss with a given confidence level σ for a certain period of time in the future, namely

$$\mathrm{Prob}(\Delta V > \mathrm{VaR}_{\sigma}) = 1 - \sigma \tag{3.33}$$

where σ is the confidence level and ΔV is the loss in the given period. We take fund KY as an example to explain how the VaR is computed. As shown in Fig. 3.16, for $\sigma = 0.95$, the VaR of fund KY on July 23, 2004 was 5 million, which means that it is guaranteed that the loss of this fund is below 5 million with probability 95%.

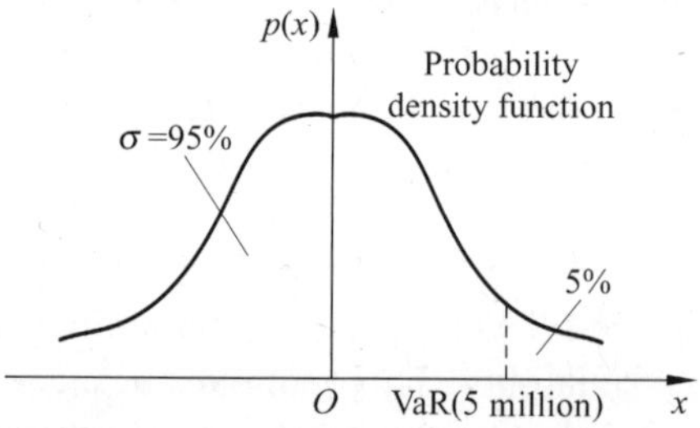

Figure 3.16 Distribution of fund KY

When the probability density function $p(x)$ is known, we have the relationship

$$\sigma = \int_{-\infty}^{\mathrm{VaR}_{\sigma}} p(x)\mathrm{d}x \tag{3.34}$$

which says that for a given value of VaR_{σ}, one can compute the confidence level σ. When we are given a certain value of σ, the VaR_{σ} can be calculated as follows. First, choose two values $\mathrm{VaR}_{\sigma 1}$ and $\mathrm{VaR}_{\sigma 2}$ that make the confidence value satisfy $\sigma_1 < \sigma < \sigma_2$. Then calculate the VaR_{σ} through golden section method or dichotomy.

The VaR is easy to understand and calculate, and can be used to evaluate the risks of different systems. In addition, the VaR combines together the scale of loss together with the estimated probability. By adjusting the confidence level σ, the value of the VaR changes accordingly. However, the VaR has two shortcomings[39]. One is that it does not satisfy the Coherent Axiom[40,41]; and the other is that the information about the tails is not sufficient statistics. Since the VaR is a chosen point at a certain confidence level in nature, it dose not provide information about the loss beyond the VaR, especially when the probability distribution function has a fat tail as shown in Fig. 3.17.

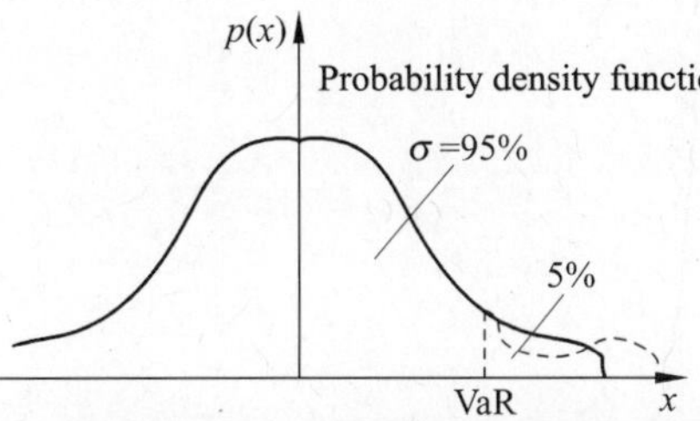

Figure 3.17 Different tail distributions with the same VaR

Obviously, the risk associated with the dashed line is greater than that associated with the solid line, but the VaR at the confidence level σ for these two curves are the same.

3.6.2 CVaR

To overcome the shortcomings of VaR, Rockafeller and Uryasev proposed the CVaR in 2000[39], which is the abbreviation for the "conditional value at risk", and is also called the mean excess loss, the mean shortfall or the tail VaR. It is the conditional mean with a certain confidence coefficient that is defined by

$$\text{CVaR} = \frac{1}{1-\sigma}\int_{\text{VaR}_\sigma}^{\infty} xp(x)\text{d}x$$

where x is the loss scale and $p(x)$ is the probability density function of the loss.

The VaR and CVaR provide unified and normalized indices to evaluate the security level of different systems.

Remark 3.4 Since σ must be given before calculations, $\dfrac{1}{1-\sigma}$ is a constant. So for convenience in this book except for Chapter 12, we use the following formula to calculate CVaR instead

$$\text{CVaR} = \int_{\text{VaR}_\sigma}^{\infty} xp(x)\text{d}x \tag{3.35}$$

3.6.3 Blackout Risk Assessment Based on Historical Data

The key step in using the VaR and CVaR in power systems is to determine the probability density function for blackouts. One can use either historical data or simulation data to approximate the probability density function. We use the blackout data in North America from 1996 to 2002 as an example, for which the probability density curve is shown in Fig. 3.18.

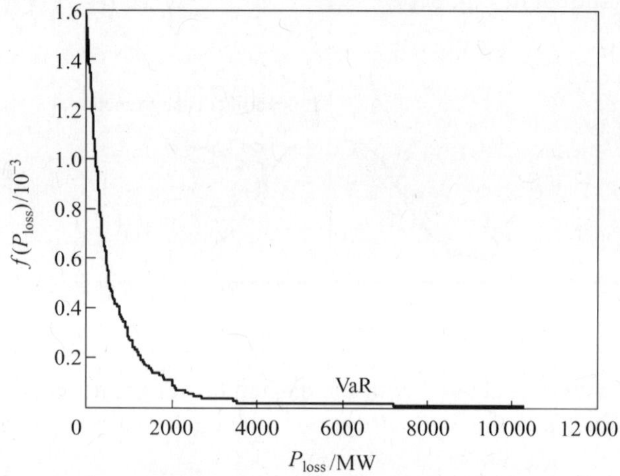

Figure 3.18 Probability density curve of North America blackout from 1996 to 2002

When $\sigma = 0.95$, according to Eqs. (3.34) and (3.35), we have

$$\begin{cases} \mathrm{VaR}_{\sigma=0.95} = 5840\ \mathrm{MW} \\ \mathrm{CVaR}_{\sigma=0.95} = 385\ \mathrm{MW} \end{cases} \tag{3.36}$$

which implies that with 95% confidence one may believe that a cascading failure cannot result in a blackout whose loss scale is greater than 5840 MW and the expected blackout loss beyond 5840 MW is 385 MW. Therefore, the VaR and CVaR are complementary to each other and a coordinated use of both of them can improve the assessment of power system blackout risks.

One can further normalize the blackout scales, namely dividing the power loss by the total load to describe the relative blackout scale. Then the VaR and CVaR can compare the risks of different systems or the same system under different operation conditions.

The computations of the VaR and CVaR are not too hard, but the real difficulty lies in acquiring the blackout data and thus the blackout probability distribution. In the following chapters of this book, we propose several blackout models based on SOC theory, which can be used to simulate cascading processes and the long-term evolution of power systems in general. In particular, after recording the blackouts on each day during the simulated evolution, one can obtain the corresponding blackout probability distribution and then compute the VaR and CVaR. Since the VaR and CVaR are taken as the system-level risk indices, they can be utilized to be the characteristics of the SOC behavior of the system and indicate the system's security level. We will talk about this in detail in the following chapters.

3.6.4 Discussion on the VaR and CVaR

(1) Although both the VaR and CVaR can be used to evaluate risks in power systems and identify the systems' critical states more accurately, each of them has its own advantages and disadvantages. The VaR is easier to calculate, and its physical meaning is clearer. The CVaR is defined upon the VaR, and its physical meaning is more abstract, which has a more transparent interpretation when compared to the definition of the expectation of a continuous random variable. Setting $\sigma = 0$, and because of the fact that the blackout scales have to be positive and in view of Eqs. (3.34) – (3.35), we have

$$\begin{cases} \mathrm{VaR}_{\sigma=0} = 0 \\ \mathrm{CVaR}_{\sigma=0} = \int_0^{\infty} xp(x)\mathrm{d}x \end{cases} \tag{3.37}$$

Note that now the CVaR is in fact the expectation of the scales of all the blackouts. Also note that some researchers define

$$\mathrm{Risk} = \int_0^{\infty} xp(x)\mathrm{d}x \tag{3.38}$$

as the risk of the system. Hence, the CVaR is a more general form of the Risk while the latter reflects the overall risk for blackouts of all possible scales and the former focuses more on blackouts of larger scales. Compared with the Risk, the practical implication of the CVaR is not that straightforward if it is used alone. Therefore, the VaR and CVaR are complementary to each other and should be used jointly to better assess the reliability of power systems.

(2) As discussed before, the blackout probability distributions show the characteristics of the power law distribution, but not the normal distribution. Similar to the "fat tail" case, the VaR cannot provide sufficient statistics for the power law tails as shown in Fig. 3.19.

From Fig. 3.19, one can see that the risk of the power law distribution is greater than that of the normal distribution. But the VaR for these two curves are the same.

(3) The information about the VaR can also be inferred from the complementary cumulative distribution function (also called the survivor function) that has been used by different researchers when studying blackouts.

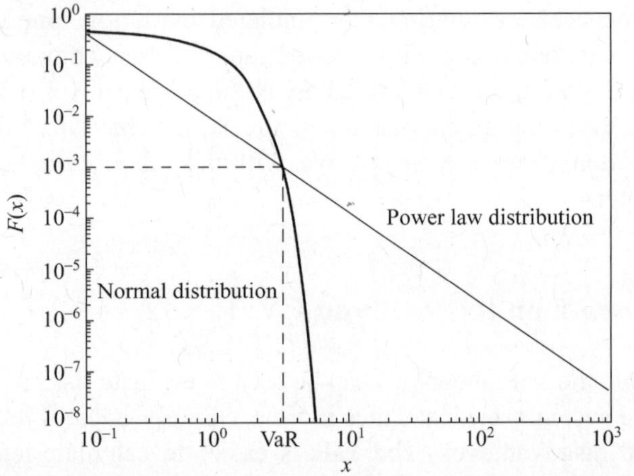

Figure 3.19 Comparison between the power law and normal distributions with the same VaR

(4) We also want to point out one difficulty in computing the CVaR. This is potentially a limitation for the application of this index and, when not being taken into account, can lead to wrong conclusions. The CVaR is the (conditional) expected value of a distribution that is likely to have a heavy tail. If the size of the tail is even greater than the scale of the system, the mean may have an infinite value. The expected values of these heavy tailed distributions, even when being truncated at the system's scale, are likely to have poor statistical properties. For example, in a heavy tailed distribution with an infinite mean, the law of large numbers does not apply and the sample mean can have a variance that does not decrease with the sample size. Even for truncated distributions with finite means, one can reasonably expect poor convergence properties of the estimators as the sample size increases.

References

[1] Warfield J N. An Introduction to System Science. Hackensack, NJ: World Scientific, 2006

[2] Ashby W R. Principles of Self-Organization. Oxford: Pergament Press, 1962

[3] Prigogine I. From Being to Becoming: Time and Complexity in the Physical Science. W H Freeman & Co (Sd), 1981

[4] Haken H. Advanced Synergeties. Berlin: Springer-Verlag, 1983

[5] Eigen M, Schuster P. The Hypercycle, a Principle of Natural Self-Organization. Berlin: Springer-Verlag, 1979

[6] Kauffman S A. The Origins of Order. New York: Oxford University Press, 1993

[7] Kauffman S A, Johnsen S. Coevolution of the edge of chaos: coupled fitness landscapes, poised states, and coevolutionary avalanches. Journal of Theoretical Biology, 1991, 149: 467 – 505

[8] Langton C G. Artificial life. In Langton C G, editor. Artificial Life, Volume VI of SFI Studies in the Sciences of Complexity, pages 1 – 47, Redwook City: Addison-Wesley, 1989
[9] Holland J H. Emergence: From Order to Chaos. Boston: Addison-Wesley, 1997
[10] Haken H. Synergetics, an Introduction: Nonequilibrium Phase Transitions and Self-Organization in Physics, Chemistry, and Biology. Berlin Heidelberg: Springer-Verlag, 1978
[11] Haken H. Synergetics Computer and Cognition: A Top-down Approach to Neural Nets. New York: Springer-Verlag, 1991
[12] Haken H. Information and Self-Organization: A Macroscopic Approach to Complex Systems. Berlin Heidelberg: Springer-Verlag, 1988
[13] Jensen H J. Self-Organised Criticality. Cambridge University Press, 1998
[14] Binney J J, et al. The Theory of Critical Phenomena. New York: Oxford University Press, 1992
[15] Jantsch E. Self Organizing Universe: Scientific and Human Implications (Systems Science and World Order Library. Innovations in Systems Science). Pergamon Press, 1980
[16] Christensen K, Moloney N R. Nicholas R. Complexity and criticality. Imperial College London, UK, 2005
[17] Guo L, Xu X M. Complex Networks. Shanghai: Shanghai Scientific & Technological Education Publishing House, 2006 (in Chinese)
[18] Rothman D H, Grotzinger J P, Flemings P. Scaling in turbidite deposition. Journal of Sedimentary Research, 1994, 64(1a): 59 – 67
[19] Diodati P, Marchesoni F, Piazza S. Acoustic emission from volcanic rocks: An example of self-organized criticality. Physical Review Letters, 1991, 67: 2239
[20] Gutenberg B, Richter C F. Seismicity of the Earth. Princeton, NJ: Princeton University Press, 1949
[21] Johnston A C, Nava S. Recurrence rates and probability estimates for the new Madrid seismic zone. Journal of Geophysical Research, 1985, 90(B8): 6737 – 6753
[22] Sornette D, Sornette A. General theory of the modified Gutenberg-Richter law for large seismic moments. Bulletin of the Seismological Society of America, 1999, 89(4): 1121 – 1130
[23] Burridge R, Knopoff L. Model and theoretical seismicity. Bulletin of the Seismological Society of America, 1967, 57: 341 – 371
[24] Zipf G K. Human Behavior and the Principle of Least Effort. Cambridge MA: Addison-Wesley, 1949
[25] Wang X F, Song Y H, Irving M. Modern Power System Analysis. New York: Springer, 2008
[26] Kundur P. Power System Stability and Control. New York: McGraw-Hill, 1994
[27] Ni Y X, Chen S S, Zhang B L. Theory and Analysis of Dynamic Power System. Beijing: Tsinghua University Press, 2002 (in Chinese)
[28] Sood V K. HVDC and FACTS Controllers: Applications of Static Converters in Power Systems. Boston: Kluwer Academic Publishers, 2004
[29] Bhatt N B. August 14, 2003 US-Canada blackout. IEEE PES General Meeting, Denver, CO, 2004

[30] Union for the coordination of electricity transmission (UCTE). UCTE releases detailed interim report on the disturbances of 4 Nov. http://www.ucte.org. Accessed Nov. 30, 2006

[31] Hu W, Mei S W, Shen C, et al. Northeast China power grid hybrid automatic voltage control. The 4th International Conference on Power Transmission and Distribution Technology, 2003: 1215 – 1220

[32] Mei S W, Xie B P, et al. Hybrid automatic voltage strategy and its application to Northeast China 500 kV power grid. European Transactions on Electrical Power, 2009, 19(3): 355 – 367

[33] Mei S W, Guan T Q, Lu Q, et al. Nonlinear disturbance attenuation control of static phase shifter in sense of L2-gain. European Transactions on Electrical Power, 2003, 13(2): 113 – 118

[34] Liu F, Mei S W, Lu Q, et al. The nonlinear internal control of STATCOM. International Journal of Electrical Power and Energy System, 2003, 25(6): 421 – 430

[35] Carreras B A, Newman D E, Dobson I, et al. Initial evidence for self-organized criticality in electric power system blackouts. Proceedings of Hawaii International conference on System Science, Maui, Hawaii, USA, Jan. 2000

[36] Weng X, Hong Y, Xue A, et al. Failure analysis on China power grid based on power law. Journal of Control Theory and Application, 2006, 4(3): 1 – 4

[37] Hull J C. Options, Futures and Other Derivatives (5th Edition.). Saddle River, New Jersey: Prentice Hall, 2003

[38] Morgan J P. Risk Metrics: Technical Document(4th Edition). 1996. http://www.riskmetrics.com

[39] Rockafeller T, Uryasev S. Optimization of conditional value-at-risk. Journal of Risk, 2000, 2(3): 21 – 42

[40] Artzner P, Delbaen F, Eber J M, et al. Thinking coherently. Risk, 1997, 10(11): 68 – 71

[41] Artzner P, Delbaen F, Eber J M, et al. Coherent measures of risk. Mathematical Finance, 1999, 9(3): 203 – 228

Chapter 4 Power Grid Growth and Evolution

This chapter models and analyzes in detail the slow dynamics in power grids. The driving forces for power grid evolutions are identified and then utilized to construct the general evolution model for complex power grids. In addition, the evolution principles are delineated for the growth of power grids. We develop a small-world power grid evolution model to analyze how the topology of a power grid evolves when the grid itself grows dynamically. We consider various factors that affect power grids' evolutions, such as the energy resource distributions, load growth patterns and power gird parameters.

From the first DC transmission line installed by Thomas Edison in 1882 to nowadays super large-scale complex AC-DC hybrid power grids, power systems have evolved with the progress of human industrial civilization. As typical man-made complex networks, the evolution of power grids involves a great deal of carefully implemented optimization design. In order to investigate the characteristics of self-organized criticality in cascading failures in power systems, we have discussed in the previous chapters the power grid evolution models that include the transient, fast and slow dynamics on three different time scales[1–5]. The mathematical description of the slow dynamics is, however, not completely consistent with the real practice in power systems because we have simply taken the power grid planning and construction as the gradual increase in electric generation capacities, power transmission capacities and load levels without considering the impact of newly constructed power plants, and the locations and connections of substations. As a result, we have ignored how a power grid's topological structures change with time. We have studied the power network complexity focusing on static snapshots of the dynamic network topologies without paying too much attention to the characteristics of power grids' long-term growth and evolution. In fact, since the birth of power grids, they have always been growing and evolving with complex dynamics that are affected by various factors, such as the construction of new power plants and substations and the addition and upgrade of transmission lines. When considering the new construction, we need to take into account factors like choices of locations, rated capacities and ways of interconnections. Obviously, the study on the complexity of power grid evolution is an important component in the framework for the study of power grid complexities, and may lead to new insight into the complicated process of the development of complex power systems. This is especially relevant to the long-term planning and scheduling for power systems in the future. To this end, we construct in this chapter a small-world power grid evolution model and analyze the influences of many factors on the evolution of power grids.

4.1 Driving Mechanism of Power Grid Evolution

As shown in Fig. 4.1, power grid evolution is a complicated dynamic process. Social economic development leads to increasingly demanding power load until at a certain stage the demand exceeds the supply, which deteriorates the reserve ratio of the whole system and requires the construction of more new power plants. At the same time, load density also increases with the demand, and in order to serve the load locally, more new substations are required to be built. When the transmitted power of the whole system increases, stricter requirements for the transmission capacities have to be satisfied. Towards this end, either new transmission lines are built or FACTS devices are utilized so that the electric energy can be transmitted more efficiently and the phenomenon of "local power surplus due to congestion" can be avoided. This is a microscopic self-evolution process in power grids. Of course, the evolution of power grids is also affected by related energy policies, which are regulated and adjusted by the corresponding central or regional governmental agencies in view of the macroscopic status of the regional economy.

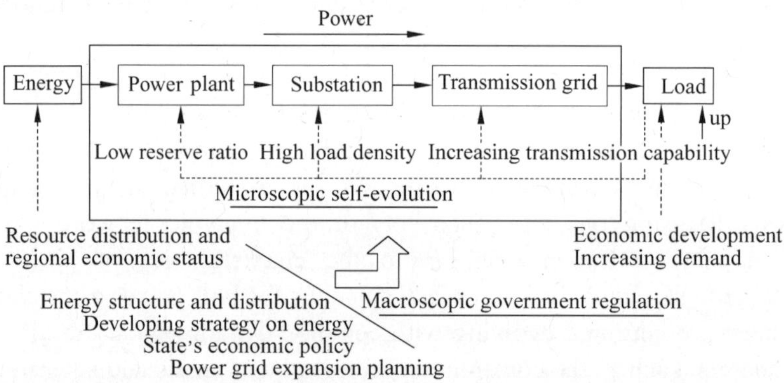

Figure 4.1 Driving mechanism of the power grid evolution

So the growth and evolution of power grids are complicated processes, which are driven both locally and globally by forces of the regional economic development and power load growth. In this chapter, we only consider the power grid evolution modeled by the microscopic self-evolution process.

4.2 General Power Grid Growth and Evolution Model

4.2.1 Local-World Network Evolution Model

In order to reveal the relationship between network structures and network behaviors and to further design control strategies to improve network performances, researchers have proposed various network structure models including the small-world model[6], the scale-free model[7], the local-world evolution model[8] and so on. Here, we list the evolution process of the local-world model as follows:

Step 1 The initial network consists of m_0 nodes and e_0 edges. Set the iteration index $t=1$.

Step 2 Choose M nodes randomly, which make up the "local world" L for the nodes to be added into the network.

Step 3 Add a new node and m edges ($m \leqslant M$), and the m edges are connecting the new node to m nodes in the local world L.

Step 4 If the network is still evolving, set $t=t+1$ and return to Step 2. Otherwise, stop.

It should be noted that in Step 3 the m nodes in L are chosen according to the connection probability P_i defined below:

$$P_i = P_i' \frac{k_i}{\sum_{j \in L} k_j}, \quad i \in L \tag{4.1a}$$

$$P_i' = \frac{M}{m_0 + t} \tag{4.1b}$$

where k_i is the degree of node i, which is the number of edges that are adjacent to i, M is the number of nodes in the local-world, and m_0 is the number of nodes in the initial network.

Using the steps above, when new nodes are added into the network, m nodes are chosen from the local world L with the given probabilities instead of being chosen from the whole network. By doing so, we have considered the local world properties in practical networks, and so it is convenient to distinguish the prioritized connections inside the local world and the weak connections among different local worlds. The way to construct local worlds relies on the specific properties of the given practical network.

4.2.2 Design of Power Grid Evolution Model

The power grid evolution includes the following six aspects:

(1) Construction of new power plants and substations corresponding to adding new nodes(in power system, nodes are called bus as well) in the network;

(2) Building new lines to connect new edges to the existing network;

(3) Removing old power plants and substations, corresponding to deleting nodes from the network;

(4) Removing aging transmission lines, corresponding to deleting edges from the network;

(5) Upgrading the capacities of power plants and substations, corresponding to changing some nodes' attributes in the network;

(6) Upgrading the line transmission capacities, corresponding to changing the edge attributes in the network.

In power systems, power plants, substations and transmission lines have relatively long service lives, so in the power grid evolution model discussed in this section, we are mainly concerned with the first two of the above six aspects, namely the construction of power plants, substations and lines. In particular, we consider the site selection, the capacity setup and the connection modes of power plants and substations, where the connection mode is defined to be the way of connecting power plants and substations to existing nodes in the network by adding new lines. We want to point out that the voltage levels of the newly added nodes may not always be the same as those in the local worlds that they are connected to.

1) Power plant construction

We consider the following aspects in the construction of power plants: construction time, location, capacity, voltage level and connection mode. The detailed process is shown in Fig. 4.2.

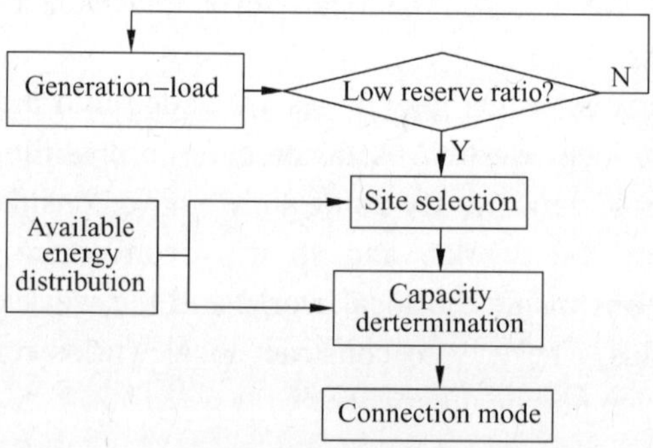

Figure 4.2 Power plant construction flowchart

(1) Construction time

To satisfy the requirement on power grid backup capacities, new power plants are built when the difference between the total generation capacity and the total load demand is less than a certain threshold.

(2) Construction site

We need to take into account various factors such as energy distribution and transportation. The acquisition of these data is closely related to practices in real networks. For simplicity, we use in this section the available energy distribution function P_s (see Remark 4.1 for the design of P_s), which is generated randomly during the initialization of the program. We choose the location where the energy is the most abundant as the construction site of new power plants according to the energy distribution function P_s.

(3) Capacity and voltage level

To make good use of the available energy at the chosen location, we set the generation capacity S_g of the power plant to be the one that is described by the following piecewise constant function determined by the available energy distribution function P_s in this area, namely

$$S_g = f(P_s(x,y)) = \begin{cases} S_1, & P_s(x,y) \geqslant \lambda \\ S_2, & \lambda > P_s(x,y) \geqslant 0.5\lambda \\ S_3, & 0.5\lambda > P_s(x,y) \geqslant 0.1\lambda \end{cases} \tag{4.2}$$

where λ is the parameter determined by the specific types of different energy sources.

In the computation, because of practical engineering consideration, an upper limit $P_s^{\max}$ is needed for the energy distribution function. We then determine the power plant's voltage level V_g according to its capacity S_g. Here we choose $P_s^{\max} = 10$ and $\lambda = 6$ and use the three voltage levels: 500 kV, 220 kV and 110 kV. The details are listed in Table 4.1.

Table 4.1 Power plant's capacity and voltage level

Energy distribution $P_s(x,y)$	Power plant capacity S_g /100 MW	Voltage level V_g/kV
$[\lambda, P_s^{\max})$	6	500
$[0.5\lambda, \lambda)$	2	220
$[0.1\lambda, 0.5\lambda)$	0.6	110

(4) Connection mode

There are two aspects when considering how a power plant is connected to the power grid. First, we need to determine its corresponding local world. The substations that have the same voltage level as that of the newly constructed power plant are chosen to be the local world of this power plant. If such a local world is empty, we choose the substations whose voltage are one level lower than that of the newly established power plant to be the local world.

Second, we need to choose the nodes to be connected to in the local world. To lower the investment cost, we choose the nodes in the local world with the shortest paths.

Remark 4.1 Design of the available energy distribution function P_s.

Usually we take P_s to be the following function of two variables

$$S = P_s(x, y) \tag{4.3}$$

where x, y denote the coordinates of the geographical locations. Thus, S is the amount of the available energy at the location whose geographical coordinates are (x, y).

Strictly speaking, to determine the value of S in a real power grid, we need to investigate comprehensively the available energy distributions and their associated geographical information in the whole area. In addition, the function should be modified when the technology for energy exploitation develops and when requirements for environmental protection become stricter. However, in the cases like the simulation examples in this section, one is not concerned with a specific practical power grid, and so a random matrix can be generated according to some probability distribution at the beginning of the computation. The elements of this randomly generated matrix can be used to describe the available energy distribution for simulation purposes.

2) Substation construction

Similar to the construction of power plants, the substation construction also needs to consider the following aspects: time, location, capacity, voltage level and connection mode. The detailed description is shown in Fig. 4.3.

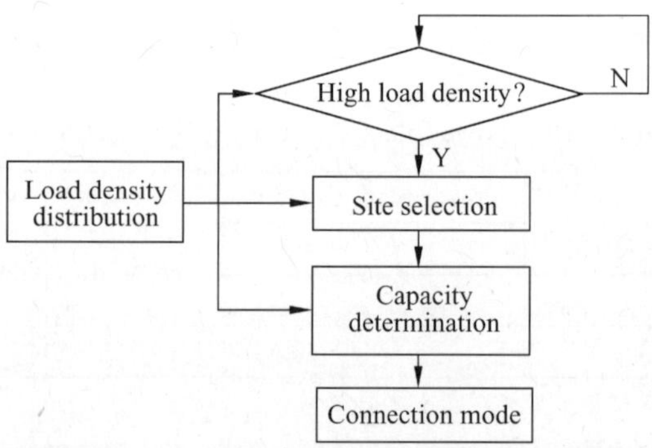

Figure 4.3 Substation construction flowchart

(1) Construction time

When the load demand exceeds the capacity of the local substation, new substations should be constructed.

(2) Construction location

We assume here that being close to the load center is a basic requirement for choosing a substation's location. Similar to the use of available energy distribution function P_s in choosing the locations for power plants, we use the load distribution function P_D as the reference for choosing the locations for substations (see Remark 4.2 for the design of P_D). If the load demand at a certain location is especially high, we build new substations at this location.

(3) Capacity and voltage level

The capacity of the new substation should be determined by the load demand at this place. So we define the substation's capacity S_t to be a piecewise constant function in the load demand P_D, namely

$$S_t = g(P_D(x,y)) = \begin{cases} S_1, & P_D(x,y) \geqslant \mu \\ S_2, & \mu > P_D(x,y) \geqslant 0.5\mu \\ S_3, & 0.5\mu > P_D(x,y) > 0 \end{cases} \tag{4.4}$$

where μ is a parameter that is determined by the specific condition of the load area. We also need an upper limit $P_D^{\max}$ of the load demand in the computation. We assume that the substation's voltage level V_t is determined by its capacity S_t. In this section, we take $P_D^{\max} = 3$ and $\mu = 2$. We consider three different voltage levels: 500 kV, 220 kV and 110 kV. The design is shown in Table 4.2.

Table 4.2 Principles for determining the substation's capacity and voltage level

Load demand distribution $P_D(x,y)$/100 MW	Substation capacity S_t/100 MW	Voltage level V_t/kV
[μ, $P_D^{\max}$]	10	500
[0.5 μ, μ]	3	220
(0, 0.5 μ)	1	110

(4) Connection mode

Similar to the connection modes of power plants, the connection mode of the newly constructed substation also relies on the following two factors:

First, we need to choose its corresponding local world. Power grids usually have substations spreading out from the center with voltage levels from high to low. So here we choose the substations with the voltage that is one level higher than that of the new substation to be the local world. In the special case when the new substation is with the highest voltage level, the substations with the same voltage level are chosen to be the local world.

Second, we need to choose the nodes to be connected to in the local world. To save investment cost, we choose those nodes in the local world that have the shortest paths to the new substation.

In summery, in the process of the growth and evolution of the power grid, the newly added components have the priority to choose the nearest nodes in the local world to be their connecting points in order to save construction cost in addition to the consideration of different voltage levels. When the local world of the new substation is empty, other local worlds can be chosen to implement the connections between different local worlds.

Remark 4.2 Design of the load demand distribution function P_D.

Similar to the available energy distribution function P_s, P_D can be written as the following function with two variables

$$d = P_D(x, y) \tag{4.5}$$

where x, y are the geographical coordinates as before. Then d is the load demand at the place whose geographical coordinates are (x, y).

The specific form of P_D can be determined by the load distribution in the area with its associated geographical information. In addition, P_D should also be modified when the environment in the area changes, such as the cases when the regional economy develops and the industrial structures are adjusted. When not referring to any specific real power grid, we utilize random matrices generated according to a certain probability distribution to describe the load distribution characteristics. The elements of the matrix can be used as the values for $P_D(x, y)$ to simulate load demand distributions.

Remark 4.3 The settings of λ, P_s, μ and $P_D^{\max}$ are based on practical experiences and not on strict theoretical basis.

4.2.3 Simulation Results and Analysis

We use the evolution process of the following 100-bus network as an example to validate the effectiveness of the proposed model. In the beginning, there are only two buses and one line in the network. Then the network evolves by itself into the 100-bus system using the rules we have discussed before. For simplicity, we divide the process into three stages, which are the 40-bus, 70-bus and 100-bus stages. The results at different stages are showed in the following figures.

The *X*-axis and *Y*-axis in the figures denote the coordinates of the geographic locations of the buses. The power plants and substations are denoted by ○ and □ respectively, whose sizes from large to small correspond to voltage levels of 500 kV, 220 kV and 110 kV respectively.

The following conclusions can be drawn from Fig. 4.4:

(1) The power grids generated using the proposed evolution model usually have multi-center radiating structures in which the central nodes are the substations at higher voltage levels. This is consistent with the real situations in practical power grids.

(2) The power grid evolution model constructed in this section is based on the local world evolution theory for complex networks, and consequently the

construction, connection and utilization of power plants and substations have been able to be considered comprehensively; in other words, we have considered the time-varying characteristics of the power grids' topological structures. Hence, the simulation for slow dynamics and the study on the power system's SOC behaviors are closer to engineering practices. However, we also want to point out that more factors can be considered in the proposed model to further improve the model. For example, one may want to consider the time-varying characteristics of the distribution functions of the available energy supplies and the load demands, and the existence of super-long transmission lines between certain pairs of buses.

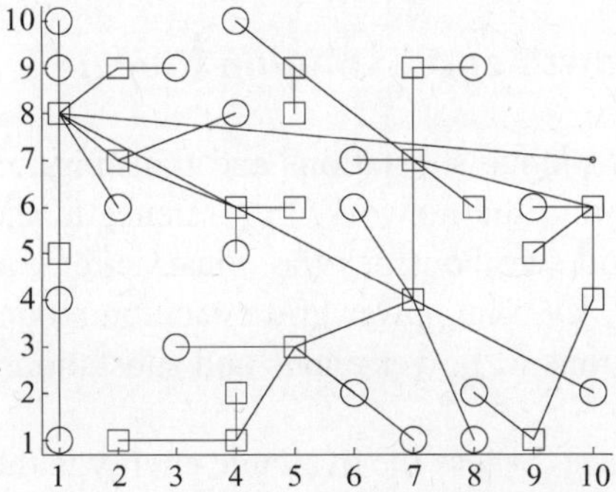

(a) The first stage of the power grid evolution

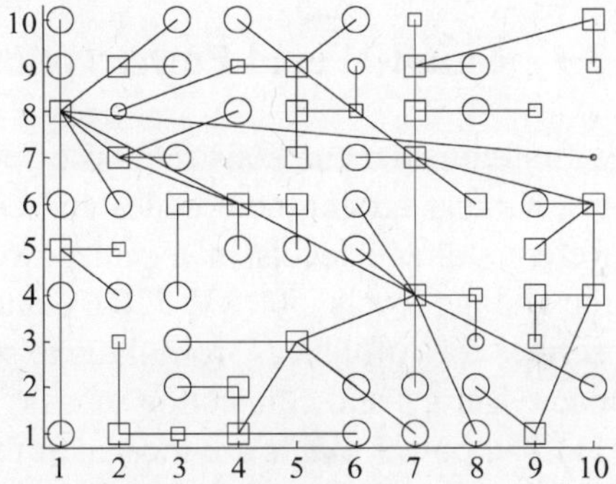

(b) The second stage of the power grid evolution

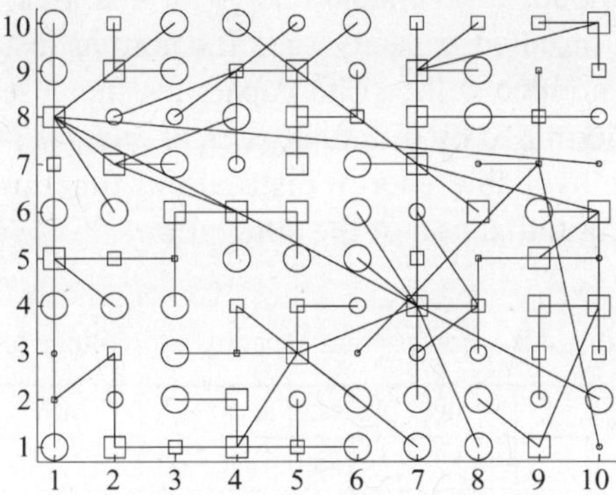

(c) The third stage of the power grid evolution

Figure 4.4 Power grid evolution process

4.3 Small-World Power Grid Evolution Model and Its Simulation Analysis

Recently the study on the complexity of interconnected power grids using tools from complex network theory has attracted more and more attention. The notion of small-world networks[6] was first proposed by Watts and Strogatz in 1998. It has been verified that many power grids including the American Western Power Grid[9], the Brazilian Power Grid[10] and the Chinese Northern Power Grid[11], all exhibit small-world characteristics. The power grid evolution is always under the driving forces of long-term self-selection and man-made optimization design. The small-world structural characteristics in many power grids are not just incidental, but related to decisive evolutionary mechanism. In this section, we discuss the related power grid evolution mechanism, taking into account the principles in the planning for real power grids. We establish a small-world power grid evolution model, for which we carry out simulation analysis.

4.3.1 Small-World Power Grid Growth and Evolution Model

We first consider the construction of power plants, substations and transmission lines, namely adding new nodes and edges into the network. In reference to the different voltage levels in a real power grid, we consider the constructions at three voltage levels, 500 kV, 220 kV and 110 kV. The power grid evolution model describes the influences from the site selections of power plants and substations, capacity setups and connection modes[12].

(1) The power plants are located at the places where the available energy is the most abundant. The installed capacity and voltage level of a new power plant are determined by the amount of available energy in this area. We list the details in Table 4.3, where the installed capacity satisfies normal distribution $N(S_g, 0.1S_g)$ with mean S_g and variance $0.1S_g$. The capacities increase at an annual rate α. The value of λ is determined by different types of energy. The energy distribution is represented by the available energy distribution function P_s that is generated randomly during the initialization of the simulation.

Table 4.3 Power plant capacity and voltage level

Energy distribution $P_s(x,y)$	Power plant capacity S_g /100 MW	Voltage level U_g /kV
$P_s \geqslant \lambda$	$N(6,0.6)\,(1+\alpha)^t$	500
$0.5\lambda \leqslant P_s < \lambda$	$N(3,0.3)\,(1+\alpha)^t$	220
$0.1\lambda \leqslant P_s < 0.5\lambda$	$N(1,0.1)\,(1+\alpha)^t$	110

(2) Substations should be constructed according to the distribution of load demands. In this section, we generate randomly the load density distribution function P_D during the initialization of the simulation and increase the loads in a chosen fashion during the system's evolution. As the loads grow, we choose the load center where load density is the highest as the site for the construction of substations, and the substations' capacities and voltage levels are determined by the load demand in that area. We list the details in Table 4.4, where the

substations' capacities satisfy the normal distribution $N(S_t, 0.2S_t)$ with mean S_t and variance $0.2S_t$. This capacity increases annually with a rate β. The value of μ in Table 4.4 is determined by practical considerations.

Table 4.4 The principles for the determination of the substations' capacities and voltage levels

Load demand distribution $P_D(x,y)$	Substation capacity S_t/100 MW	Voltage level U_t/kV
$P_D \geqslant \mu$	$N(10,2)\cdot(1+\beta)^t$	500
$0.5\mu \leqslant P_D < \mu$	$N(5,1)\cdot(1+\beta)^t$	220
$0 \leqslant P_D < 0.5\mu$	$N(2,0.4)\cdot(1+\beta)^t$	110

(3) When connecting new power plants and substations to a power grid, we always connect them to existing substations. Here, connecting power plants in parallel are not considered. We follow the three rules below.

Rule 1 The new power plants and substations are connected with priority to the substations with the same voltage level. If there is no such substations in the existing grid, they are connected to the substations whose voltage is one level lower than the desired one.

Rule 2 Among all the substations chosen according to Rule 1, we connect the new power plants or substations to those that are closest.

Rule 3 For all the substations chosen according to Rule 1, the second closest to the new power plants or substations is connected to the new power plants or substations with probability p to simulate new transmission lines.

The flowchart for the power grid growth and evolution algorithm is shown in Fig. 4.5 and the main steps are as follows:

Step 1 Define the range of the area $Area = [N, N]$, in which the power grid develops. Define the coordinate of bus i to be $v_i(x_i, y_i)$, where x_i, y_i are integers, $i = 1, 2, \cdots, N$. Determine the initial energy distribution P_s and load distribution P_D.

Step 2 Suppose there are m_0 buses and e_0 transmission lines in the initial power grid. At each time step, add a power plants and b substations to the existing grid at those locations which are optimal according to the site selection rules.

Step 3 Set capacities and voltage levels for the new power plants and substations according to Tables 4.3 and 4.4.

Step 4 Connect the new power plants and substations to the existing system. We connect a new power plant or substation with probability one to those nearest buses with the same voltage level and with probability p to other nearby buses with the same voltage level. The distance l_{ij} between buses i and j is defined to be

$$l_{ij} = \sqrt{(x_i - x_j)^2 + (y_i - y_j)^2} \tag{4.6}$$

After t steps, this procedure produces a grid of $n = (a+b)\cdot t + m_0$ buses and $l = (1+p)\cdot t + e_0$ lines.

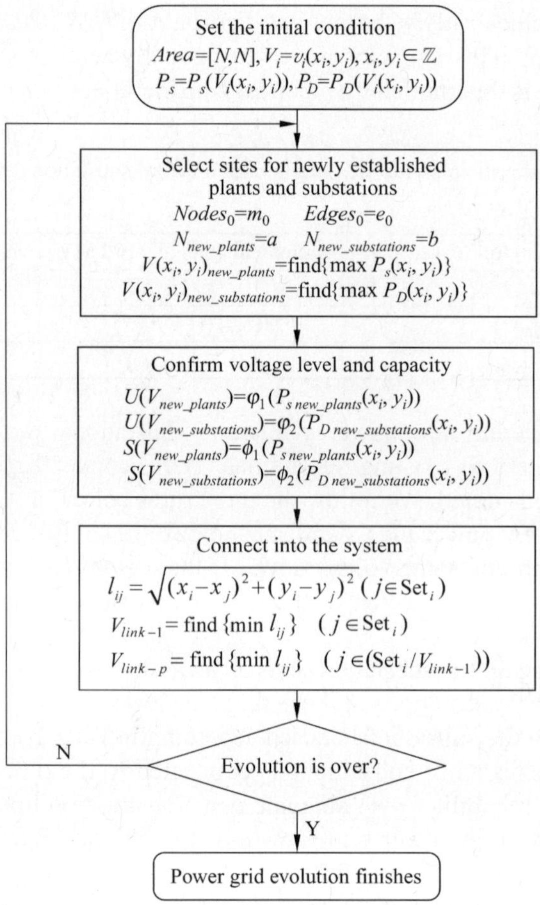

Figure 4.5 Flowchart for the power grid evolution algorithm

4.3.2 Simulation of Power Grid Evolution

In an area with the size 200 km-by-200 km, we assume that the initial energy distribution P_s and the load distribution P_D are given. Assume that within a construction cycle, e.g. one year, one power plant is built together with five substations, namely $a=1$, $b=5$. Set the connection probability for the second nearest buses to be $p=0.1$.

Remark 4.4 The assignments of a, b and p are chosen according to experimental experiences.

We simulate an evolution period of 50 years for the power grid in this area using the power grid evolution model just discussed. We take snapshots at the starts of every ten years and show the results in Fig. 4.6, where the circles denote the power plants and the squares denote the substations. Different sizes of the circles and squares and different thicknesses of the lines represent different voltage levels with the biggest circles and squares and the thickest lines corresponding to the 500 kV voltage level, the lesser ones 220 kV and the rest 110 kV.

After 50 years of evolution, the power grid in this area has developed into a 300-bus transmission power system with 50 power plants, 250 substations and 331 transmission lines. The system's total installed capacity is 6.68×10^4 MW and

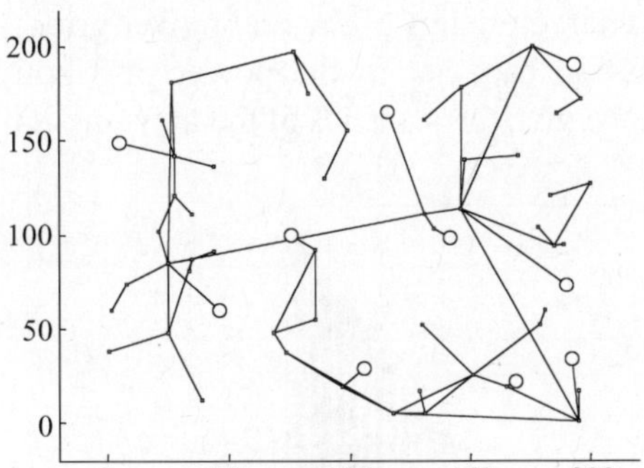

(a) Snapshot at the 10th year of the power grid evolution

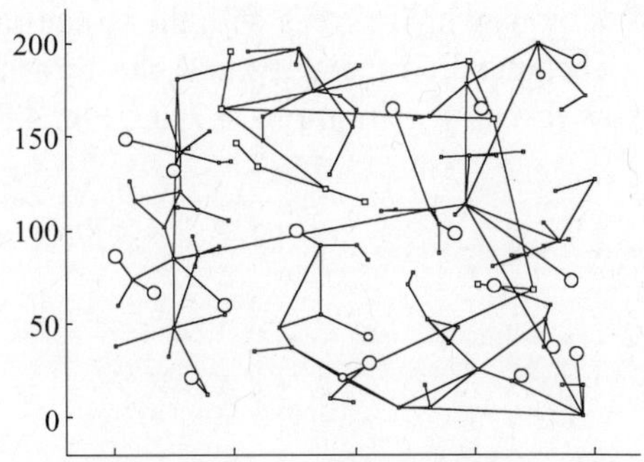

(b) Snapshot at the 20th year of the power grid evolution

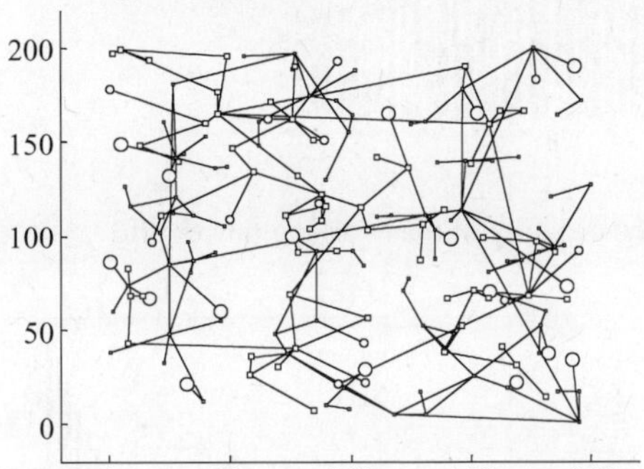

(c) Snapshot at the 30th year of the power grid evolution

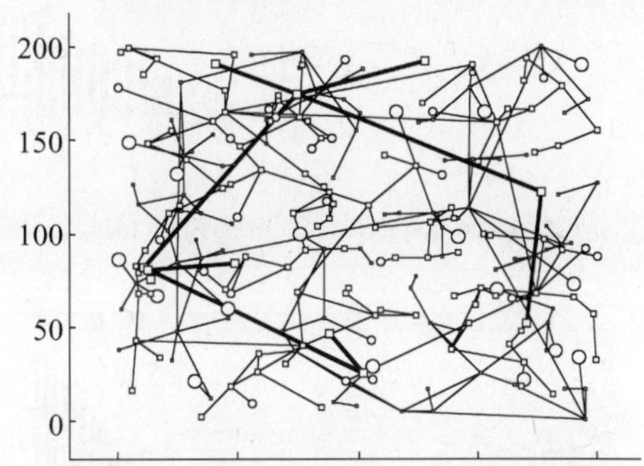

(d) Snapshot at the 40th year of the power grid evolution

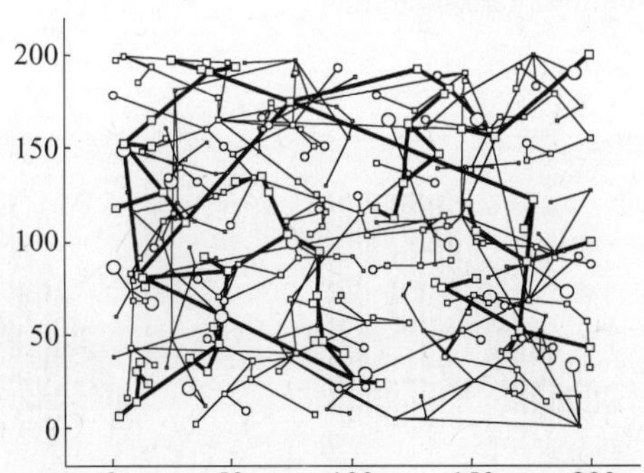

(e) Snapshot at the 50th year of the power grid evolution

Figure 4.6 Schematic drawing of the 50-year power grid evolution

the total power transformation capacity is 1.78×10^5 MW. There are 61 substations at 500 kV level, 99 substations at 220 kV level and 90 substations at 110 kV level. The total length of the 331 transmission lines is 7492.2 km, including sixty-four 500 kV lines of 1766.2 km, one hundred and forty-seven 220 kV lines of 2926.6 km, and one hundred and twenty 110 kV lines of 2799.4 km. It is clear from the figure that the 500 kV backbone framework contains the radiating topologies with local ring structures, which agrees with the structural characteristics of the real power grids.

The installed capacity and the transmission capacity in the power grid both grow fast as shown in Fig. 4.7 and Fig. 4.8. We give the statistics of the transmission

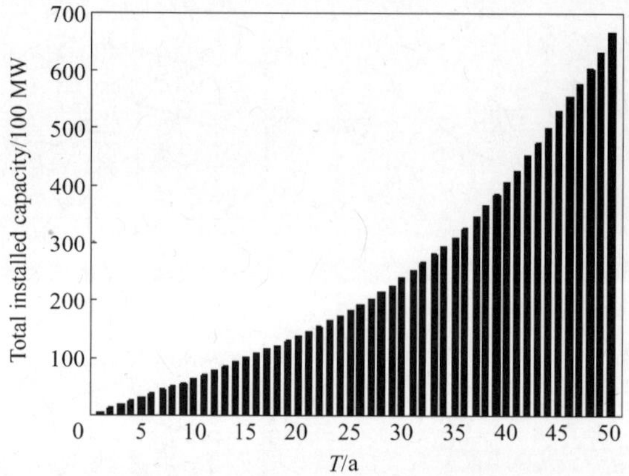

Figure 4.7 The trend of the total installed capacity of the evolving power grid

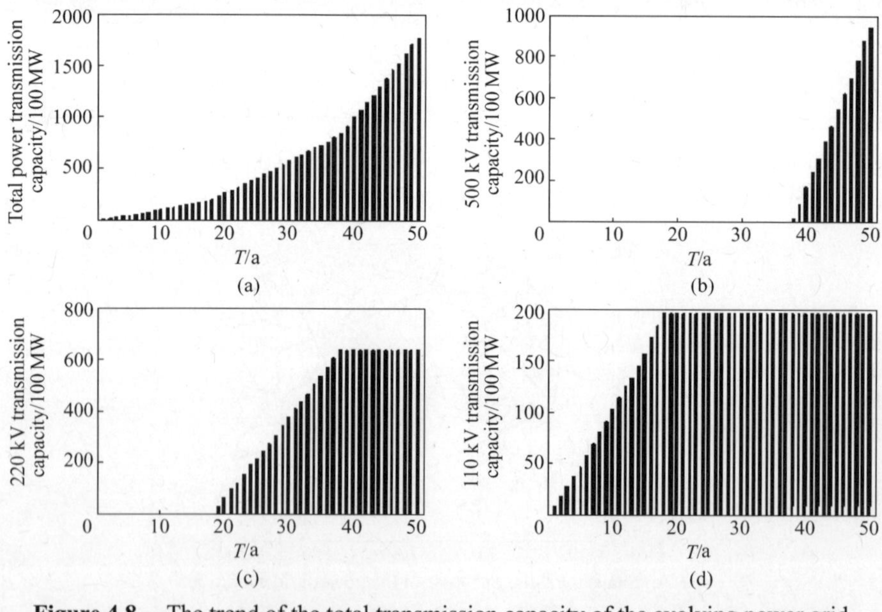

Figure 4.8 The trend of the total transmission capacity of the evolving power grid

development in Table 4.5. When the power grid evolves into the 19th year, the first 220 kV transmission line is built and the first 500 kV line is built in the 39th year. This is consistent with the experiences in the practical power grid construction.

Table 4.5 Transmission lines in the power grid

Duration/a	Total number of lines	Total length /km	The number of 500 kV lines	The length of 500 kV line/km	The number of 220 kV lines	The length of 220 kV lines/km	The number of 110 kV lines	The length of 110 kV lines/km
10	66	2082.4	—	—	—	—	66	2082.4
18	118	2764.9	—	—	—	—	118	2764.9
19	125	3204.4	—	—	5	404.9	120	2799.4
20	131	3414.5	—	—	11	615.0	120	2799.4
30	199	4837.6	—	—	79	2038.1	120	2799.4
38	254	5574.3	—	—	134	2774.8	120	2799.4
39	261	6101.2	6	519.7	135	2782.0	120	2799.4
40	267	6276.2	11	679.4	136	2797.3	120	2799.4
50	331	7492.3	64	1766.2	147	2926.6	120	2799.4

4.3.3 Analysis of Evolution Characteristics of Power Grid Topology Parameters

The degree distributions in the power grid evolution processes are shown in Fig. 4.9 and Fig. 4.10, where d denotes the degree and $P(d)$ is its distribution. We choose snapshots at the 20th year, 30th year, 40th year and 50th year of the power grid evolution to present the degree distribution statistics. The degree distribution in the log-log scale is shown in Fig. 4.9. Only the distribution of the 30th year snapshot is roughly a straight line, i.e. its degree distribution agrees with the power law distribution; the power grid at this moment is a typical scale-free network. However, in the other snapshots, there are obvious turning points in the range of high degrees, and so the degree distributions do not agree with power law distributions and the corresponding grids are not scale-free networks. Hence, we know that the power grid degree distribution under this evolution model does not have the scale-free characteristic consistently.

The degree distributions in logarithmic scales are shown in Fig. 4.10. They present no obvious linear characters and thus the degree distributions are not exponential.

In Fig. 4.11, we present the changes in the clustering coefficients C and average path length L during the power grid evolution. In the initial stage, since there are only a few nodes, C is zero. In the later evolution stages, C is stabilized around 0.07 (the order of magnitude is 10^{-2}). L increases gradually as time

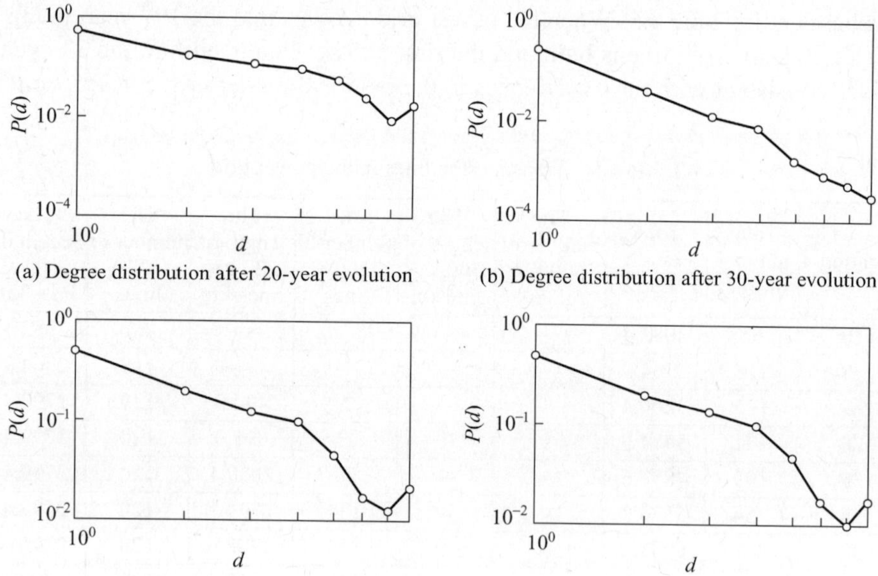

(a) Degree distribution after 20-year evolution
(b) Degree distribution after 30-year evolution
(c) Degree distribution after 40-year evolution
(d) Degree distribution after 50-year evolution

Figure 4.9 Degree distribution diagrams in log-log scales

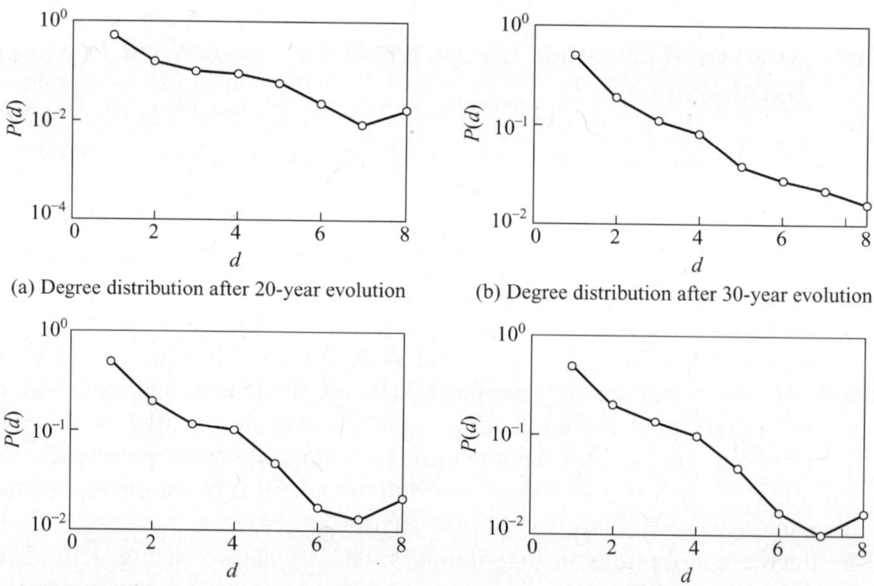

(a) Degree distribution after 20-year evolution
(b) Degree distribution after 30-year evolution
(c) Degree distribution after 40-year evolution
(d) Degree distribution after 50-year evolution

Figure 4.10 Degree distribution diagrams in logarithmic scales

evolves. In Fig. 4.12, we show the changes in the average node degree $\overline{d}$ during the evolution. In the first few years, $\overline{d}$ increases rapidly, and becomes stable around 2.2 after about ten years.

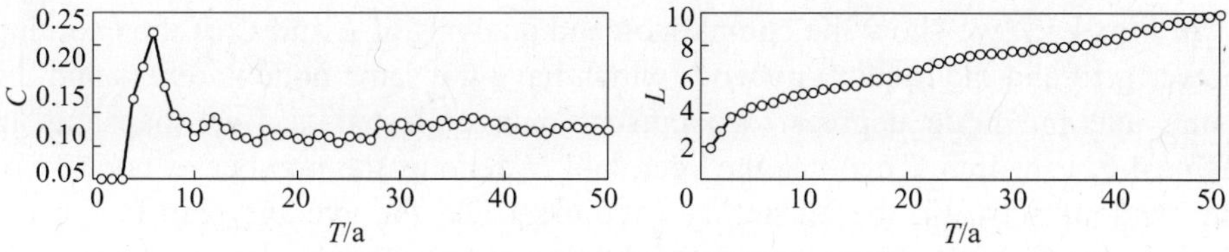

(a) (b)

Figure 4.11 The clustering coefficient and average path length during the evolution of the power grid

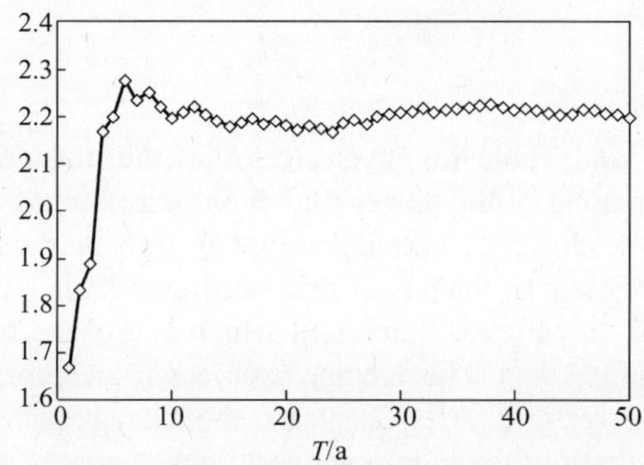

Figure 4.12 Average node degree during the evolution of the power grid

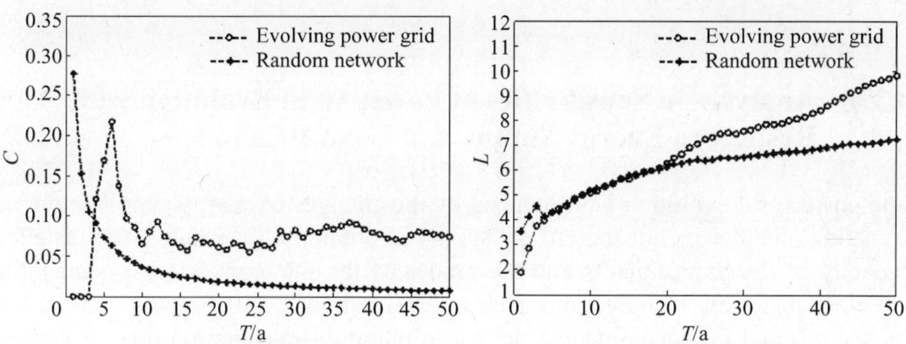

(a) Clustering coefficients (b) Average path lengths

Figure 4.13 The comparison of the evolving power grid and the random network

Table 4.6 The comparison of the evolving power grid and the random network

T	N	$\overline{d}$	L	C	L_r	C_r
10	60	2.2000	5.1000	0.0649	5.1929	0.0367
20	120	2.1833	6.2875	0.0610	6.1311	0.0182
30	180	2.2111	7.5992	0.0828	6.5444	0.0123
40	240	2.2167	8.4127	0.0790	6.8852	0.0092
50	300	2.2000	9.8204	0.0742	7.2341	0.0073

In Fig. 4.13, we show the comparison and analysis of L and C of the evolving power grid and the random network which have the same node numbers and the same average node degrees. We present data collected at five snapshots in Table 4.6, in which T denotes the year, and N denotes the number of nodes. For the random network, the clustering coefficient and the average path length are denoted by C_r and L_r, which are defined to be

$$L_r = \frac{\ln N}{\ln \overline{d}} \tag{4.7}$$

$$C_r = \frac{\overline{d}}{N} \tag{4.8}$$

From Fig. 4.13 and Table 4.6, it is clear that the clustering behavior is not obvious at the beginning of the power grid evolution because of the small number of nodes and thus the clustering coefficients are all zero. At the later evolution stage, the clustering coefficient of the power grid becomes stable around 0.07 (the order of magnitude is 10^{-2}), which is much larger than that of the random network (the order of magnitude is 10^{-3}). The average path length increases gradually with the evolution of the power grid. After 20 years, the average path length of the power grid is larger than that of the random network, which agrees with the criterion for small-world networks. So the power grid evolves towards becoming a small-world network.

4.3.4 Analysis on Sensitivities of Power Grid Evolution with Respect to Energy Supply and Load Demand

The power grid evolution is influenced by the changes of energy supply and load demand. Whether or not the energy supply is abundant determines the installed capacity of the power plants and the grades of the interconnection systems. On the other hand, the load demand determines the grades of the newly constructed substations and further influences the construction of transmission lines at various voltage levels. In this section, we consider the influence of energy supply and load demand on the evolution of power grids.

(1) Power grids' development under different supply and demand conditions

In Table 4.7, we give different supply-demand modes under which the power grid evolves for 50 years. To implement different scenarios of various amounts of energy supplies and load demands in the simulation, we use different amplitudes for the energy distribution function P_S and the load distribution function P_D. The power grid evolutions with their topological characteristics under the four different modes are shown in Fig. 4.14 – Fig. 4.17. The power grids under the four modes all exhibit small-world structural characteristics in theirs network topologies when they reaches stable states.

Table 4.7 Power grid evolutions under different supply and demand modes

Mode	Energy supply	Load demand	Characteristics of the evolution of the power grid	Is it a small-world network?
1	Abundent	Large	Power plants are usually with large capacities. There are many 500 kV substations and 500 kV backbone frameworks with local ring networks and distinct layers.	Yes
2	Deficient	Small	The majority of the power plants are small ones. There are mainly 110 kV and 220 kV substations and 220 kV backbone frameworks with multiple centers and radiating structures. There are no obvious layers.	Yes
3	Abundent	Small	There are relatively more large power plants. The substations are mainly at 110 kV and 220 kV levels. The 220 kV backbone framework has multiple centers with radiating structures.	Yes
4	Deficient	Large	There are mainly small power plants. The substations are mainly at the 500 kV level. The backbone framework is at the 500 kV level. There are many long-distance transmission lines and relatively large-scale local ring networks.	Yes

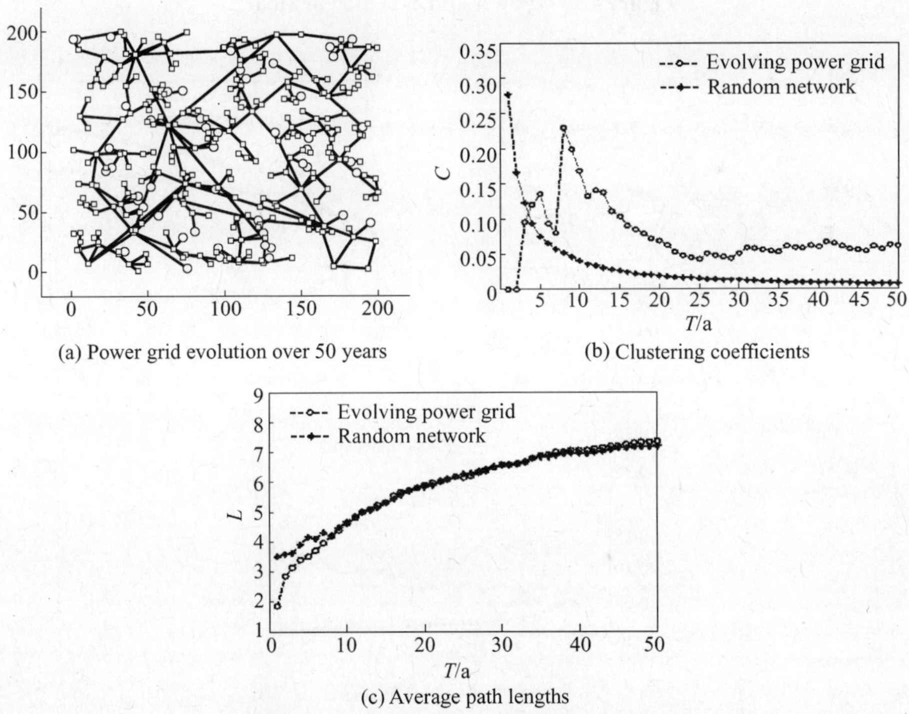

(a) Power grid evolution over 50 years

(b) Clustering coefficients

(c) Average path lengths

Figure 4.14 Power grid evolution in Mode 1

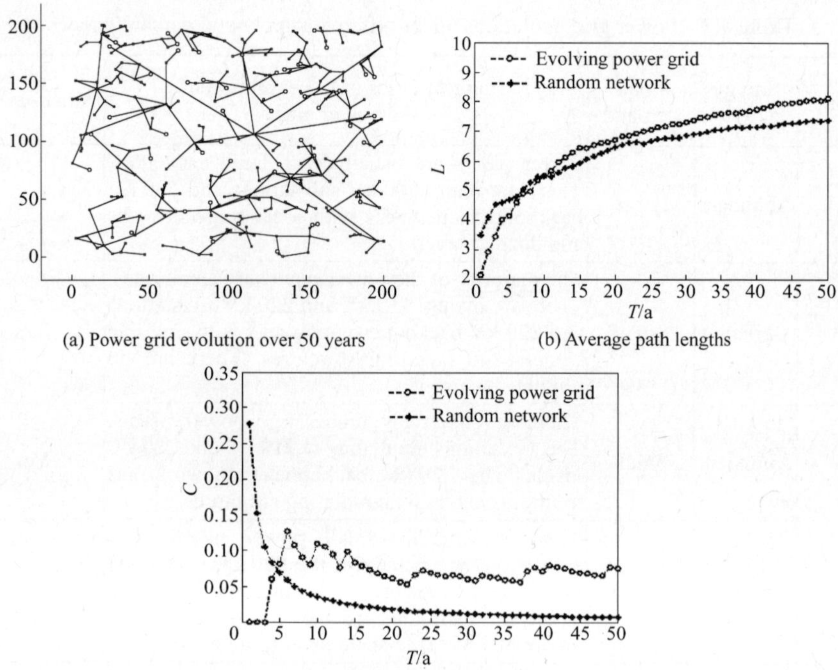

(a) Power grid evolution over 50 years

(b) Average path lengths

(c) Clustering coefficients

Figure 4.15 Power grid evolution in Mode 2

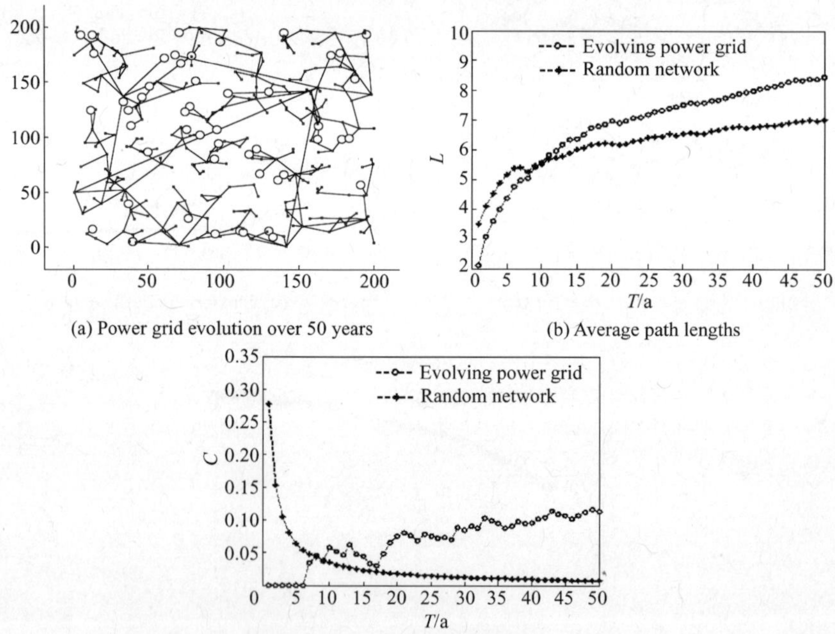

(a) Power grid evolution over 50 years

(b) Average path lengths

(c) Clustering coefficients

Figure 4.16 Power grid evolution in Mode 3

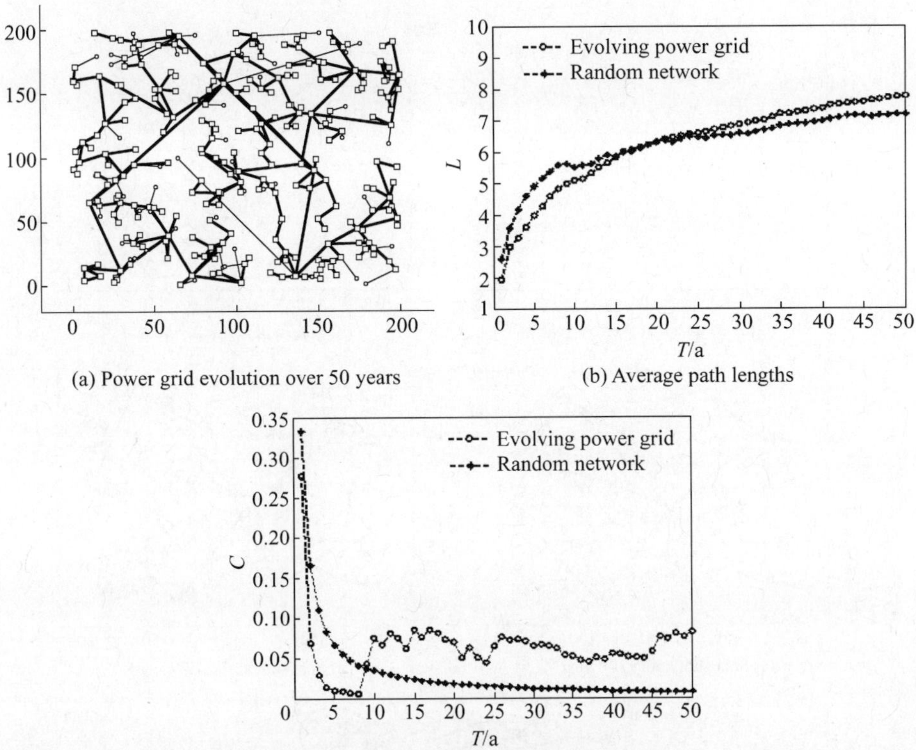

(a) Power grid evolution over 50 years (b) Average path lengths

(c) Clustering coefficients

Figure 4.17 Power grid evolution in Mode 4

(2) Power grid evolution characteristics under different geographic distributions for supply and demand and different load growth modes

Energy supply and load demand have different distributions in different geographical regions in the development of practical power grids. In addition, load growth modes in different regions are also different. For example, the central and southwest regions of China have developed into energy bases because of the abundant hydroelectric resources while the areas near Beijing, Shanghai and Guangzhou have become the main load centers whose load growth speeds are much faster than that in the inner land. In Fig. 4.18, we show the power grid evolution processes in different regional supply and demand patterns and different load growth modes. In the end, the power grid develops into two energy bases S1, S2 and three load centers D1, D2 and D3, which agrees with the practical development of the power grid. In Fig. 4.19, we show the comparision results of L and C between the evolving power grid and the random network.

From these figures, it is clear that the power grid still evolves into small-world networks even after taking into consideration the regional differences of energy supply and load demand and different modes of load growth.

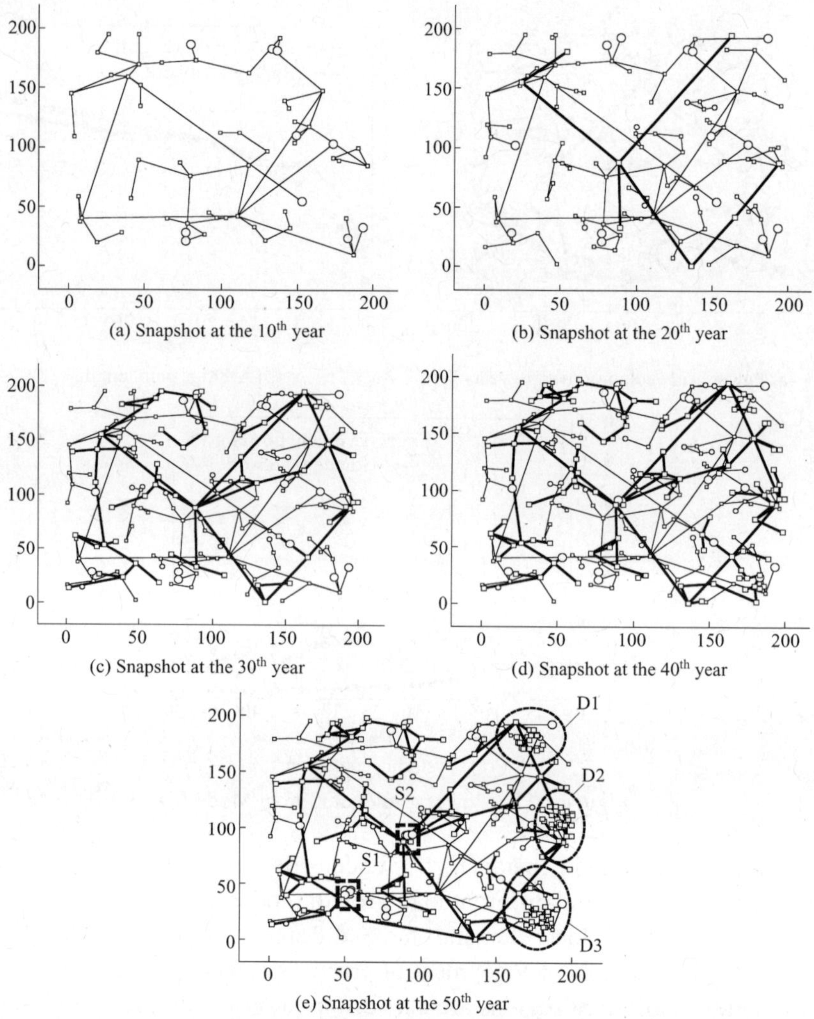

(a) Snapshot at the 10th year (b) Snapshot at the 20th year (c) Snapshot at the 30th year (d) Snapshot at the 40th year (e) Snapshot at the 50th year

Figure 4.18 Power grid evolution taking into account regional differences

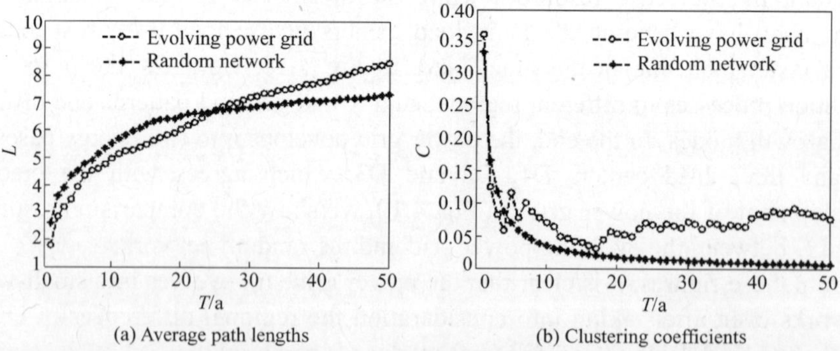

(a) Average path lengths (b) Clustering coefficients

Figure 4.19 Comparison between the evolving power grid and the random network

4.3.5 Saturation Characteristics of Power Grid Evolution

In order to study the saturation characteristics of the evolution of power grids, we extend the simulated period of the power grid evolution to 160 years. The simulation results are shown in the following figures.

From Fig. 4.21 and Fig. 4.22, we can see that the clustering coefficient, the

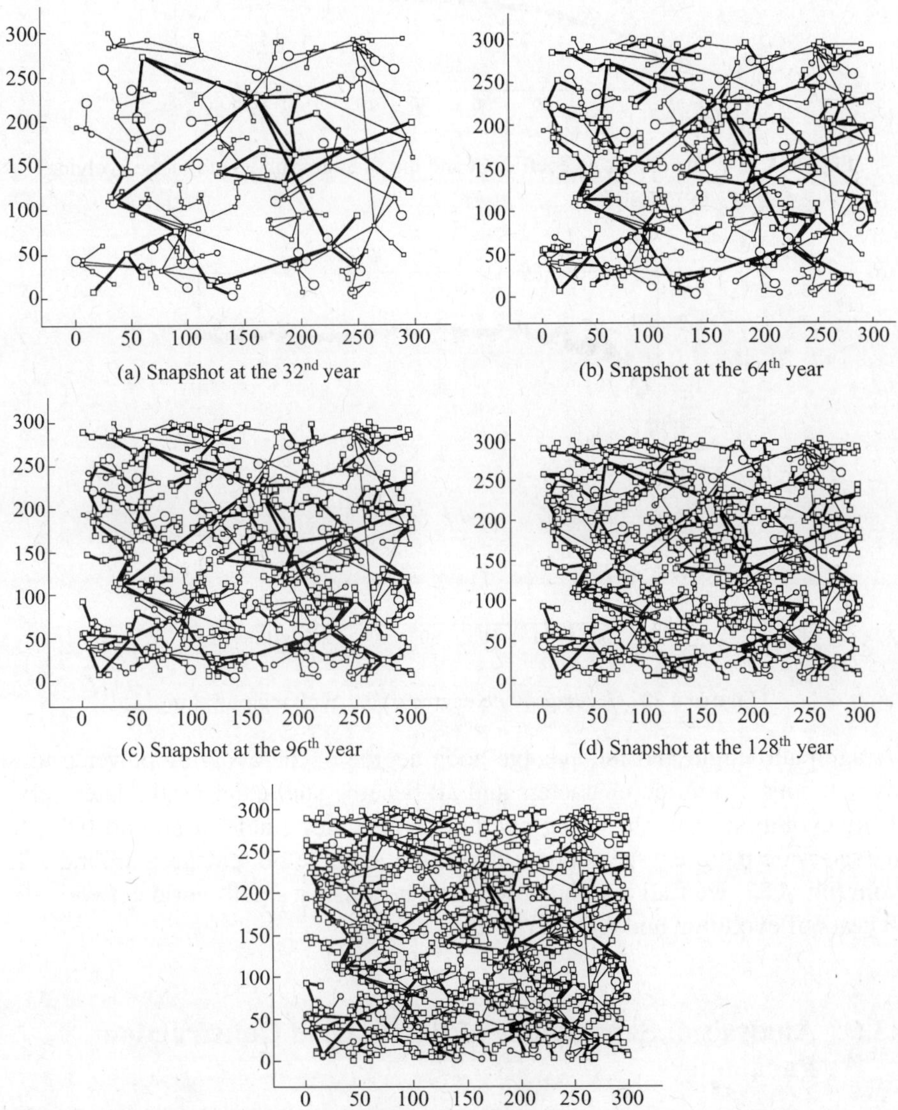

(a) Snapshot at the 32nd year

(b) Snapshot at the 64th year

(c) Snapshot at the 96th year

(d) Snapshot at the 128th year

(e) Snapshot at the 160th year

Figure 4.20 The evolution of the power grid over a period of 160 years

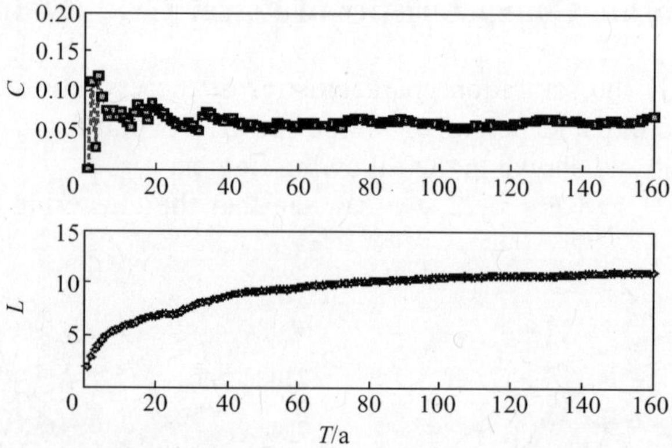

Figure 4.21 The clustering coefficient and the average path length of the evolving power grid

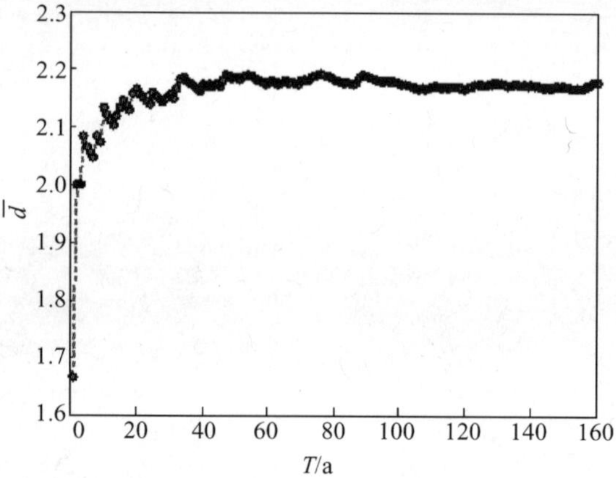

Figure 4.22 Average node degree of the evolving power grid

average path length and the average node degree of the evolving power grid all have obvious saturation characters and all become stable during the later period of the evolution. The clustering coefficient becomes stable at around 0.07, the characteristic path length around 10, and the average node degree around 2.15. From Fig. 4.23, we find that the power grid becomes a small-world network after 20 years of evolution and remains so afterwards.

4.3.6 Analysis of Sensitivities of Power Grid Construction Parameters

In the power grid evolution model, we have used the connection probability p to simulate building new lines to connect power plants and substations to the nearby

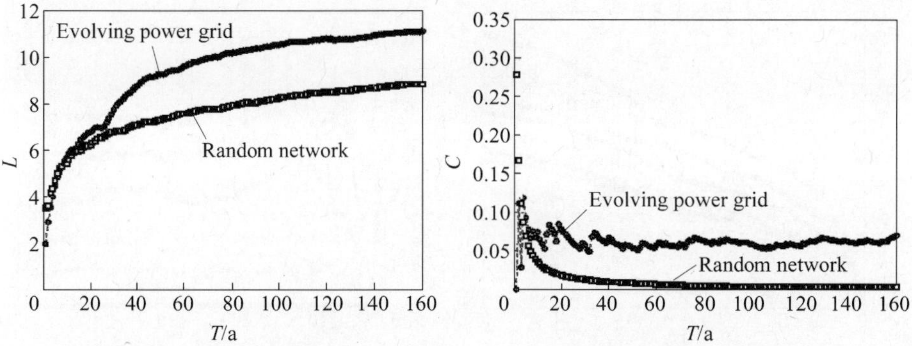

(a) Characteristic path lengths (b) Clustering coefficients

Figure 4.23 Comparison between the evolving power grid and the random network

buses in the system. In order to study the influence of the parameter p on the power grid evolution, we simulate a period of 50 years using the power grid evolution mode with different p.

In Fig. 4.24, we give 3D drawings for the evolution of the topological parameters with different p. In Fig. 4.25 and Fig. 4.26, we present changes in the topological parameters from the perspectives of p and the evolution of the grid respectively. It can be seen that as p grows, the characteristic path length decreases and the clustering coefficient increases. These are exactly the small-world networks' characters of large clustering coefficients and small characteristic path lengths. So the increase of p directly accelerates the evolution of the power grid into a small-world network. Said differently, when p is large, the construction scale in the power grid is also large, and thus the grid evolves into a small-world network at an earlier time.

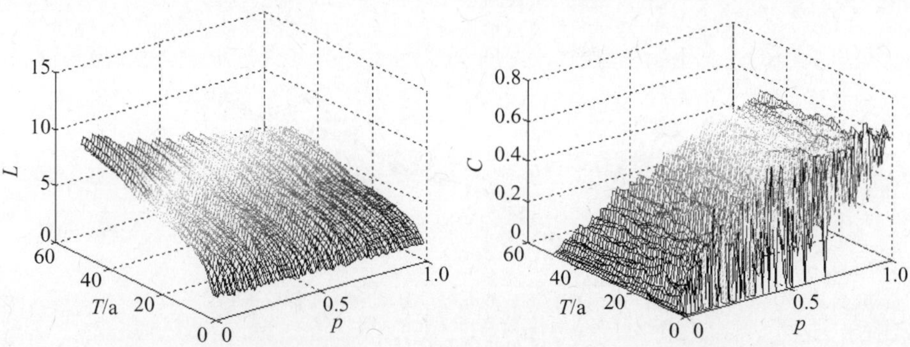

(a) Characteristic path length (b) Clustering coefficient

Figure 4.24 Relationships between the topological characteristic parameters of the evolving power grid and the connection probability p

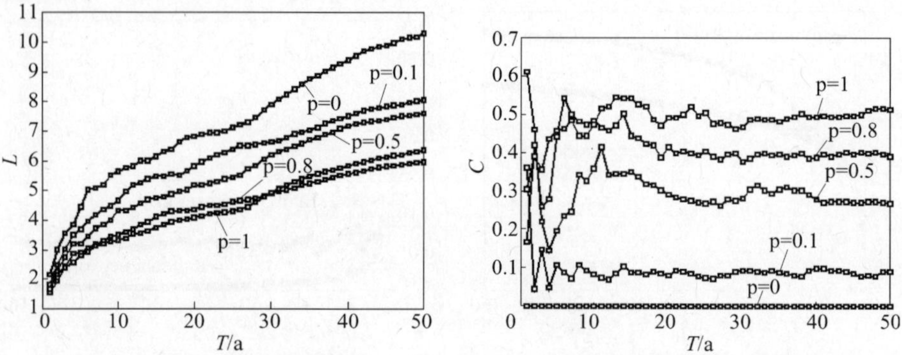

(a) Characteristic path lengths (b) Clustering coefficients

Figure 4.25 Topological parameters of the evolving power grid with different connection probability *p*

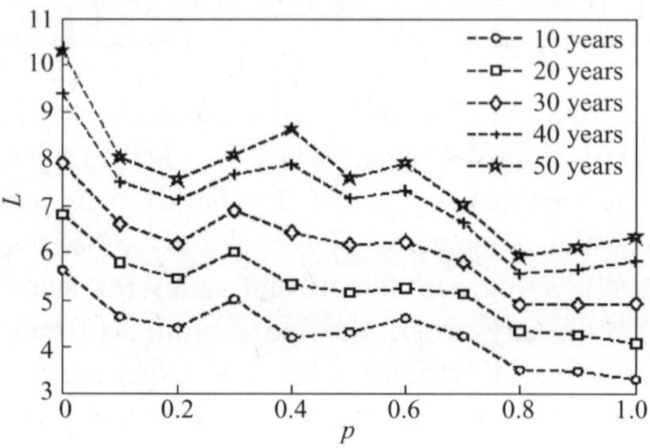

(a) Characteristic path lengths

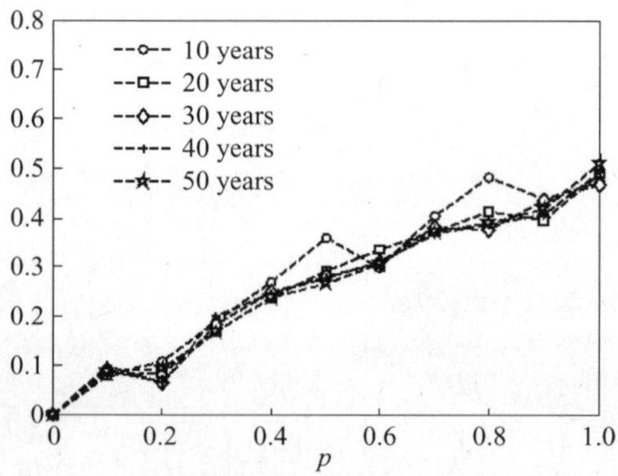

(b) Clustering coefficients

Figure 4.26 Relationship between topological parameters of the evolving power grid and the connection probability *p* at different times of the evolution

4.3.7 Analysis of the Mechanism for the Evolution

It has been shown that power grids behave differently in different situations of energy supply and load demand. However, their topological structures always have the small-world characteristics in the end. When we consider the differences among regional energy and load distributions and among regional load increase modes in the simulation, the grid starts to have distinct energy bases and regional load centers during the evolution, which agrees with the situations in practical power systems. Even in this more complicated situation, the power grid still develops into a small-world network. Increasing the connection probability p accelerates the evolution of the power grid into a small-world network. The following power grid evolution mechanism that has been discussed is the determinant factor for the observed behaviors. Firstly, we have required that when adding new buses to the grid, they must be connected to those with the same voltage level, and thus the buses at the same voltage level are more tightly interconnected, which leads to a larger clustering coefficient of the power grid. Secondly, we have also required that the new buses have to be connected to the nearby buses and this results in shorter characteristic paths. Simulations have validated the effectiveness of the studied evolution mechanism under this mechanism, the power grid always evolves towards a small-world network and the small-world evolution model discussed in this chapter is more faithful to the real situations in practical power systems compared to the general power grid evolution model discussed in Section 4.2.

4.4 Further Discussion

(1) The evolution of a practical power system is a controlled evolving process, and not an autonomous process. How to introduce control and decision-making variables into the power grid evolution is a question that deserves to be further studied. In fact, feedback mechanism needs to be introduced to realize the power grid's adaptive and learning capabilities. This gives the power grid some "intelligence" in its evolution process.

(2) We also want to give some additional remarks on the driving mechanism of the power grid evolution. The regional economy development and load demand growth are the direct driving forces for the power grid evolution. We can further consider the feedback effects of the constructions of power plants and substations. Obviously, the construction of a power plant helps to meet the demands from the local loads, and this can be taken as an adjusting feedback effect. When the load demand increases, more substations need to be built rapidly and when the gap of the backup capacity is large, more power plants need to be built. So a quantified function of the load and the installed capacity can be defined in order to introduce the feedback mechanism to stimulate the development of the network. In other

words, one can analyze the evolution of power grids from the perspective of control system theory while explaining the thrust of the power grid evolution using social development theory.

(3) We can give more attention to the situations in a practical power grid. To do this, we need to further compare the model discussed in this chapter with the real situations in practical power grids.

References

[1] Dobson I, Chen J, Thorp J S, et al. Examining criticality of blackouts in power system models with cascading events. Proceedings of the 35th Annual Hawaii International Conference on System Sciences, Jan. 2002: 7 – 10

[2] Carreras B A, Lynch V E, Dobson I, et al. Critical points and transitions in an electric power transmission model for cascading failure blackouts. Chaos, 2002, 12(4): 985 – 994

[3] Mei S W, He F, Zhang X M, et al. An improved OPA model and blackout risk assessment. IEEE Transactions on Power Systems, 2009, 24(2): 814 – 823

[4] Chen J, Thorp J S, Dobson I. Cascading dynamics and mitigation assessment in power system disturbances via a hidden failure model. International Journal Electrical Power and Energy Systems, 2005, 27(4): 318 – 326

[5] Kirschen D S, Jayaweera D, Nedic D P, et al. A probabilistic indicator of system stress. IEEE Transactions on Power Systems, 2004, 19(3): 1650 – 1657

[6] Watts D J, Strogatz S H. Collective dynamics of "small-world" networks. Nature, 1998, 393(6684): 440 – 442

[7] Barabasi A L, Albert R. Emergence of scaling in random networks. Science, 1999, 286: 509 – 512

[8] Shi D H, Chen Q H, Liu L M. Markov chain-based numerical method for degree distribution of growing networks. Physical Review E, 2005, 71, 036140-1: 9

[9] Watts D J. Small World—The Dynamics of Networks Between Order and Randomness. Princeton(NJ): Princeton University Press, 1998

[10] Surdutovich G, Cortez C, Vitilina R, et al. Dynamics of "Small World" Networks and Vulnerability of the Electric Power Grid. VIII Symposium of Specialists in Electric Operational and Expansion Planning, Brazil, 2002

[11] Meng Z W, Lu Z X, Song J Y. Comparison analysis of the small-world topological model of Chinese and American power grids. Automation of Electric Power Systems, 2004, 28(15): 21 – 29 (in Chinese)

[12] Mei S W, Ni X P. Several new progresses on the complexity research of the interconnected power network. Journal of Changsha University of Science & Technology (Natural Science), 2008, 5(2): 1 – 8 (in Chinese)

Chapter 5 Complex Small-World Power Grids

The electric features of small-world power grids are the focus of this chapter. The DC power flow model is utilized to analyze the power flow characteristics and the synchronization capabilities of small-world power grids. We then perform the assessment of the grid's structural vulnerability when examining the key buses of power plants, substations and important tie lines.

The model of small world networks is an important network model in complex network theory[1]. Many power grids, such as the American Western Interconnected System, the Brazilian Power Grid and the Northern Power Grids of China, exhibit small-world structural characteristics[2−4], including small characteristic path lengths and large clustering coefficients. As a further study on such power grids, we discuss in this chapter their electric characteristics, and thus provide critical references for the system planning and secure and stable operation of large-scale power grids.

5.1 Power Flow and Synchronization in Small-World Power Grids

One of the main principles in complexity theory is that network functionalities are determined by network structures[5]. In long-term evolutions, power grids are subject to self-selection and man-made developments that aim to implement optimization objectives, and thus their small-world characteristics shown in their topological structures have their intrinsic trend evolving towards optimality. In this context, we study the characteristics of power flow and synchronization in small-world power grids, by looking into how the connectivity of a grid evolves with changing transmission lines and fixed numbers of generator buses and load buses (see Fig. 5.1). From the angle of spatial geometry, we analyze how one can map topological space into electric performance space and thus infer the electric characteristics, such as power flow performances and synchronization capabilities from the topological features (see Fig. 5.2).

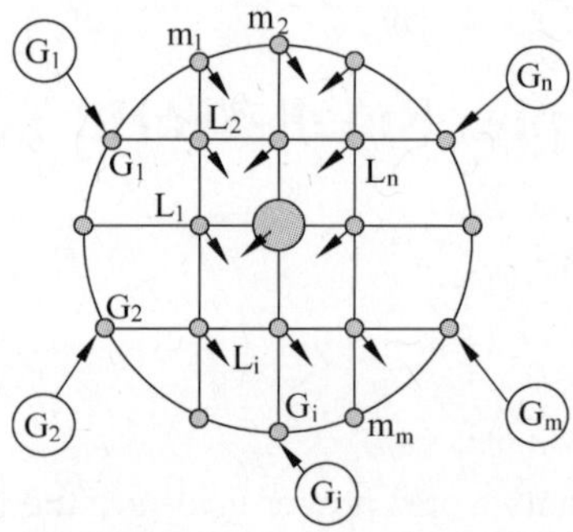

Figure 5.1 Power grid's topological structures and electric characteristics

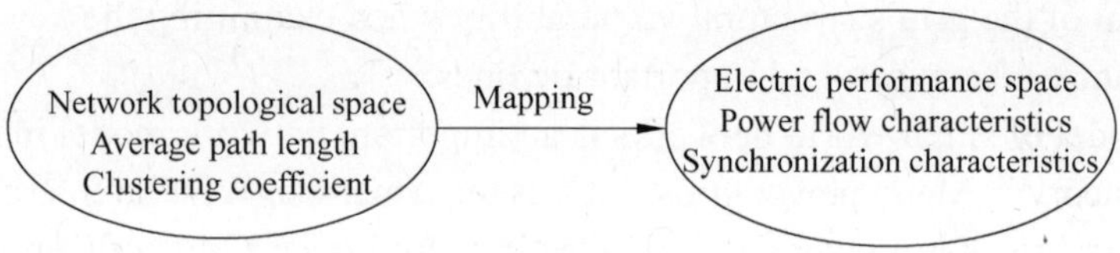

Figure 5.2 Mapping from network topologies into power grid performances

5.1.1 Analysis of Power Flow Characteristics in Small-World Power Grids

In order to realize various network topologies, we utilize a typical small-world network generator here to obtain power grids with different structural characteristics, such as clustering coefficients and characteristic path lengths, by adjusting the reconnection probability p. The DC power flow model[6] is adopted as well. We use the following notations: A is the node-branch incidence matrix, C is the branch admittance matrix, $B = A \cdot C \cdot A^{\mathrm{T}}$ is the nodal admittance matrix, and P is the injected power at the buses. In view of the DC power flow algorithm, we write the branch power flow equation as

$$f = C \cdot A^{\mathrm{T}} \cdot B^{-1} \cdot P = C \cdot A^{\mathrm{T}} (A \cdot C \cdot A^{\mathrm{T}})^{-1} P \tag{5.1}$$

From Eq. (5.1) we know that when the network's generator buses and load buses are fixed and the transmission line parameters and the injected powers at the buses are given, the power flows in transmission lines are closely related to the node-branch incidence matrix A. Since in addition A describes the topological properties of the network, it changes and thus affects the distribution of power flows when the reconnection probability p changes leading to the changes in network topologies. To evaluate the system's power flow performances, we define two assessment indices, which are the average power flow level F and the number of overloaded lines NL. The flowchart of the algorithm is shown in Fig. 5.3.

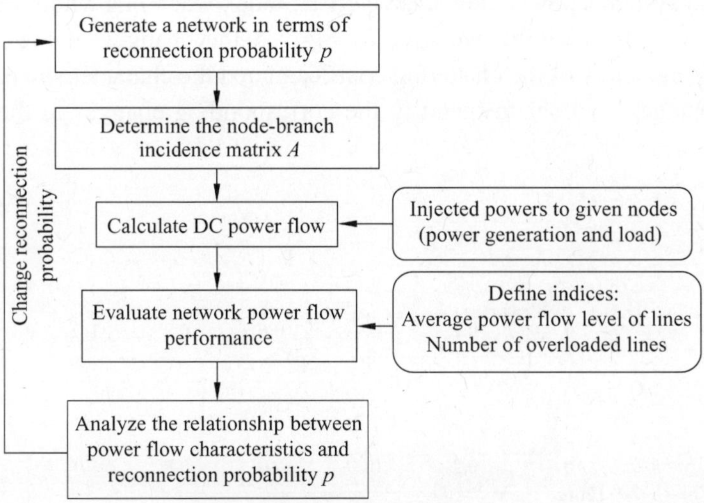

Figure 5.3 Algorithm flowchart

Now consider the 300-bus system generated by the power grid evolution model discussed before, which contains $G=50$ generator buses and $L=250$ load buses. Assume the reactance of every line is 1 for the sake of simplicity and the initial load (p.u.) at every load bus satisfies the uniform random distribution over the interval (0, 1). We set generator bus N to be the slack bus, and assume that it balances 10% of the total load. We choose the line overload threshold to be 2.0. When the reconnection probability p changes from 0 to 1, we calculate the mean F and the variance X of the system's power flow and count the number of overloaded lines NL. For each p, we run the simulation for 10 times, the results of which are used to compute the average as the final statistic index.

Since the parameters of the lines are assumed to be identical, the network has a better transmission performance and the system's power flow is better distributed when the average power flow level is low, its variance is small and the number of overloaded lines is small. And consequently, when the load increases or the operation mode changes, the network can deliver greater power supplies and is thus more robust. When the system's average power flow is low, the power flow's distribution may still be unbalanced with overloaded lines running close to their upper limits and under-loaded lines that are far below their transmission capacities. Hence, we introduce the system's power flow variance index X to better evaluate the system's power flows.

From Fig. 5.4, it is clear that as p increases, the system's power flow improves with decreasing average power flow levels and decreasing numbers of overloaded lines. The most obvious change occurs when p is within $0-0.1$, and the change becomes slower when p is between 0.1 and 1. In Table 5.1, we show the detailed data for the system's power flow characteristic index when p is within the interval of $0.001-0.1$. In addition, we give in Fig. 5.5 the changes of the network's structural parameters of the clustering coefficient and the characteristic path length when p changes in order to quantify the corresponding changes in the network topology.

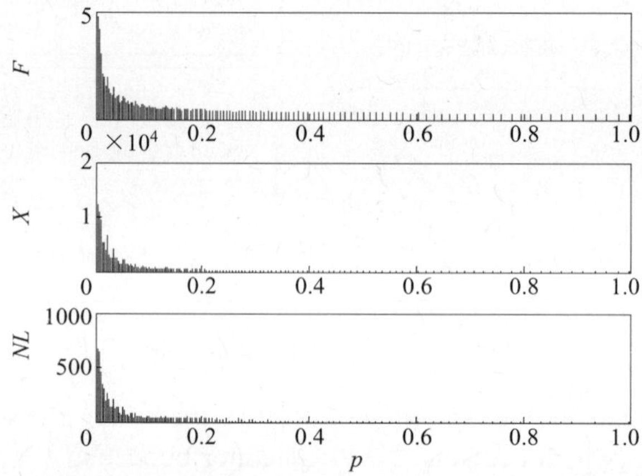

Figure 5.4 Relationship between the system's power flow characteristics and the reconnection probability p

Table 5.1 Relationship between the system's power flow characteristics and p

p	F	X	NL
0.001	3.6503	9251.4097	546
0.01	1.9242	5303.2148	291
0.02	1.4656	3681.3496	195
0.03	0.9940	1885.6954	115
0.04	0.8744	1583.8785	92
0.05	0.7521	1113.3259	69
0.06	0.7268	1144.3162	67
0.07	0.6424	905.2641	49
0.08	0.6291	778.2622	48
0.09	0.5840	670.9929	41
0.1	0.5599	620.9348	39

From Table 5.1 and Fig. 5.5, one can see that the network has the largest clustering coefficient and average path length when $p = 0.001$ with the corresponding average power flow level being 3.6503. When $p = 0.05$, the network has a relatively large clustering coefficient and a small average path length with the average line power flow level being 0.7521, which is 79.4% lower than that in the former case. Thus, with the same network buses and injected powers, the power flow varies greatly under different network topological structures. So a small variation in p changes the network topology and at the same time changes the power flow characteristics greatly.

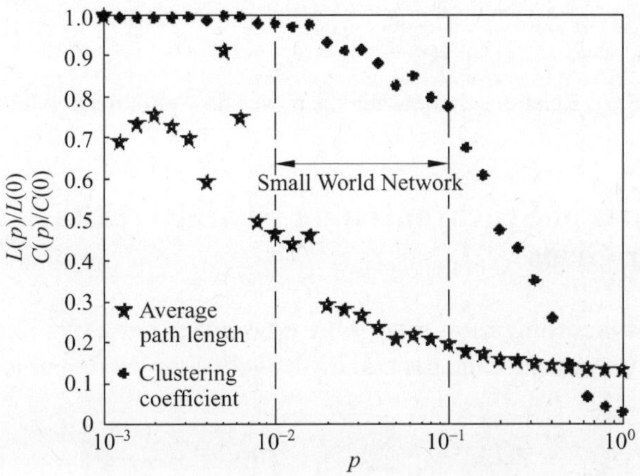

Figure 5.5 Relationships between network structural parameters and p

From Table 5.1 we know that the average power flow and the number of overloaded lines are both at a relatively low level and thus the systems has a better power flow performance when p varies within 0.01 – 0.1. From Fig. 5.5, it is clear that when p is within 0.01 – 0.1, the network has greater clustering coefficient and shorter average path length, which indicates that the system has small-world structure. Hence, power grids with small-world network structures have better power flow performances.

The relationship between the optimization level of power flow performance and p is shown in Fig. 5.6, in which we normalize the indices using the reduced amount when p changes from 0.001 to 0.01 as the base. From the figure, one can see clearly during the transition into small-world networks, the power flows are improved the most. When the network has been transformed into a small-world network, the change in the power flow performances becomes slow. This is because although p keeps increasing, the basic features of the network almost remain the same, and thus the characteristics of the power flows do not change very much.

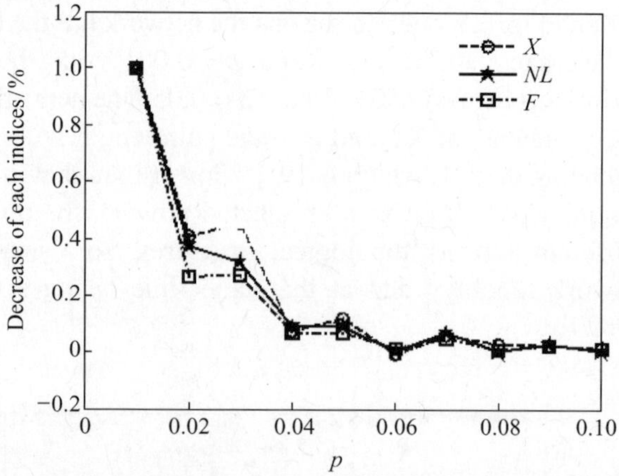

Figure 5.6 Relationship between the power flow optimization level and p

5.1.2 Analysis of Synchronization Characteristics in Small-world Power Grids

From network synchronization theory[7] for complex networks and also stated in Section 2.14.1, we consider the general diffusively coupled dynamical networks

$$\dot{x}_i = f(x_i) + c\sum_{j=1}^{N} a_{ij}\Gamma(x_j), \quad i = 1,2,\cdots,N \tag{5.2}$$

where $A = [a_{ij}] \in \mathbb{R}^{N\times N}$ denotes the network's topological structures and is called the external coupling matrix , which satisfies the diffusive coupling condition

$$\sum_{j} a_{ij} = 0, \quad i = 1,2,\cdots,N \tag{5.3}$$

If there is a connection between nodes i and j, then $a_{ij} = a_{ji} = 1$; otherwise, $a_{ij} = a_{ji} = 0\ (i \neq j)$. The diagonal elements are $a_{ii} = -\sum_{\substack{j=1\\ j\neq i}}^{N} a_{ij}$.

Since $\sum_{j} a_{ij} = 0\ \ (i = 1,2,\cdots,N)$, A must have a zero eigenvalue. And since $a_{ij} = a_{ji}$ we know A is symmetric and its eigenvalues are all real. Without loss of generality, we assume the eigenvalues are

$$0 > \lambda_2 \geqslant \cdots \geqslant \lambda_N \tag{5.4}$$

It has been shown[7] that for general diffusively coupled complex dynamical networks, a network's synchronization domain is determined by the dynamics

$f(\cdot)$ at isolated nodes, the coupling strength c, the external coupling matrix A and the internal coupling matrix $\Gamma(\cdot)$. The network synchronization capability is strong when λ_2 or λ_N / λ_2 is small. So to investigate the synchronization capability of a complex network, one can mainly look at λ_2 or λ_N / λ_2 of the external coupling matrix.

For the example in 5.1.1, we study the relationship between the network synchronization capability and its topological structure by taking $S = \lambda_N / \lambda_2$ as the index for a variety of networks generated by different p. As shown in Fig. 5.7, the system's synchronization capability changes the most when p is within 0 – 0.1 and it hardly changes when p is within 0.1 – 1.

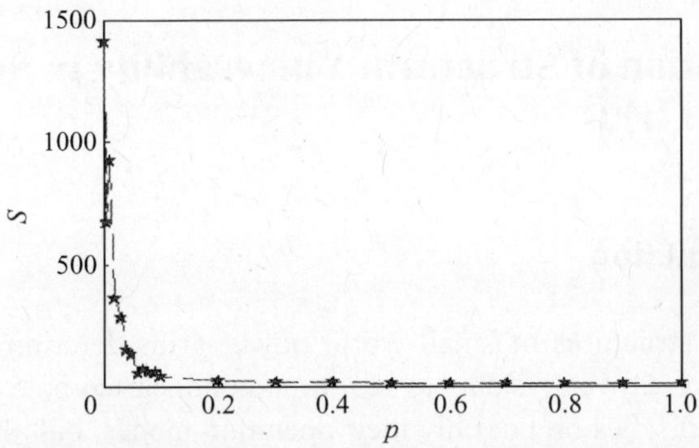

Figure 5.7 Relationship between the system's synchronization capability and its topological structure

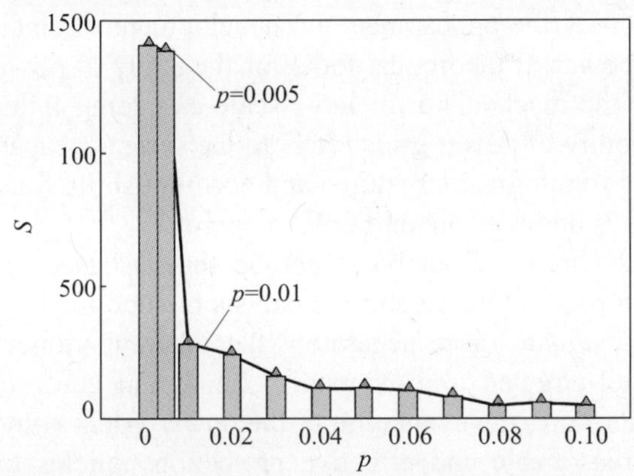

Figure 5.8 Network synchronization characteristic for small-world power grids

As shown in Fig. 5.8, when p changes within 0.01 – 0.1, the network synchronization performance index S is relatively small, and thus the system has a stronger synchronization capability. At the same time, as shown in Fig. 5.5, the network has small-world structures. When p changes from 0.005 to 0.01, the network transforms from a regular network into a small-world network, S changes obviously, and the synchronization capability is improved greatly.

In short, under similar operation conditions, the small-world power grids have stronger synchronization capabilities and are more robust when facing external disturbances like load fluctuations. It needs to be clarified that the network synchronization capability discussed here is different from the classical notion of the transient power angle synchronization in power systems. It evaluates the network's synchronization capability in terms of topological structures and describes the system's robustness and the capability to reject small exogenous disturbances.

5.2 Evaluation of Structural Vulnerability in Small-World Power Grids

5.2.1 Introduction

The topological structures of small-world power grids determine the network's vulnerability level. Power systems as real-time nonlinear complex networks, their vulnerability level relies on not only their operation modes, but also the network's structures. In particular, the structural characteristics are a power grid's inherent properties and once the network's structure is set up, external influences can hardly change such characteristics fundamentally. Progress in complex network theory[8–10], especially the appearance and development of small-world network theory, provides powerful theoretical tools for the study of power grid structural vulnerability and the mechanism for large-scale cascading failures. Usually, the structural vulnerability of power grids refers to the system's capability to maintain the integrity of its topological structures and accomplish its functionalities when some node or line is under attack and fails to operate.

Compared to the transient analysis method that is based on the balance of power flows, the structural vulnerability analysis method for large-scale complex power grids can evaluate more accurately the inherent vulnerability in power grids, identify well-targeted improvement actions, and guide to strengthen the power grid. Furthermore, the evaluation of the power grid's vulnerability helps to determine more reasonable and effective protection policies to protect critical components in the system and prevent disastrous consequences caused by cascading failures. It also serves as references for power grid planning and development to make the power grid framework more efficient and robust.

In this section, we establish the small-world power grid model using complex network theory. We propose the network failure modes and the network performance evaluation indices from the perspective of the topologies of complex networks in order to study the network's performances under different attacks and evaluate structural vulnerability of small-world power grids.

5.2.2 Topological Models for Small-World Power Grids

To study the power grid characteristics using complex network theory, we have to first simplify such characteristics in view of the real power grids' behaviors to establish topological models. Here we use sparsely connected graphs to denote the topological structures of power grids, where the nodes represent power plants and substations in practical power systems, and the edges represent high-voltage transmission lines. Assume that the weights of all edges are identically one. The following simplification principles are used.

(1) We consider only the transmission and distribution networks that are above 110 kV and take the other low-voltage distribution networks as simplified equivalent loads.

(2) We merge the buses in the same substation with the same voltage level as a single bus.

(3) We take the transmission lines as undirected edges and ignore the orientations.

Then the power grid can be described by an undirected connected graph without weights that contains N nodes and K edges. We separate the nodes into generator buses and load buses. We use the node adjacency matrix A to describe the connections between nodes. If there is a connection between nodes i and j, then $a_{ij} = a_{ji} = 1$; otherwise, $a_{ij} = a_{ji} = 0 \, (i \neq j)$.

In this model, the transmission lines are all the same and the electric energy is transmitted along the shortest path. So we define the distance between any pair of distinct nodes in the network to be the number of edges contained in the shortest path connecting the two nodes. A node's betweenness is defined to be the times that the node is traversed by the shortest paths between all pairs of generator buses and load buses in the network. Correspondingly, an edge's betweenness is defined to be the times that the edge is traversed by the shortest paths between all pairs of generator buses and load buses. A node's degree is defined to be the number of edges that are adjacent to it. The degree and betweenness of a node reflect to a certain extend the node's importance in the network and the edge betweenness reflects how actively the corresponding transmission line is involved in electric energy transmission process and its influence.

5.2.3 Evaluation Algorithms for Power Grid Structural Vulnerability

The evaluation of power grid structural vulnerability is to evaluate a power grid's performances when it is under different attacks. Small-world power grids have their own network structural characteristics, and to study the structural vulnerability of small-world power grids, we consider the following network failure attack modes and the corresponding network performance evaluation indices.

1) Network failure attack modes

Two attack modes with distinct features are considered here, which include the random attack mode and the deliberate attack mode. The latter refers to the attack that is intentionally carried out targeting at a particular property of the network, which has a stronger destructive effect than the former.

We further classify the attack modes into four sub-categories. For network nodes, we consider:

(1) Random node attack: attack a node randomly and then gradually attack more nodes randomly.

(2) Node degree static attack: deliberately attack the node with the largest node degree and then attack the nodes according to the ordering of their node degrees.

(3) Node degree dynamic attack: deliberately attack the node with the largest node degree, recalculate the nodes' degrees and then repeat the attacking and calculation process.

(4) Node betweenness attack: deliberately attack the node with the largest node betweenness and then attack the nodes according to the ordering of their betweenness.

For network edges, we consider:

(1) Random line attack: attack a line in the network randomly and then gradually attack more lines randomly.

(2) Line betweenness attack: deliberately attack the line with the largest line betweenness and then attack the lines according to the ordering of the betweenness.

2) Network performance evaluation indices

Since small-world power grids have small characteristic path lengths and large clustering coefficients, they usually have key buses and long-range tie lines. Once these buses and tie lines are under attack, the electric transmission paths may be altered dramatically and the shortest paths between pairs of generator and load buses grow greatly, which affects the network transmission performances.

We define the network transmission efficiency index E to be the sum of all the reciprocals of the lengths of the shortest transmission paths and the sum is further normalized using the network transmission efficiency before the attack as the base, namely

$$E = \frac{\sum\limits_{i \in G, j \in D} \frac{1}{d_{ij}}}{\sum\limits_{i \in G^0, j \in D^0} \frac{1}{d_{ij}^0}} \tag{5.5}$$

where G is the set of generator buses, D is the set of load buses, and d_{ij} is the length of shortest transmission path from the generator bus i to the load bus j; G^0, D^0 and d_{ij}^0 are the corresponding parameters for the power grid before the attack.

When the islanding occurs after the system is attacked and then cascading failures develop, it is an important index to evaluate whether the system can maintain its normal power supplies and keep the system's integrity. Here, we use the system connectivity index to assess the influences of the attacks on power grids. To be more specific, we count the number of buses in the largest connected component in the network after each attack and take this number to be the connection index Q for the largest connected network

$$\begin{cases} Q = \max\{\text{sum}(\text{Set}_m)\}, \quad m = 1,2,\cdots,k \\ H = \bigcup_{m=1}^{k} \text{Set}_m \end{cases} \tag{5.6}$$

where Set_m is the set of buses of the sub-network m after the splitting, k is the number of sub-networks, and H denotes the overall network after the attack.

One of the fundamental functions of power grids is to deliver electric powers to satisfy the load demands, and so one of the basic requirements for power systems is to supply powers reliably. Hence, we take R, the maximum amount of load that can be supported by the system as an important index to evaluate the vulnerability of large-scale power grids:

$$\begin{cases} R = \dfrac{\text{Load}_{\max}}{\text{Load}_0} \\ \text{Load}_{\max} = \sum_{m=1}^{k} \min(\alpha \cdot \sum_{i \in \text{Set}_m} P_{Gi}, \sum_{j \in \text{Set}_m} P_{Lj}) \\ H = \bigcup_{m=1}^{k} \text{Set}_m \end{cases} \tag{5.7}$$

where Load_0 denotes the system's total initial load and $\text{Load}_{\max}$ denotes the maximum amount of load that can be supported by the system after the attack, which is the sum of the maximum capacities of all the sub-networks after the islanding. When calculating the maximum load capacity in each sub-network, we take the generator sets' adjusting coefficients to be $\alpha = 1 + 5\%$, which implies that because of the generation-load balancing principle and the adjustability of generators' outputs, the generators can increase their outputs by 5% with respect to their present outputs.

5.2.4 Simulation Analysis

1) Vulnerability evaluation on evolving small-world power grids

Using the small-world power grid evolution model discussed in Section 4.3, we study the 300-bus small-world network for a simulated evolution period of 50 years. We take the power plants as the generator buses and the substations as the load buses. The injected powers at the load buses are given in Gaussian distributions according to their voltage levels, namely

$$P_{Di} = \begin{cases} N(5,0.5), & i \in 500\text{ kV} \\ N(2,0.2), & i \in 220\text{ kV} \\ N(1,0.1), & i \in 110\text{ kV} \end{cases} \tag{5.8}$$

The injected power at generator buses are given according to the voltage levels at the places where the generator buses are connected and the generation-load balancing principle, namely

$$\begin{aligned} \lambda_{Gj} &= \begin{cases} N(5,0.5), & j \in 500\text{ kV} \\ N(2,0.2), & j \in 220\text{ kV} \\ N(1,0.1), & j \in 110\text{ kV} \end{cases} \\ P_{Gj} &= \frac{\lambda_{Gj}}{\sum_j \lambda_{Gj}} \sum_i P_{Di} \end{aligned} \tag{5.9}$$

(1) Attacks on buses

From Fig. 5.9, the small-world power grid shows different features under random attacks and deliberate attacks including the node degree static attack, the node degree dynamic attack and the node betweenness attack. The horizontal ordinate is attack number n. The grid is more robust to random attacks, which means that its various network performance indices decrease only slightly when its buses are under continuous random attacks. However, those indices decrease substantially especially at the beginning when the network is under deliberate attacks, which indicates that the small-world power grid is vulnerable to deliberate attacks.

As to the network transmission efficiency E and the maximum amount of load R supported by the network, they change in a similar fashion under three deliberate attacks, but behave very differently under random attacks. The node betweenness attack is more destructive when evaluated using the maximum connected component index Q. In Fig. 5.10, we use black frames to denote the buses that are under attack in the three deliberate attack modes, in which the smaller ones are the 220 kV substation buses and the bigger ones are the 500 kV substation buses. From Fig. 5.10, it can be seen that the attacked buses are mainly important substations of 500 kV and 220 kV. Especially in the modes of the node degree dynamic attack and the node betweenness attack, more 500 kV substations are attacked, and the corresponding damages are more serious. These substations play important roles in the grid, and large-scale blackouts caused by cascading failures take place easily if these buses are attacked and thus out of service.

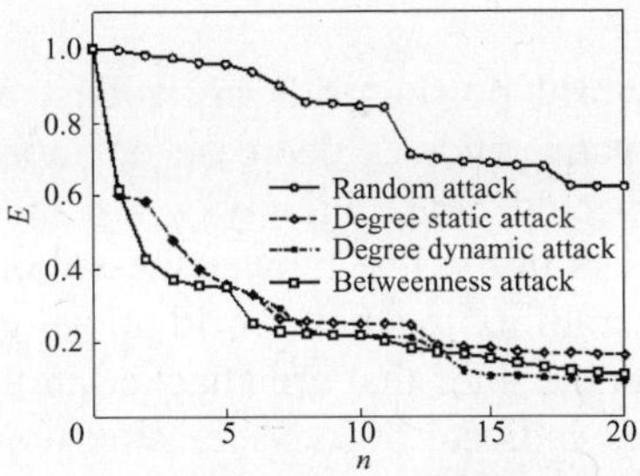

(a) Network transmission efficiency E

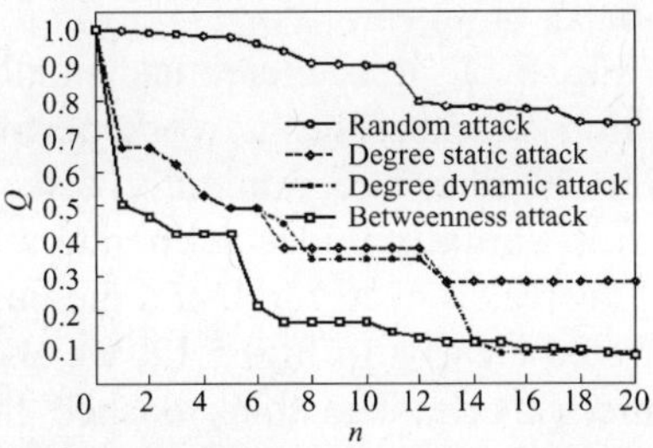

(b) Network largest connected component Q

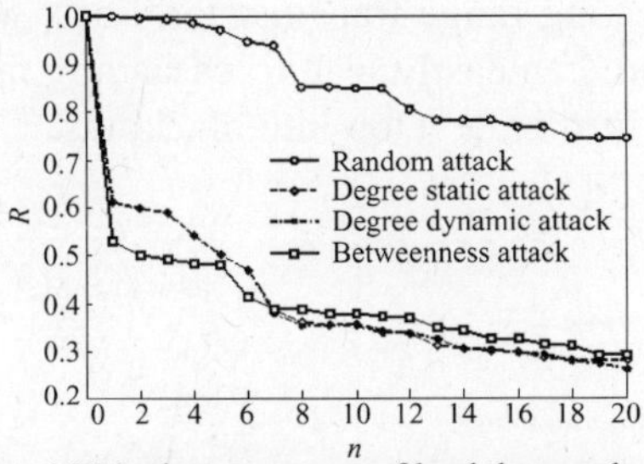

(c) Maximum amount of load that can be supported by the network R

Figure 5.9 Vulnerability indices for the 300-bus small-world power grid under node attacks

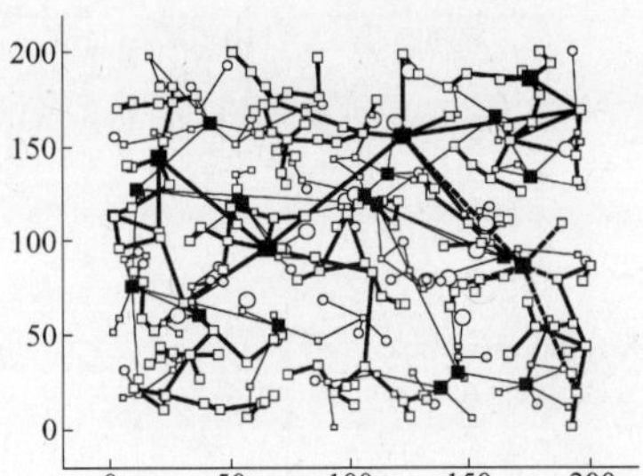

(a) Attacked buses in the node degree static attack mode

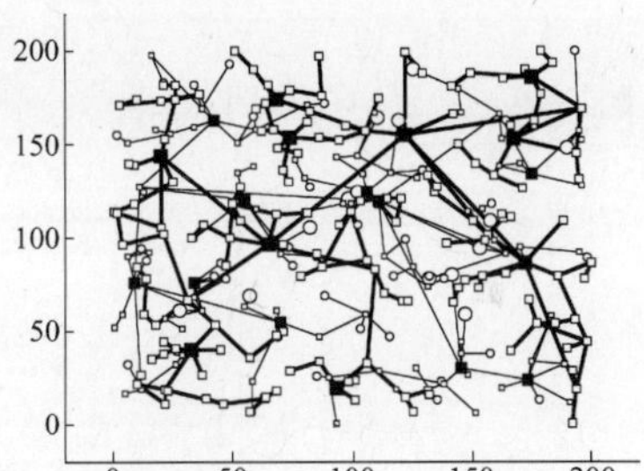

(b) Attacked buses in the node degree dynamic attack mode

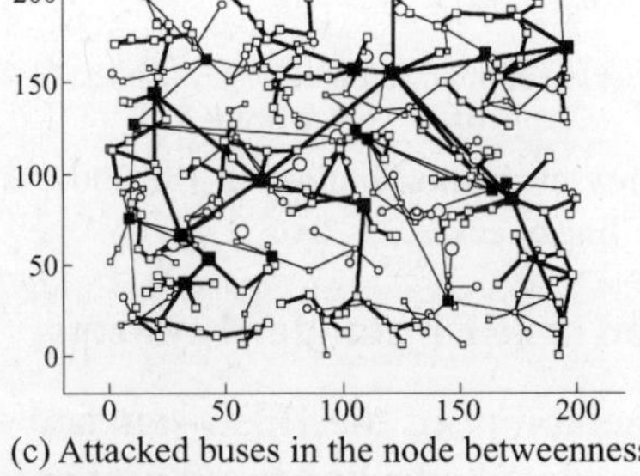

(c) Attacked buses in the node betweenness attack mode

Figure 5.10 Node selection in the 300-bus small-world power grid in different attack modes

(2) Attacks on lines

From Fig. 5.11, it is clear that small-world power grids are robust against random line attacks. The network performance indices decrease no more than 10% after 20 lines are randomly disconnected consecutively while the same indices drop abruptly under deliberate attacks in the line betweenness mode, for which Q decreases nearly to 0 and the vulnerability in small-world power grids is demonstrated clearly. In Fig. 5.12, we show the lines that are attacked in the line betweenness mode. The thick dashed lines indicate 500 kV transmission lines and the thin dashed lines indicate 220 kV lines. Obviously, these lines are all important tie lines and long-range transmission lines, which are key lines in the power grid. Their failure can easily lead to cascading failures in the small-world power grid, which can further develop into blackouts.

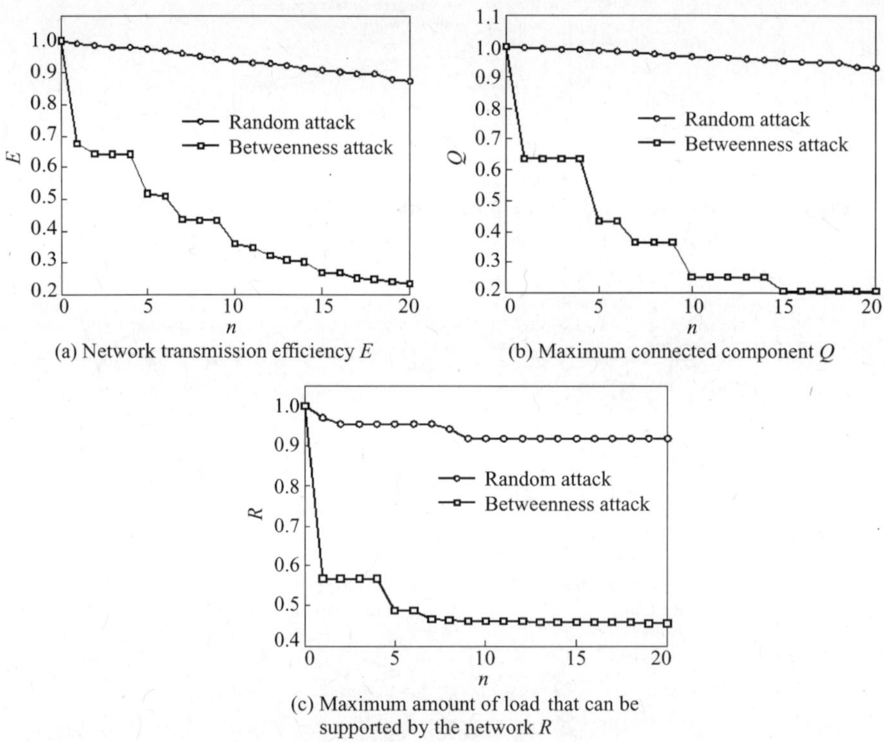

Figure 5.11 Vulnerability evaluation indices of the 300-bus small-world power grid under line attacks

2) Vulnerability evaluation in IEEE standard systems

Now we evaluate the vulnerability of the IEEE standard systems with different topological features. We pay special attention to the network transmission efficiency index E and the maximum connection component index Q. It is shown in [11] that the IEEE 118-bus system has the typical small-world characteristics, and the

IEEE 30-bus system is not a small-world power grid, as shown in Table 5.2. In this section, we carry out the structural vulnerability assessment of the two standard systems.

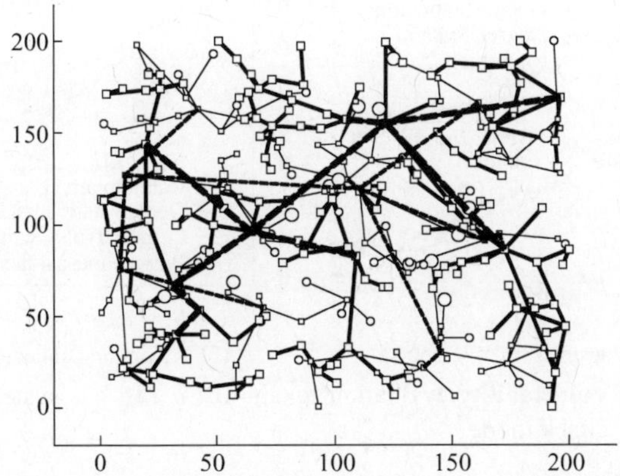

Figure 5.12 Lines that are attacked in the line betweenness mode

Table 5.2 Comparisons of IEEE standard power systems[11]

N	$\bar{d}$	L	C	L_r	C_r
IEEE 5-bus system	2	1.6	0.73	2.32	0.4
IEEE 9-bus system	2	2.33	0	3.17	0.22
IEEE 10-bus system	1.8	2.71	0	3.92	0.18
IEEE 11-bus system	2	2.87	0	3.46	0.18
IEEE 13-bus system	2	3.51	0	3.7	0.15
IEEE 14-bus system	2.86	2.37	0.37	2.51	0.20
IEEE 30-bus system	**2.74**	**3.31**	**0.23**	**4.27**	**0.09**
IEEE 39-bus system	2.36	4.7	0.04	4.27	0.06
IEEE 43-bus system	1.95	5.75	0	5.63	0.05
IEEE 57-bus system	2.74	4.95	0.12	4.01	0.05
IEEE 118-bus system	**3.02**	**6.33**	**0.16**	**4.32**	**0.03**
IEEE 145-bus system	5.82	4.39	0.54	2.83	0.04
IEEE 162-bus system	3.46	5.66	0.10	4.10	0.02
IEEE 300-bus system	2.73	9.94	0.09	6.68	0.01
WSCC 4941-bus system	2.67	18.7	0.08	8.66	0.0005

The vulnerability evaluation results under the bus attack are shown in Fig. 5.13 and Fig. 5.14, and those under the line attack are shown in Fig. 5.15 and Fig. 5.16.

Comparing Fig. 5.13 – Fig. 5.16, one can see that the main differences of structural vulnerability between small-world power grids and the other power grids lie in the following aspect. The small-world power grids perform differently under random attacks and deliberate attacks. They are more robust to the former but more vulnerable to the latter while in comparison the other power grids do not differ too much under these two classes of attacks.

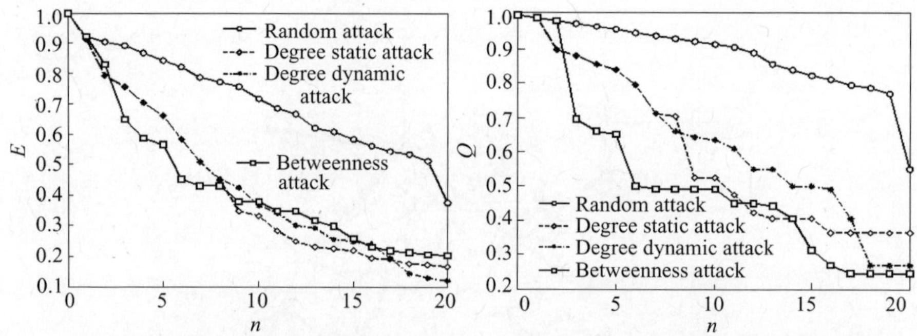

(a) Network transmission efficiency E (b) Maximum connected component Q

Figure 5.13 Vulnerability evaluation for the IEEE 118-bus system in the bus attack mode

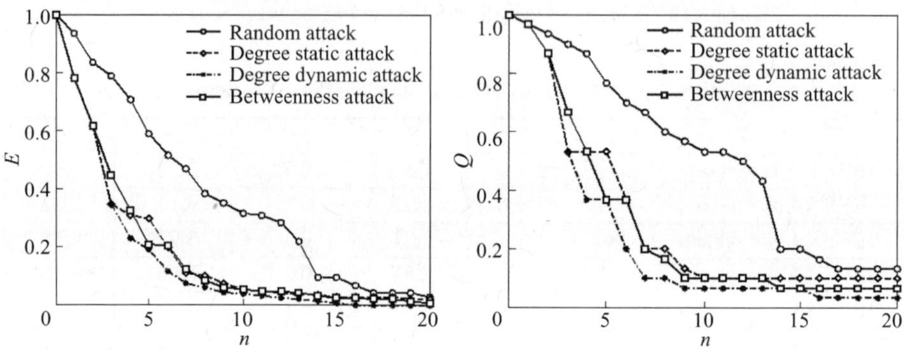

(a) Network transmission efficiency E (b) Maximum connected component index Q

Figure 5.14 Vulnerability evaluation for the IEEE 30-bus system in the bus attack mode

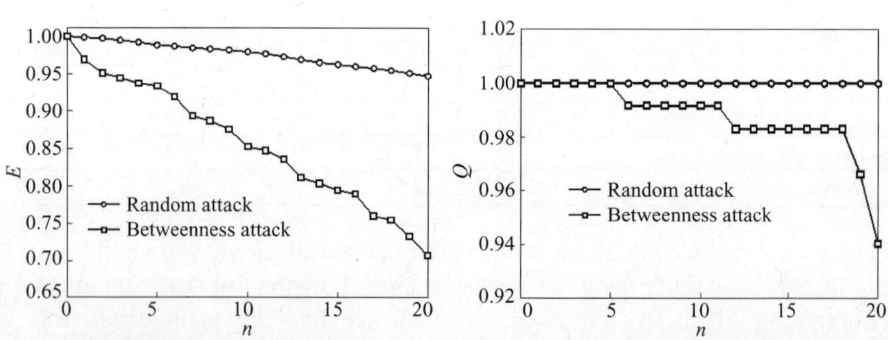

(a) Network transmission efficiency E (b) Maximum connected component index Q

Figure 5.15 Vulnerability evaluation for the IEEE 118-bus system in the line attack mode

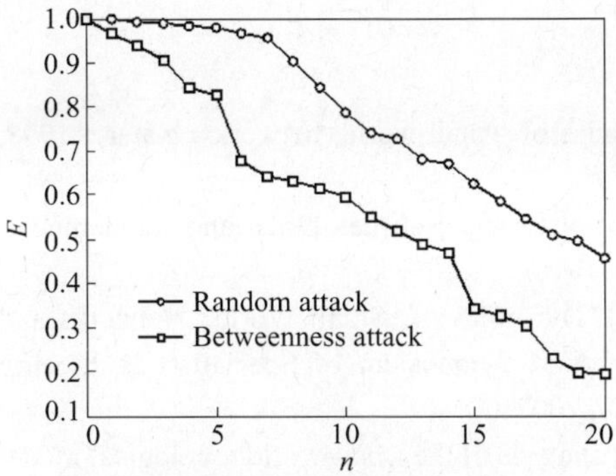

(a) Network transmission efficiency E

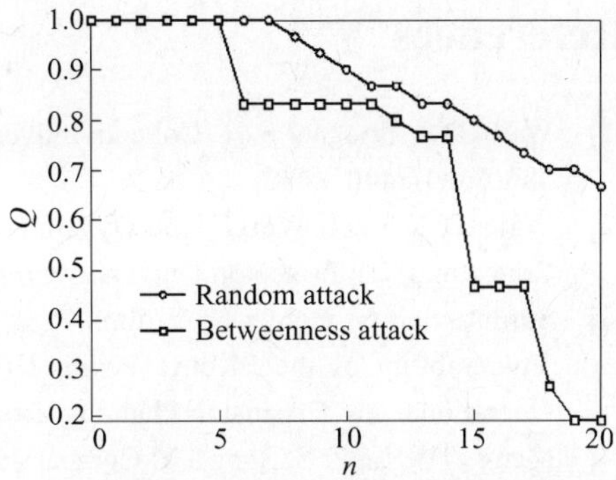

(b) Maximum connected component index Q

Figure 5.16 Vulnerability evaluation for the IEEE 30-bus system in the line attack mode

5.3 Summary and Discussion

(1) The study in this Chapter shows that in the long-term stable operation of power grids, the small-world structures are superior for power flows and network synchronization and thus small-world power grids are robust against changes in load demands and operation modes. This can be explained by the fact that small-world power grids have shorter characteristic paths and closer generator-load pairs. From the engineering point of view, small-world power grids can deliver electric energy more effectively and have better power flow performances.

(2) Generally speaking, since the small-world power grids have small characteristic path lengths and large clustering coefficients, they are prone to cascading failures. The clustering coefficient corresponds to the width that failures may spread out and the average path length indicates the depth that failures may propagate to lower hierarchies. The small-world networks have both considerable width and depth, and so the faults' spreading speed and its effective range in small-world power grids are far greater than those of regulate networks and random networks. Small-world power grids always have long-range tie lines and key buses, such as the power plants with lots of outgoing lines and the tie lines connecting different regions. When faults take place on these lines and at these power plants, cascading failures usually appear and can further develop into large-scale blackouts. In this respect, small-world power grids have intrinsic structural vulnerability.

(3) For the external factors, the development in regional economy and the increase in load demand are the direct driving forces for the growth in power grids. For internal factors, the conflict between power flow and synchronization performances and the structural vulnerability is the driving force that pushes the small-world power grid further into its evolution. This is exactly the mechanism for the growth and development of small-world power grids.

References

[1] Watts D J, Strogatz S H. Collective dynamics of 'small-world' networks. Nature, 1998, 393(6684): 440 – 442

[2] Watts D J. Small World—the Dynamics of Networks Between Order and Randomness. Princeton (NJ): Princeton University Press, 1998

[3] Surdutovich G, Cortez C, Vitilina R, et al. Dynamics of "Small World" Networks and Vulnerability of the Electric Power Grid. VIII Symposium of Specialists in Electric Operational and Expansion Planning, Brazil, 2002

[4] Meng Z W, Lu Z X, Song J Y. Comparison analysis of the small-world topological model of Chinese and American power grids. Automation of electric power systems, 2004, 28(15): 21 – 29 (in Chinese)

[5] Newman M E J. The structure and function of complex networks. SIAM Review, 2003, 45: 167 – 256

[6] Wang X F, Song Y H, Irving M. Modern Power System Analysis. New York: Springer, 2008

[7] Guo L, Xu X M. Complex Networks. Shanghai: Shanghai Scientific & Technological Education Publishing House, 2006

[8] Albert R, Jeong H, Barabasi A L. Error and attack tolerance in complex networks. Nature, 2000, 406: 378 – 382

[9] Freeman L C. A set of measures of centrality based upon betweenness. Sociometry, 1997, 40: 35 – 41

[10] Holme P, Kim B J, Yoon C N, et al. Attack vulnerability of complex networks. Physical Review E, 2002, 65: 056109

[11] Holme P. Edge overload breakdown in evolving networks. Physical Review E, 2002, 66(2): 036119

Chapter 6 Decomposition and Coordination of Static Power Grids

This chapter proposes a new algorithm to realize the control-oriented reactive power partitioning and key-bus selection. The first step is to decompose a network into several communities in order to decouple partitioning from control. By using the centrality degree index, the second step is to compare nodes' capabilities to affect the whole system's operation. Such a comparison provides information about how effective it will be to control the system at each node and hence helps us to simplify the control strategy to be running the key buses at their target operation modes. The effectiveness of the proposed algorithm is validated through simulations at the end of this chapter.

The control of reactive power and voltage is an important component of the control for a power system, which is especially critical to accomplish the system's stable and economic operation[1–6]. Nowadays, multi-layer voltage control usually includes three layers, the structure of which is shown in Fig. 6.1.

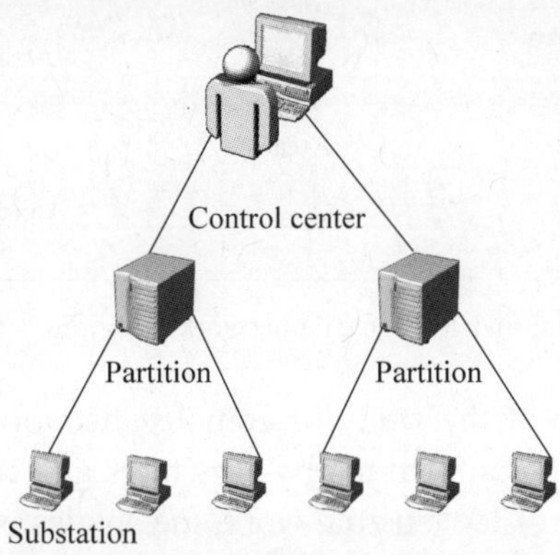

Figure 6.1 Three-layer structure for voltage control

The first layer is the control center whose control target is the efficiency and safety of the whole grid. It utilizes the optimal power flow or the computation of eigenvalues of the power flow matrix to obtain the desired power flow operation point, which is then handed down to the lower two layers to realize. As the scales of networks grow, the local balancing feature of reactive power has to be exploited to control the reactive power flows with respect to different local regions[2,7]. So the control in the second layer is implemented in regional networks where all the buses are forced to satisfy the requirements from the first layer. Since the number of controllable equipments within each region is usually less than that of the buses that need to be controlled, most of the time one cannot fully realize the goal delineated by the first layer, and consequently we need to identify key buses in each region. We adjust the voltages of the generator buses to control the voltages of the key buses and further influence the voltages of all the buses in the region. In this way, one can approximately realize the control objective handed down from the first layer. The third layer is the place where the internal automatic control devices in generators or transformers implement the voltage settings determined by the second layer.

The spatial distributions of a network's regions and key buses are shown in Fig. 6.2, where "△" denotes the key buses in the corresponding region. Following the two key steps, which are the partitioning of reactive power networks and the identification of key buses, the three-layer control can realize the decomposition and coordination control of complex static power grids. In this chapter, we analyze and study in detail these two key steps.

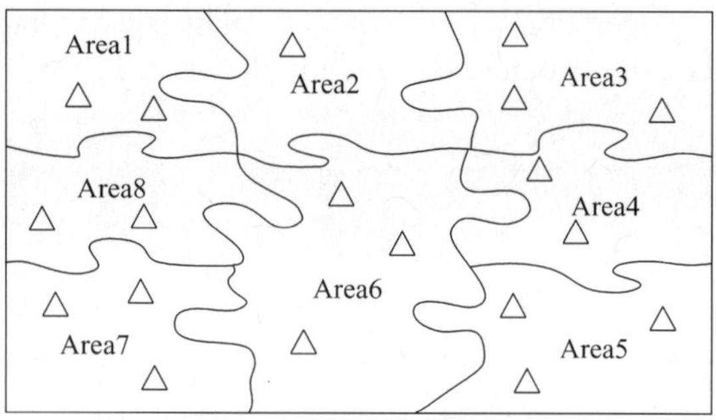

Figure 6.2 Spatial distribution of a network's regions with their key buses

From the perspective of the study of complex networks, these two key steps are related to the research on community structures and center nodes in complex network theory. In most of the existing work, the studies on community structures and the index for the centrality degree are independent, and these studies are usually based upon the network's topology or information flow to analyze the network's static characteristics and have not been applied into the control of complex networks.

The main approach in this chapter is as follows. First, we exploit the community structure characteristics of the reactive power flow to divide the network into several relatively simple regions, and such partitioning algorithms ensure that the regions are decoupled from one another to reduce the control complexity. Then we use the centrality degree index to determine key buses that are critical for the control objective. Such key buses are well controlled and monitored to achieve the control objective over the whole network. We want to emphasize here that if the key buses are selected at a scale of the entire network, then their distribution may not be consistent with the distribution of the geographic regions and hence it is difficult to guarantee the effects of monitoring and controlling each region independently. For this reason, we select key buses in each region separately. Following the above approach, we propose a new control-oriented reactive power partitioning and key bus selection algorithm in this chapter.

6.1 Network Partitioning, Pilot Buses and Community Structures

6.1.1 Network Partitioning for Reactive Power/Voltage Control

Generally speaking, network partitioning is to decouple large-scale power grid geographically to make the information within each partitioned region independent from that of the other regions. And consequently, the dimension of the searching space for control strategies is greatly reduced and the control for reactive power/voltage becomes more robust[2]. From the view of complexity science, network partitioning is the optimization and identification of the community structures of complex power systems, and is the foundation for the optimal reactive power operation.

Before going into the details for reactive power partitioning, one first needs to answer the following three questions.

(1) Is the network reactive power partitioning necessary?

The answer is "Yes". In the hierarchical structure for multi-layer voltage control, the network partitioning can decompose the global objective in the first layer. After decoupling the information and control among different regions, the system reliability is improved and the computation burden is reduced. So network partitioning is the bridge between the first and the second layers, and is key for realizing the global reactive power/voltage control.

(2) Is the network partitioning feasible?

It is desirable for all large-scale control systems to decompose the system-level control objective, but the decomposition can only be achieved under certain conditions. In reactive power/voltage control, the fact that reactive power is balanced locally is the physical basis for the partitioning. However, for an arbitrary power grid, its distribution of reactive power flows may not be suitable for network partitioning and control. So we evaluate the suitability for decoupling the reactive power flows in a power grid by using the weighted modularity index Q. The theoretical justification for this is that statistical physicsists have discovered that when the modularity index Q for an optimal partitioning scheme in a weighted network is greater than 0.3, the network exhibits a clear separable structure.

In the network partitioning problem for reactive power/voltage control, we usually use the transmitted reactive power as the weights for the corresponding lines. When evaluating different partitioning schemes for the IEEE 39-bus system and the IEEE 118-bus system using the weighted modularity index Q, we find that the values of Q for the optimal partitioning schemes are both greater than 0.4. This is consistent with the intuition gained from electrical engineers' experience that reactive power flows can be decoupled. However, there are exceptions. The value of Q for different partitioning schemes in the IEEE 14-bus system is always less than 0.3, which implies that the IEEE 14-bus system is not suitable for partitioning in the reactive power/voltage control. This shows from a different angle the importance for the research on how to decouple reactive power flows.

(3) How can one partition a network?

In the existing reactive power partitioning algorithms, a weighted network is usually constructed by taking the sensitivities of voltages with respect to reactive power at all the buses as the weights for transmission lines. There are mainly two classes of such algorithms. The first is the α-nested decomposition algorithm based on graph theory[8]. It makes use of the sensitivity matrix and has the advantage of being simple and straightforward. However, it also has the disadvantage that the number of partitioned regions has to be set manually beforehand and it ignores the local balance of reactive power in each region. The second is the hierarchical clustering algorithm[2]. It takes into account the generators' control capabilities and determines automatically the number of portioned regions, but cannot guarantee the local balance of reactive power in each region, and in addition, the number of computational iterations is equal to the number of the PQ buses in the system, which can be huge in large-scale systems.

The reactive power partitioning algorithm that we propose to use takes into account both the sensitivity distribution to adjust conveniently the reactive power and voltages in each region, but also the power distribution in the present operation mode. We pay special attention to the distribution of reactive power in order to utilize adequately the characteristic of their local balance. In short, we consider jointly the sensitivity relationships among buses and the actual distribution of reactive power such that new reactive power partitioning algorithm can be constructed successfully.

6.1.2 Choosing Pilot Buses

After the static power network is partitioned into several regions using the algorithm we just discussed, we need to choose representative controllable buses as pilot buses because in each region the number of buses that need to be controlled is much smaller than the number of buses that can be controlled. The pilot buses are well monitored and controlled in order to realize the coordination of the whole system.

The principles for choosing pilot buses are as follows. We first apply some disturbances at load buses, usually in the form of random Gaussian disturbance, whose mean is zero and standard deviation is proportional to the load's reactive power before the disturbance. Then we adjust the system's reactive power supply to guarantee the voltage deviations at certain buses to be zero. If as a result we are able to minimize $I(C)$, the expected weighted squared sum of the voltage deviations of all buses, then the buses that are chosen to have zero deviations are identified to be the pilot buses. The value of $I(C)$ is then called the stabilization effect of the pilot buses. We give a more rigorous definition as follows[9]:

$$\begin{aligned} I(C) &= E\{\Delta V_L^{\mathrm{T}} Q_X \Delta V_L\} \\ &= E\{[(I - BFC)M\Delta Q_L]^{\mathrm{T}} Q_X [(I - BFC)M\Delta Q_L]\} \\ &= \mathrm{trace}\{P_L (I - BFC)^{\mathrm{T}} Q_X (I - BFC)\} \\ &= \mathrm{trace}\{P_L Q_X\} - \mathrm{trace}\{(2H_1 - H_2 H_3^{-1} H_4) H_3^{-1}\} \end{aligned} \tag{6.1}$$

where

$$\begin{aligned} H_1 &= C P_L Q_X B B^{\mathrm{T}} C^{\mathrm{T}} \\ H_2 &= C B B^{\mathrm{T}} Q_X B B^{\mathrm{T}} C^{\mathrm{T}} \\ H_3 &= C B B^{\mathrm{T}} C^{\mathrm{T}} \\ H_4 &= C P_L C^{\mathrm{T}} \end{aligned} \tag{6.2}$$

$$\begin{aligned} P_L &= M D_{LL} M^{\mathrm{T}} \\ D_{LL} &= E\{\Delta Q_L^{\mathrm{T}} \Delta Q_L\} \\ M &= S_{LL}^{-1} \\ B &= -S_{LL}^{-1} S_{LS} \end{aligned} \tag{6.3}$$

Here, trace$\{\cdot\}$ denotes the trace of a matrix, which is the sum of the diagonal elements; $E\{\cdot\}$ denotes the expected value; ΔQ_L is the change of the reactive power at the load buses; Q_X is the reactive power vector for all the load buses; S_{LL} and S_{LS} are sensitivity matricies that will be defined in more detail in the next section; C is the matrix corresponding to the choice of pilot buses and D_{LL} is the covariance matrix.

Mathematically speaking, to solve for the matrix C is an integer programming problem. Such a large-scale combinatorial problem can easily lead to the "curse of dimensionality". Up to now, there are mainly two algorithms, which are the searching algorithm and the greedy algorithm. The former can be further classified into the local searching algorithm and the global searching algorithm according to the searching space. The performances of such algorithms depend on the initial conditions and the execution of such algorithms also relies on the pre-determined number of pilot buses. The performance of the greedy algorithms is affected by the size of the data set. As the size grows, the computation time grows dramatically and the accuracy also degenerates. The advantage of the greedy algorithm is that it does not depend on the initial solutions and does not require the knowledge of the number of pilot buses in advance.

Neither of the above two algorithms considers the characteristics of the pilot buses, so they do not have a clear objective and are not suitable for on-line applications. From physical consideration, the choice of pilot buses needs to take into account two factors, namely the monitoring and control effects in the corresponding region. During a random disturbance, the pilot buses should be chosen in such a way that they reflect the status of all the load buses in the region. By looking at only the information about the pilot buses, we can infer the operation status of the entire region. When facing random disturbances, the choice of the pilot buses should guarantee that by maintaining the voltages of the pilot buses at a given level, the voltage of the whole region is kept at a desirable level.

6.1.3 Structure Information of Static Power Grids

In order to partition the network to implement the decoupled control actions, we class all the buses into two categories, namely the set of controlling buses S and the set of controlled buses L. Then for a network with $G = g(V, E, W)$, we have $V = \{S, L\}$. Suppose the total number of buses is N_{ND}. We denote $S = \{s_i\}$, $i = 1, 2 \cdots, N_{CN}$, which include all the buses that are connected to generators, phasers and other devices that can adjust reactive power flows as reactive power sources. We also denote $L = \{l_i\}$, $i = 1, 2, \cdots, N_{BCN}$, and $N_{BCN} = N_{ND} - N_{CN}$.

The linearized model of the system is

$$\begin{bmatrix} \Delta Q_S \\ \Delta Q_L \end{bmatrix} = \begin{bmatrix} S_{SS} & S_{SL} \\ S_{LS} & S_{LL} \end{bmatrix} \begin{bmatrix} \Delta V_S \\ \Delta V_L \end{bmatrix} \tag{6.4}$$

where ΔQ_S and ΔV_S are the changes of reactive powers and voltages of the generator buses respectively; ΔQ_L and ΔV_L are the changes of reactive powers and voltages of the load buses respectively; S_{SS}, S_{SL}, S_{LS} and S_{LL} are the sensitivity matrices, which are the parts in the Jacobian matrix that are related to reactive power/voltage and can be replaced by the imaginary parts of the admittance matrix.

Assume that the injected reactive powers at load buses are constant, i.e., $\Delta Q_L = 0$, then

$$\Delta V_L = -S_{LL}^{-1} S_{LS} \Delta V_S = S_V \Delta V_S \tag{6.5}$$

where the sensitivity matrix S_V reflects the control capability of generator buses with respect to the voltages of the load buses.

Assume that the voltages of the controlling buses are constant, i.e. $\Delta V_S = 0$, then

$$\Delta V_L = S_{LL}^{-1} \Delta Q_L = S_Q \Delta Q_L \tag{6.6}$$

where the sensitivity matrix S_Q reflects the influence of load fluctuations on the voltages at load buses.

In short, in a complex power system the controlling buses have the self-adjustment capabilities, and the states of the controlled buses are affected by the settings of the controlling buses. When some controlled buses are disturbed, the states of all the controlled buses are affected, i.e.

$$\begin{aligned} \Delta V_L &= S_V \Delta V_S \\ \Delta V_L &= S_Q \Delta Q_L \end{aligned} \tag{6.7}$$

where ΔV_L are the changes in the states of the controlled buses, ΔV_S are the changes in the settings of the controlling buses, and ΔQ_L are the disturbances at the controlled buses; S_V and S_Q are sensitivity matrices, representing the influences of ΔV_S and ΔQ_L upon ΔV_L.

6.2 Network Partitioning Algorithm Based on Community Structures

6.2.1 Analysis of Decoupling Reactive Power Flows

We analyze the reactive power flows in the IEEE 39-bus system and the IEEE 118-bus system using the modularity index, and find that for most of the partitioning schemes, the values of Q are greater than 0.3. This implies that the community structures of reactive power are very clear in these two systems. Compared to other natural or man-made systems that have been commonly studied using complex system theory, the distribution of reactive power flows in power systems are easier to be decoupled.

On the other hand, if we use the modularity index Q directly, it is possible to encounter negative effects. The values of the index for different partitioning schemes are too close to compare the performances; even when we are able to identify the optimal partitioning scheme, it usually returns too many divided regions to be controlled effectively. This is because the modularity index Q is designed for the networks whose community structures are not so obvious and thus whose optimal partitioning scheme is chosen among several schemes of similar mediocre performances. In other words, the evaluation is not refined and cannot compare different schemes all with satisfying performances. To better serve for the need in the reactive power partitioning in power systems, we make the following improvements. (1) We pay more attention to the lines between different communities; and (2) we modify the weights of inner attraction to distinguish them from those for decoupling. These two improvements decrease the weights for the lines between communities, thus reduce the necessity to decouple the network into more regions and lead to smaller number of the divided regions. The improved index is

$$Q_n = \sum_r \left[\frac{e_{rr}}{2m} - \frac{1}{2}\left(\frac{a'_r}{2m} \right)^2 \right], \quad 0 \leqslant Q_n \leqslant 0.5 \tag{6.8}$$

where a'_r is the degree of the rth community, which equals two times the sum of the number of the lines within the rth community and the number of the lines interconnecting the rth community to other communities.

Using the new index, if the whole system is taken as a single region, then the value of its index Q_n is 0.25. So if the values of Q_n for most of the regions are greater than 0.25, then we say the reactive power flow in the system can be decoupled.

Now we study the network partitioning problem using both the division algorithm and the integration algorithm, and the main idea is shown in Table 6.1. First, we use the division algorithm to tentatively divide the reactive power network into several regions. In this process we use the sensitivity matrix to describe the control capabilities and reduce the number of iterations. Then we use the weighted modularity index Q_n to guide the integration of different areas and the choice of the optimal number of divided regions. In this process, it is ensured that the reactive powers are balanced locally in each region.

Table 6.1 Main ideas of the network partitioning algorithms

	Decomposition	Integration
Objective	Ensure the control capability in each region	Ensure the local balance of reactive power in each region
Utilized information	Sensitivity matrix	Weighted modularity index

The flow chart is shown in Fig. 6.3.

6.2.2 Partitioning Algorithm Based on Sensitivity Matrix

The aim for dividing the network is to ensure that there is at least one generator

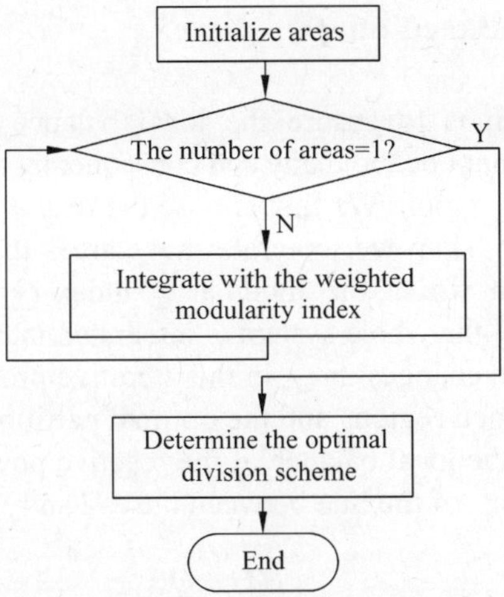

Figure 6.3 Flow chart of the proposed network partitioning algorithm

in each region, and the generators in each region are capable to control the states of the load buses. We first compute the sensitivity matrix for the relationship between the voltages at the load buses and those at the generator buses. Then we obtain the initial partition according to the values of the sensitivities. After the division, the number of initial regions is equal to the number of generators, and each load bus belongs to one and only one region.

To be more specific, we construct N_{CN} initial regions, where the jth region Area_j^0 contains only the jth generator bus s_j. Then we further divide the regions according to the control capabilities of the generators and the controlled buses are assigned to different initial regions. If the voltage at load bus l_i has the highest sensitivity with respect to the voltage at the jth generator bus, then l_i is assigned to the jth initial region Area_j^0. So we have guaranteed that each load bus belongs to one and only one initial region, and the number of initial regions is equal to the number of the generator buses N_{CN}. We further list the main steps below.

Step 1 Compute the sensitivity matrix S_V

$$S_V = -S_{LL}^{-1} S_{LS} \tag{6.9}$$

Step 2 Obtain the initial regions using S_V

$$\begin{aligned}
& A_j^0 = \{s_j\}, \qquad j = 1, 2, \cdots, N_{CN} \\
& \text{For } \ i = 1 : N_{BCN} \\
& \qquad \text{If } \ S_{V ij} == \max_{j=1,2,\cdots,N_{CN}} (S_{V ij}) \\
& \qquad\qquad A_j^0 = l_i \cup A_j^0 \\
& \qquad \text{End} \\
& \text{End}
\end{aligned} \tag{6.10}$$

6.2.3 Integration Based on Q_n

The aim of integration is to ensure the local balance of reactive power. In addition, the regions cannot be too many and consequently there are usually several generator buses in each region. We first choose the reactive power transmitted by each line as its weight. Then we integrate those areas that share common lines along the directions in which the modularity index Q_n increases the most or decreases the least, until the whole system is integrated into a single region. Based on this, we examine the changes in Q_n in this iterative process and determine the optimal number of divided regions and the optimal partitioning schemes.

In order to evaluate the local balance of the reactive power within each region, we choose the weight w_{ij} of the line between buses i and j to be

$$w_{ij} = \frac{|Q_f + Q_t|}{2S_r} \tag{6.11}$$

where Q_f is the reactive power at the sending end of the line, Q_t is the reactive power at the receiving end, and S_r is the transmission capacity of the line.

Using the improved modularity index Q_n, we evaluate the weighted network to check how balanced each region is in terms of reactive powers. The main steps are as follows.

First, we compute the weighted modularity index for the initial regions, which is denoted by Q_n^0. Then we compute the modularity index Q_{nTxy} for the integration of any arbitrary two connected areas x and y. If the maximal weighted modularity index $\max(Q_{nT})$ is achieved after the integration of areas j and k, then we combine these two areas permanently and set $Q_n^m = \max(Q_{nT})$, where m is the number of iterations. We repeat this integration process until the whole system is combined into a single region. If the modularity index Q_n^t after the tth iteration is the maximal $\max\limits_{m=0,1,2,\cdots,N_{CN}-1} Q_n^m$ in the whole iteration process, we then conclude that the corresponding partition scheme is optimal.

In the general integration algorithms, the regions are combined according to the distances between different regions[2], and the distances can be chosen from the minimum distance, the maximum distance, the average distance, and the central distance. Such choices of distances focus on the closeness of the buses' geographical locations and the control capabilities of the generators. Here we have focused on the local balance of reactive powers. Given the initial partitioning, we integrate connected regions along the directions in which the modularity index Q_n increases the most or decreases the least. We list the steps in detail as follows.

Set the iteration number to be $m = 0$, the number of regions $N_{\text{AREA}} = N_{CN}$, and compute the initial modularity index, which is denoted by Q_n^0.

Step 1 Construct the line weighted matrix W

$$w_{ij} = \frac{|Q_f + Q_t|}{2S_r} \tag{6.12}$$

Step 2 Compute the weighted modularity index Q_{nT}

$$\begin{aligned}
&\text{If area } x \text{ is not connected to area } y \\
&\quad Q_{nT\,xy} = 0 \\
&\text{Else} \\
&\quad Q_{nT\,xy} = \sum_r \left[\frac{e_{rr}}{2m} - \frac{1}{2}\left(\frac{a'_r}{2m}\right)^2 \right] \\
&\text{End}
\end{aligned} \tag{6.13}$$

where $x = 1, 2, \cdots, N_{\text{AREA}} - 1$; $y = x+1, x+2, \cdots, N_{\text{AREA}}$.

Step 3 Combine the neighboring two areas

$$\begin{aligned}
&\text{If } \; Q_{nT\,jk} == \max(Q_{nT}) \\
&\quad A_j^{m+1} = A_j^m \cup A_k^m; \quad A_k^{m+1} = \phi \\
&\text{End}
\end{aligned} \tag{6.14}$$

Step 4 Update the variables

$$\begin{cases} N_{\text{AREA}} = N_{\text{AREA}} - 1 \\ m = m + 1 \\ Q_n^m = \max(Q_{nT}) \end{cases} \tag{6.15}$$

Step 5 Return to Step2 until $N_{\text{AREA}} = 1$.

Step 6 Determine the optimal partitioning scheme.

If $Q_n^t = \max\limits_{m=0,1,\cdots,N_{GEN}-1} Q_n^m$ for some t in the integration process, then the optimal partitioning scheme is that after t times of integration.

Hence, from the two procedures of dividing and integrating, we have taken into consideration separately the two requirements for reactive power network partitioning regarding the control capabilities for generators and the local balance of reactive powers. We use the sensitivity matrix to describe the control capabilities of generators with respect to the load buses and divide the network into N_{CN} initial regions. Then we use the modularity index Q_n to ensure the local balance of reactive powers in each region, combine neighboring regions, and determine the optimal number of divided regions and the optimal partitioning scheme. This algorithm takes care of the two characteristics of reactive power network partitioning by utilizing the sensitivity matrix and the modularity index respectively, which preserves the clear physical intuition on the one hand, and guarantees the algorithm's feasibility and simplicity on the other.

6.3 Choosing Pilot Buses Using the Control Centrality Index

To implement the decoupled control in each region, we choose pilot buses for each region seperately. Suppose the optimal partitioning scheme has divided the system into N_{AREA} regions. Denote the sets of buses, controlled buses and controlling buses in the ith region by V_i, L_i and S_i respectively, where $V_i = L_i \cup S_i$. Suppose in the ith region there are N_{BCNi} controlled buses, then the enumeration method needs to compute for N_{ai} times the stabilization effects $I_i(C)$ to select the pilot buses in the ith region, where N_{ai} is determined by

$$N_{ai} = \sum_{i=1}^{N_{BCNi}} C^i_{N_{BCNi}} = 2^{N_{BCNi}} - 1 \tag{6.16}$$

To reduce the computational burden, we propose a new algorithm for choosing pilot buses. We first choose candidate pilot buses (CCNs) from those controlled buses (BCNs) in each region using the control centrality index (CCI). Then we enumerate all the possible combinations of the candidate pilot buses, and determine the set of pilot buses (CTNs) according to the stabilization effect of the candidate pilot buses. In the end, we compute the stabilization effect of the pilot buses for the overall system. Figure 6.4 shows the relationship among CNs, BCNs, CCNs, CTNs, and here CNs denote controlling buses.

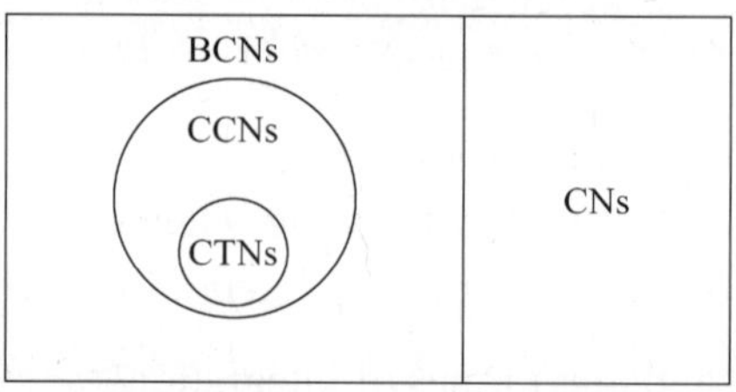

Figure 6.4 The relationship among CNs, BCNs, CCNs and CTNs

Using this algorithm, we only need to compute N_{bi} times the stabilization effect $I_i(C)$ for the ith region before the pilot buses are selected for the region. Here, N_{bi} is computed by

$$N_{bi} = \sum_{i=1}^{N_{CCNi}} C^i_{N_{CCNi}} = 2^{N_{CCNi}} - 1 \tag{6.17}$$

where N_{CCNi} is the number of candidate pilot buses in the ith region, which is much smaller than the number of the controlled buses in the region.

The flowchart for choosing pilot buses is shown in Fig. 6.5.

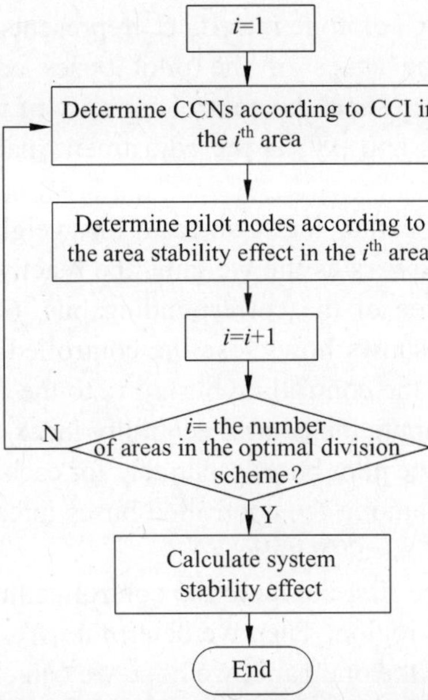

Figure 6.5 Flowchart for the new algorithm of choosing pilot buses

6.3.1 Choosing Candidate Pilot Buses Using the Control Centrality Index

The pilot buses should have the following properties during the control process. Once the information about these buses is collected, the whole system's status can be deduced. And once the states of these buses are kept at the desired level, the whole system is guaranteed to operate in desirable states. We want to point out that for the pilot buses (a) the buses themselves deserve great attention; (b) they are the "center" of the controlled buses and reflect the status of all the controlled buses; and (c) they are also the "center" of the controlling buses and can propagate the control effects.

In view of the definitions for center nodes in complex network theory, we propose the following control centrality indices

$$
\begin{aligned}
I_{ccp} &= \prod_{i=1}^{3} C_i \\
C_1 &= Q_p \\
C_2 &= \sum_{l \in L_i} \left\| S_{Qpl} \times \Delta Q_l \right\|^2 \\
C_3 &= \sum_{s \in S_i} \left\| S_{Vls} \times \Delta V_s \right\|^2
\end{aligned}
\tag{6.18}
$$

where C_1 is the importance of node i itself, C_2 represents the extent to which the controlled bus affects the states of the pilot buses and the variation ΔQ_l is proportional to Q_p, C_3 represents the control capacity of the controlling bus with respect to the pilot buses and ΔV_s is the adjustment margin for the controlling buses.

In fact, C_1 is a degree center index since in the weighted networks that take the transferred reactive powers as the weights, the reactive power load Q_p at the load bus equals the degree of the corresponding bus. C_2 and C_3 are closeness center indices. Here, C_2 shows how close the controlled buses are to the center and C_3 shows how close the controlling buses are to the center.

In this chapter, we compute the control centrality index for the controlled buses and the number of candidate pilot buses separately for each region. We then choose the candidate pilot buses among the controlled buses in each region according to the values of the control centrality indices.

To be more specific, we first compute the control centrality indices for all the controlled buses in the ith region. Then we determine N_{CCNi}, the number of pilot buses in the ith region. On the one hand, we hope we can choose all the controlled buses as candidate buses to identify the optimal combination; on the other, we are constrained by the computational capacity and the allowable time for screening the candidate buses. So we instead choose N_{CCNi} to be the smaller one of the number of controlled buses N_{BCNi} and the constant N_{CMAX} that is jointly determined by the system's scale, the computational capacity, the allowable time and the precision needed. In the end, we order from the largest to the smallest the control centrality indices of all the controlled buses in the ith region and denote the jth controlled bus by p_{ij}. The top N_{CCNi} controlled buses are chosen to be the candidate pilot buses for the ith region.

The process is summarized as follows.

Step 1 Compute the control centrality index for each controlled bus

$$I_{ccp} = Q_p \times \sum_{l \in L_i} \left\| S_{Qpl} \times \Delta Q_l \right\|^2 \times \sum_{s \in S_i} \left\| S_{Vls} \times \Delta V_s \right\|^2 \tag{6.19}$$

Step 2 Determine the number of the candidate pilot buses

$$N_{CCNi} = \min(N_{\text{CMAX}}, N_{BCNi}) \tag{6.20}$$

where $i = 1, 2, \cdots, N_{\text{AREA}}$.

Step 3 Order the control centrality indices for each region.

Step 4 Determine the candidate pilot buses for each region

$$C_{CNi} = \bigcup_{j=1}^{N_{CCNi}} p_{ij} \tag{6.21}$$

6.3.2 Choosing Pilot Buses for Each Region

To evaluate the effectiveness of the chosen pilot buses, we need to compute the stabilization effect $I_i(C)$ for each region. A smaller value of $I_i(C)$ implies that the control effect is better and thus the choice of the pilot buses is more effective[9].

Suppose there are N_{CCNi} candidate pilot buses in the ith region. Then as shown in Eq. (6.17), there are N_{bi} possible combinations of candidate pilot buses. We denote the set of the controlled buses in the jth combination by $C_i(j)$. In order to determine the pilot buses among the N_{bi} combinations, we propose the following algorithm. We first check the feasibility of the combinations. For each region to be controllable, its number of controlled variables cannot be greater than the number of control variables. If in the jth combination the number of controlled buses $N_{Ci}(j)$ is greater than the number of controlling buses N_{CNi}, then the combination is not feasible. If the opposite is true, we then calculate the stabilization effect $I_i(\text{C})$ for the combination of candidate pilot buses.

We calculate the $I_i(C)$ for all the combination. The buses in the combination with the optimal stabilization effect are chosen to be the region's pilot buses. The set of the pilot buses of all the regions make up the system's set of pilot buses. We list the steps below.

Step 1 Enumerate all the combinations of controlled buses in the ith region, and record the corresponding stability effects.

$$\begin{aligned}&\text{For } j=1:N_{bi}\\&\quad \text{If } N_{Ci}(j)\leqslant N_{CNi}\\&\qquad I_{ij}(C)=E\{\Delta V_{L_i}^{\text{T}} Q_{L_i}\Delta V_{L_i}\}\\&\quad \text{End}\\&\text{End}\end{aligned}\tag{6.22}$$

Step 2 In the ith region, the controlled buses in the best combination are chosen to be the pilot buses for the region

$$\begin{aligned}&\text{If } I_{ij}(C) == \min_{t=1,2,\cdots,N_{bi}} (I_{it}(C))\\&\quad C_i = C_i(j)\\&\text{End}\end{aligned}\tag{6.23}$$

Step 3 The set of the pilot buses in all the regions is the set of the system's pilot buses

$$C_s = \cup C_i \tag{6.24}$$

where $i=1,2,\cdots,N_{\text{AREA}}$.

Generally speaking, the proposed algorithm utilizes the control centrality index to choose pilot buses for each region separately, which reduces significantly the computation cost. It also covers all the combinations of candidate pilot buses to determine the pilot buses in each region, which ensures the effectiveness of the chosen pilot buses.

6.4 Stabilization Effect of Pilot Buses

According to Eq. (6.1), we compute the effectiveness of choosing the pilot buses, which is defined to be the stabilization effect $I_{\text{sys}} = E\{\Delta V_L^{\mathrm{T}} Q_X \Delta V_L\}$ of the pilot buses with respect to the whole system. Here, I_{sys} represents the capability of pilot buses to reflect and control the system's operation, ΔV_L is the change in the states of all the controlled buses in the system, and Q_X denotes the reactive power load at all the controlled buses.

In short, the goal of choosing pilot buses is to reduce the number of the controlled buses and guarantee that the numbers of the controlling buses and the controlled buses match each other. The key is to choose those buses that can monitor and control the performance of the whole network. The existing algorithms do not take enough consideration of the physical characteristics of the network and search for controlled buses randomly. We have proposed to simplify the global controlling and monitoring objective by controlling and monitoring within individual divided regions. Hence, we have introduced the control centrality index to reduce the dimension of the search space. We want to emphasize again that the relationship between the regions and the whole network is determined by the division of reactive power distribution and the reactive power partitioning algorithm ensures that different regions are decoupled.

6.5 Simulation and Analysis

In power systems, partitioning reactive power networks and choosing pilot buses are important components in reactive power/voltage control, which are especially important for the decoupled control of divided regions and thus ensuring stable and economic operations and the voltage quality of the system.

Here we use the IEEE 14-bus system, the IEEE 39-bus system, the IEEE 118-bus system and the Shanghai power grid as examples to analyze whether a system can be decoupled. For those systems that can be decoupled, we further apply the new algorithms for partitioning reactive power networks and choosing pilot buses in order to realize the decoupled and coordinated control.

6.5.1 IEEE 14-Bus System

There are 5 generator buses and 9 load buses in the IEEE 14-bus system as shown in Fig. 6.6.

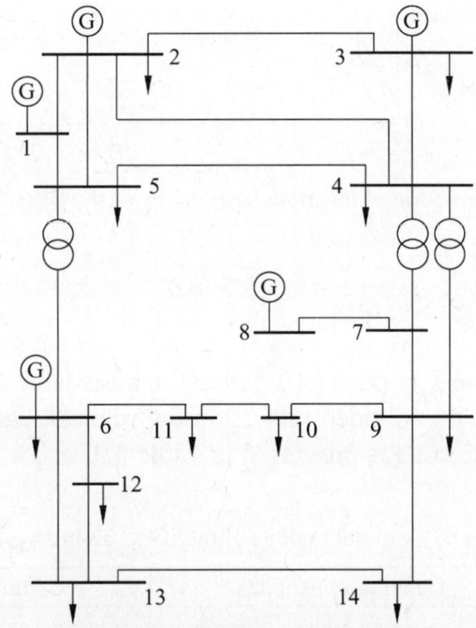

Figure 6.6 IEEE 14-bus system

The system is divided into 5 initial regions using the sensitivity matrix as shown in Table 6.2.

Table 6.2 Initial regions of the IEEE 14-bus system

Area index	Generator index	Load index
1	1	
2	2	4,5
3	3	
4	6	9–14
5	8	7

The initial regions are combined along the directions in which the modularity index increases the most or decreases the least. We show the evolution of the modularity index in Fig. 6.7, whose values in the three partitioning schemes are less than 0.25, the value for the whole system as a single region. So the distribution of the reactive power flows in the IEEE 14-bus system does not have an obvious community structure, and is thus not suitable for decoupled control.

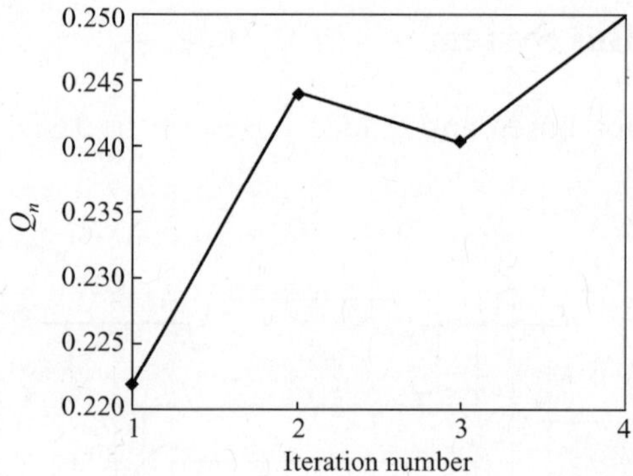

Figure 6.7 Evolution of the modularity index of the IEEE 14-bus system

6.5.2 IEEE 39-Bus System

The IEEE 39-bus system contains 10 generator buses and 29 load buses. Using the sensitivity matrix, it is divided into 10 initial regions, each of which contains a generator. The partitioning is presented in Table 6.3.

Table 6.3 Initial regions of the IEEE 39-bus system

Area index	Generator bus index	Load bus index
1	30	2,3,18
2	31	5,6,7,8
3	32	4,10,11,12,13,14
4	33	19
5	34	20
6	35	15,16,17,21,22,23,24
7	36	
8	37	25
9	38	26,27,28,29
10	39	1,9

The iterative process for combining different regions using the modularity index is shown in Fig. 6.8, where the regions are represented by the index of its generator bus. In this process, the evolution of the modularity index Q_n is shown in Fig. 6.9.

From the evolution of Q_n, we determine that the optimal partitioning scheme appears after the 5th iteration, and as shown in Fig. 6.10, at that time there are 5 regions and the value of Q_n is 0.4079. In [10], a related research on the partitioning of the IEEE 39-bus system has been performed using the slow coherency method, and different partitioning schemes have been obtained. We present

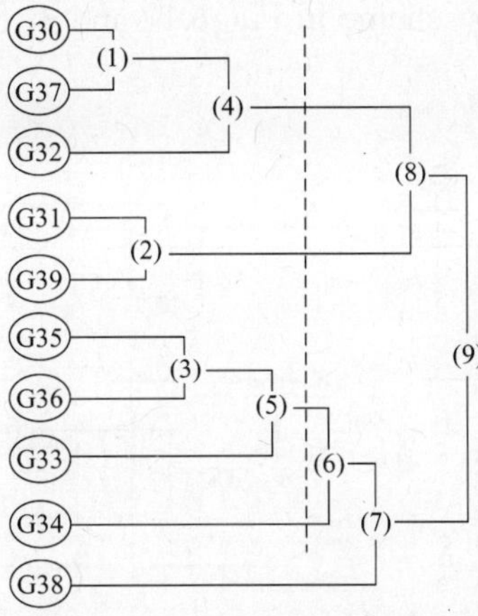

Figure 6.8 Combining regions in the IEEE 39-bus system

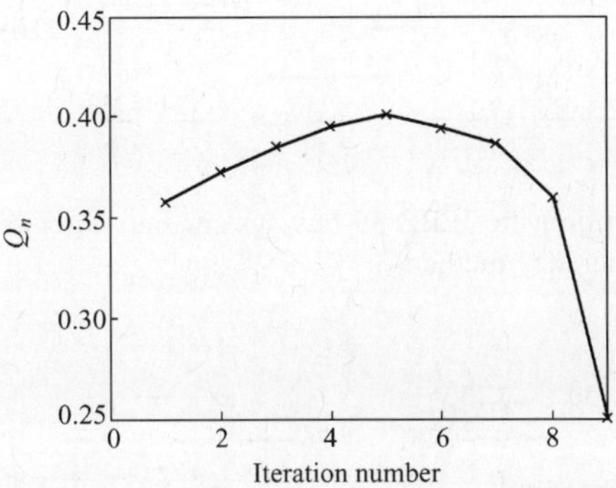

Figure 6.9 Evolution of the modularity index in the IEEE 39-bus system

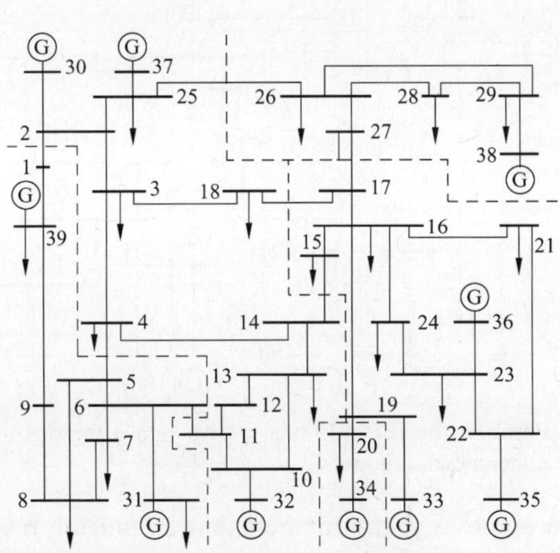

Figure 6.10 Partitioning scheme for the IEEE 39-bus system using the proposed algorithm

their 6-region partitioning scheme in Fig. 6.11 and 3-region partitioning scheme in Fig. 6.12.

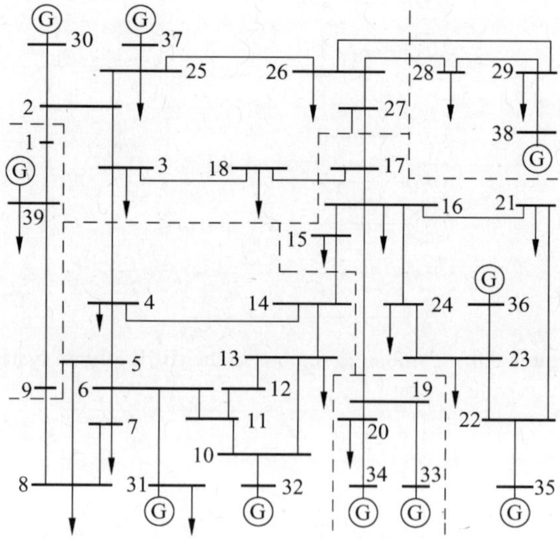

Figure 6.11 Partition the IEEE 39-bus system into 6 regions using the slow coherency method

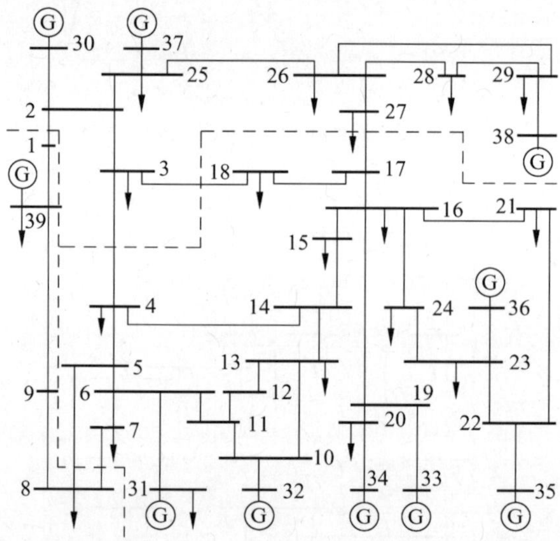

Figure 6.12 Partition the IEEE 39-bus system into 3 regions using the slow coherency method

We compare the above 3 partitioning schemes, and the results are shown in Table 6.4. Obviously, the proposed algorithm corresponds to the largest modularity index, which implies that the proposed algorithm is more effective to ensure smaller couplings between the divided regions while preserving the control effects in the regions.

Table 6.4 Comparison of the three partitioning schemes for the IEEE 39-bus system

Algorithm	Number of partitioned regions	Modularity index Q_n
Proposed agorithm	5	0.4079
Slow coherency method	6	0.3699
Slow coherency method	3	0.3220

Now we set $N_{CMAX} = 4$ and choose candidate pilot buses for each region. The sets of candidate pilot buses in the regions are {12,4,25,18}, {8,7}, {15,21}, {23,24}, {20}, and {26,27,28,29} respectively. We enumerate all the combinations of the candidate pilot buses in each region, and choose the pilot nodes as shown in Fig. 6.13, for which the stabilization effect $I(C)$ is 0.0335.

In comparison, the set of pilot buses chosen in [9] is {18,20,22,26}, with $I(C)$ being 0.2419. So the proposed algorithm is more effective for controlling the voltages at the load buses.

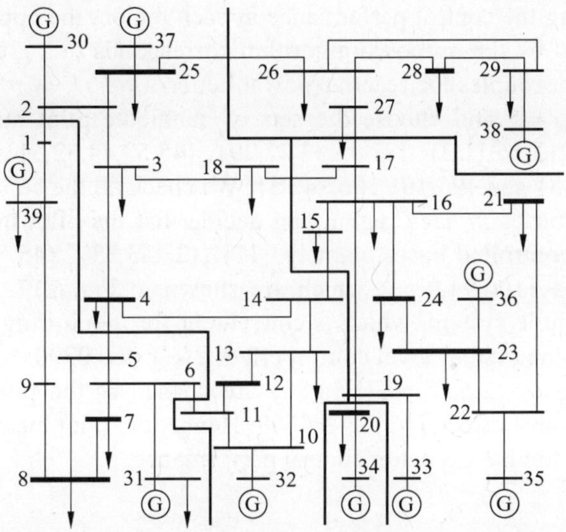

Figure 6.13 Optimal partitioning scheme and pilot buses for the IEEE 39-bus system using the proposed algorithm

6.5.3 IEEE 118-Bus System

The IEEE 118-bus system contains 54 generator buses and 64 load buses. Using the proposed algorithm, the evolution of the modularity index Q_n in the integration process is shown in Fig. 6.14. So the optimal number of regions is 6 with Q_n being 0.4137, and the corresponding partitioning result is shown in Fig. 6.15.

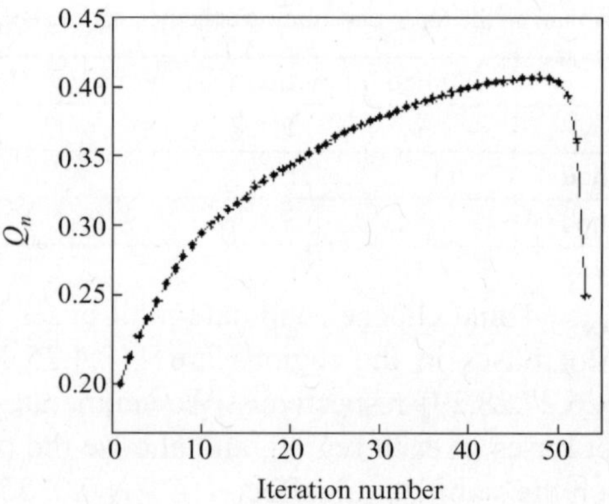

Figure 6.14 Evolution of Q_n in the IEEE 118-bus system

In comparison, the system is divided into 6 regions in [10], which is shown in Fig. 6.16, and the corresponding modularity index $Q_n = 0.3629$.

While ensuring the control performance in each region, the optimal partitioning scheme obtained by the proposed algorithm corresponds to a greater modularity index and thus decouples the reactive power better.

We set $N_{\mathrm{CMAX}} = 5$ and choose the sets of candidate pilot buses in different regions as {117,13,16,11,2}, {21,43,33,22,20}, {45,53,44,52,51}, {67,60,59,61}, {79,82,86,118,78} and {95,101,106,94,93}. We check all the combinations of the candidate pilot buses in each region and decide that the pilot buses are a combination of 15 controlled buses, namely {11}, {21,43,33}, {45,53,51}, {67,60}, {79,86,118} and {106,94,93}}, which are shown in Fig. 6.17. They distribute evenly in the whole system, which is convenient for monitoring and controlling the system, and the system-level control effect $I(C)$ is 0.0200.

In [9], the buses {2,5,28,50,71,84,88} are chosen as the pilot buses, whose system-level control effect $I(C)$ is 0.0608. Hence, the pilot buses chosen by the proposed algorithm have a better control performance.

6.5.4 Shanghai Power Grid

The Shanghai power grid is a typical heavy-load network, and the local balance of its reactive powers is critical for the system's stability. Since the network is complicated and no sophisticated theoretical analysis has been applied, the grid is now divided into 7 fixed regions according to its geographical features, which is shown in Fig. 6.18.

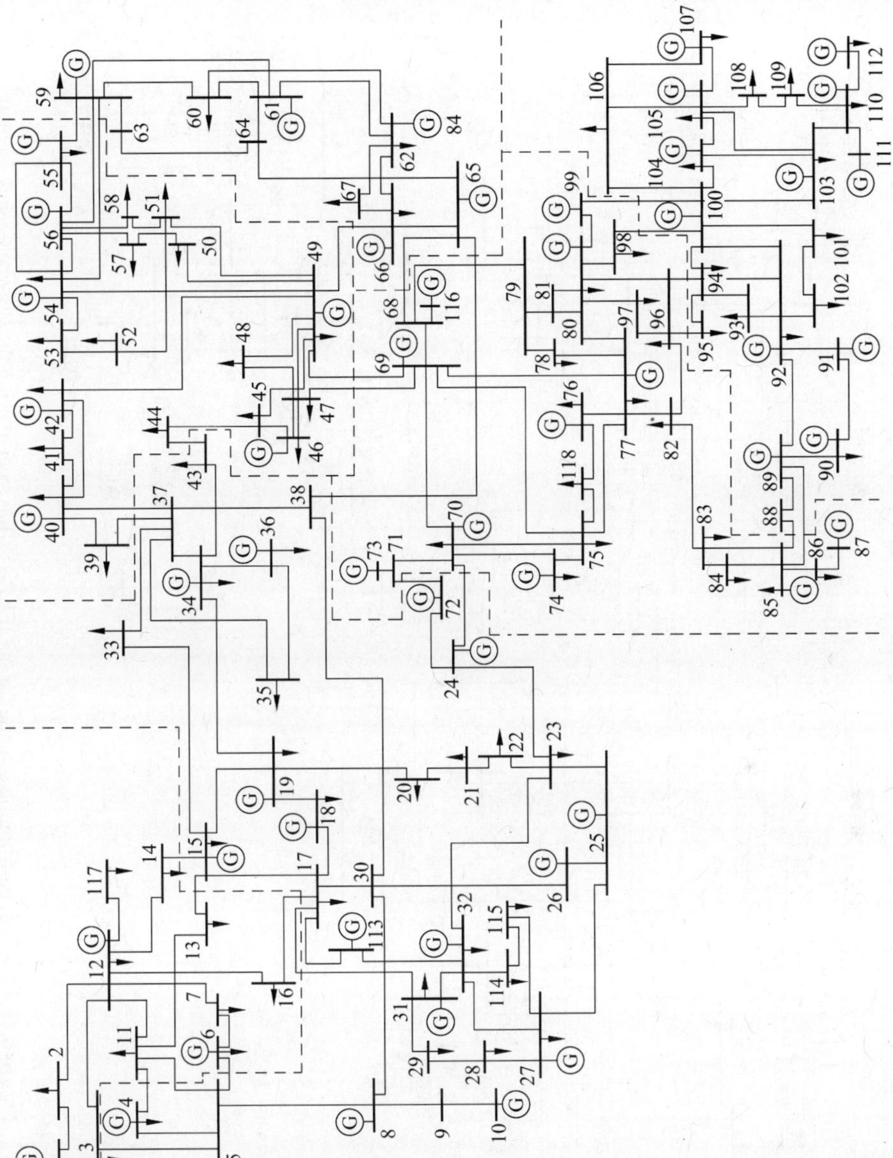

Figure 6.15 Partitioning scheme of the IEEE 118-bus system using the proposed algorithm

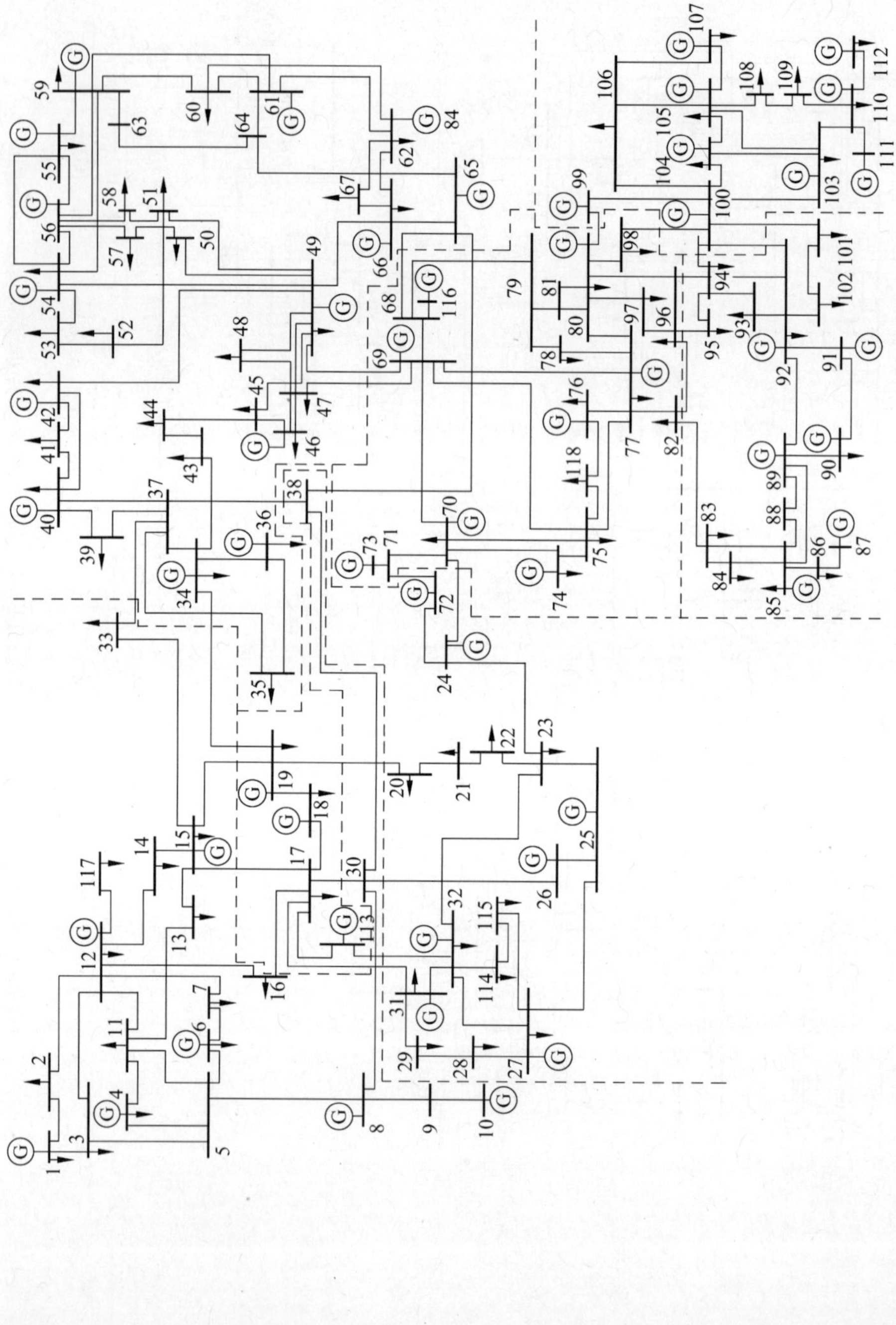

Figure 6.16 Optimal partitioning scheme for the IEEE 118-bus system using the slow coherency method

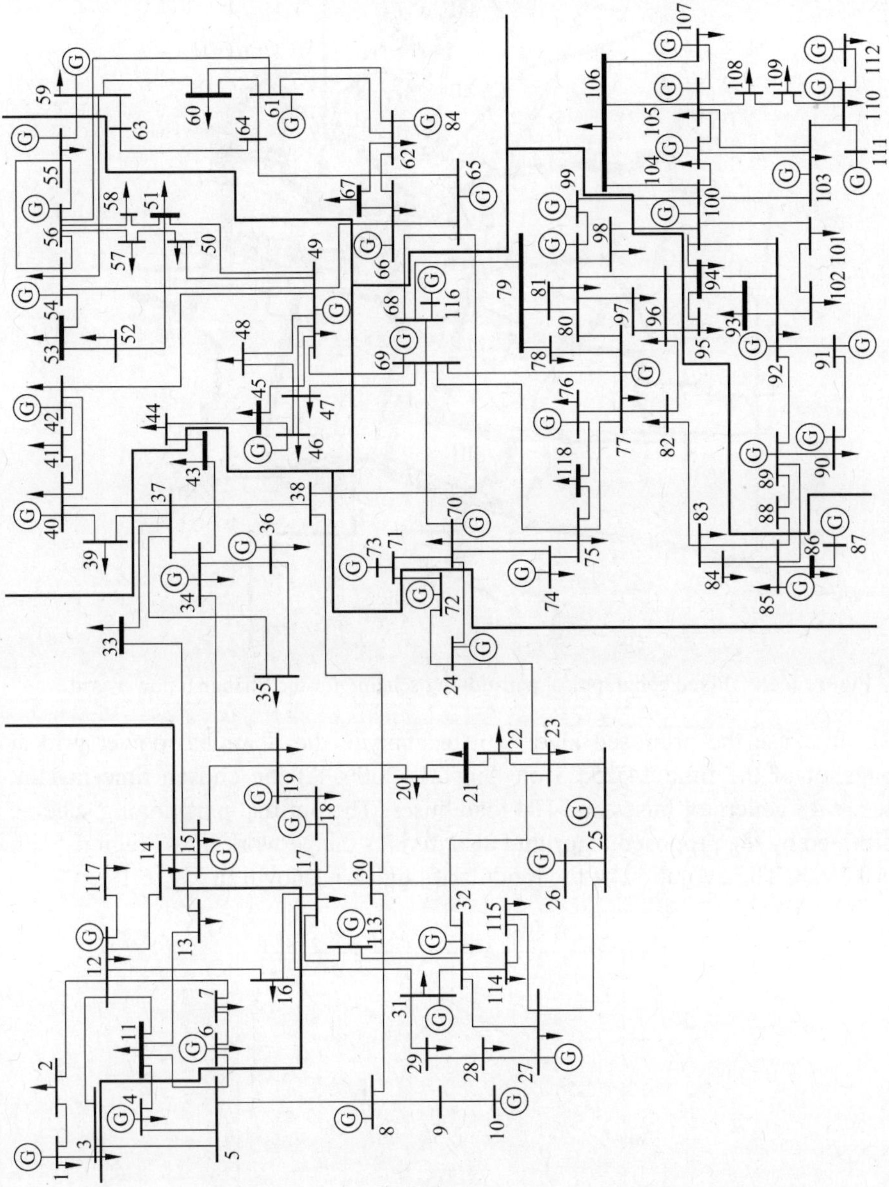

Figure 6.17 Optimal partitioning scheme pilot buses in the IEEE 118-bus system using the proposed algorithm

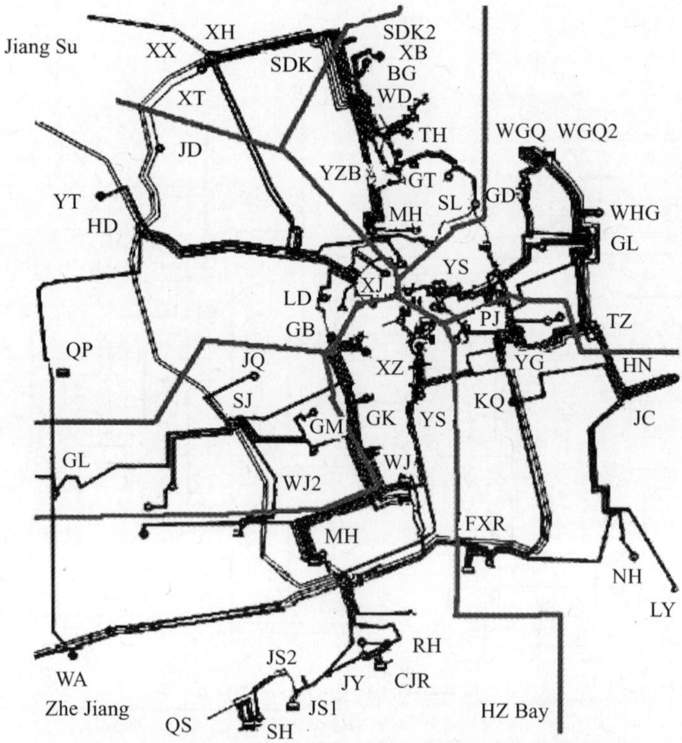

Figure 6.18 Fixed geographical partitioning scheme for the Shanghai power grid

We now use the proposed algorithm to analyze the Shanghai power grid at a snapshot of the time 14:25:23 on Aug 21, 2006. At the chosen time instant, there are 45 generator buses and 164 load buses. The optimal partitioning scheme determined by the proposed algorithm also divides the network into 7 regions with $Q_n = 0.3928$. The evolution of the modularity index is shown in Fig. 6.19.

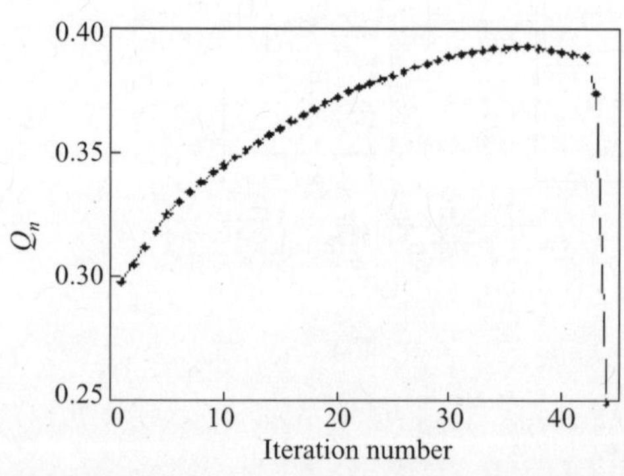

Figure 6.19 Evolution of the modularity index in the Shanghai power grid

We compare the two partitioning schemes in Table 6.5. It is clear that the proposed algorithm is more capable to balance the reactive powers locally and realize the multi-layer reactive power/voltage control.

Table 6.5 Comparison of two partitioning schemes for the Shanghai power grid

Algorithm	Number of regions	Q_n
Proposed algorithm	7	0.3928
Fixed division scheme	7	0.3754

Comparing Fig. 6.18 with Fig. 6.20, we find the following differences between the fixed geographical partitioning scheme and the division scheme obtained by the proposed algorithm.

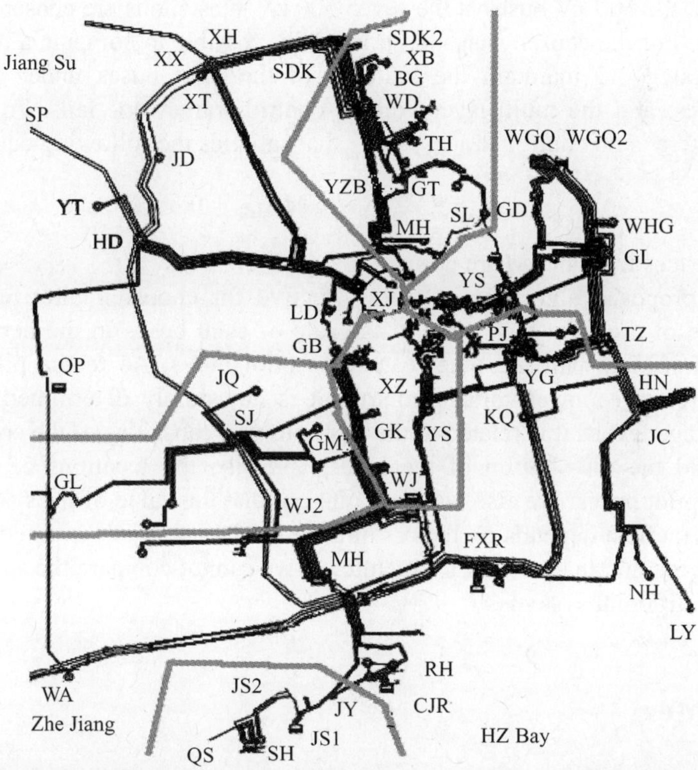

Figure 6.20 Partitioning scheme for the Shanghai power grid using the proposed algorithm

(1) The proposed algorithm combines the pair of substations HD and XH, and the pair of substations NQ and YG respectively. This is because the reactive power transmitted by the double-circuit transmission lines from the HD 500 kV bus to

the XH 500 kV bus are 168.28 Mvar and 174.62 Mvar respectively. In comparison, the reactive power transmitted by the double-circuit transmission lines from the HD 500 kV bus to the SJ 500 kV bus are 64.06 Mvar and 65.47 Mvar respectively. This implies that the two substations HD and XH are closely coupled and should be combined into one region. Similar argument holds for NQ and YG substations.

(2) The WJ generation plant, JS thermal power plant and their nearby loads are separated from the NQ and YG region. This is because these power plants have little control over the other loads in the region, and their reactive powers are mainly consumed by their nearby loads and not exchanged with the other part of the region.

We set $N_{\mathrm{CMAX}} = 5$, and choose the optimal combination of the pilot buses, which contains 9 controlled buses including three 500 kV buses (HD, SJ and NQ) and six 220 kV buses (XB, ZB, YSP, WZ, XZ and JY). The system's stabilization effect $I(C) = 6.9392$. In the present fixed partitioning scheme in the Shanghai power grid, the 500 kV buses at the seven 500 kV substations are chosen to be the pilot buses. For the chosen snapshot, it is hardly possible to compute a reasonable control strategy to maintain the voltages of the pilot buses under stochastic disturbances, and the multi-layer voltage control framework fails. To be more specific, there is no control strategy ΔV_G that satisfies the following equation

$$\Delta V_p = CB\Delta V_G + CM\Delta Q_L = 0 \tag{6.25}$$

where matrices C, B and M are defined in Eq. (6.1).

So the proposed algorithm is more effective for choosing pilot buses. The importance of the pilot buses lies in the role of such buses in the network and their relationships with the other candidate pilot buses. As to the pilot buses' effect on reactive power/voltage control, it is not simply determined by these buses' voltage levels, it is related to the positions and capacities of the controllable devices and the distribution of the loads, and also the locations of the other candidate pilot buses. We also want to point out that the value of the system-level stabilization effect depends on factors like the total amount of the reactive power in the network and the network's structure, so we cannot compare the stabilization effects for different systems.

6.6 Notes

In order to effectively control complex power grids, we have followed an approach to cooperatively optimize the global performance by executing local decoupled control strategies in different regions. We have proposed new algorithms to partition reactive power distributions and choose pilot buses. Firstly, we take steps that can be roughly summarized as "division first, integration later". We have considered the control capabilities and the system's present operating state, and also utilized the modularity index to evaluate the effectiveness for the partitioning schemes. Secondly, we have used the control centrality degree to select candidate pilot buses in each region, and by checking all the combinations of the candidate pilot buses, we have chosen in the end the pilot buses. Simulations of the IEEE 14-bus system, the IEEE 39-bus system, the IEEE 118-bus system and the Shanghai power grid have proved the effectiveness of the proposed algorithm.

It should be noted that the proposed algorithm for identifying community structures and choosing pilot buses in order to obtain coordinated control strategies can be generalized to the application in other practical complex networks. Hence, the discussion in this chapter can also be viewed as a contribution for a class of engineering systems to solve the challenging topic to design decoupled and coordinated controls of complex systems.

References

[1] Wang X F, Song Y H, Irving M. Modern Power System Analysis. New York: Springer, 2008

[2] Guo Q L, Sun H B. Power network partitioning based on clustering analysis in Mvar control space. Automation of Electric Power Systems, 2005, 29(10): 36 – 40 (in Chinese)

[3] Van C T, Vournas C. Voltage Stability of Electric Power Systems. Norwell, MA: Kluwer, 1998

[4] Wang W, Ejebe G C, Tong J, Waight J G. Preventive/corrective control for voltage stability using direct interior point method. IEEE Transactions on Power Systems, 1998, 13(3): 878 – 883

[5] Morison G K, Gao B, Kundur P. Voltage stability analysis using static and dynamic approaches. IEEE Transactions on Power Systems, 1993, 8: 1159 – 1171

[6] Gao B, Morison G K, Kundur P. Voltage stability evaluation using modal analysis. IEEE Transactions on Power Systems, 1992, 7: 1529 – 1542

[7] Lagonotte P, Sabonnadiere J C. Structural analysis of the electrical system: application to secondary voltage control in france. IEEE Transactions on Power Systems, 1989, 4(2): 478 – 486

[8] Li X. Uniform synchronous criticality of diversely random complex networks. Physica A, 2005

[9] Sancha J L, Fernandez J L, Cartes A, et al. Secondary voltage control: analysis, solutions and simulation results for the Spanish transmission system. IEEE Transactions on Power Systems, 1996, 11(2): 630 – 638

[10] Yusof S B, Rogers G J, Alden R T H. Slow coherency based network partitioning including load buses. IEEE Transactions on Power Systems, 1993, 8(3): 1375 – 1382

Chapter 7 Vulnerability Assessment of Static Power Grids

This chapter develops a vulnerability evaluation method for transmission lines in static power grids using vulnerability theory. It first proposes an integrated vulnerability index considering both average electric transmission distances of active powers over the entire network and local balances of reactive powers. Then it assesses the vulnerability of transmission lines, which examines, from both the active and reactive power aspects, the system's stability after a line fault based on the vulnerability index.

In recent years, "vulnerability" has been a hot topic in the study of complex networks[1–4]. The degree of the vulnerability of a node or line is defined to be the decrease in the level of network performances after the node or line is removed intentionally from the network. This notion has its origin in the research on computer networks[5], and has been introduced into the study on complex networks to assess from different angles the consequences of attacks using various complex network models[6]. The aim is to identify the most critical faults threatening the operation of the system, and thus to take effective actions for preventions and adjustments. In this chapter, we focus on the line vulnerability of power networks and its related assessment methods.

7.1 Overview

There are in general three categories of indices for line vulnerability.

(1) The shortest path as a global index

Generally speaking, after a fault takes place in a line, the system is considered to be more vulnerable as a result of the fault if the length of the shortest path between any pair of nodes increases significantly. However, in a real network, the propagation of power and information has to abide by certain physical rules (for example, Kirchhoff's law for the circuit networks), and is not determined by the shortest path between nodes. So it is hardly possible to describe accurately the impact of a fault simply by inspecting the change in the length of the shortest path.

(2) Connectivity as a local index

This class of indices is used as the basic tool to investigate the impact of a fault in a line on the local connectivity of a network. However, in real networks, the transmission of power and information generally does not depend on the direct connection between any two nodes, so we need to use instead indices that reflect the local transmission imbalance or congestion, such as the load level of the nodes around the faulted line. After the fault, if the power and information that are supplied or transmitted to the nearby nodes increase greatly, a local failure is more likely to happen.

(3) The size of the largest connected subnet

This index examines the impact of a fault in a line on the connectivity of the whole network. However, in power networks, important tie lines are all double lines, and the cut of one line of the two usually does not lead to large-scale load loss.

In vulnerability theory, the importance of the lines that are experiencing faults in the network is evaluated by the comparison of the values of the average path length, clustering coefficient and some other indices before and after the faults. Thus, the lines that are the most seriously faulted are not those that transmit the largest amount of power and information in normal operation conditions, but those the existence of which affects significantly the system's global or local performances. As a typical class of complex networks, power systems deliver simultaneously active and reactive powers and exhibit the characteristics of complex networks at both the global and local levels. On the one hand, the active powers flow from generators to loads through the routes determined by Kirchhoff's law, but not necessarily the shortest paths. So faults in lines affect the average electric distance for the delivery of active power over the entire network. On the other hand, the reactive powers are decoupled at regional levels and the effects of line faults on reactive powers are usually confined within the neighborhood of the faults. Hence, the existing indices cannot reflect the above characteristics of electric networks.

In order to evaluate efficiently the impact of line faults on the system's functionalities and thus identify the most severe faults, power system researchers usually adopt two classes of algorithms in the fault screening modules for static safety analysis[7–11]: The first class of algorithms uses one (or a set of) scalar performance index to assess the local impact of line faults on generators and loads. Such algorithms are computationally efficient, but ignore the influence of lines on active power transmission, and are less useful when one is interested in having a comprehensive description. The second class of algorithms compute load margins in certain directions of load growths. If the load margin is small after a fault, the system is considered to be fragile under the fault. Such algorithms can compare the influences of different faults with the same scale, but generally lead to different computational results when the directions of load growths are chosen differently.

In fact, the first class of algorithms only considers the impact of line faults on the delivery of reactive powers, while the second class of algorithms when combined with the continuation power flow method may also find the maximum active power load limit at the point when voltage collapse occurs. The latter observation is a result of the fact that the collapse in the continuation power flow method corresponds to the active power and reactive power limits. In this chapter, based upon the first class of algorithms we consider in addition the impact of line faults on the average transmission electric distances of active powers and the local balance of reactive powers. So we preserve the fast computation speed of the first class of algorithms and at the same time are able to detect globally the faulted lines that may cause the static instability in the system.

Now we take the 8/14 blackout failure in the US as an example to analyze the impact of the transmission distances of active power on power systems' static stability. As shown in Fig. 7.1, the New York ISO is a hub of the North American power grid. The numbers listed in the figure are the maximum loads in each network.

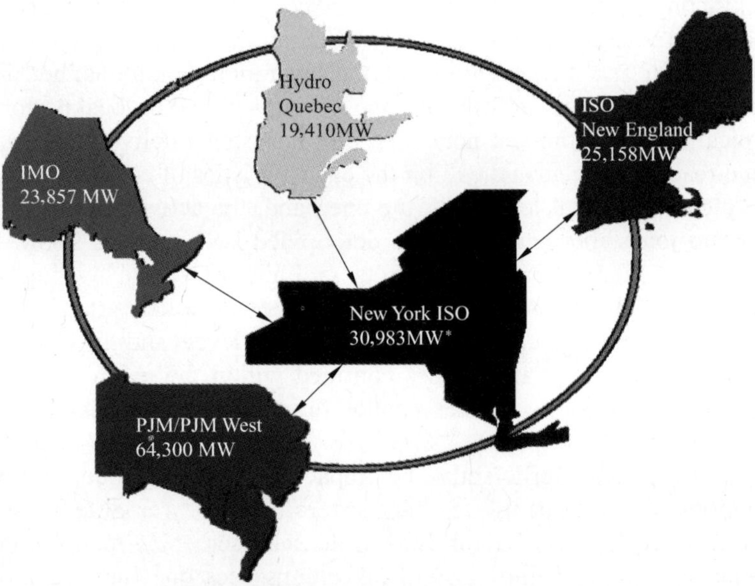

Figure 7.1 North American power grid

The development of the 8·14 blackout can be divided into three stages according to the reports by the ITC and the NERC:

The first stage is the evolution before the blackout, corresponding to the time interval 3:06 – 4:10 pm. This stage lasted for one hour and 5 minutes, during which 14 lines and 10 generator sets were cut off and 2000 MW power flows were redistributed over a large area. This process was the quasi-steady state evolution of power flows before the outage, and the corresponding transient process could be neglected. The main destructive effect on the system's static stability was caused by the redistribution of power flows.

The second stage is the dramatic change period during the blackout from 4:10 to 4:25 pm. In this stage of just a few minutes, nearly 100 generator units were in outage and the voltage fell as low as 50% but the frequency did not fluctuate too much.

The third stage is the extremely long recovery process after the power outage that lasted until August 17. After the fault, 2.2% of the power supply was restored after 3 hours and 20 minutes, 34.5% after 6 hours and 50 minutes, 66.5% after 12 hours and 50 minutes, and 76.8% after 18 hours and 50 minutes. New York city restored its power supply after 29 hours and 20 minutes, and regional mandatory regulation was carried out after 42 hours and 50 minutes. The complete restoration was realized after 48 hours and 50 minutes when 21 nuclear power units were still in outage.

In this chapter, we are mainly concerned with the first stage, namely the steady-state evolution process with wide-range power flow redistribution before the blackout. In Fig. 7.2, the 0's denote the generators in outage, the x's denote the tripped lines and the arrows denote the transmission routes of active powers.

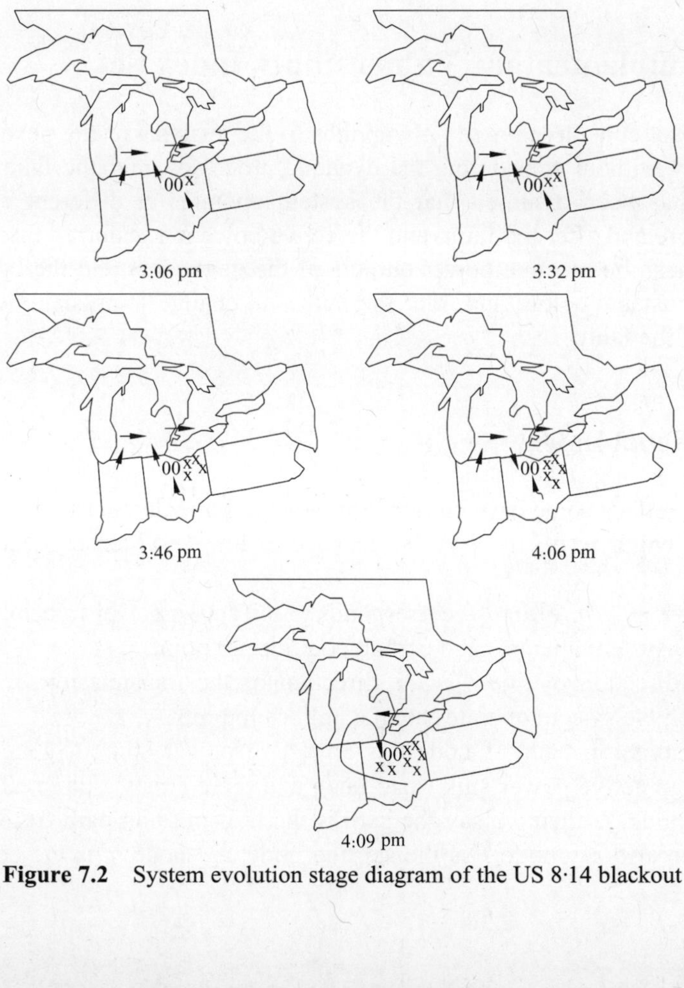

Figure 7.2 System evolution stage diagram of the US 8·14 blackout

From the process of the 8·14 blackout, it is easy to see the influence of the average active power transmission distance on the system's static stability. As the lines were tripped, a large amount of powers redistributed over a very large area and consequently the electric distances for active power transmissions increased greatly, and in the end the system collapsed gradually.

In order to overcome the shortcomings in the existing methods to evaluate transmission lines' vulnerability, in this chapter we combine the complex network theory with the physical characteristics of power networks. We propose a set of complementary vulnerability indices that assess both globally and locally the impact of line faults on the average active power transmission distance and the local balance of reactive powers. Thus we are able to quickly evaluate the seriousness of the faults on transmission lines. Furthermore, we normalize the complementary index set, propose a comprehensive vulnerability index, and then present a transmission line vulnerability assessment method. We carry out simulations on the IEEE 30-bus and the IEEE 118-bus systems to validate the effectiveness of the evaluation index set and the assessment methods.

7.2 Complementary Vulnerability Index Set

The index set considered here only applies to the analysis of the power network static safety without considering the dynamic processes after the faults. In other words, it has been assumed that the system operates at different equilibrium points before and after the fault, and the active powers transferred through transmission lines, the reactive power outputs of the generators and the bus voltages all keep constants before and after the fault and change impulsively only at the moment of the fault.

7.2.1 Global Index

We first introduce some basic notions for weighted directed networks.

(1) The length w_l of line l is the distance of line l, which corresponds to its reactance.

(2) The weight p_l of line l corresponds to the active power transmitted in line l when the system operates at a certain equilibrium point.

(3) The direction of line l is the direction of the transmission of the active power when the system operates at an equilibrium point.

(4) Transmission path: if node i is an active power source (generator bus), node j is an active power sink (load bus), and there exists a directed path from node i to node j, then we say the path is the transmission path from node i to node j. We also say node i is the starting node and node j is the ending node

of the transmission path. In power systems, there may exist multiple transmission paths from a generator node i to a load node j.

Figure 7.3 is a directed weighted graph G consisting of 5 nodes and 5 edges.

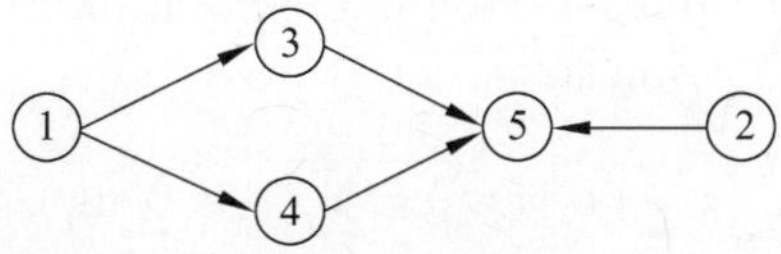

Figure 7.3 A five-node directed weighted graph

In Fig.7.3, the transmission path from node 1 to node 3 is 1-3, which means that the active power is transmitted from node 1 to node 3; the transmission paths from node 1 to node 5 are 1-3-5 and 1-4-5 respectively; and there is no transmission path from node 2 to node 3, which means that no active power is transmitted from node 2 to node 3.

(5) The length w_r of a transmission path r is the sum of the lengths of the edges in the transmission path r, namely

$$w_r = \sum_{l(r)} w_l \tag{7.1}$$

where $l(r)$ is the set of all the edges in the transmission path r. In Fig. 7.3, the length of the transmission path 1-3-5 is $w_{13} + w_{35}$, where w_{13} and w_{35} are the lengths of edges 1-3 and 3-5 respectively.

(6) The direction of the transmission path r is the direction of the active power transmitted through the transmission path r.

(7) The weight of the transmission path r is the active power supplied by the starting node to the ending node. The active power supplied by generator i to the network is the sum of the weights of all the transmission paths starting at node i, and the active power received by load j is the sum of the weights of all the transmission paths ending at node j. The weight of an edge is the sum of the weights of all the transmission paths containing this edge, namely

$$p_l = \sum_{r(l)} p_r \tag{7.2}$$

where $r(l)$ is the set of all the transmission paths containing edge l.

Using the above basic concepts, we propose the weighted total transmission path length index L_T and the average transmission distance index L_a, and the latter can be used to evaluate globally the average electric transmission distance for the transmission of active powers when the system is at a given equilibrium point. We now define these two indices.

(1) Index L_T

The weighted total transmission path length index L_T is defined to be the sum of the products of the weights and lengths of all the transmission paths. In power systems, generators usually supply active powers to several loads, and loads usually receive active powers from more than one generator, so the weight of a transmission path might be difficult to determine[12,13]. In order to circumvent this difficulty, we change the weighted total transmission path length to the sum of weighted edge lengths by taking the following steps:

$$\begin{aligned} L_T &= \sum_r (p_r \times w_r) = \sum_r (p_r \times \sum_{l(r)} w_l) \\ &= \sum_r \sum_{l(r)} (p_r \times w_l) = \sum_l \sum_{r(l)} (p_r \times w_l) \\ &= \sum_l (w_l \times \sum_{r(l)} p_r) \\ &= \sum_l (w_l \times p_l) \end{aligned} \tag{7.3}$$

where the subscript r is the index of an arbitrary transmission path, the subscript l is the index of an arbitrary edge, $l(r)$ denotes all the edges that the transmission path r contains, and $r(l)$ is the set of all the transmission paths that contain l. Here, p_r is the active power transferred along the transmission path r, and p_l is the active power traversing through edge l, which is the sum of the active power transferred through all the transmission paths containing l.

From the derivation above, it is clear that the weighted total transmission path length is equal to the sum of the products of the edge weights and edge lengths. It needs to be pointed out that active powers cannot be transferred bi-directionally in power systems. If one wants to apply this index to other systems, the bi-directional power and information flows cannot be canceled out.

(2) Index L_a

The average electric transmission distance L_a is defined to be the ratio of the weighted total transmission path length over the total transferred active power, namely

$$L_a = \frac{L_T}{\sum_{v \in V} P_v} = \frac{\sum_l (w_l \times p_l)}{\sum_{v \in V} P_v} \tag{7.4}$$

where p_l is the weight of edge l, which is the active power delivered through the edge, w_l is the length of edge l, which is the reactance of the edge, P_v is the active power received by node v, and V is the set of all the nodes. Note that if several mutually isolated connected sub-networks appear after a fault, we only consider the average transmission electric distance L_a of the largest connected sub-network.

The computation of L_a circumvents the problem of tracing the power flows and evaluate directly the transmission distance for a unit of active power from its source node to its destination node. So it can describe conveniently the influence of network structures on the transmission of active powers. When the occurrence of a fault leads to a large increase in L_a, we know that there are no alternative active power transmission paths near the faulted line, and the active powers have to be redistributed over a large area in the system. Therefore, the larger the L_a is, the more vulnerable the system is. A comparison between the L_a's before and after the fault reveals the influence of the fault on the overall performance of the network.

7.2.2 Local Index

In order to assess the influence of a fault in a transmission line on the local balance of reactive powers in the system, we need to calculate the reactive power outputs of the generators and the voltages at the load buses after the fault [12]. Towards this end, we define the following indices.

(1) Index D_v

D_v is determined by the change in local voltages, namely

$$D_V = \sum_{s \in S(l)} |U_s - U_{s0}| \tag{7.5}$$

where U_{s0} and U_s are the magnitudes of voltages (p.u.) at the load bus s before and after the fault respectively and $S(l)$ is the set of all the load buses that are significantly affected by the fault in l. If after the fault, the voltage at a load bus s drops by an amount that is more than α_1, then the load bus is considered to be affected to a significant extent, which in turn implies that

$$S(l) = \{s \mid U_{s0} - U_s > \alpha_1\}.$$

(2) Index D_q

D_q is determined by the change in local reactive powers namely

$$D_q = \sum_{g \in G(l)} \frac{Q_g - Q_{g0}}{Q_{g\max}} \tag{7.6}$$

where $G(l)$ is the set of the generator buses that are significantly affected by the fault in l. Here, we use Q_{g0} and Q_g to denote the reactive power outputs of generator g before and after the fault respectively and $Q_{g\max}$ is the reactive power capacity of generator g. If the reactive power output of a generator increases by at least α_2 folds after the fault in l, then this generator is considered to be significantly influenced, and thus we can determine

$$G(l) = \{g \mid (Q_g - Q_{g0})/Q_{g\max} > \alpha_2\}.$$

In this chapter, we use the indices D_v and D_q to measure the influence of the fault on the nearby load buses and generator buses. If a fault results in a large increase in D_v or D_q, then the voltages of the load buses near the fault fall greatly or the reactive powers of the generators increase greatly, and thus the local reactive-power balance is severely damaged.

7.2.3 Normalizing Process

The indices of L_a, D_v and D_q defined before can evaluate comprehensively various influences of faults, and they together form a set of complementary indices. We now normalize these indices in order to get rid of the effects of different units.

$\tilde{L}_a$ is obtained by normalizing L_a, namely

$$\tilde{L}_{ai} = \frac{L_{ai} - L_{a\_\min}}{L_{a\_\max} - L_{a\_\min}} \tag{7.7}$$

where

$$L_{a\_\min} = \min_{i=1,2,\cdots,N_{\text{line}}} (L_{ai}) \tag{7.8}$$

$$L_{a\_\max} = \max_{i=1,2,\cdots,N_{\text{line}}} (L_{ai}) \tag{7.9}$$

Here, $\tilde{L}_{ai}$ and L_{ai} are the corresponding indices after the fault in line i, $L_{a\_\min}$ denotes the minimum value of L_a after all the failures, N_{line} is the number of transmission lines, and $L_{a\_\max}$ denotes the maximum value of L_a after all the failures.

From Eq. (7.7), one can see that $0 \leqslant \tilde{L}_{ai} \leqslant 1$. If $L_{ai} = L_{a\_\min}$, then $\tilde{L}_{ai} = 0$ and if $L_{ai} = L_{a\_\max}$, then $\tilde{L}_{ai} = 1$. Similarly, one can normalize the indices of D_v and D_q to obtain $\tilde{D}_v$ and $\tilde{D}_q$, which satisfy $0 \leqslant \tilde{D}_{vi} \leqslant 1$ and $0 \leqslant \tilde{D}_{qi} \leqslant 1$.

7.2.4 Comprehensive Vulnerability Index

We define the comprehensive vulnerability index to be the weighted sum of the normalized indices, which can be computed by

$$I_{Vi} = \frac{1}{2} \times \left(\tilde{L}_{ai} + \frac{1}{2} \times (\tilde{D}_{vi} + \tilde{D}_{qi}) \right) \tag{7.10}$$

where the subscript i denotes the ith line. From the equation, we know that I_{Vi} incorporates various effects of the fault in a line on the system's functions and can be computed straightforwardly, so it can be used online in real time to evaluate the vulnerabilities of different lines in the present operation mode. However, since I_{Vi} has been normalized ($0 \leqslant I_{Vi} \leqslant 1$), it cannot be used to compare the vulnerability of the same line under different operation conditions.

7.3 Assessment Methods for the Vulnerability of Transmission Lines

In the previous section, we have introduced a set of complementary indices that evaluate the influence of a fault in a transmission line on the delivery of active powers through the average electric transmission distance index and evaluate the influence on the balance of local reactive powers through the local variable indices. We have also normalized the set of indices to obtain the comprehensive vulnerability index, which can examine the impact of the fault on the system's functionalities from both the global and local perspectives and with respect to both the active and reactive powers.

Now we use the comprehensive vulnerability index to identify the most severe faults. The advantage of this approach is that the computation does not rely on the directions of load growth and the computation speed is fast. As a result, this method can be applied in real time to identify the most serious fault under the present working condition. The flowchart of this method is shown in Fig. 7.4, and we list its main steps in detail as follows:

Step 1 Initialize $i = 1$.

Step 2 Estimate the equilibrium point E_{pi} after the fault in line i.

Two iterations of computation are carried out using the fast power flow decomposition method[12] to quickly estimate E_{pi}.

Step 3 Calculate L_{ai} according to E_{pi}.

In power grids, the distribution of active powers is mainly determined by the reactances of transmission lines, which are used here as the lengths of the transmission lines. We also use the active power transferred by a line after the fault to be the weight of that line. Then L_{ai} can be calculated using Eq. (7.4), and we further analyze the impact of the fault on the transmission of active powers.

Step 4 Calculate D_{vi} and D_{qi}.

We calculate D_{vi} and D_{qi} according to Eqs. (7.5) and (7.6) using the information about E_{P0} and E_{pi}, where E_{P0} is the system's equilibrium point before the fault and can be obtained by using the measurement data from the system. From engineering practical experiences, α_1 in Eq. (7.5) is usually set to be 0.02p.u., and α_2 in Eq. (7.6) is set to be 0.1. It needs to be pointed out that the values of α_1 and α_2 do not need to change for grids with different voltage levels, because they represent the changes in the relative values of the voltages and reactive powers that are caused by line failures.

Step 5 $i = i + 1$.

If $i > N_{\text{line}}$, go to Step 6; otherwise, go to Step 2. Here, N_{line} is the number of transmission lines.

Step 6 Normalize L_a, D_v and D_q, and calculate I_V.

Step 7 Pick out the serious faults.

If $I_{Vi} > \alpha_3$, line i is taken as under a serious fault. The choice of the threshold value α_3 will be discussed in detail in the simulation analysis.

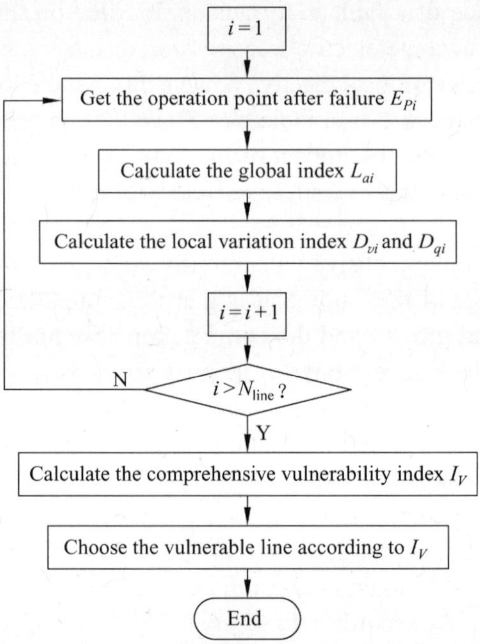

Figure 7.4 Flowchart for the assessment of the transmission line vulnerability

7.4 Discussions on Computational Complexity

In real practice, a small change in power network topological structures can lead to significant variation in the system's vulnerability. Therefore, it is necessary to study the vulnerability of transmission lines in real time. In order to guarantee the feasibility of online evaluation, we need to analyze the computational complexity of the proposed line vulnerability evaluation algorithm.

The main computation demand comes from two steps in the algorithm. (1) After a fault in a transmission line happens, we need to tell whether the network is still connected. (2) When necessary, we need to estimate the power flow in the largest island after the fault. For the former, we use the depth first search method, whose computational complexity is $O(N)$; and for the latter, we perform two iterations of power flow computations for the largest island to estimate the operation point of the power flow, so the computation is to solve a set of linear equations for two times, whose complexity is $O(\text{iteration time} \times N^2)$.

Hence, we conclude that the computation speed of the proposed line vulnerability evaluation algorithm is fast enough to satisfy all the online assessment requirements.

7.5 Simulation Results

We use the IEEE 30-bus system, the IEEE 118-bus system and the Shanghai power grid as test systems to validate the effectiveness of the algorithm just described.

7.5.1 IEEE 30-Bus system

We first look at the IEEE 30-bus system, which is shown in Fig. 7.5. The system contains six generator buses, 24 load buses and 41 transmission lines (including four transformers). The system's initial load is 283.4 MW.

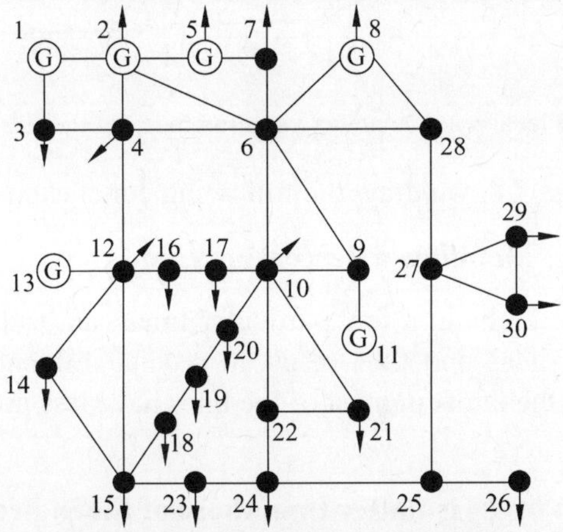

Figure 7.5 The IEEE 30-bus system

First, we calculate the set of complementary vulnerability indices L_{ai}, D_{vi} and D_{qi}. In Fig. 7.6 – Fig. 7.8, we list the top 15 lines that have the largest values for the indices. The horizontal coordinate is the line number n.

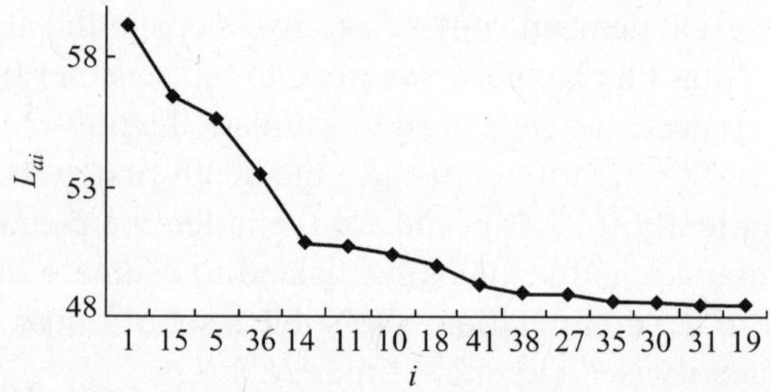

Figure 7.6 The average electric transmission distance index of the IEEE 30-bus system

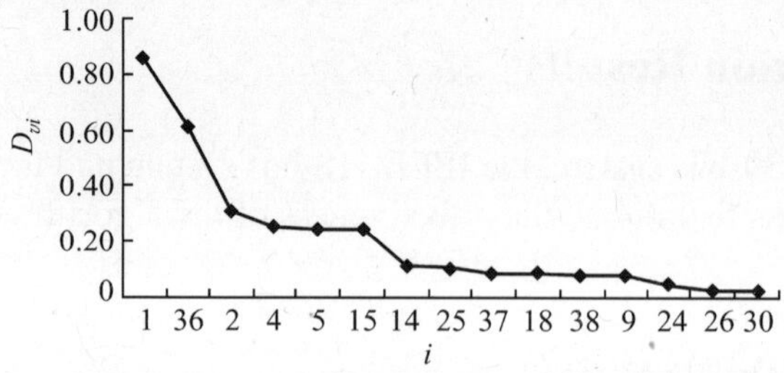

Figure 7.7 The local voltage variation index of the IEEE 30-bus system

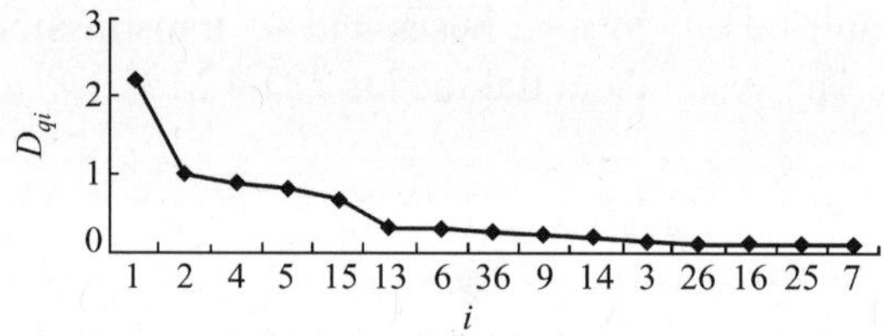

Figure 7.8 The local reactive power variation index of the IEEE 30-bus system

From Fig.7.6 – Fig. 7.8, we draw the following conclusions:

(1) The three indices all follow a decreasing trend

This implies that the faults in a few particular lines can lead to large changes in the values of the indices and thus result in crucial influences on the system's functionalities while the faults in most of the lines have less impact on the system's performances.

(2) The distribution of L_{ai} is flatter than those of the indices D_{vi} and D_{qi}

This is because L_{ai} is a global index while D_{vi} and D_{qi} are local indices. Even extremely serious faults can hardly bring significant influences on global properties.

Now we calculate the comprehensive vulnerability index I_{Vi} after the normalization of the set of complementary indices. The horizontal coordinate is the ranking of the comprehensive vulnerability index $I_{Vi,R}$

The distribution of the I_{Vi} in a decreasing order is shown in Fig. 7.9. The index satisfies $0 \leqslant I_{Vi} \leqslant 1$. Only a few lines correspond to larger values of I_{Vi}, and the majority of lines correspond to smaller I_{Vi}. This implies that the faults in a few particular lines can bring serious impact on the system's performances while faults in most of the lines have little influence. In view of the distribution of I_{Vi}, we usually set $0.1 \leqslant \alpha_3 \leqslant 0.2$ because I_{Vi} decreases rapidly at first and then declines slowly after some critical point.

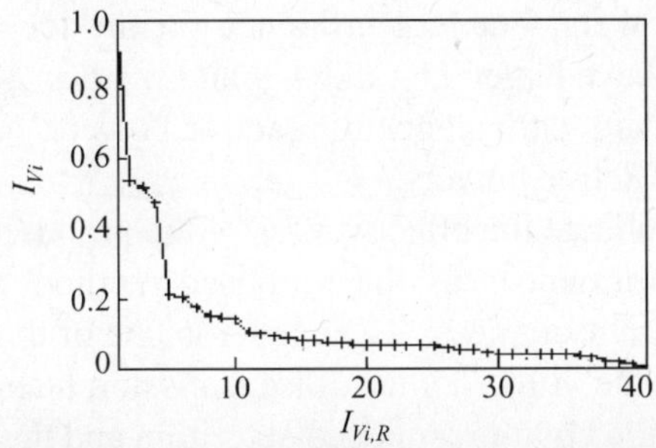

Figure 7.9 Distribution of the vulnerability index of the IEEE 30-bus system

In view of the number of the total transmission lines in the IEEE 30-bus system, we set $\alpha_3 = 0.15$ and pick nine most fragile lines. According to the ordering of the comprehensive vulnerability index, we list the results in Table 7.1.

Table 7.1 Line vulnerability results of the IEEE 30-bus system

Ranking of I_{Vi}	Line index	I_{Vi}	$\tilde{L}_{ai}$	$\tilde{D}_{vi}$	$\tilde{D}_{qi}$
1	1	1.0000	1.0000	1.0000	1.0000
2	15	0.5374	0.7770	0.2877	0.3079
3	5	0.5202	0.7075	0.2919	0.3739
4	36	0.4744	0.5310	0.7205	0.1150
5	14	0.2093	0.3104	0.1364	0.0800
6	2	0.2040	0.0033	0.3612	0.4483
7	4	0.1719	0.0000	0.2973	0.3902
8	11	0.1497	0.2994	0.0000	0.0000
9	18	0.1448	0.2385	0.1022	0.0000

In Table 7.1, the first column lists the ranking of the comprehensive vulnerability index I_{Vi}, the second lists the line indices, the third shows the value of I_{Vi}, and the fourth to the sixth columns correspond to the normalized indices $\tilde{L}_{ai}$, $\tilde{D}_{vi}$ and $\tilde{D}_{qi}$ respectively. We draw the following conclusions from Table 7.1.

1) The three evaluation indices share some common features

For example, lines 1, 5 and 15 all correspond to larger $\tilde{L}_{ai}$, $\tilde{D}_{vi}$ and $\tilde{D}_{qi}$; line

36 has greater values of $\tilde{L}_{ai}$ and $\tilde{D}_{vi}$. This implies that serious fault may deteriorate both global and local performances.

In Table 7.1, line 1 has the highest vulnerability with I_{Vi} being 1.0000, while line 15 is the second highest with I_{Vi} being only 0.5374. The indices $\tilde{L}_{ai}$, $\tilde{D}_{vi}$ and $\tilde{D}_{qi}$ of line 1 are all 1.0000, which indicates that a fault in line 1 can attack intensively on various aspects of the system.

2) The three indices used in the evaluations are complementary to one another

In Table 7.1, lines 14, 18 and 11 have larger $\tilde{L}_{ai}$ and smaller $\tilde{D}_{vi}$ and $\tilde{D}_{qi}$, implying that these lines have greater influence on the average electric transmission distances of active powers, but less influence on the local balance of reactive powers. Lines 2 and 4 have larger $\tilde{D}_{vi}$ and $\tilde{D}_{qi}$ and smaller $\tilde{L}_{ai}$, indicating that they influence directly on the local balance of reactive powers, but have less influence on the transmission of active powers.

In order to further validate the effectiveness of the proposed method, we analyze and compare the performances of the proposed method with the load margin method based on sensitivity indices[8]. The latter is one of the most typical existing methods for assessing the vulnerability of transmission lines. Table 7.2 shows the evaluation results obtained by the proposed algorithm and the load margin method.

Table 7.2 Vulnerability result comparison among different algorithms for the IEEE 30-bus system

I_{Vi} in decreasing order	Line index	Vulnerability ranking using the load margin method		
		Load growth direction Ⅰ	Load growth direction Ⅱ	Load growth direction Ⅲ
1	1	1	1	1
2	15	6	5	5
3	5	2	4	2
4	36	3	2	4
5	14	5	23	7
6	2	7	6	6
7	4	4	3	3
8	11	16	8	26
9	18	10	7	27

In Table 7.2, the first column lists the ranking of the comprehensive vulnerabilities of the lines, the second column lists the indices of the lines, and the third to the fifth columns are the ranking of the vulnerabilities obtained by the load margin method with three different load growth directions. The higher the ranking is, the less load growth the system can endure, and thus the more vulnerable the system is. Here, the load growth direction Ⅰ corresponds to the growth with constant power factors, the growth direction Ⅱ corresponds mainly to the increase of reactive power loads, and the growth direction Ⅲ mainly relates to the growth in active power loads. Obviously, although the load margin method can return quickly the estimated load margin and choose severe fault locations, the results depend on the choice of the load growth direction. However, since it is hard to estimate the correct load growth direction during the online vulnerability assessment, the load margin method usually misses some of the most vulnerable lines. As shown in Table 7.2, if we choose the estimated growth direction to be direction Ⅰ or Ⅲ, then lines 18 and 11 will be overlooked by the load margin method, which are most vulnerable lines under load growth direction Ⅱ with rankings 7 and 8 respectively in Table 7.2; when the estimated growth direction is chosen to be direction Ⅱ, the load margin method fails to identify line 14, which is a vulnerable line in the directions of I and Ⅲ. In comparison, the proposed indices do not depend on the load growth directions, and the top nine lines with the highest values of I_{Vi} contain all the top seven lines in the three load growth directions.

We have developed computing algorithms using PSAT on the Matlab platform[14]. The main computation operations include the $2 \times N_{\text{line}}$ power flow iterations, namely solving a set of $2 \times N_{\text{line}}$ linear equations. The calculation time of the IEEE 30-bus system is 2.87 s, and correspondingly the analysis of a line takes 0.07 s on average. Hence, the algorithm is not demanding in computational power and is suitable for online applications.

7.5.2 IEEE 118-Bus System

The IEEE-118 system has 54 generator buses, 64 load buses, and 186 lines (including seven transformers). To examine the proposed algorithm's performances when the system has certain heavy loads, we increase the load level proportionally until we reach the high load level of 4743 MW. The comprehensive vulnerability index I_{Vi} of the 186 lines are shown below.

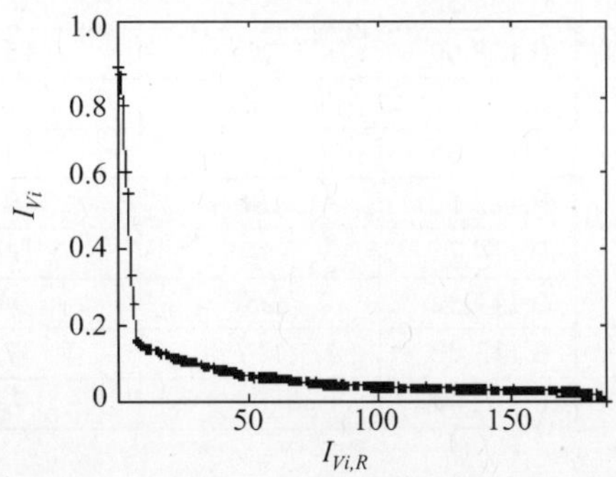

Figure 7.10 The distribution of the vulnerability index for the IEEE 118-bus system

The distribution of I_{Vi} is similar to that in Fig. 7.9. Considering the total number of lines in the system, we choose $\alpha_3 = 0.11$, and pick out 24 transmission lines with the most serious faults, which are lines 3, 4, 7-9, 21, 29, 33, 36, 38, 51, 93, 94, 96, 97, 104, 107, 108, 116, 118, 163, 174, 185 and 186.

We compute the transmission line vulnerability using the load margin method[8] in three different load growth directions. It is clear that the top 15 lines with the worst fault are all elements of the set of lines selected by the proposed method. We present the details in Table 7.3.

Table 7.3 Comparison of vulnerability among different methods for the IEEE 118-bus system

I_{Vi} ranking	Line index	I_{Vi}	Vulnerability ranking obtained by the load margin method		
			Load growth direction I	Load growth direction II	Load growth direction III
1	7	0.877,35	1	1	1
2	9	0.858,26	2	2	2
3	8	0.775,73	3	3	3
4	51	0.600,36	6	6	6
5	96	0.543,24	5	8	5
6	38	0.330,01	13	31	11
7	36	0.252,24	15	27	14
8	118	0.160,47	7	5	7
9	97	0.156,9	10	11	10
10	116	0.147,16	11	10	15
11	185	0.147,01	4	4	4
12	4	0.137,24	26	18	28
13	93	0.136,38	14	16	16
14	104	0.136,07	9	33	8
15	29	0.135,8	18	58	70
16	21	0.134,56	32	38	30
17	33	0.128,69	21	15	19
18	108	0.126,03	12	12	12
19	94	0.123,66	184	13	144
20	3	0.122,59	186	49	13
21	174	0.114,75	34	14	142
22	186	0.114,32	25	9	61
23	107	0.113,02	177	37	17
24	163	0.112,44	8	7	9

Using the Matlab simulation platform that we have constructed, the computation for the IEEE 118-bus system takes 18.53 s, and the analysis for each fault in a different line takes 0.1 s on average. This shows that the proposed algorithm achieves fast computation speed.

7.5.3 Shanghai Power Grid

Now we use the proposed algorithm to analyze a snapshot of the Shanghai Power Grid at 14:25:23 on August 21, 2006. At this snapshot, the system contains 45 generator buses, 164 load buses and 317 transmission lines. The distribution of the comprehensive vulnerability index I_{Vi} for the transmission lines is shown in Fig. 7.11.

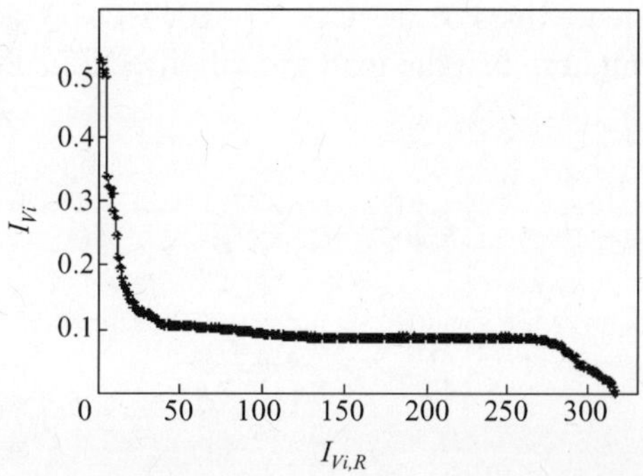

Figure 7.11 Vulnerability index distribution for the Shanghai power grid

The distribution of I_{Vi} is similar to that in Fig. 7.9. Note that the IEEE 30-bus system and the IEEE 118-bus system have the vulnerability indices as high as 0.87 and 1.00 respectively, while the highest vulnerability index of the Shanghai power grid is only 0.52. This is because that the Shanghai power grid is required to satisfy the $N-1$ regulation in its operation, and as a result a random fault in a line has less impact on the system's static stability and the system in the snapshot is less vulnerable.

In view of the total number of lines in the system, we choose $\alpha_3 = 0.12$ and select 32 lines with the most serious faults, which are listed below:

(1) Some lines in the 500 kV backbone network

HD – XH 500 kV tie lines (double lines): One of the double lines transfers the active powers of up to 860.58 MW and the reactive power of up to 181.08 Mvar. If this line is out of service, it will bring great influence on the transmission of active powers and the local balance of reactive powers.

HD – SJ 500 kV tie lines (double lines): One of the lines transfers the active power of up to 803.8 MW and the reactive power of 66.593 Mvar. The fault on this line will affect greatly the transmission of active powers.

(2) Outgoing lines of generators at strategic locations

The power plants, including the WJ second power plant, the WJ power plant and the MH power plant, are all at strategic locations. If their power supplies cannot be delivered to the system, the transmission of active powers and the support for local voltages will be affected greatly.

(3) Transformers in the substations at strategic locations

These transformers include the Sijing 500 kV substation and the Yanghang 500 kV substation.

(4) Important tie lines between some substations

These lines include the tie lines of WJ-CC, XJ-GB, HD-JD, and JD-XT.

We want to point out here that the load growth directions in the simulations are chosen to be

$$\begin{cases} P_{inri} = \lambda_P \times \dfrac{P_{Li}}{\sum\limits_i P_{Li}} \\ Q_{inri} = \lambda_Q \times \dfrac{Q_{Li}}{\sum\limits_i Q_{Li}} \end{cases} \tag{7.11}$$

where P_{inri} is the stepsize for the growth of the active power at load node i, P_{Li} is the initial active power at bus i, $\sum\limits_i P_{L_i}$ is the sum of the initial active power of all the load buses, λ_P denotes the growth coefficient for the active power loads; similarly, Q_{inri} is the stepsize for the growth of the reactive power at load bus i, Q_{Li} is the initial reactive power at bus i, $\sum\limits_i Q_{L_i}$ is the sum of the initial reactive power of all the load buses, and λ_Q denotes the growth coefficient for the reactive power loads.

The increased load has to be balanced by increasing the outputs of the generators in service in proportion to their initial outputs. If the fault in a line results in the islanding of the network, we then analyze the largest connected component. In order to ensure the balance of generation and load in this maximum connected sub-network, we rearrange the growth directions for load and generation according to the principle just described. We consider three load growth directions as shown in Table 7.4.

Table 7.4 Three load growth modes in the simulation

Index	Notes	λ_P	λ_Q
Load growth direction Ⅰ	Active and reactive power growth with equal proportion	1	1
Load growth direction Ⅱ	Mainly reactive power growth	0.2	1
Load growth direction Ⅲ	Mainly active power growth	1	0.2

7.6 Notes

In complex network theory, global indices based on the shortest path and local indices based on connectivity are usually used to assess the vulnerability of connections in the network. In power systems, faults in transmission lines affect both the average electric transmission distances of active powers over the entire network and the local balance of reactive powers. For this reason, we have proposed a set of complementary vulnerability indices and the comprehensive vulnerability index. We have also proposed methods to assess the vulnerability of transmission lines, which examines the system's stability after a fault in a transmission line from both the active and the reactive power aspects. In particular, the average electric transmission distance index has been used to evaluate the influence of the fault on the system's global performances, which has a clear physical interpretation and the potential to be applied to other complex networks, like traffic networks and the World Wide Web. In addition, we have analyzed the IEEE 30-bus system, the IEEE 118-bus system and the Shanghai power grid as testing systems to validate the facts that the proposed indices complement one another and the proposed evaluation method is effective.

It needs to be pointed out that the set of complementary vulnerability indices has been discussed closely in the context of power system engineering and cannot be directly applied to other practical networks. Nevertheless, we have performed a sequence of steps for analyzing the vulnerability of complex systems, which include studying the influence of a fault from both local and global perspectives, defining a complementary index set, and normalizing the indices to identify the lines with serious faults. Such steps are also potentially useful for the study of other complex networks.

References

[1] Holme P, Kim B J, Yoon C N, et al. Attack vulnerability of complex Networks. Physical Review E, 2002, 65(1): 056109

[2] Zhao L, Park K, Lai Y C. Attack vulnerability of scale-free networks due to cascading breakdown. Physical Review E, 2004, 70(3): 035101

[3] Albert R, Albert I, Nakarado G L. Structual vulnerability of the North American power grid. Physical Review E, 2004, 69(2): 025103

[4] Latora V, Marchiori M. Vulnerability and protection of intrastucture networks. Physical Review E, 2004, 71(1): 015103

[5] Barabasi A L, Albert R, Jeong H. Scale-free characteristics of random networks: the topology of the world wide web. Physica A, 1999, 27(2): 173 – 187

[6] Albert R, Jeong H, Barabasi A L. Error and attack tolerance in complex networks. Nature, 2000, 406: 378 – 382

[7] Chiang H D, Flueck A J, Shah K S, et al. CPFLOW: A practical tool for tracing power system steady state stationary behavior due to load and generation variations. IEEE Transactions on Power Systems, 1995, 10(2): 623 – 634
[8] Greene S, Dobson I, Alvarado F L. Sensitivity of the loading margin to voltage collapse with respect to arbitrary parameters. IEEE Transactions on Power Systems, 1997, 12(1): 262 – 272
[9] Ejebe G C, Wollenberg B F. Automatic contingency selection. IEEE Transactions on Power Apparatus and Systems, 1979, PAS-98(1): 97 – 109
[10] Ejebe G C, Irisarri G D, Mokhtari S, et al. Methods for contingency screening and ranking for voltage stability analysis of power systems. IEEE Transactions on Power Systems, 1996, 11(1): 350 – 356
[11] Galiana F D. Bound estimates of the severity of line outages in power system contingency analysis and ranking. IEEE Transantions on Power Apparatus and Systems, 1984, 103(9): 2612 – 2624
[12] Ni X P. Studies on the Complexity of the Interconnected Power Network [D]. Beijing: Tsinghua University, 2008 (in Chinese)
[13] Bialek J. Tracing the flow of electricity. IEE Proceedings on Generation, Transmission and Distribution, 1996, 143(4): 313 – 320
[14] Federico M. The power system analysis toolbox (PSAT). http: //www.power.uwaterloo.ca/~fmilano/psat.html, 2008

Chapter 8 Simplification, Equivalence, and Synchronization Control of Dynamic Power Grids

This chapter studies the dynamic performances of power systems. It introduces the simplification and equivalence methods for dynamic power systems. In the proposed simplification method, the Laplacian matrix of a dynamic power system is used to evaluate the synchronization capacities among generators, and is utilized as the main tool for the coherence equivalence analysis. It also discusses the synchronization problem for complex power grids. The synchronization control method discussed in this chapter chooses the optimal tripping locations using a linear model of power networks. After a tripping strategy is determined, the effectiveness of the strategy is checked using a nonlinear model.

In the previous chapters, we have analyzed static power systems. However, buses, like generator buses and load buses, have their own dynamic characteristics. The couplings between these buses are influenced by how power transmission and network structures change with time, so they exhibit dynamic characteristics as well. Hence, it is of great importance to study dynamic performances of power systems. Towards this end, we introduce in this chapter the simplification and equivalence methods for dynamic power networks, and then discuss the synchronization control method in power systems.

8.1 Overview

The power gird in China is developing rapidly with the implementation of the west-east electricity transmission project, the north-south power exchange project and the nation-wide interconnection project. The system becomes larger and more complex. As a result, a tremendous amount of transient dynamics simulations are needed to evaluate the system's stability after disturbances to enhance reliability and efficiency. In this context, simplification and equivalence are necessary to reduce the sometimes extremely heavy computational burden[1, 2]. The coherency-based dynamic equivalence algorithm is a classical dynamic equivalence method, which has been used extensively because it has fully utilized the available transient simulation software.

Coherency-based dynamic equivalence algorithms can be divided into three components[2].

(1) Area separation

According to the degrees to which generators are affected by a fault, the system is divided into the study area and the external area. Researchers have proposed several indices[3] to inspect comprehensively the electric distances between a generator and the fault, and between a generator and the inertia of generators. If the value of a generator's index exceeds a given threshold, this generator is then considered to be in the study area. The indices used here usually have clear physical meanings and are less demanding for calculation capacities. However, the choice of the indices and their threshold values are usually subjective, and consequently there are no systematic procedures to determine the study area's boundaries and scopes.

(2) Identification of coherent generator groups

The external area is further divided into several equivalent zones by identifying coherent generator groups. There are two classes of methods to identify coherent generator groups. One class is based on linearized system models, which includes the slow coherency method[4], Epsilon method[5] and weak coupling method[6]; and the other class is based on generators' trajectories after disturbances, which includes the swing trajectory aggregation method, frequency aggregation method[7] and Prony analysis method[8]. The first class does not require simulations at a system level, but needs to compute several eigenvalues and set manually the thresholds to determine the number of equivalence zones. The second class, on the other hand, requires the whole-system level simulation, and the amount of the calculations needed is huge.

(3) Generator dynamic aggregation and network simplification

While keeping the study area fixed, we can aggregate the generators in the same equivalence zone in the external area into a single equivalent generator. In this way, the system model is simplified and at the same time the influence from the external area to the study area is preserved to an acceptable extent.

From the viewpoint of complex networks, the study of area separation and coherent generator identification belongs to the topic of identifying "community structures" in a complex network[9,10]. The complex network theory of community structures is based on the study of the weighted Laplacian matrices, through which the system is divided automatically into two or more components using spectral bisection method[11,12], agglomerative method or divisive method[13]. Such methods do not construct distance indices, nor set critical or threshold values for the given indices, and it is unnecessary to decide the number of zones. Hence, they break new ground for area separation and coherent generator identification, and in particular have the potential to overcome the limitations of the existing methods in this field. It is thus the aim of this chapter to discuss the novel coherency-based equivalence methods that can determine automatically the range of the study area and the number of equivalent zones, and ensure the tight couplings within a zone and the weak couplings between zones.

Moreover, in this chapter we also study the synchronization control of power grids, which is related to the study of the power-angle stability in multi-generator systems. The power system security control systems consist of three components, namely preventive control, emergency control and recovery control[14]. When a system enters an emergent state that is caused by large disturbances, various emergency control actions need to be taken to stabilize the system. Among these control actions, tripping generators can change the inertia and the kinetic energy and redistribute power flows, and is thus considered to be one of the most important emergency control methods.

The three key issues of emergency control are timing, location and intensity. Generally speaking, the time for control actions should be the earliest permissible moment in practice, which can be determined by the searching policy-table or calculating the execution time of control devices with the associated delays. The intensity is inherently related to the location, and so the determination of the location is the most important among the three.

There are currently two main methods to determine the locations for tripping generators. One is to trip the generator with the maximum acceleration or kinetic energy at the moment when the fault is cleared or the generator with the maximum angular speed when the system losses stability. This method is usually based on time-domain simulations and engineering experiences. The other is built upon stability analysis and related to sensitivity analysis. The first method comes from engineering experiences, which is effective for certain systems and certain faults, but cannot provide an ordering for the locations where generators are tripped and lacks a systematic procedure for calculation. The second method relies on stability analysis; however, tripping generators might change the unstable modes of the system and affect the accuracy of the analysis results. To make it worse, the amount of the generation to be tripped is related to the location. Hence, the location and intensity for tripping generators are strongly coupled in stability analysis, which adds up the computational complexity. In addition, the stability margin indices and the control variables usually have relationships that are strongly nonlinear, so it is difficult to determine the location and intensity for tripping generators using only local sensitivity information.

In this chapter, we analyze the coupling of dynamic power grids and define the diagonal elements of the system state matrix to be the synchronization capacity coefficients for the corresponding generators to reflect the coupling strengths between the generator groups and the network. We analyze the changes in the synchronization capacity coefficients before and after a fault (or a control action). Then, we are able to evaluate the influence of faults and controls on the synchronization capacities. We further propose an index to evaluate the locations for tripping generators, which compares different choices using the network structure directly, not relying on stability analysis. Thus, we decouple location, control action and stability analysis, so the calculation as a result is greatly simplified. Furthermore, we discuss strategies for tripping generators. Simulations are performed to validate the effectiveness of our strategies.

8.2 State Matrix and Laplacian Matrix of Dynamic Power Grids

When studying the characteristics of static power grids, the network can be described by a weighted network $G = g(V, E, W)$, where $V = \{v\}$ is the set of buses, containing generator buses and load buses; $E = \{e\}$ is the set of branches, including transmission lines and transformers; and $W = \{w\}$ is the set of weights, describing the impendence or power flow of transmission lines and transformers. Both buses and branches represent physical devices. The degrees of buses are usually between 2 and 4.

When studying the synchronization problem in electric networks, we need to consider at the same time both the dynamic characteristics of generators and the connectivity of power grids. Hence, dynamic electric networks can be described as follows by a set of differential equations representing the dynamic characteristics of generators and a set of algebraic equations representing the network topological structures[2]:

$$\dot{x} = F(x, y) \tag{8.1a}$$

$$0 = G(x, y) \tag{8.1b}$$

Here, when generators are described by the second order model with constant E'_q, we have the state $x = [\delta_1, \delta_2, \cdots, \delta_n, \omega_1, \omega_2, \cdots, \omega_n]$, where δ_i is the angle of the ith generator, ω_i is the angular speed of the ith generator, and n is the number of generators. The state in the algebraic equations is $y = [V_1, V_2, \cdots, V_N, \theta_1, \theta_2, \cdots, \theta_N]$, where V_i and θ_i are the magnitude and angle of the ith bus respectively, and N is the total number of buses that have no dynamic characteristics.

When the system runs at a given operating point, the state of Eq. (8.1) is known, so state y can be solved using Eq. (8.1b), which can be substituted into Eq. (8.1a) to obtain

$$\dot{x} = F_2(x) \tag{8.2}$$

where x is the same as that in Eq. (8.1).

This process can be explained physically. We keep the state x in the model to preserve the internal-potential buses of all the generators; by eliminating the state y, we eliminate all the buses that have no dynamic characteristics, such as load buses, tie-line buses and generator terminal buses. It needs to be pointed out that in the reduced network after eliminating y, the internal-potential buses of generators are linked to each other.

8.2.1 State Matrix

When a generator's inertia is large, and the deviation of its angular speed from its rated value is small, one can take the rated value of its torque approximately to be the rated value of its power[2]. Under this assumption and ignoring the damping, one can write the dynamics of the ith generator as

$$\begin{cases} \dot{\delta}_i = \omega_0(\omega_i - 1) \\ \dot{\omega}_i = \dfrac{1}{M_i}[P_{mi} - P_{ei}(\delta)] \end{cases} \tag{8.3}$$

where δ_i and ω_i are the angle and angular velocities of the rotor respectively, $\omega_0 = 2\pi f_0 \approx 314\text{rad/s}$, M_i is the moment of inertia of the ith generator, P_{mi} = const is the mechanical power of the ith generator, P_{ei} is the electrical power of the ith generator, which is a function of the vector $\delta = [\delta_1, \delta_2, \cdots, \delta_n]^{\mathrm{T}}$, and n is the number of generators. All the ω_i, P_{mi} and P_{ei} are in per-unit values.

Linearizing the above equations, we have

$$\begin{cases} \Delta\dot{\delta}_i = \omega_0 \Delta\omega_i \\ \Delta\dot{\omega}_i = -\dfrac{1}{M_i}\sum_{j=1}^{n}\left[\dfrac{\partial P_{ei}(\delta)}{\partial \delta_j}\Delta\delta_j\right] \end{cases} \tag{8.4}$$

namely the linearized rotor angle equation of the ith generator is

$$\Delta\ddot{\delta}_i = -\frac{\omega_0}{M_i}\sum_{j=1}^{n}\left[\frac{\partial P_{ei}(\delta)}{\partial \delta_j}\Delta\delta_j\right] \tag{8.5}$$

Combining the rotor angle equations of all the n generators, we obtain

$$\ddot{x}_2 = -Ax_2 \tag{8.6}$$

$$A_{ij} = \frac{\omega_0}{M_i}\frac{\partial P_{ei}(\delta)}{\partial \delta_j} \tag{8.7}$$

where

$$A_{ij} = \frac{\omega_0 E_i E_j}{M_i}(g_{ij}\sin\delta_{ij} - b_{ij}\cos\delta_{ij}), \quad i \neq j \tag{8.8}$$

$$A_{ii} = -\sum_{\substack{j=1 \\ j\neq i}}^{n} A_{ij} \tag{8.9}$$

Here, $x_2 = [\Delta\delta_1, \cdots, \Delta\delta_n]^T$, $\Delta\delta_i$ is the difference between the rotor angle of the ith generator and its steady state value, n is the number of generators, which is also the number of buses in the reduced network; A is the system state matrix, representing the connectivity of the buses in the reduced network, which determines the coherence characteristic among generator groups and serves as the foundation for synchronization analysis; M_i is the inertia of the ith generator; E_i denotes the transient potential of the ith generator; and g_{ij} and b_{ij} are the real and imaginary parts of the corresponding elements of the admittance matrix of the reduced network; $\delta_{ij} = \delta_i - \delta_j$ is the rotor angle difference between the ith and j^{th} generators.

In fact, Eq. (8.6) is the linearized model of the electric network after being reduced to generator internal-potential buses. The reduced network has n buses corresponding to the n generators. The matrix A denotes the connectivity between buses, which is a complete matrix implying that the buses are fully connected to each other and the generators' power-angle characteristics are pairwise correlated.

8.2.2 Laplacian Matrix

The state matrix A of the dynamic power grid is the main tool used in coherency analysis because it reflects, from different aspects including electric distances, generator inertia distributions and operation points, how generators are inter-connected. However, the matrix is not symmetric, and thus the workload to compute its eigenvalues and eigenvectors is comparatively large. To deal with this difficulty, we construct the Laplacian Matrix L of dynamic power grids:

Step 1 Construct the diagonal matrix T

$$T_{ij} = \begin{cases} M_i^{-\frac{1}{2}}, & i = j, \ i = 1, 2, \cdots, n \\ 0, & i \neq j \end{cases} \tag{8.10}$$

where M_i is the inertia of the ith generator and n is the number of generators.

Since M_i satisfies

$$M_i > 0, \quad i = 1, 2, \cdots, n \tag{8.11}$$

It must be true that T is nonsingular and

$$[T^{-1}]_{ij} = \begin{cases} M_i^{\frac{1}{2}}, & i = j, i, j = 1, 2, \cdots, n \\ 0, & i \neq j \end{cases} \tag{8.12}$$

Step 2 Construct the Laplacian matrix

$$L = T^{-1} A T \tag{8.13}$$

where A is the state matrix of the power system at a given time instant and L is the corresponding Laplacian matrix.

From the above steps, one can see that the matrices L and A are similar matrices and L is symmetric. We expand our discussion further as follows:

(1) The Laplacian matrix *L* and the state matrix *A* are similar

Let λ be one of the eigenvalues of L with the associated eigenvector v. Then

$$Lv = \lambda v \tag{8.14}$$

Substituting Eq. (8.13), we get

$$T^{-1}ATv = \lambda v \tag{8.15}$$

After arranging the terms in the above equation, we have

$$A(Tv) = \lambda(Tv) \tag{8.16}$$

So λ is also an eigenvalue of A with associated eigenvector Tv. From Eq. (8.16) we know that the eigenvalues of L are the same as those of A, the components of the eigenvectors of L have the same sign as those corresponding components of the eigenvectors of A because the elements of the diagonal matrix T are all positive.

(2) The Laplacian matrix *L* is symmetric

From Eq. (8.10) and Eq. (8.13), we know that the elements of L satisfy

$$L_{ij} = \sqrt{\frac{M_i}{M_j}} A_{ij}, \quad i \neq j \tag{8.17}$$

Substituting Eq.(8.8), we obtain

$$L_{ij} = \frac{\omega_0 E_i E_j}{\sqrt{M_i M_j}} (g_{ij} \sin\delta_{ij} - b_{ij} \cos\delta_{ij}), \quad i \neq j \tag{8.18}$$

When neglecting the effects of phasers, the imaginary part of the system admittance matrix is symmetric, namely $b_{ij} = b_{ji}$. In addition, in transmission systems, g_{ij} can be approximately taken to be 0, namely $g_{ij} = 0$. Then L must be symmetric.

So the Laplacian matrix L has the following properties:

(1) L is real and symmetric.

So all the eigenvalues are real as well.

(2) L has a zero eigenvalue.

From Eq. (8.9) and Eq. (8.13) we know that the row sums of L are all zero, so L has zero as an eigenvalue.

(3) L is similar to the state matrix.

As we have explained, the two matrices have the same eigenvalues, and the components of the corresponding eigenvectors have the same signs. Hence, comparing with carrying out analysis directly on the state matrix, the analysis of L instead can lower the requirement for computational power. We also want to point out that the system can be divided into two different regions using the eigenvector associated with the minimum nonzero eigenvalue λ_2 via the spectral bisection method, and the magnitude of λ_2 can be used to indicate the effectiveness of the division.

(4) L is time-varying

The occurrence and clearance of faults and the change of the operating point all lead to variations in matrices A and L. This implies that the state matrix and the Laplacian matrix are both time-varying.

Comparing the two matrices A and L, we want to further emphasize that

a. A is not symmetric. A_{ij} and A_{ji} represent the weights of branch i to j and branch j to i respectively in the bidirectional weighted networks.

b. L is symmetric. $L_{ij} = L_{ji}$ represents the weight of the branch between bus i and bus j in an undirected weighted network.

c. The two matrices, A and L, are similar, and consequently the network divisions obtained from each of them using the spectral bisection method must be the same.

8.3 New Coherency-Based Equivalence Algorithm

The algorithm discussed here consists of three parts. First, we separate the reduced power grid into two areas by the spectral bisection method. Then we choose one area as the study area and the other as the external area according to the capability for synchronization. In the end, we divide the external area into several equivalent zones using the agglomerative algorithm and the modularity index.

8.3.1 Separation into Two Areas

After a fault, L's second smallest eigenvalue undergoes a sudden decrease. This implies that under the influence of the fault, the connection between certain buses and the rest of the network gets weaker abruptly, and the network experiences the strong tendency to be separated into two areas. Hence, here we compute $L(t_f)$ at time t_f right after the fault, and then divide the network into two areas by the spectral bisection method.

Firstly, we compute the state matrix $A(t_f)$ and the corresponding Laplacian matrix $L(t_f)$. Then we calculate the second smallest eigenvalue λ_2 and its associated eigenvector v_2. In the end, we put the buses corresponding to the positive elements of v_2 into one area and the rest of the buses into the other area.

It should be pointed out that the division of the two areas via the spectral bisection method has the following characteristics:

(1) $L(t_f)$ can be calculated directly from the network structure, the system's initial operating point and the fault location without time-domain simulations.

(2) L is real symmetric for which various mature algorithms exist to calculate its eigenvalues and eigenvectors. Furthermore, one only needs to know the signs of the elements of v_2 to do the separation and there is no need to compute precisely the magnitudes of the elements. When iterative algorithms are adopted, the precision requirement is low and the calculation complexicity is also low.

(3) The spectral bisection method can better determine the scope and the boundary of the study area, which does not require to set up the thresholds manually.

8.3.2 Choice of Study Area

The spectral bisection method can only divide a network into two areas, but is not capable to determine which of the two is the study area that has been affected by the fault more seriously. Thus, we introduce the index for synchronization capability to determine the study area from the two areas.

The diagonal elements $L_{ii}(t_0)$ of L represents the attraction from the network acting on generator i with unit inertia at time t_0. The larger the $L_{ii}(t_0)$ is, the greater the synchronization capability generator i has. The changes in L_{ii} before and after the fault reflect the fault's influence on the generator set's synchronization capability. So we propose to use the synchronizability index F_i

$$F_i = \frac{L_{ii}(t_f)}{L_{ii}(t_s)}, \quad i = 1, 2, \cdots, n \tag{8.19}$$

where $L_{ii}(t_f)$ and $L_{ii}(t_s)$ are the diagonal elements of L at time t_s before the fault and t_f after the fault respectively. From the equation we know that the smaller the F_i is, the greater influence generator i suffers from the fault. The area containing the generator with the smallest F_i is affected by the fault the most severely and can thus be chosen as the study area.

8.3.3 External Area and Equivalent Zones

As the study area is affected by the fault the most severely, its detailed models have to be used in the transient simulation. The external area is less affected by the fault and can be simplified to reduce the demand for transient simulation. We use the agglomerative method with the modularity index to divide the external area into several equivalent zones in order to ensure that the attraction within a zone is strong and those between zones are weak.

We want to clarify that since the external area is less influenced by the fault, we can use $L(t_s)$ to describe synchronization capabilities between the generator sets. Moreover, since the distinction between different zones in a dynamic power grid is less obvious than those of the distribution of reactive power in a static power grid, we use the weighted modularity index Q in Eq. (2.216) to compute the iteration directions.

First, we take every generator itself in the external area as a zone, and calculate its modularity index Q^0. Then we merge the zones along the positive or negative gradient of Q. In other words, for any arbitrary two zones x and y, we calculate the modularity index Q_{Txy} after merging them together. If the zone obtained by merging zones j and k has the maximum modularity index $\max(Q_T)$, we merge these two zones and set $Q^m = \max(Q_T)$ where m is the number of iterations. Repeating this process until the external area is merged into a single zone. If the modularity index Q^t after the tth iteration is equal to the maximum modularity index $\max\limits_{m=0\cdots n_{\text{ex}}-1} Q^m$ in this iterative process, then the pattern and the number of different zones at iteration t is the optimal zoning choice[15]. The steps of this process are listed below.

Step 1 Initialization

Set the iteration number $m = 0$ and the number of zones $N_{\text{AREA}} = n_{\text{ex}}$, where n_{ex} is the number of generators in the external area. Calculate the modularity index Q^0 of the initial zoning scheme.

Step 2 Calculate the modularity index matrix Q_T

$$
\begin{aligned}
&\text{If}\quad \text{zone } x \text{ is not connected to zone } y\\
&\quad Q_{Txy} = 0\\
&\text{Else}\\
&\quad Q_{Txy} = \frac{1}{2m}\sum_{i=1}^{n}\sum_{\substack{j=1\\ j\neq i}}^{n}\left\{\left[(-L_{ij}(t_s)) - \frac{L_{ii}(t_s)L_{jj}(t_s)}{2m}\right]\delta(i,j)\right\}\\
&\text{End}
\end{aligned}
\tag{8.20}
$$

where $x = 1, 2, \cdots, N_{\text{AREA}} - 1$, and $y = x+1, \cdots, N_{\text{AREA}}$

Step 3 Merge two zones

$$
\begin{aligned}
&\text{If}\quad Q_{Tjk} == \max(Q_T)\\
&\quad A_j^{m+1} = A_j^m \cup A_k^m;\quad A_k^{m+1} = \phi\\
&\text{End}
\end{aligned}
\tag{8.21}
$$

Step 4 Update variables

$$\begin{cases} N_{\text{AREA}} = N_{\text{AREA}} - 1 \\ m = m + 1 \\ Q^m = \max(Q_T) \end{cases} \tag{8.22}$$

Step 5 Go back to Step 2 until $N_{\text{AREA}} = 1$

Step 6 Determine the best zoning scheme

If $Q^t = \max\limits_{m=0,1,\cdots,n_{\text{ex}}-1} Q^m$, then the best zoning scheme has been obtained in the tth iteration.

Using the modularity index Q, the algorithm just described ensures the strong coupling within a zone and the decoupling between zones. It determines the optimal number of zones and the optimal zoning scheme. This algorithm can be implemented efficiently with fast convergence speed. The number of zones is determined automatically and it thus circumvents the difficulty of setting threshold values.

8.4 Generator Synchronizability Coefficient

The elements in the state matrix A (Eq. (8.6)) reveal the couplings between generators. The diagonal elements are the sums of opposites of the corresponding off-diagonal elements, each of which represents the attraction experienced by each generator with unit inertia. We improve in this chapter the generator synchronizability coefficient discussed in [16]. We consider the influence of the generator inertia by utilizing directly the diagonal elements of A to describe the interaction between the generators and the grid. We define the synchronizability coefficient of the ith generator to be

$$S_i(t) = -A_{ii}(t) \tag{8.23}$$

From the above equation, one can see that the generator synchronizability coefficient is time-varying. $A_{ii} < 0$ and $S_i > 0$ when the system operates near its stable equilibrium point, and the values of A_{ii} reveal the synchronization capability of the ith generator with respect to the grid.

8.5 Influence of Disturbance on Synchronizability Coefficient

When a disturbance takes place, the system's states cannot change instantaneously; however, the state matrix A, which denotes the couplings in the network, can vary abruptly. The change in the synchronizability coefficient S_i of each generator can describe directly the influence of the fault on the couplings in the network. Note that both faults and controls can be taken as disturbances to the system. Faults in general weaken the synchronization capabilities of the generators and can be taken to have negative effects. The objective of controls is to strengthen the synchronization capabilities, and they might have negative effects for certain generators. So the accurate evaluation of the negative effects of control actions is important for the determination of the control locations.

8.5.1 Influences of Fault

When a fault occurs, the topological connections between the nearby generators and the network are greatly weakened. We can identify the generator that is influenced by the fault the most through calculating the ratio

$$F_{if} = \frac{S_{if}}{S_{i0}} \tag{8.24}$$

where S_{if} is the synchronizability coefficient of the ith generator at the moment when the fault takes place and S_{i0} is the corresponding synchronizability coefficient before the fault. The smaller the F_{if} is, the greater influence of the fault the ith generator has experienced.

As the fault progresses, the state starts to deviate and S_i varies gradually. When the fault is cleared, the state matrix A changes abruptly again and S_i recovers to some extent. But since the operation point has shifted during the fault, S_i cannot recover to its value before the fault immediately. If we use S_{ic} to denote the synchronizability coefficient of the ith generator when the fault is cleared, then $F_{ic} = S_{ic} / S_{i0}$ can be used to indicate the recovered synchronization capabilities after the fault. The smaller the F_{ic} is, the greater accumulated influence generator i has experienced during the fault. Note that because of the system's nonlinearity, the generator with the minimum F_{if} at the moment of fault may not be the one with the minimum F_{ic} after the fault.

8.5.2 Effects of Control

Since the dynamic power system is a fully coupled network, tripping one generator somewhere affects the synchronization capabilities of all the generators. To deal with this, we propose to use the assessment index to evaluate different locations for tripping generators by inspecting how the other generators are influenced after tripping one generator.

First, assume that the generation at location p is cut off by 40%. We calculate the state matrix $A(t_p^+)$ at this moment, where t_p denotes the time of tripping the generator, and we further solve for the synchronizability coefficients

$$S_{ip} = -A_{ii}(t_p^+), \quad i = 1,2,\cdots,n \tag{8.25}$$

Then we compute the recovery ratios of the synchronizability coefficients

$$R_{ip} = \frac{S_{ip}}{S_{i0}}, \quad i = 1,2,\cdots,n \tag{8.26}$$

where S_{i0} is the synchronizability coefficient of generator i before the fault. R_{ip} is the ratio of the synchronizability coefficient S_{ip} after the fault to S_{i0} before the fault. It describes the combined influence of the fault and the tripping of generators.

We further compute the minimum recovery ratio of the synchronizability coefficients among all the generators except for the tripped generator. We define this minimum value to be the effectiveness index for location p

$$E(p) = \min_{i=1,2,\cdots,n;\ i \neq p} R_{ip} \tag{8.27}$$

$E(p)$ can be used to characterize the influence of the tripping on the other generators. The larger the $E(p)$ is, the more effective the tripping action is on the other generators' recovery capabilities, namely the better the tripping location p is.

We take the same amount of control actions at different p, calculate their influences on the other generators and then obtain the effectiveness indices $E(p)$, based on which the effectiveness of tripping locations can be evaluated.

It is worth mentioning that the value of $E(p)$ depends on how much of the generation has been tripped, which is not true for the ordering of $E(p)$. So when evaluating the effectiveness of tripping locations the proposed method works as long as the same amount of generation is tripped at the tested locations. This property enables us to decouple the evaluation of tripping locations from the determination of how much generation is needed to be tripped.

8.6 Calculation of Tripping Generators

The tripping location index evaluates the effectiveness of different tripping locations, and is independent of the stability assessment and control action computation. So the computation for generator tripping strategies has been greatly simplified. Hence, in this section we propose the following procedure for tripping generators that consists of three parts: stability assessment, evaluation of tripping locations and calculation of control. We use the flowchart below to explain.

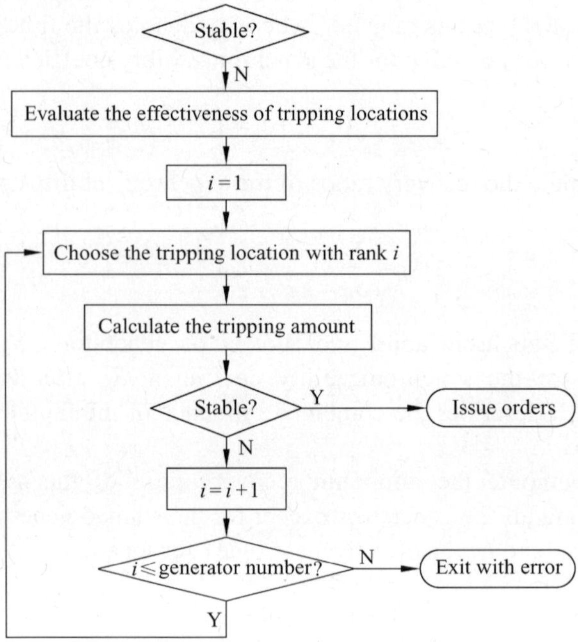

Figure 8.1 Flowchart of tripping strategy

(1) Assessment of stability

Since the three parts are completely decoupled, any stability assessment method applies here. Since the location and amount of the generation to be tripped can be determined accurately and quickly using the tripping location evaluation index together with the tripping amount calculation method, the number of iterations needed is small, and so here we use time domain simulations to assess stability.

(2) Evaluation of tripping locations

We calculate the effectiveness of different tripping locations using Eqs. (8.25) – (8.27), and then order the computation results. According to the ordering, we compute the tripping amount at each location until a stabilization strategy is found that makes the system satisfy the stability condition. From the simulation in this chapter, we know that the system can be stabilized after tripping appropriate amounts of generation at the top two locations in the ordering.

(3) Calculation of controls

At the same location, as the tripping amount increases, the trajectories of a generator set's imbalanced power exhibit some similarities. Zhang et al.[17] proposed a method to calculate the tripping amount using predication functions. Here we follow this approach to compute the controls.

8.7 Simulation and Analysis

8.7.1 Simulation Results of the Simplification and Equivalence Methods

In this section, we take the IEEE 39-bus system (Fig. 8.2) as an example to validate the effectiveness of the simplification and equivalence methods that we have proposed. We divide the system into the study area and the external area, and then devide the external area into several equivalent zones. Then we simulate separately in time domain the system's behaviors before and after the equivalence.

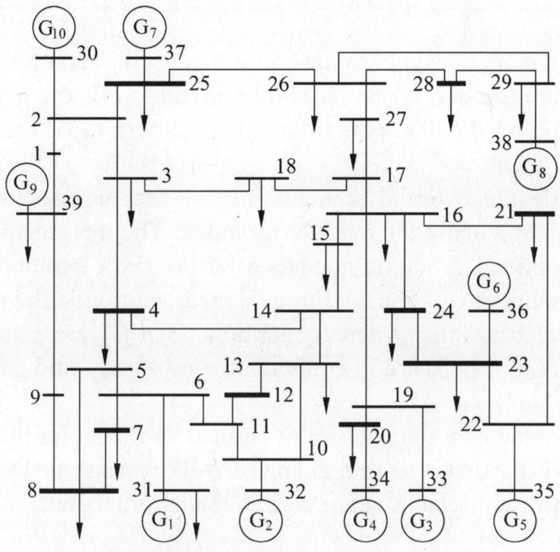

Figure 8.2 IEEE 39-bus system

1) Fault at generator terminal buses

Consider a three-phase short circuit fault takes place at the terminal Bus 36 of generator G_6 at 0.1 s, which is cleared at 0.2 s.

First, we examine the fault's influence on the division of the two areas. Before the fault, the second smallest eigenvalue λ_2 of $L(t_s)$ is 0.1664 and after the fault, λ_2 of $L(t_f)$ decreases to 0.0224. When λ_2 is smaller, the two-area system structure is more obvious. In other words, the fault has significantly weakened the connection of some buses and strengthened the two-area separation. So we compute the eigenvector v_2 associated with the λ_2 of $L(t_f)$, and divide the system into two areas according to the signs of the elements of v_2 using the spectral bisection method. As shown in Table 8.1, the generator G_6 corresponds to a positive element in v_2 and is in area D_1, and the other generators are in D_2.

Table 8.1 Zone division when the fault occurs at Bus 36

Generator	Component of v_2	Zone	F_i
G_1	−0.0399	D_2	0.8898
G_2	−0.0423	D_2	0.8827
G_3	−0.0690	D_2	0.8171
G_4	−0.0471	D_2	0.7942
G_5	−0.0311	D_2	0.5088
G_6	0.9824	D_1	0.0287
G_7	−0.0709	D_2	0.9255
G_8	−0.1109	D_2	0.8565
G_9	−0.0667	D_2	0.9805
G_{10}	−0.0412	D_2	0.8493

Second, we calculate the synchronization index F_i for each generator to choose the study area from the two areas. As shown in Table 8.1, the generator with the minimum F_i is G_6, so the area containing G_6 is chosen to be the study area, and D_2 is the external area.

Then we divide the external area D_2 into several equivalent zones via the agglomerative method using the modularity index. The merging process is shown in Fig. 8.3. The optimal grouping appears after the sixth iteration in the process, and the optimal number of zones is three. Hence, we divide the external area D_2 into three equivalent zones, where generators $G_3 - G_5$ are in equivalent zone Area_1, generators G_1, G_2, G_9 are in equivalent zone Area_2, and generators G_7, G_8, G_{10} are in equivalent zone Area_3.

Now we have obtained a four-generator equivalent model for the original system using the method discussed in this chapter. We have one generator (G_6) in the study area and three equivalent generators in the external area.

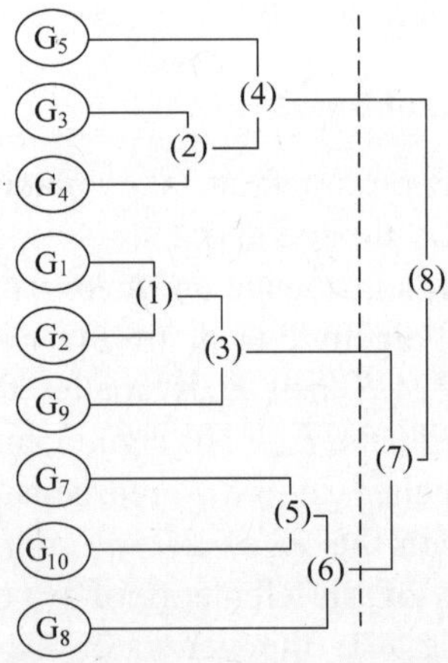

Figure 8.3 Merging process of the external zones (The fault occurs at Bus 36)

In comparison, the system is divided into three coherent areas using the slow coherency method[4]. The model contains eight generators, which includes six generators G_1-G_6 in the fault area and two equivalent generators in the external area.

Now we carry out the time domain simulations using the detailed model and the two equivalent models (The details of the coherency model can be found in [18]). The swing curves of G_6 are shown in Fig. 8.4, from which one can see that the system loses stability at 0.24 s according to the detailed model, at 0.23 s according to the equivalent model proposed in this chapter and at 0.19 s according to the coherency equivalent model. So the proposed model is closer to the detailed model and performs much better than the coherency equivalent model.

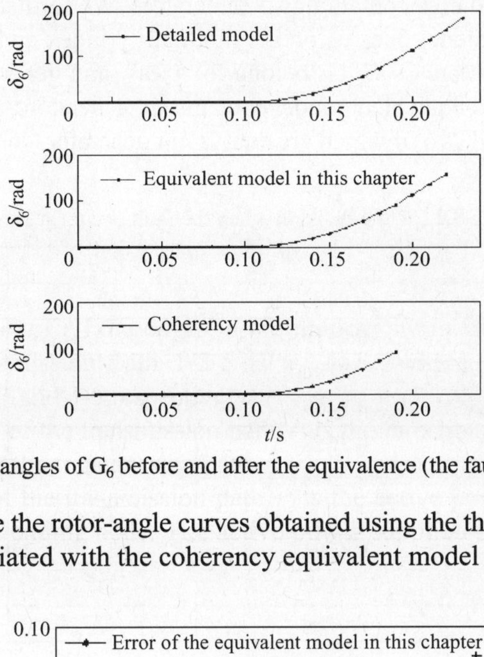

Figure 8.4 Rotor angles of G_6 before and after the equivalence (the fault occurs at Bus 36)

Now we compare the rotor-angle curves obtained using the three models, We find that the error associated with the coherency equivalent model is larger (Fig. 8.5).

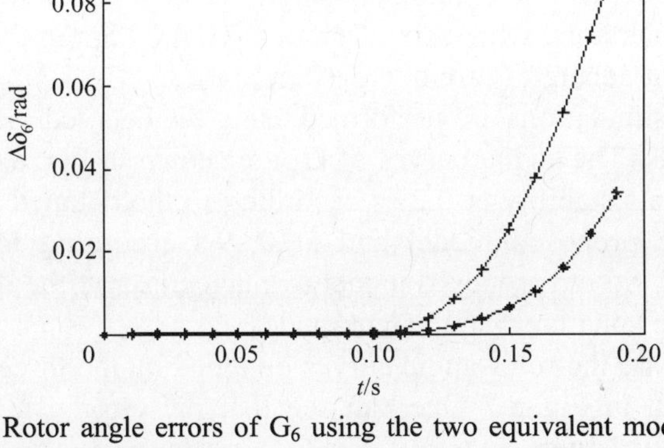

Figure 8.5 Rotor angle errors of G_6 using the two equivalent models (the fault occurs at Bus 36)

2) Fault at the tie line

Consider that a three-phase short circuit fault takes place at Bus 21 at 0.1 s, and is cleared at 0.2 s.

At the moment when the fault occurs, the second smallest eigenvalue λ_2 of $L(t_s)$ decreases from 0.1664 to 0.0865, which implies that the fault makes the two-area characteristic more obvious. We calculate the eigenvector v_2 associated with λ_2 of $L(t_f)$, and divide the system into two areas according to the signs of the components of v_2, as shown in Table 8.2, generators G_1, G_2, G_7, G_8, G_9 and G_{10} are in area D_1, and the other generators are in D_2. The generator with the minimum F_i is G_5, so D_2 is chosen as the study area that contains four generators and D_1 is the external area containing six generators. We further divide the external area into two equivalent zones using the modularity index and the agglomerative method. Then generators G_1,G_2,G_9 belong to Area_1, and generators G_7,G_8,G_{10} are in Area_2. Hence, the equivalent model contains six generators, four of which are in the study area D_2 and two of which are equivalent generators in the external area.

Table 8.2 Zone division when the fault occurs at Bus 21

Generator	Component of v_2	Zone	F_i
G_1	−0.0474	D_1	0.7551
G_2	−0.0440	D_1	0.7417
G_3	0.2180	D_2	0.6128
G_4	0.1409	D_2	0.5618
G_5	0.7558	D_2	0.2648
G_6	0.2548	D_2	0.3017
G_7	−0.1707	D_1	0.8272
G_8	−0.4893	D_1	0.6881
G_9	−0.1182	D_1	0.9286
G_{10}	−0.0988	D_1	0.6762

A coherent model can be obtained through the slow coherency method[4] that contains eight generators, where six generators $G_1 - G_6$ are in the study area and the two equivalent generators are in the external area.

Time-domain simulations are performed using the detailed model and the two equivalent models. The swing curves of G_5 are shown in Fig. 8.6. In this figure, the system loses its stability at 0.27 s according to the detailed model, at 0.25 s according to the proposed model and at 0.19 s according to the coherency equivalent model. So the proposed model is much closer to the detailed model in comparison to the coherency equivalent model.

We then compare the rotor-angle curves obtained by the three models and the results are shown in Fig. 8.7 One can see that the error using the coherency equivalent model is larger.

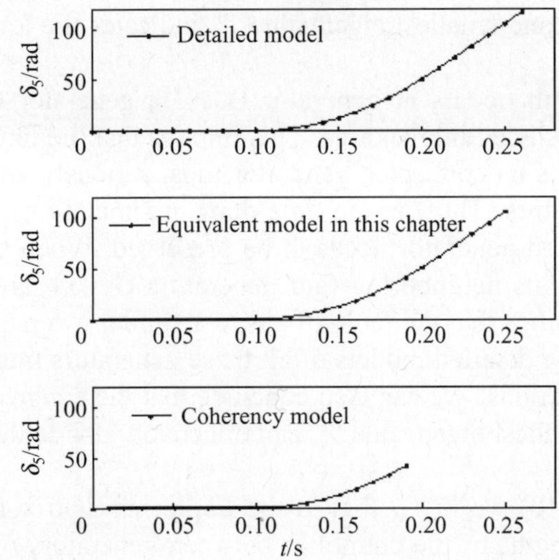

Figure 8.6 Rotor angles of G_5 before and after the equivalence (the fault occurs at Bus 21)

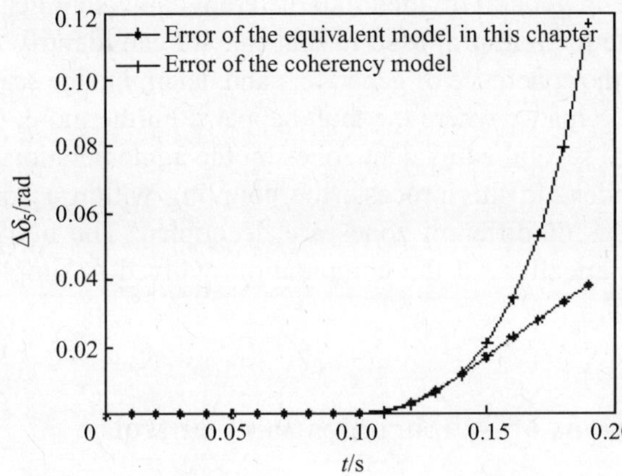

Figure 8.7 Rotor angle errors of G_5 using the two equivalent methods (the fault occurs at Bus 21)

3) Analysis of simulation results

From the above two simulation results, we draw the following conclusions.

(1) In the IEEE 39-bus system, the second smallest eigenvalue λ_2 of $L(t_s)$ is 0.1664 before the fault. If the fault takes place at generator Bus 36, then the λ_2 of $L(t_f)$ after the fault decreases to 0.0224, and if the fault occurs at the tie-line Bus 21, the λ_2 of $L(t_f)$ after the fault decreases to 0.0865. So the connections of some buses have been weakened because of the faults and the bipartition structure of the system is more obvious since the λ_2 of $L(t_f)$ decreases significantly. Generally speaking, the fault at a generator bus is more serious than the fault at a tie-line bus with a much smaller λ_2 after the fault. This observation

implies that the second smallest eigenvalue λ_2 indicates the level of the severity of a fault.

(2) When the fault occurs at generator Bus 36, generator G_6 near the fault location alone is assigned into an area. This implies that the fault at a generator's terminal only affects its connected generator most seriously and has less impact on the other generators. Thus, in the time domain simulations, only the detailed model of the affected generator needs to be preserved. When the fault occurs at the tie-line Bus 21, its neighboring four generators $G_3 - G_6$ are divided into the same area. This implies that the fault affects to a similar extent several generators near the fault, so the detailed models of all these generators must be preserved in time-domain simulations. We can also conclude that the eigenvector v_2 associated with the second smallest eigenvalue λ_2 can effectively divide the area affected by the fault.

(3) From Eq. (8.18) we know that in the Laplacian matrix L, the element L_{ij} describing the strengths of the couplings between generators i and j is inversely proportional to the inertias of these two generators. This explains that in the external area, the connection between two generators with large inertias is usually weak and they are in general divided into different equivalent generator groups.

In summary, the method proposed in this chapter can identify the influence of fault locations on the coherence of generators and determine the scale and boundary of the study area no matter where the fault happens. Furthermore, the external area can be divided into several equivalent zones by the agglomeration algorithm using the modularity index. In this process, the coupling within a zone is preserved while the dynamics of different zones are decoupled. The number of zones is determined automatically. All the computation evolved is simple and has clear physical meanings.

8.7.2 Simulations of Synchronization Control

We apply the synchronization control method to the IEEE 9-bus system and the IEEE 39-bus system in this section. We analyze the influence of fault and generator tripping on the generators' synchronization capabilities and evaluate the effectiveness of locations for generator tripping. The simulation results validate the effectiveness of the index for tripping generators and the calculation for tripping strategies.

1) IEEE 9-bus system

The IEEE 9-bus system (Fig. 8.8) contains 3 generator buses, 3 load buses and 3 tie-line buses as shown in Fig. 8.8. We assume that each generator bus (Bus 1 – Bus 3) is connected to ten identical generators, which means that the minimum tripping ratio is 10%. Since the ten generators at the generator bus are identical, they are considered to remain synchronous throughout the fault process and after the fault.

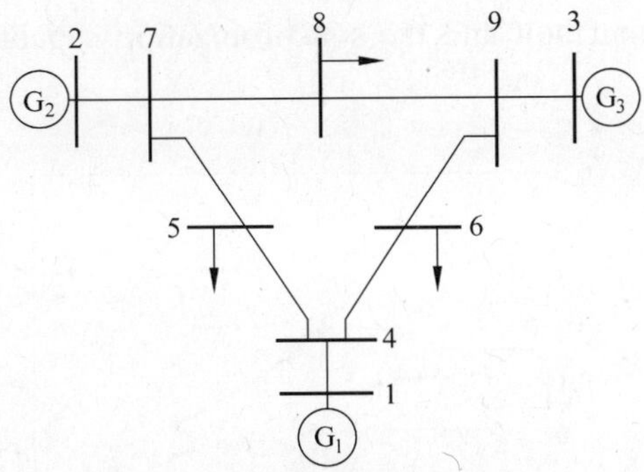

Figure 8.8 IEEE 9-bus system

Consider the case in which a three-phase short circuit occurs at Bus 2 at 0.1 s, and the fault is cleared at 0.32 s. The system will lose its stability at 0.69 s if no control actions are taken. The trajectories of the system states are shown below.

In Fig. 8.9, x_1, x_2 and x_3 denote the rotor-angle curves at Bus 1, Bus 2 and Bus 3 respectively. One can see that from the moment when the fault occurs, the three rotor-angle curves diverge gradually and the difference between x_1 and x_2 is greater than π at 0.69 s, after which the system loses its stability.

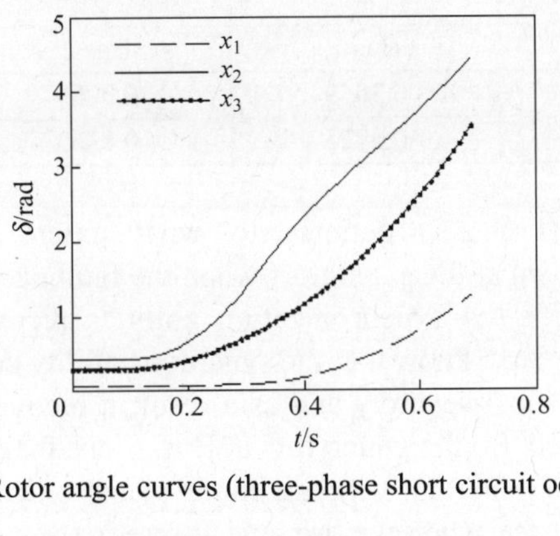

Figure 8.9 Rotor angle curves (three-phase short circuit occurs at Bus 2)

(1) The influence of the fault on the synchronizability coefficient

The changes in the synchronizability coefficients of the generators are shown in Fig. 8.10.

Here, S_1, S_2 and S_3 denote the synchronizability coefficients of Bus 1, Bus 2 and Bus 3 respectively. One can see from the figure that the synchronizability coefficients of all the three generator buses are affected severely when the fault occurs. During the evolution of the fault, their synchronization capabilities decrease to a certain extent but not significantly. After the fault is cleared, the synchronizability coefficients of all the generator buses tend to recover. However, since the system's operation point has shifted, the coefficients cannot recover to the level before the fault and the synchronization capabilities vary with the operating point.

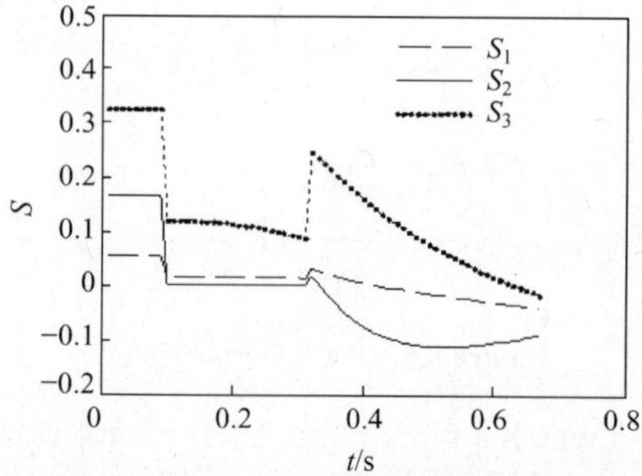

Figure 8.10 The coefficient curves (the three-phase short circuit occurs at Bus 2)

We list in Table 8.3 the influences of the fault on the synchronization capabilities of the generators when the fault happens and after the fault is cleared.

Table 8.3 The fault's influence on the synchronizability coefficients

i	1	2	3
Generator bus	Bus 1	Bus 2	Bus 3
F_{if}	0.3039	0.0099	0.3729
F_{ic}	0.5725	0.1263	0.7631

In Table 8.3, F_{if} $(i=1,2,3)$ denotes to what extent the synchronization capabilities are preserved at the generators when the fault occurs and F_{ic} $(i=1,2,3)$ denotes to what extent the synchronization abilities recover after the fault is cleared. From Table 8.3 we know that the synchronizability coefficient at generator Bus 2 is affected the most severely and as a result it recovers to the least extent after the fault is cleared. So the generators at Bus 2 lose their stability first.

(2) The influence of generator tripping on the synchronizability coefficients

Now consider the case when we are able to predict that the system will lose stability and decide to take control actions at 0.02 s after the clearance of the fault. We trip 40% of the generation outputs at all the generator buses, and the control effects are listed below.

In Table 8.4, p denotes the candidate location for tripping generators, R_{ip} is the extent to which the synchronizability coefficient at bus i has recovered after 40% of the generation outputs had been tripped, and $E(p)$ is the effectiveness index for location p.

As shown in Table 8.4, after tripping 40% of the generation output at Bus 2, the effectiveness index $E(p)=0.4443$, which ranks at the top and implies that the system recovers to the best extent.

Table 8.4 Effectiveness evaluation of tripping location

The index of candidate tripping location	p	R_{1p}	R_{2p}	R_{3p}	$E(p)$
1	Bus 1	/	−0.1113	0.6409	−0.1113
2	Bus 2	0.4443	/	0.6084	0.4443
3	Bus 3	0.3240	−0.2246	/	−0.2246

Using the correction-prediction function[17], we find that the appropriate ratio of the generator tripping at Bus 2 is 30%.

We trip 30% of the generation output at Bus 2 at 0.02 s after the fault. The trajectories of the system states are shown in Fig. 8.11, and the system is stable. If we trip 30% of the output of generators at Bus 1 or Bus 3, the system loses its stability.

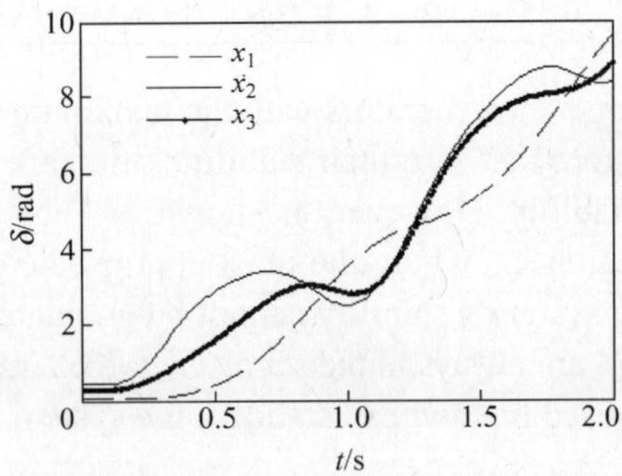

Figure 8.11 Rotor angle curves (30% of the generation output is cut off at Bus 2 when the short circuit occurs at Bus 2)

To validate the effectiveness of the proposed method in more detail, we place faults at each bus respectively in the IEEE 9-bus system and then calculate the effectiveness of each tripping location to choose the optimal tripping location. Consider the case when the fault occurs at 0.1 s, and the generator tripping actions are taken 0.02 s after the fault is cleared.

In Table 8.5, the first column lists the indices of all the calculations; the second column lists all the buses where the faults occur; the third column t_{cl} denotes the time, at which faults are cleared; the fourth column G_a lists generator buses, for which the acceleration is the maximum at the moment when the fault is cleared; the fifth column G_K presents the buses where the generators are with the maximum kinetic energies at the moment when the fault is cleared; the sixth column G_x lists the buses where the generators have the maximum angular speeds at the moment when the system loses its stability if no control actions are taken; the seventh column G_E gives the most effective locations for tripping generators that are computed by the proposed method; the eighth column G_r gives the locations of those generators, which are tripped to stabilize the system according to the time-domain simulations.

Table 8.5 Tripping location chosen by different algorithms

Index	Fault bus	t_{cl} /s	G_a	G_K	G_x	G_E	G_r
1	Bus 1	0.41	Bus 3	Bus 1	Bus 2	Bus 1	Bus 1
2	Bus 2	0.32	Bus 2	Bus 1	Bus 2	Bus 2	Bus 2
3	Bus 3	0.36	Bus 3	Bus 1	Bus 3	Bus 3	Bus 3
4	Bus 4	0.38	Bus 3	Bus 1	Bus 2	Bus 1	Bus 1
5	Bus 5	0.44	Bus 2	Bus 1	Bus 2	Bus 2	Bus 2
6	Bus 6	0.46	Bus 3	Bus 1	Bus 2	Bus 1	Bus 1 and Bus 2
7	Bus 7	0.33	Bus 2	Bus 1	Bus 2	Bus 2	Bus 2
8	Bus 8	0.37	Bus 2	Bus 1	Bus 2	Bus 2	Bus 1 and Bus 2
9	Bus 9	0.35	Bus 3	Bus 1	Bus 3	Bus 3	Bus 3

In engineering practice, the generators with the maximum acceleration or kinetic energy and the generators that lose their stabilities first are usually tripped off to improve the system stability. However, as shown in Table 8.5, these generators may not be at the location G_r where the most appropriate generator sets need to be cut off and thus the system's stability cannot be guaranteed. At the same time, the chosen locations G_E are always in the set of G_r, which implies that the locations calculated by the proposed method can stabilize the system.

2) Simulation result for the IEEE 39-bus system

The IEEE 39-bus system contains 10 generator buses as shown in Fig. 8.2.We also assume that each generator bus (Bus 30 – Bus 39) is connected to ten identical generators.

Now we consider three-phase short circuit faults at the tie-line buses and generator buses to check the effectiveness of the synchronization control method proposed in this chapter.

(1) Three-phase short circuits at tie-line buses

We consider a three-phase short circuit takes place at Bus 4 at 0.1 s, which is cleared at 0.18 s. The system loses its stability at 0.51 s if no control actions are taken as shown in Fig. 8.12.

In this figure, x_1 – x_9 denote the rotor-angle curves of the generators at Buses 31 – 39 respectively, x_{10} is the rotor-angle curve of the generator at Bus 30.

Table 8.6 presents the effectiveness for each tripping location. Different effectiveness indices $E(p)$ are obtained when cut off 30% and 60% of each generator's output while the ranking of $E(p)$ remains the same. Thus, it is clear that the tripping locations are independent of the tripping amounts, and the choices for these two variables are decoupled. According to the ranking of $E(p)$, Bus 32 is the most effective tripping location.

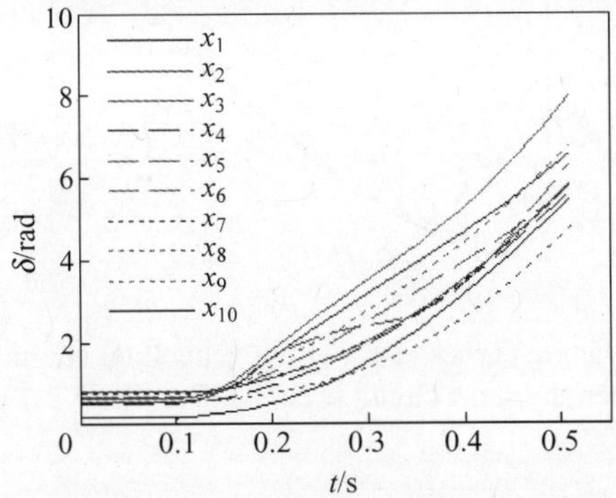

Figure 8.12 Rotor angle curves (the three phase short-circuit occurs at Bus 4)

Table 8.6 The effectiveness of tripping locations and its ranking when the short circuit occurs at Bus 4

Candidate tripping location	Ranking of effectiveness	$E(p)$ (tripping 30%)	$E(p)$ (tripping 60%)
Bus 31	10	0.0687	−0.3783
Bus 32	1	0.3073	−0.0675
Bus 33	5	0.1343	−0.2064
Bus 34	3	0.1402	−0.1902
Bus 35	7	0.1278	−0.2298
Bus 36	8	0.1111	−0.2680
Bus 37	6	0.1304	−0.2179
Bus 38	4	0.1381	−0.1972
Bus 39	9	0.1025	−0.2684
Bus 30	2	0.1506	−0.1815

Using the prediction function[17], 30% of the outputs of the generators at Bus 32 need to be tripped. We do so at 0.2 s. The rotor-angle curves of the generators are shown in Fig. 8.13 and the system is stable.

(2) Three-phase short circuits at generator buses

We consider a three-phase short circuit takes place at the generator Bus 33 (G_3) at 0.1 s, which is cleared at 0.22 s. The system loses its stability at 0.51 s if no control actions are taken as shown in Fig. 8.14.

In Fig. 8.14, $x_1 - x_9$ denote the rotor-angle curves of the generators at Buses 31 – 39 respectively, x_{10} denotes the rotor-angle curve of the generators at Bus 30.

Table 8.7 shows the effectiveness of each tripping location.

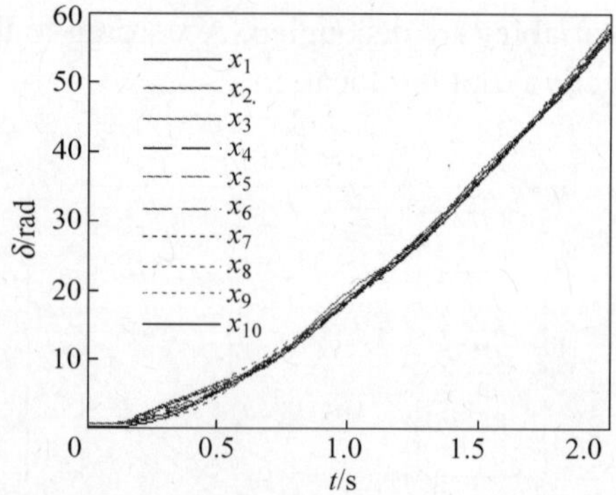

Figure 8.13 Rotor-angle curves (30% of the generation output are tripped at Bus 32 when the short circuit occurs at Bus 4)

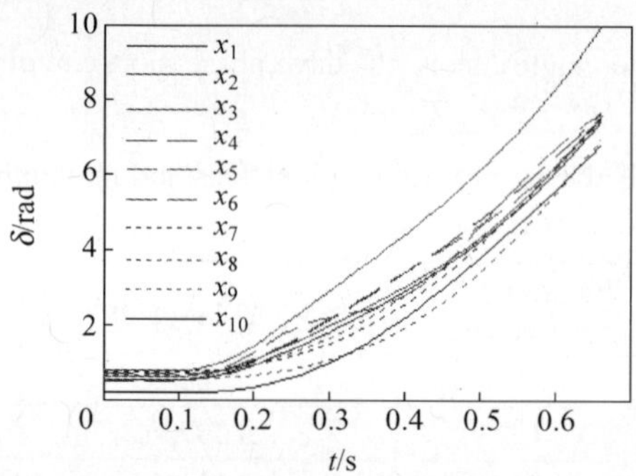

Figure 8.14 Rotor-angle curves (the three-phase short circuit occurs at Bus 33)

Table 8.7 The effectiveness indices of tripping locations and their ranking when the short circuit occurs at Bus 33

Candidate tripping location	Ranking of effectiveness	$E(p)$ (tripping 30%)	$E(p)$ (tripping 60%)
Bus 31	6	0.0661	0.0545
Bus 32	7	0.0569	0.0349
Bus 33	1	0.3771	0.3283
Bus 34	10	–0.0022	–0.0936
Bus 35	8	0.0416	0.0023
Bus 36	9	0.0241	–0.0426
Bus 37	5	0.0686	0.0599
Bus 38	4	0.0716	0.0664
Bus 39	3	0.0760	0.0783
Bus 30	2	0.0834	0.0913

Similar to Table 8.6, different effectiveness indices $E(p)$ are obtained when we trip 30% and 60% of the generators' outputs while the ranking of $E(p)$ remains the same. Bus 33 is the most effective tripping location according to the ranking of $E(p)$, and Bus 30 is the second most effective one.

Using the prediction function[17], 60% of the outputs of the generators at Bus 33 need to be tripped. We do so at 0.24 s, and the rotor-angle curves of the generators are shown in Fig. 8.15. The system is obviously unstable.

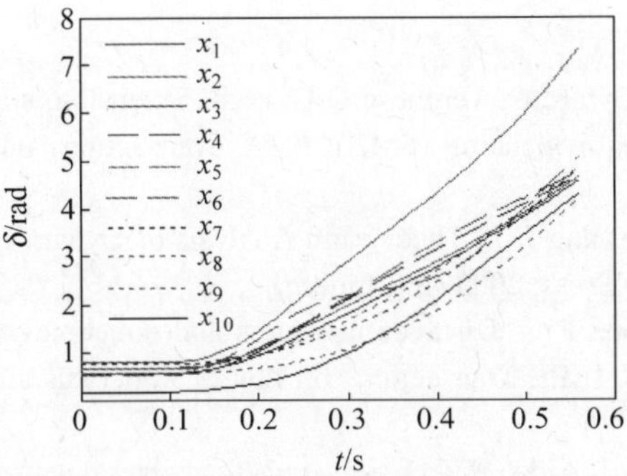

Figure 8.15 Rotor-angle curves (60% of generation outputs are cut off at Bus 33 when the short circuit occurs at Bus 33)

The first computation iteration fails to solve for the tripping strategy. We choose Bus 30 to be the tripping location according to the effectiveness index ranking, and according to the computation result of the prediction function, 90% of its outputs need to be tripped. So we cut off 90% output of the generation output at Bus 30 at 0.24 s. The time-domain simulations of the generators' rotor-angel curves are shown in Fig. 8.16, and the system is stable. Note that the power is not balanced because of the tripping, and the frequency declines. But since the tripped generation capacity is only a small fraction of that of the overall system, the generation outputs will be controlled in the secondary frequency regulation after the system recovers to its stable states.

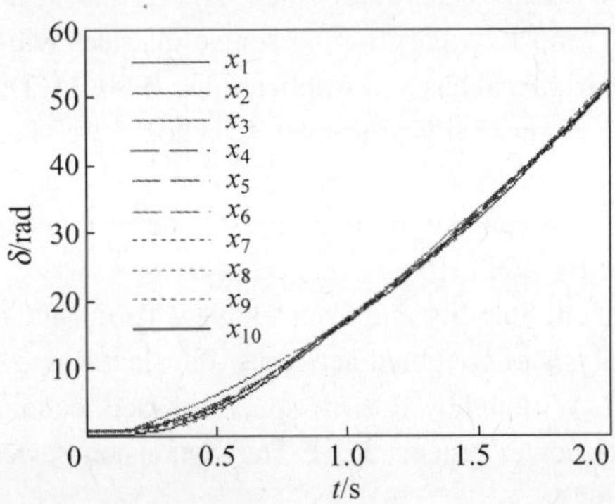

Figure 8.16 Rotor angle curve (90% of output are cut off at Bus 30 when short circuit occurs at Bus 33)

One can see from the above process that, Bus 33 corresponds to the tripping location with the maximum effectiveness index, but we cannot stabilize the system after tripping at this location, so we instead choose Bus 30 with the second largest effective index to be the tripping location, and a tripping strategy is successfully obtained.

References

[1] Ramaswamy G N, Evrard C, Verghese G C, et al. Extension, simplification, and tests of synchronic modal equivalencing (SME). IEEE Transactions on Power Systems, 12(2): 896 – 905

[2] Ni Y X, Chen S S, Zhang B L. Theory and Analysis of Dynamic Power System. Beijing: Tsinghua University Press, 2002 (in Chinese)

[3] Lee S T Y, Schweppe F C. Distance measures and coherency recognition for transient stability equivalents. IEEE Transactions on Power Apparatus and Systems, 1973, 92(4): 1550 – 1557

[4] Yusof S B, Rogers G J, Alden R T H. Slow coherency based network partitioning including load buses. IEEE Transactions on Power Systems, 1993, 8(3): 1375 – 1382

[5] Gacia N, Zecevic A I, Siljak D D. Coherency recognition using epsilon decomposition. IEEE Transactions on Power Systems, 1998, 13(2): 314 – 319

[6] Nath R, Lamba S S, Rao K S P. Coherency based system decomposition into study and external areas. IEEE Transactions on Power Apparatus and Systems, 1985, 104(6): 1443 – 1449

[7] Lo K L, Qi Z Z, Xiao D. Identification of coherent generators by spectrum analysis. IEE Proceedings on Generation, Transmission and Distribution, 1995, 142(4): 367 – 371

[8] Hauer J F, Demeure C J, Scharf L L. Initial results in prony analysis of power system response signals. IEEE Transactions on Power Systems, 1990, 5(1): 80 – 89

[9] Newman M E J, Girvan M. Finding and evaluating community structure in networks. Physical Review E, 2004, 69(2): 026113

[10] Girvan M, Newman M E J. Community structure in social and biological networks. Proceedings of the National Academy of Sciences, USA, 2002, 99: 7821 – 7826

[11] Pothen A, Simon H, Liou K P. Partitioning sparse matrices with eigenvectors of graphs. SIAM Journal on Matrix Analysis and Applications, 1990, 11(3): 430 – 452

[12] Fiedler M. Algebraic connectivity of graphs. Czechoslovak Mathematical Journal, 1973, 23(98): 298 – 305

[13] Wang X F, Chen G. Synchronization in complex dynamical networks. Journal of Systems Science and Complexity, 2003, 16: 1 – 14

[14] Kundur P. Power System Stability and Control. New York: McGraw-Hill, 1994

[15] Newman M E J. Analysis of weighted networks. Physical Review E, 2004, 70: 056131

[16] Chow J H, Cullum J, Willoughby R A. A sparsity-based technique for identifying slow-coherent areas in large power systems. IEEE Transactions on Power Apparatus and Systems, 1984, 103(3): 463 – 473

[17] Zhang X M, Mei S W, et al. Transient and emergency control. Science in China, Series E: 2009, 52(2): 420 – 428

[18] Hu J,Yu Y X. A practical method of parameter aggregation for power system dynamic equivalence. Power System Technology, 2006, 30(24): 26 – 30 (in Chinese)

Chapter 9 Blackout Model Based on DC Power Flow

This chapter first reviews briefly how to compute DC power flows. Then the blackouts and SOC phenomena in the Northeast Power Grid of China are discussed so that one can apply the blackout model to this power grid based on DC power flows. Consequently, we are able to identify the relationship between certain power system parameters and the blackouts and find the probability distribution of faults. At the same time, indices for blackout risks are defined and computed.

As shown in previous chapters, power systems evolve with typical SOC characteristics, the insight into which will have significant impact on raising the security level for the operation of power systems. It has been discussed in Chapter 3 that two classes of methods have been used to study SOC, namely the priori and the posteriori methods. When the posteriori method is adopted, the critical characters of a real power system, such as the power law correlations and the long-range correlations, are obtained by statistical analysis of a sufficiently large amount of data collected from the system, and correspondingly the future dynamical behavior of the system is predicted in a statistical sense. Though widely utilized, the a posteriori method has some limitations that are difficult to overcome. For example, blackout data need to be analyzed in order to reveal the mechanism of SOC; however, these data are difficult to be collected and cannot be validated by replaying the cascading process in the same real system. Thus in this context, it is preferable to use the a priori method, with which SOC is studied via constructing blackout models and implementing simulation platforms. Since engineers usually use power flows to describe the delivery of electric power in a power system and cascading failures always happen with power flow redistribution, it is natural to take the a priori approach to build the cascading model upon power flow equations and consequently use this model to study the start and development of a cascading blackout.

In a typical power system with hundreds of transmission lines, to determine the power flows requires dealing with an extremely large set of simultaneous nonlinear algebraic equations. On the one hand, such a process of finding a precise solution can easily become too complicated to proceed; on the other hand, in some cases, such as real-time operation and planning, the precision of a power flow solution is less important compared to the convergence speed of the computation process itself. Hence, the DC power flow has been introduced which can be computed efficiently by solving a set of linear equations. It is because of the guaranteed fast convergence of the computation of DC power flows that the blackout model based on DC power flows becomes advantageous to study cascading failures. For the same reason, in this chapter we utilize DC power flow models to simulate the long-term evolution of power systems and analyze the mechanism of blackouts.

9.1 DC Power Flow and Related Optimization

9.1.1 DC Power Flow

In various scenarios of the operation of power systems, engineers are mainly concerned with the distribution of active power, in which case it is less important to compute bus voltages. This is where the computation of DC power flows is directly related[1]. In what follows, we review the steps to compute DC power flows.

The following observations hold when a power system operates in a steady state: the magnitude of the voltage at each bus is around its nominal value; the phase angle difference between the two ends of a line is comparatively small; the resistance of an extra high voltage transmission line is also comparatively small. So for a line connecting bus *i* and bus *j*, we may assume that $U_i \approx U_j \approx 1$, $\sin\theta_{ij} \approx \theta_{ij}$, $\cos\theta_{ij} \approx 1$, $r_{ij} \approx 0$, and that its shunt branch can be ignored. Then the power flow equation of line (i, j) is

$$\begin{aligned} P_{ij} &= (U_i^2 - U_i U_j \cos\theta_{ij}) g_{ij} - U_i U_j b_{ij} \sin\theta_{ij} \\ &\approx (\theta_i - \theta_j)/x_{ij} \end{aligned} \tag{9.1}$$

where for the line (i, j), g_{ij} is the conductance; b_{ij} is the susceptance and x_{ij} is the reactance.

Comparing with Ohm's law for DC circuits, we can take P_{ij} as the DC current, θ_i and θ_j as the bus voltages, and x_{ij} as the line resistance. Then applying Kirchhoff's current law to bus *i*, we have the current balance equation

$$P_i^{SP} = \sum_{j \neq i} P_{ij} = \sum_{j \neq i} (\theta_i - \theta_j)/x_{ij}, \quad i = 1, 2, \cdots, n \tag{9.2}$$

where P_i^{SP} is the injected active power at bus *i*, and *n* is the total number of buses. Eq. (9.2) can be further written into its matrix form as

$$P^{SP} = B_0 \theta \tag{9.3}$$

where P^{SP} and θ are n-dimensional vectors and B_0 is the $n \times n$ bus admittance matrix constructed by using the line admittances $1/x_{ij}$. Since the shunt resistance has been neglected, it holds that for line (i,j)

$$P_{ij} + P_{ji} = 0 \tag{9.4}$$

Again from Kirchhoff's current law, we know that

$$\sum_{i=1}^{n} P_i^{SP} = 0 \quad \text{or} \quad P_n^{SP} = -\sum_{i=1}^{n-1} P_i^{SP} \tag{9.5}$$

So the sequence $\{P_i^{SP} \mid i = 1, 2, \cdots, n\}$ is not independent. Set the phase angle of bus n to be zero, namely $\theta_n = 0$. Then delete the components corresponding to bus n from the vectors P^{SP} and θ, and delete the column and row corresponding to bus n from the matrix B_0. As a result, the power flow equation becomes

$$\tilde{P}^{SP} = \tilde{B}_0 \tilde{\theta} \tag{9.6}$$

where $\tilde{P}^{SP}$ and $\tilde{\theta}$ are $(n-1)$-dimensional vectors, and $\tilde{B}_0$ is $(n-1) \times (n-1)$ matrix with elements

$$\begin{cases} \tilde{B}_0(i,i) = \sum_{j \neq i} 1/x_{ij} \\ \tilde{B}_0(i,j) = -1/x_{ij} \end{cases} \tag{9.7}$$

We follow the convention to use F to denote the line flows, which are calculated by

$$\begin{aligned} F &= M_f \theta \\ &= [\tilde{M}_f \quad M_{fn}] \begin{bmatrix} \tilde{\theta} \\ \theta_n \end{bmatrix} \\ &= \tilde{M}_f \tilde{\theta} = \tilde{M}_f (\tilde{B}_0)^{-1} \tilde{P}^{SP} \end{aligned} \tag{9.8}$$

where M_f is the branch-bus incidence matrix, $\tilde{M}_f$ is the matrix obtained from M_f by deleting the column corresponding to bus n, and M_{fn} is the column vector corresponding to bus n. Define $A = \tilde{M}_f (\tilde{B}_0)^{-1}$, then the line flow equation becomes

$$F = A\tilde{P}^{SP} \tag{9.9}$$

In view of the DC power flow Eqs. (9.6) and (9.9), the phase angle of each bus and the line flows can be easily obtained using the active power injected at the buses.

So in the computation of DC power flows, it has been assumed that the resistances and the power losses of transmission lines are negligible, the differences in phase angles are small, and only active power flows need to be considered when setting voltages at all buses to be the same without upper and lower voltage limits. In high voltage networks, such assumptions usually hold approximately. To see this, we use the Northeast Power Grid of China as an example to check these assumptions.

We compare the computation results of DC and AC power flows in the Northeast Power Grid of China at 11 o'clock in the morning of July 11, 2005. The grid contains 568 buses and 693 transmission lines. Here we take transformers equivalently as transmission lines.

The histogram of the voltage distribution obtained from AC power flows is showed in Fig. 9.1, where the horizontal axis indicates the voltage magnitude U (p.u.) and the vertical axis indicates the number of buses N. Figure 9.1 shows that the average magnitude of voltage is 0.983 p.u. and only 8 buses' voltage magnitudes deviate by more than 0.1 p.u. from 1 p.u., the voltage magnitude in DC power flows. Note that 8 buses are only about 1.41% of the total buses.

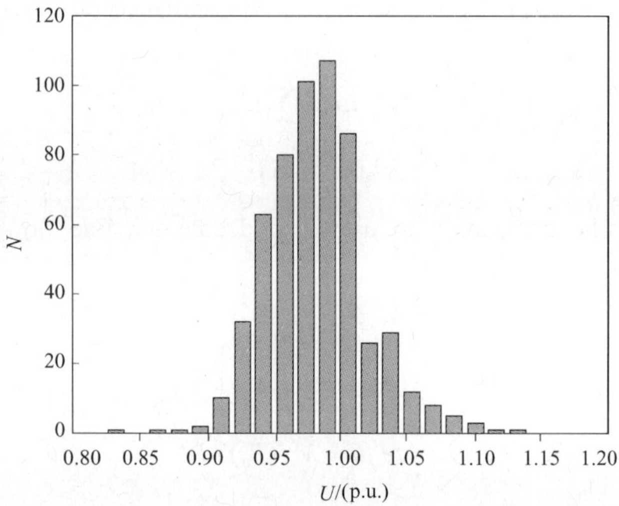

Figure 9.1 Distribution of the voltage magnitudes calculated by AC power flows

The histogram of the distribution of differences of phase angles obtained from the AC power flows is showed in Fig. 9.2, where the horizontal axis indicates the difference of phase angles θ (degree) and the vertical axis indicates the number of transmission lines M. Figure 9.2 shows that the average phase-angle difference is 2.49 degrees and there are 19 transmission lines with phase-angle differences greater than 10 degrees, which take up about 3.34% of all lines. The maximum phase-angle difference is 18 degrees. Hence, the assumption that $\sin\theta_{ij} \approx \theta_{ij}$ and $\cos\theta_{ij} \approx 1$, will only lead to very small inaccuracy in the DC power flow computation.

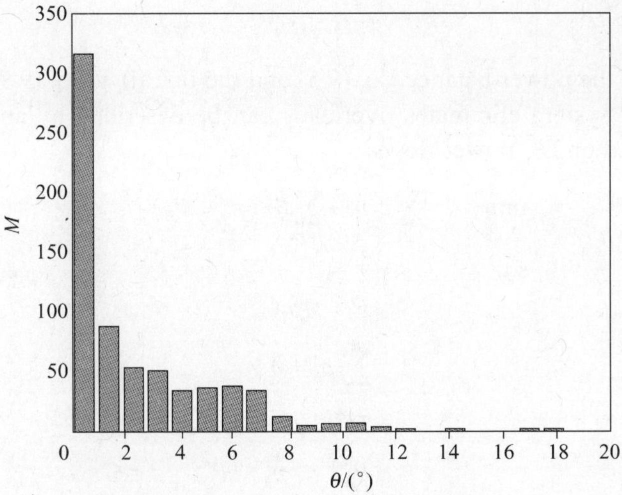

Figure 9.2 Distribution of phase-angles calculated by the AC power flows

Last but not least, the average resistance of the high voltage line is only about 0.0053 p.u. Thus, the assumption $r_{ij} \approx 0$ is also reasonable.

9.1.2 The Optimization Problem Based on DC Power Flow

When overload occurs in a transmission line or a transformer in a real power system for some reason, adjustment has to be made to change the outputs of generators or, when necessary, to shed some loads. These adjustments are done either manually according to the commands from system operators or automatically. This process can be described by the following optimization problem.

Consider a system with n buses and m lines. Let p_i be the active power injected at bus i and $p=[p_1, p_2, \cdots, p_n]^{\mathrm{T}}$. Let G denote the set of generators and D the set of loads. The maximum and minimum power limits for generator i are denoted by $P_{gi}^{\max}$ and $P_{gi}^{\min}$, then generator i's power limits are

$$P_{gi}^{\min} \leqslant p_i \leqslant P_{gi}^{\max}, \quad i \in G \tag{9.10}$$

Let P_{dj} (<0) be the demand at bus j, then the range for load p_j at bus j satisfies

$$P_{dj} \leqslant p_j \leqslant 0, \quad j \in D \tag{9.11}$$

Let L denote the set of transmission lines, F_l be the power flow in line l and $F_l^{\max}$ be the transmission capacity of line l. Let $F=[F_1, F_2, \cdots, F_m]^{\mathrm{T}}$. Taking into account the directions of the power flows, the constraint for the line transmission capacity is

$$-F_l^{\max} \leqslant F_l \leqslant F_l^{\max}, \quad l \in L \tag{9.12}$$

In view of the power-balance Eq. (9.5) and the line flow Eq. (9.9), the process that a power system eliminates overloads can be described by an optimization problem based on DC power flows:

$$\min \quad \sum_{i \in G} c_i p_i + \sum_{j \in D} W_j (p_j - P_{dj}) \tag{9.13}$$

s.t.

$$F = Ap$$
$$\sum_{i=1}^{n} p_i = 0$$
$$P_{dj} \leqslant p_j \leqslant 0, j \in D$$
$$P_{gi}^{\min} \leqslant p_i \leqslant P_{gi}^{\max}, \ i \in G$$
$$-F_l^{\max} \leqslant F_l \leqslant F_l^{\max}, l \in L$$

where p is the injected active power to be optimized, c_i is the unit generation cost for generator i, and W_j is the unit cost for shedding loads at bus j. The optimization task is to minimize the sum of costs for generation and power loss. Since power systems usually operate to meet demands from loads, the parameters W_j are in general large numbers. The constraints in this optimization problem include the power flow equation, the power balance equation, the load limits, the generator output limits and the transmission capacity limits. Once the solution of the injected power p is obtained, the overloads in transition lines or transformers can be eliminated with minimum costs by adjusting the outputs of generators or shedding chosen loads.

9.2 OPA Model

9.2.1 Introduction of OPA Model

Generally speaking, the time-scale for cascading failures differs greatly from that of power system evolutions because a cascading failure can happen within a short period of time while power system evolution may take a very long time. Thus an appropriate model to study cascading failures and blackouts in power systems must take into account both of these two time-scales. The OPA model[2–4] proposed in this context is a simplified model jointly studied by the Oak Ridge National Laboratory (ORNL), the Power System Engineering Research Centre (PSERC) at Wisconsin University and the Physics Department at Alaska University. In fact,

OPA, the name of the model, consists of the first letters of the names of the three institutes.

The OPA model is based on the DC flows for transmission lines, loads and generators. Starting from the changes in loads, the OPA model has been developed to explore the global dynamics of blackouts in power systems. The main idea is that as generations and loads increase continuously, line power flows grow accordingly, which results in breakdowns in some overloaded transmission lines that in turn lead to the power flow increase in other lines and further overloads and breakdowns until finally the cascading failure happens. Note that usually the planning department will improve the transmission capacity of the overloaded lines to increase the safety level of the transmission system.

As shown in Fig. 9.3, the OPA model contains two loops. The inner loop corresponds to the fast dynamics that simulates the cascading failure; the outer loop corresponds to the slow dynamics that simulates the upgrades and improvements in power systems including the increase of generations in response to the growing loads and the improvement in transmission capabilities.

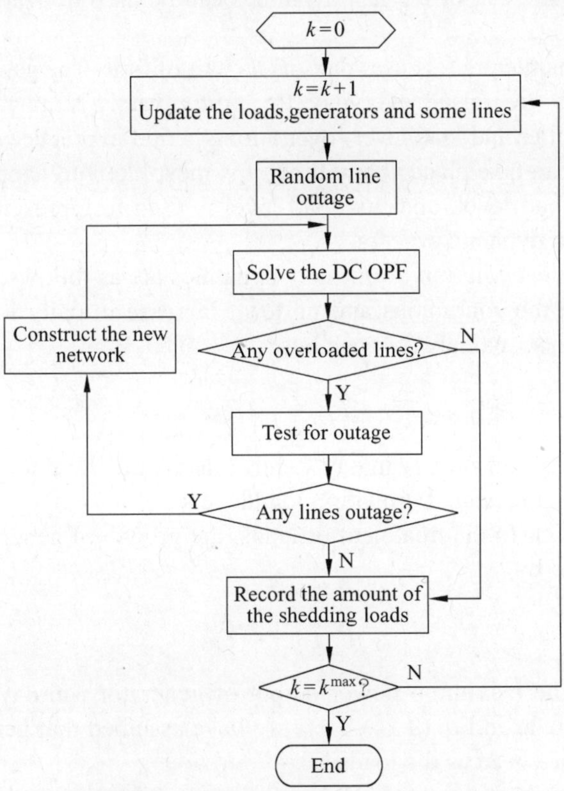

Figure 9.3 Flowchart of the OPA model

The fast dynamics take place and complete in a single day within the slow dynamics, which describes the cascading failure developing according to the following steps:

Step 1 Consider day k. The capacities and initial outputs of the generators and the load demands are determined by the slow dynamics (see slow dynamics below for details). The outage probability of every transmission line is set to be τ to simulate accidental faults.

Step 2 Calculate the generator outputs and the loads by solving the optimization problem (9.13). The power flows are calculated at the same time.

Step 3 If there are transmission lines being overloaded, namely $F_l/F_l^{\max} \geqslant \alpha$, then go to Step 4. Otherwise, exit the fast dynamics.

Step 4 For the lines that are found in Step 3, each of them is tripped with probability β to simulate the action of the relay protection. If there are lines tripped off, go to Step 2; otherwise, exit the fast dynamics.

In the fast dynamics, the parameter α can be taken as a threshold value for being overloaded. The transmission lines satisfying $F_l / F_l^{\max} \geqslant \alpha$ are considered as overloaded and will be tripped off with probability β. The amount of loads that are shed at the end of the fast dynamics can be used to evaluate the scale of the blackout.

The slow dynamic process uses days as its unit of time. The generator capacities and load levels are raised everyday to approximately simulate the jumps in generator capacities and load levels over a long period in practice. Within one day, though failure can take place at any time, it is most likely to happen at a moment with the peak load level, and this load level is used to represent the daily load level in the slow dynamic process.

The steps in the evolution of the slow dynamics are as follows:

Step 1 Use the continuous and uniform increase in daily loads to simulate the annual increase, namely for each bus

$$P_{di,k+1} = \lambda P_{di,k} \tag{9.14}$$

where $P_{di,k}$ is the load at bus i in day k and λ is the growth rate of load level that is determined by the annual growth in load levels.

Step 2 Similar to the treatment of loads, the growth of generators' capacities can be described by

$$P_{gi,k+1}^{\max} = \lambda P_{gi,k}^{\max} \tag{9.15}$$

where $P_{gi,k}^{\max}$ is the maximum power output of generator i in day k and λ is the same growth rate as in Eq. (9.14). Here we have assumed that generation outputs grow at the same speed as the loads.

Step 3 For those lines that have been tripped off in the fast dynamics, use the average improvement effect to simulate the transmission upgrades, namely

$$\begin{cases} F_{j,k+1}^{\max} = \mu F_{j,k}^{\max} \\ Z_{j,k+1} = \dfrac{1}{\mu} Z_{j,k} \end{cases} \tag{9.16}$$

where $F_{j,k}^{\max}$ is the transmission capacity of line j in day k and μ is the growth rate of the transmission capacity. And $Z_{j,k}$ is the impedance of line j in day k.

For those lines that have been working within its capacities, no improvement is made

$$\begin{cases} F_{j,k+1}^{\max} = F_{j,k}^{\max} \\ Z_{j,k+1} = Z_{j,k} \end{cases} \tag{9.17}$$

Combining the fast and slow dynamics, the overall OPA model is illustrated in Fig. 9.3 and can be summarized as the following three steps:

Step 1 $k=0$. Initialization. Determine the generation outputs, load demands and network topology.

Step 2 Fast dynamics in day k. Record the amount of the shed loads.

Step 3 Slow dynamics. Update the load levels, generation outputs and line capacities. $k = k+1$. If k reaches a pre-set limit $k = k^{\max}$, stop the simulation process; otherwise, go to Step 2.

9.2.2 SOC Analysis Based on the OPA Model

The study on the OPA model shows new directions for the application of SOC theory to the analysis of power systems. It has been recognized that the OPA model is a seminal work to introduce complex system theory to study the security issues in large power grids. In particular, the OPA model leads to the promising approach of using simulations to obtain data for system evolutions and blackouts and accordingly carrying out research on the fault distributions and their SOC characteristics. In this section, we apply this model to the Northeast Power Grid of China. To be more concrete, we simulate a 2000-day dynamical process, obtain the blackout distribution curves, and analyze the relationships between power system operating parameters and the blackout distributions. In order to evaluate the blackout risks, we further utilize indices of the VaR and the CVaR that have been discussed in Chapter 3.

1) The Northeast Power Grid of China and its simulation

The Northeast Power Grid of China consists of 4 parts, which are Heilongjiang province, Jilin province, Liaoning province and the north of the Inner Mongolia municipality. There are about 500 transmission lines, more than 300 transformers, 250 substations and about 200 generators. It covers an area of more than 1.2×10^6 km^2 and serves for more than 100 million people. In this system, most of the hydropower plants are located in the east, and most of the thermal power plants are located in the west and some port cities in the east. The main consumers are in the middle of the south. Hence, the electric power is transmitted from the west and the east to the middle and from the north to the south.

In the simulation study, we consider 500 kV and 220 kV transmission lines and substations which correspond to a 570-bus system, and the 500 kV system is shown in Fig. 9.4.

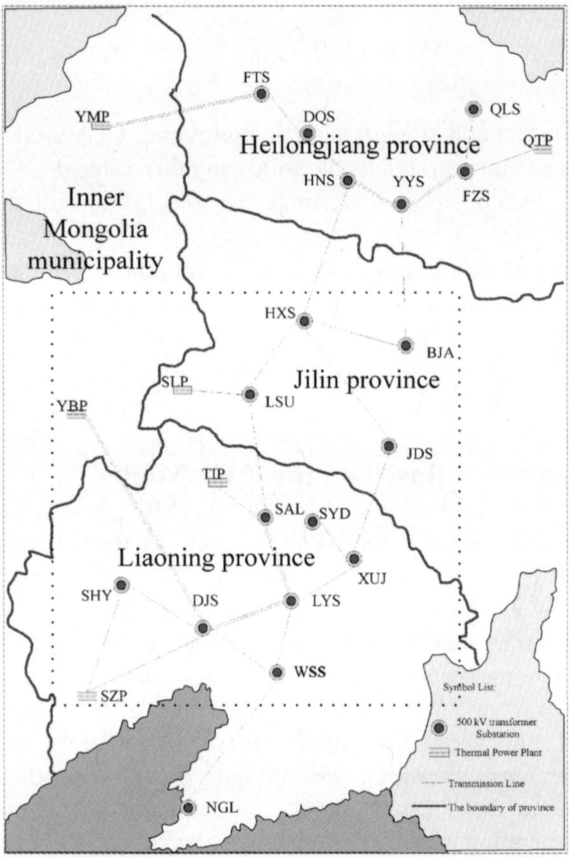

Figure 9.4 Northeast Power Grid of China (500 kV)

The parameters are collected from the statistics of the system:

(1) The generation and load growth rates λ

The load level in July 2005 is 18,394 MW[5], and that in July 2006 is 21,400 MW[6]. So the annual growth rate in load level is about 16.45%, which amounts to about 0.041% in daily growth. Thus we set $\lambda = 1.00041$.

(2) The probability τ for line outage due to natural causes like lightening

There are 178 reported faults in the grid in 2003[7]. Since there are about 700 transmission lines and transformers, in a given day and for each transmission line, the probability of line outage is $\tau = 0.0007$.

(3) The trip probability β for overloaded lines

In real systems, to guarantee the system security, overloaded lines should be tripped off with a probability greater than 99%. So in the simulation, we set $\beta = 0.999$.

(4) The growth rate μ for the improvement in transmission line capacities

This growth rate must be greater than the load growth rate λ. We set $\mu = 1.005$.

We also set $\alpha = 0.99$. For convenience, we say the chosen values for the above parameters are the parameters' initialization values.

Using the OPA model, simulation results are shown as follows for the evolution of 2000 days for the Northeast Power Grid of China. The change in daily load shedding P_{loss} with respect to day k is shown in Fig. 9.5.

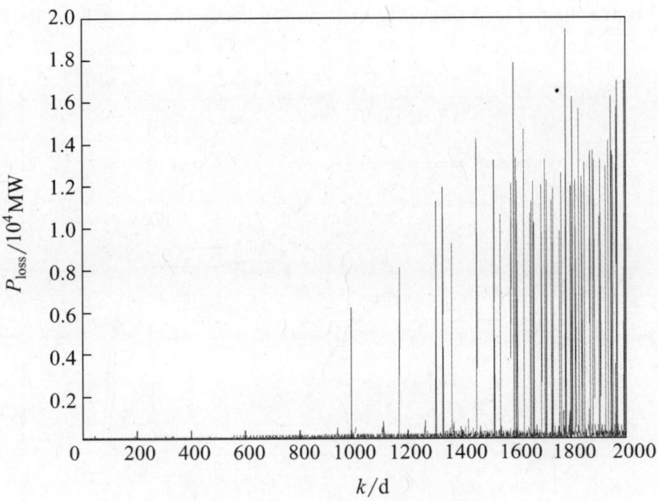

Figure 9.5 Daily load shedding

It is clear from Fig. 9.5 that the operation state of the grid varies from day to day. When a fault does happen, its scale differs greatly. We further show the probability distribution curve $F(P_{\text{loss}})$ for the scale of faults in Fig. 9.6.

Remark 9.1 For convenience, in the rest of this chapter when referring to the probability distribution curve for the scale of faults, we always use power loss P_{loss} to describe the scale of a fault and use $F(P_{\text{loss}})$ to describe its distribution.

2) The influence of the tripping probability β of overloaded lines on the blackout distribution

In Fig. 9.7, two curves are obtained when the values of β are different and the other initialization values are the same.

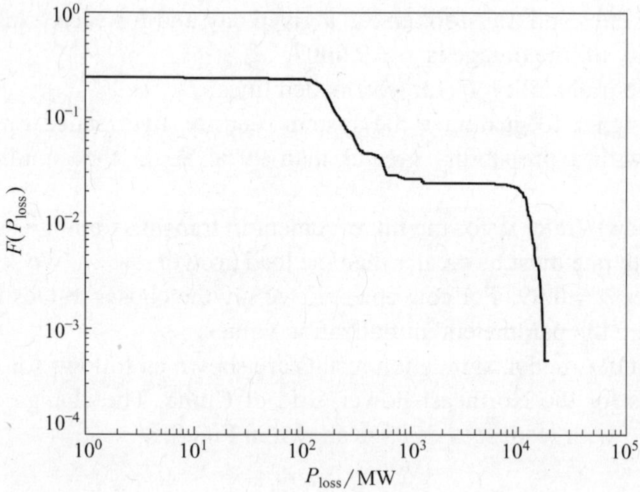

Figure 9.6 The probability distribution curve for the scale of faults

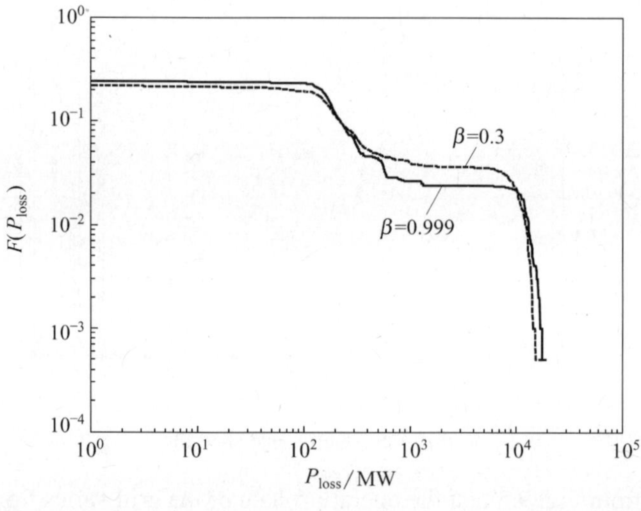

Figure 9.7 The probability distributions with different misoperation probabilities

Furthermore, let $\sigma = 0.99$, then the computation results of the VaR and the CVaR are listed in Table 9.1.

Table 9.1 The risks for blackouts with different tripping probabilities

β	VaR/MW	CVaR/MW
0.999	13,241	149.59
0.3	12,319	136.78

From Fig. 9.7 and Table 9.1 we can see that when increasing β, the tripping probability for overloaded lines, the number for large- and small-scale faults also grows, and in the mean time the number for mid-scale faults drops. When considering the blackout risks, increasing β will increase the risks evaluated by the VaR and the CVaR slightly.

The reason for these observations is rooted in the fact that the power flows in the tripped transmission lines have to be redistributed to other lines and thus may lead to cascaded faults. When a mid-scale fault has developed, the load redistribution as a result of the large trip probability helps the fault further evolve into large-scale failures. Note that in the OPA model, should an overloaded line not be tripped off, it would not be upgraded and thus the problem of insufficient capacity of that line would remain unsolved. So keeping an overloaded line can only postpone the happening of a fault, but cannot solve the system security problem in the long run.

3) The influence of the transmission capacity growth rate μ on the distribution of blackouts

The curves in Fig. 9.8 show the distributions of blackouts with the same initialization values except for the values of the growth rate μ.

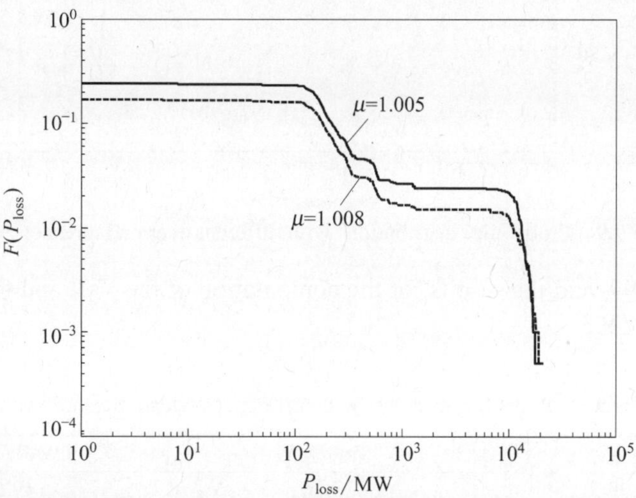

Figure 9.8 Probability distributions with different growth rates for transmission capacities

Let $\sigma = 0.99$, then the values of the VaR and the CVaR are listed in Table 9.2.

We have shown that increasing the transmission capacity growth rate can significantly reduce faults and lower the blackout risk. So upgrading transmission lines is one of the most effective methods to minimize the loss caused by faults; of course, the implementation of an upgrade must be jointly considered with its investment costs.

Table 9.2 Risks for blackouts with different transmission capacity growth rates

μ	VaR/MW	CVaR/MW
1.005	13,241	149.59
1.008	10,962	146.35

4) The influence of the overload threshold value α on the distribution of blackouts

The curves in Fig. 9.9 indicate the distributions of blackouts with the same initialization values except for the values of the overload thershold value α.

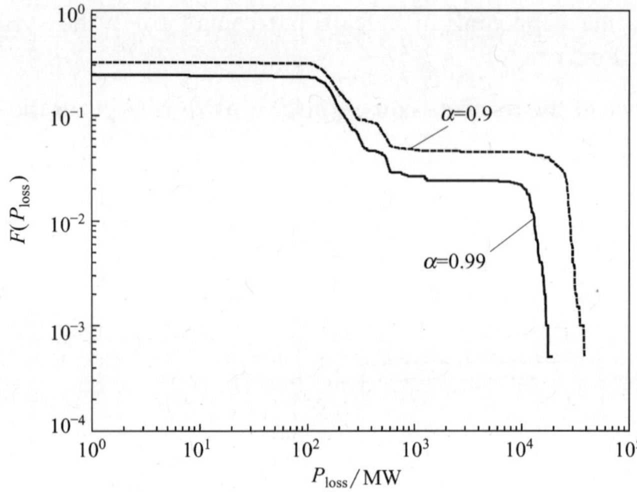

Figure 9.9 Probability distributions with different overload threshold values

Let $\sigma = 0.99$, and the results for the computation of the VaR and the CVaR are listed in Table 9.3.

Table 9.3 Risks for blackouts with different overload threshold values

α	VaR/MW	CVaR/MW
0.99	13,241	149.59
0.9	28,050	306.94

The above computation discloses the strong correlation between the overload threshold value and the risk for blackouts.

5) Discussion on SOC at a higher level

The macroscopic SOC characteristics of an interconnected power grid lie in the interaction of two conflicting forces: with the continuous social economic developments, the growth of load demands reduces the operational margin of the entire system and thus increases the risk of cascading failures and blackouts; however, at the same time the construction for new generation facilities and transmission lines improve the ability of the grid to handle increasing loads and thus lower the risk for faults. These two opposite acting forces are applied to the power system according to their inherent rules and gradually drive the power system to a critical state. To delineate indices for the macroscopic SOC phenomena, we first define two criteria that are determined by the two conflicting forces just described.

Define

$$I(k) = \frac{\sum_{i \in D} P_{i,k}}{F_k} \tag{9.18a}$$

$$\overline{P}_{\text{loss}}(k) = \frac{P_{\text{loss}}(k)}{P_{\text{demand}}(k)} \tag{9.18b}$$

where D is the set of load buses, F_k denotes the overall transmission capacity on day k, $\sum_{i \in D} P_{i,k} = P_{\text{demand}}(k)$ is the overall power demand on day k, so the indicator $I(k)$ is the ratio of the overall power demand and the overall transmission capacity, $P_{\text{loss}}(k)$ is the power loss on day k, so the indicator $\overline{P}_{\text{loss}}(k)$ denotes the blackout scale evaluated by the relative quantity of power loss.

In addition the total transmission capacity F_k can be calculated approximately by the sum of the transmission capacity of the main lines, namely

$$F_k = \sum_{j \in L} F_{j,k}^{\max} \tag{9.18c}$$

where L denotes the set of main lines that, in this book, only consists of those lines that are directly connected to generators.

Now we study the macroscopic SOC characteristics using the IEEE 30-bus system as an example whose connection diagram is shown in Fig. 9.10. There are 6 generator buses, 24 load buses and 41 transmission lines.

We utilize the OPA model to study the simulated evolution process of 5000 days (about 14 years) for the IEEE 30-bus system according to the two criteria defined in Eqs. (9.18a) and (9.18b).

Figure 9.11(a) and (b) describe how $I(k)$ and $\overline{P}_{\text{loss}}(k)$ change with time k respectively.

Figure 9.11 clearly indicates that the system evolution can be divided into two stages. In the first stage,

$$k < 450$$

$I(k)$ increases and few failures occur. In the second stage,

$$k > 450$$

as $I(k)$ goes up to a certain value, the number of faults increases, the scale of

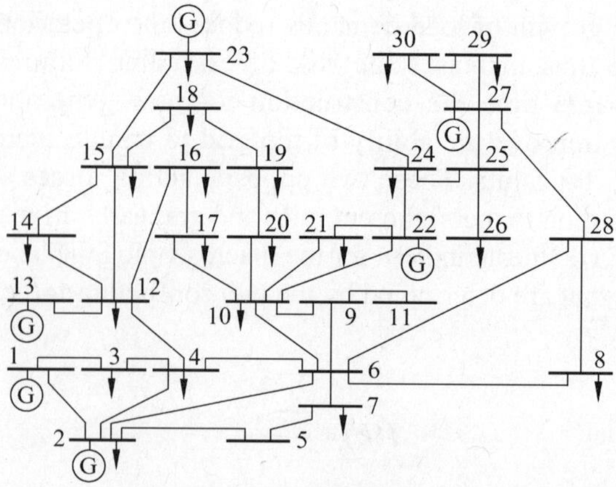

Figure 9.10 IEEE 30-bus system

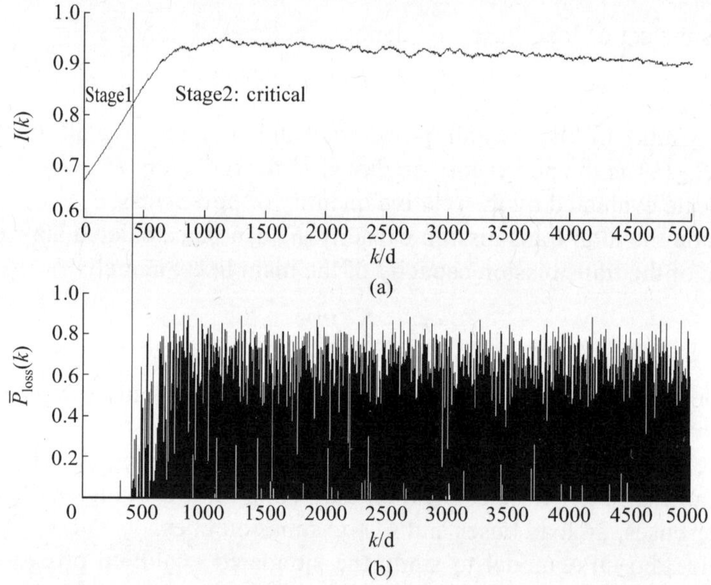

Figure 9.11 The progress for the IEEE 30-bus system evolving to a critical state

failures grows with $I(k)$, and the probability of large-scale blackouts increases rapidly.

From the viewpoint of criticality, when $I(k)$ is small, the system operates in a steady state and the probability for blackout is extremely small. As time evolves, since the increase of the transmission capacity cannot meet the demand from growing loads, the system develops into the stage where the transmission system reaches its limit. In other words, after $I(k)$ reaches its threshold value, large-scale blackouts begin to take place. So when the system evolves into its critical state, the probability for large failures becomes substantial. In fact, from Fig. 9.11, one can see that the system enters a dangerous critical state when $I(k) \approx 0.82$. Thus $I(k)$ can indeed be taken as a macroscopic criteria for SOC characterization in power grids, and so does $\overline{P}_{\text{loss}}(k)$.

9.3 Improved OPA Model

The OPA model is built upon the SOC theory. It takes account the fast and slow two types of variables that affect the evolution of power systems. When the interaction between the two types of variables create certain critical conditions, any disturbance, big or small, may cause unpredictable disastrous consequences. We have used the OPA model to identify those "critical conditions" to evaluate the safety level of power systems. However, the OPA model disagrees with two aspects of a real power system, and this to a certain extend discounts the credibility of some of the results based on the OPA model[8]. We summarize the two aspects as follows.

1) Line outage

When solving the DC power flow optimization problem (9.13) in the OPA model, the line flows are set strictly within the line capabilities. At the end of the computation, all those lines, the power flows in which are close to the line capacities satisfying $F_{jk} / F_{jk}^{\max} \geqslant \alpha$, are taken to be overloaded and tripped off. This is hardly the case in real power systems. In practice, the optimal power flow computation is related to the adjustment made by the scheduling department and tripping lines are actions taken by relay protections. To trip off a line within its limit is a false relay action which only happens in real systems with extremely low probability.

2) Updates of transmission lines

In the OPA model, only those lines that are tripped in the inner loop are upgraded. In other words, the outer loop of the OPA model will only increase the transmission capacities for those overloaded lines found in the inner loop. This underestimates the function of the planning department. In practice, the transmission lines are updated based on the load forecasting. Therefore, the effect of planning based on load forecasting should be considered in the model.

9.3.1 Design of the Improved OPA Model

Motivated by the shortcomings of the OPA model, we discuss an improved OPA model in this section. Its computation flow chart is shown in Fig. 9.12.

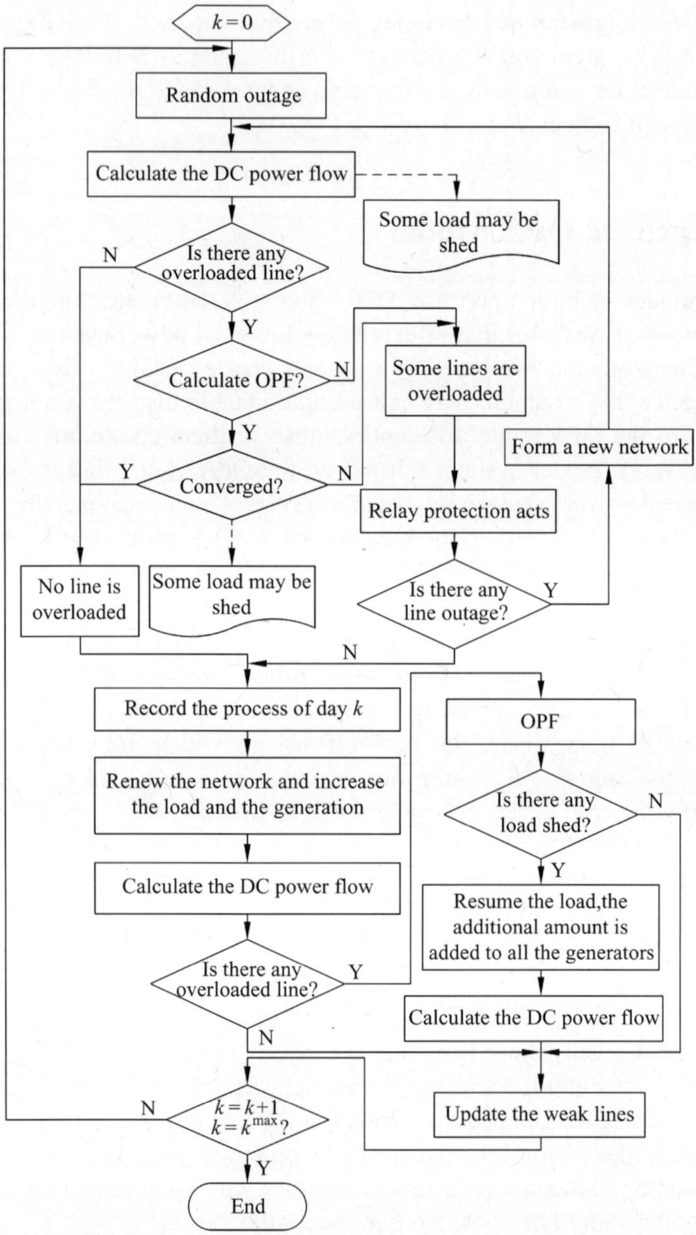

Figure 9.12 The flow chart of the improved OPA model

The improved model also has a double-loop structure. The inner loop is the fast dynamics process, which simulates power flows and cascading failures. Compared with the inner iteration of the OPA model, the improved OPA model considers the effects of dispatch, relay protection and automation, and thus reflects the practical systems more faithfully. The outer loop is the slow dynamic process, which simulates the evolution of the power system. Compared with the outer loop of the OPA model, the improved model incorporates the functions of the planning and operation departments.

The inner loop of the improved OPA model still completes within a day, while days are the unit of time used in the outer loop. The fast dynamics are implemented as follows.

Step 1 Consider day k. The capacities and initial outputs of the generators and the load demands are determined by the slow dynamics (see slow dynamics below for details). The outage probability of every transmission line is set to be τ to simulate accidental faults.

Step 2 Calculate the DC power flow according to the given generation outputs, load demands and network topology. The grid may have been separated into several islands because of line outages, and the generation and load may not be balanced in an island. This requires actions to redistribute power flows to achieve the load-generation balance. For an island, if the total generation is greater than the total load, the outputs of the generators are reduced with amounts that are proportional to their current operating states. If on the other hand, the total load is greater than the total generation, the outputs of the generators are increased in proportion. In the latter case, if the reserved generation capacities are not sufficient, besides raising the generation outputs to their limits, loads have to be lowered to an appropriate level. The DC power flow is recalculated after adjusting the generations and loads as just described.

Step 3 If there are transmission lines that are overloaded, then go to Step 4. Otherwise, exit the fast dynamics.

Step 4 Because of communication failures and the instability in the Energy Management System (EMS), the dispatch center may not be able to deal with overloads. Accordingly, it is assumed that the optimization problem (9.13) can only be solved with probability η in which case generations and loads (load shedding is possible) are determined and the line flows are computed at the same time. If the optimization problem cannot be solved because of the reasons such as communication failures or the optimization computation fails to converge, then go to Step 5. Otherwise, exit the fast dynamics.

Step 5 Check for overloaded lines. Trip the overloaded lines with probability β to simulate relay protections with misoperation probability $1-\beta$. For the lines within their operating limits, trip the lines with probability $\xi \times |F/F^{\max}|^n$ to simulate the maloperation of relay protections, where ξ is the base maloperation probability, $|F/F^{\max}|$ is the load rate of the transmission line that is the ratio of the line flow to the line capacity, and $|F/F^{\max}|^n$ is the nth power of the load rate. Note that the higher the load rate is, the larger the probability of maloperation is. If there are lines in outage, go back to Step 2; otherwise, exit the fast dynamics.

The slow dynamics of the improved OPA model are implemented in three steps.

Step 1 and **Step 2** are the same as those in the OPA model.

Step 3 According to the power demand and the generator capacity, calculate the power flow, which is the operation mode for the next day, and upgrade the transmission line which will be overloaded in the next day. Step 3 simulates the effects of scheduling and planning. This step can be executed in a subprogram as follows:

3a Calculate the DC power flow. If there is no overloaded line, go to 3e; otherwise, go to 3b.

3b Calculate the DC Optimal Power Flow in the same way as in the OPA model. Some transmission lines may not have sufficient capacities, and loads need to be shed. If loads are shed, go to 3c; otherwise, go to 3e.

3c Restore the shed loads. In order to balance the increased load, generators increase their outputs with amounts that are proportional to their rotational reserves.

3d Calculate the DC power flow.

3e Improve the lines whose values of $|F/F^{\max}|$ are greater than ε, a pre-defined constant that is the load rate of weak lines. This step simulates the actions of the department of planning. We update the transmission lines following the OPA model

$$F_{j,k+1}^{\max} = \mu F_{j,k}^{\max}$$

$$Z_{j,k+1} = \frac{1}{\mu} Z_{j,k}$$

where μ is the growth rate of the transmission capacity and $Z_{j,k}$ is the impedance of transmission line j on day k.

The overall process of the improved OPA model is summarized as follows.

Step 1 $k=0$. Initialization. Determine the generation outputs, the load demands and the network topology.

Step 2 Fast dynamics in day k. Record the amount of the shed loads.

Step 3 Slow dynamics. $k=k+1$. If k reaches the upper bound of the simulated time period, stop; otherwise, go to Step 2.

So the improved OPA model is more comprehensive in that it considers the functions of the departments of dispatch, automation, communication, relay protection, operation and planning.

9.3.2 SOC Analysis Based on the Improved OPA Model

In the previous section, we modified the OPA model to deal with some of the limitations of the OPA model. In this section, we simulate the evolution of 2000 days for the Northeast Power Grid of China to compare the improved OPA model with the OPA model. By doing so, we want to verify the correctness and effectiveness of the improved OPA model. When investigating the cascading blackouts in the grid, we establish the relationship between the blackout failures and the factors of interest, such as malfunctioning of the relay protection, growth in line capacities, failure in dispatch centers, load distribution and the load rate at weak lines. We continue to use the VaR and the CVaR to analyze the blackout risks in the grid.

1) The comparison between the OPA model and the improved OPA model

The improved OPA model has two main advantages over the OPA model[2].

(1) Flow redistribution and line outage

In the OPA model, the power flows are first determined by the DC optimal power flow program and all those lines whose power flow exceeds a certain proportion of the limit are tripped according to a given probability. In comparison, in the improved OPA model, the DC power flows are first adjusted according to the system's inherent balancing mechanism. If there are overloaded lines, the DC optimal power flows are computed in order to obtain the control strategies to maintain the power flows within the transmission limits. When the control strategies fail to be carried out due to reasons like communication failures, relay protections act to trip the overloaded lines.

(2) Line upgrade

In the OPA model, only the lines that have been tripped are upgraded, while in the improved OPA model, the lines are upgraded even before their flows reach their transmission limits. In other words, the improved OPA model better mimics the power grid constructions guided by forecasting.

In short, the improved OPA model reflects more faithfully the real situations in practical systems than the OPA model does. We find that the OPA model tends to overestimate blackout risks compared to the improved OPA model. This fact can be verified by using both the OPA model and the improved OPA model to simulate the Northeast Power Grid of China with the same set of values for the parameters. Here are some details.

① The growth rate of generation and load

As we have done in the OPA model, set $\lambda = 1.00041$.

② Accidental outage probability due to natural reasons like bad weathers

As in the OPA model, set $\tau = 0.0007$

③ The probabilities for maloperation and misoperation of relays

In real systems, strict regulations are implemented for relay protections, and thus their reliabilities can be as high as being greater than 99%. So we set the probabilities for maloperation and misoperation to be $\beta = 0.999$ and $\xi = 0.001$ respectively.

④ Other parameters

We make reasonable assumptions according to the operations in practical systems. Since it is highly likely that the dispatch center can handle faults successfully, we set $\eta = 0.95$ which means that the probability for failure is 5%. We assume that lines are upgraded when their line flows exceed 90% of their capacities, namely $\varepsilon = 0.9$. Set the growth rate for line transmission capacities to be $\mu = 1.005$. The above parameters are listed in Table 9.4.

Table 9.4 Parameters of the improved OPA model

Parameter	Value
Growth rate of generation and load (λ)	1.000,41
Growth rate of line trainsmission capacity (μ)	1.005
Accidental outage probability (τ)	0.0007
Maloperation probability (ξ)	0.001
Misoperation probability ($1-\beta$)	0.001
Failure probability in dispatch centers ($1-\eta$)	0.05
Load rate at weak lines (ε)	0.9

In the sequel, we call the values of the parameters in Table 9.4 the initialization values for the simulation of the Northeast Power Grid of China. Using these parameters, the probability distribution curves for load shedding calculated by the improved OPA model and the OPA model are shown respectively in Fig. 9.13.

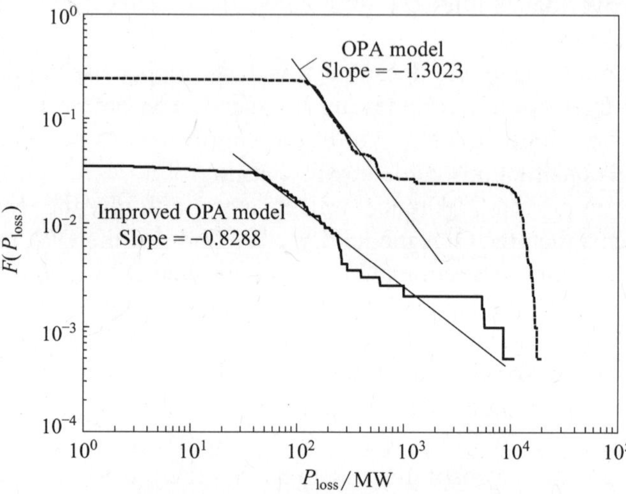

Figure 9.13 PDF of the load shedding

From Fig. 9.13, one can conclude that the number of blackouts obtained from the OPA model is much larger than that from the improved OPA model. In the simulated first 600 days, there are 22 times of actions of shedding loads in the OPA model while only 14 times in the improved OPA model. This is mainly due to the different power flow redistribution strategies and line tripping algorithms in the two models. In the last 1400 days, there are 463 times of blackouts in the OPA model while only 56 times in the improved OPA model. This is mainly because of the different line upgrade policies in the two models.

The improved model is more convincing compared to the OPA model because in reality, no more than 10 blackouts happen in this grid within a year[9], which is much closer to the outcome from the improved model. In addition, the blackout size distribution of the improved model agrees approximately with a power law distribution with the exponent being about –0.8288, which is close to –0.8641[10], the value obtained through the statistical analysis of the historical data.

2) Simulations for cascading failures using the improved OPA model

As disclosed in SOC theory, when a complex system evolves into its critical state, the influence of disturbances becomes unpredictable because both small and large "avanlaches" are possible[11]. Next using the improved OPA model, we simulate a large-scale cascading blackout in the Northeast Power Grid of China.

(1) Cause of the blackout

The triggering event of the simulated large-scale cascading failure is the outage of a transformer in the Liao Yang Substation (LYS in Fig. 9.14) in the middle of Liaoning Province.

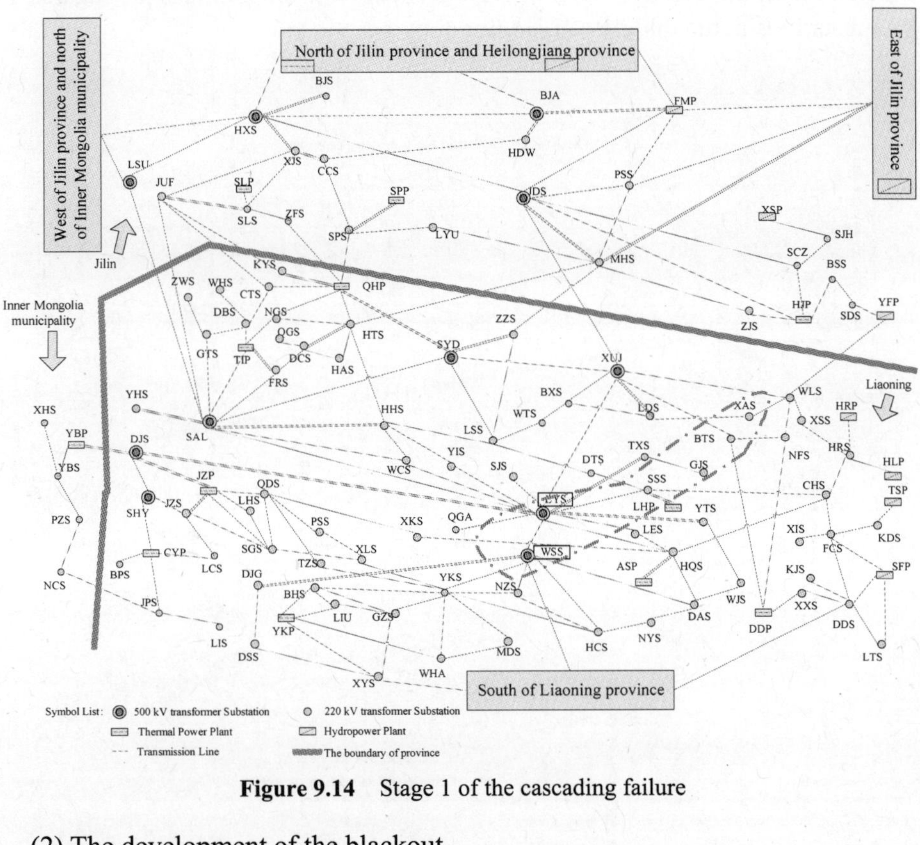

Figure 9.14 Stage 1 of the cascading failure

(2) The development of the blackout

Stage 1 The outage of the transformer at LYS overloads the transmission lines from BTS to GJS and from BTS to XAS in the encircled area in Fig. 9.14, which further overloads two other transformers at LYS and two transformers in the Wang Shi Substation (WSS in Fig. 9.14). All the substations whose names are framed in the figures are those that are affected by the fault. Since for some reason the dispatch centre does not notice these abnormal behaviors and thus takes no action, all the overloaded lines and transformers are tripped off by the relay protections. In this stage, the fault is still restricted in the area near LYS in the middle of the Liaoning Province.

Stage 2 The tie lines between the middle and the north of Liaoning province are overloaded, and consequently several transformers in XUJ and SAL as shown in Fig. 9.15 are overloaded. The situation is further worsened when two transformers in XUJ, four transformers in SAL and the lines LYS→WCS, SAL→WCS, BXS→LDS, CHS→HRS, DTS→LSS, DTS→SSS and HHS→YIS are all overloaded. At the end of the second stage, the Deng Ta Substation (DTS) has been isolated from the system because 10 out of the 12 overloaded lines and transformers are tripped. So all the 196 MW load of DTS is shed. Now the fault has propagated to 17 substations in the middle of Liaoning province.

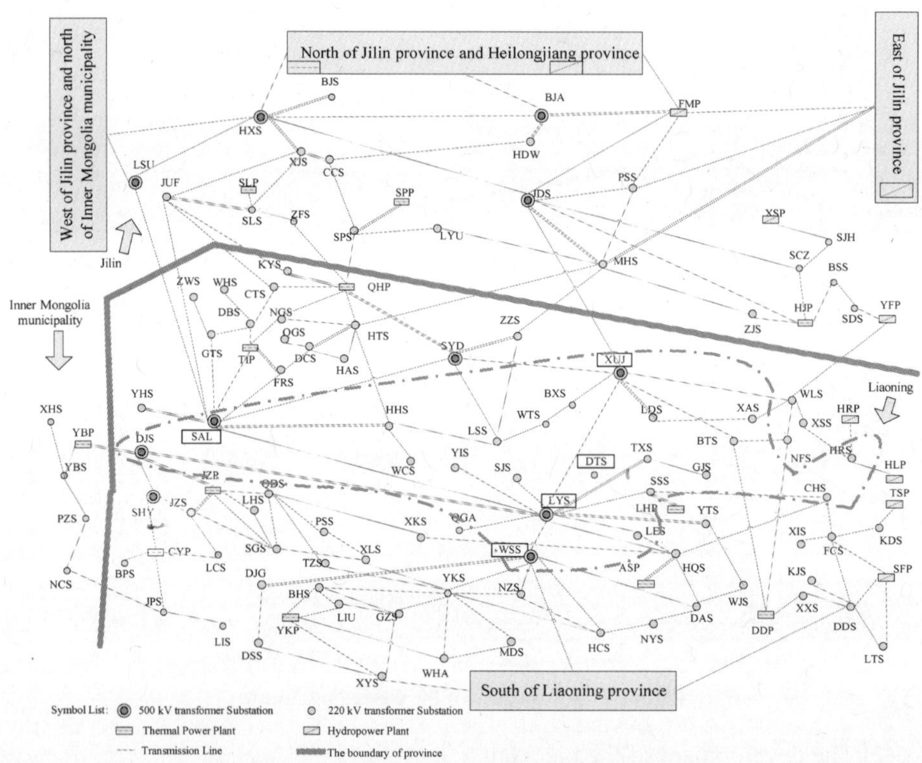

Figure 9.15 Stage 2 of the cascading failure

Stage 3 The fault spreads to the middle and north of Liaoning province and the south of Jilin province. There are 54 lines and transformers that are overloaded. As shown in Table 9.5, now 33 lines and transformers are tripped by their relay protections and as a result, loads up to 5905 MW at 67 locations in the middle and the north of Liaoning province are shed. At the end of this stage, the blackout has affected almost the whole the Liaoning province and parts of Jilin province.

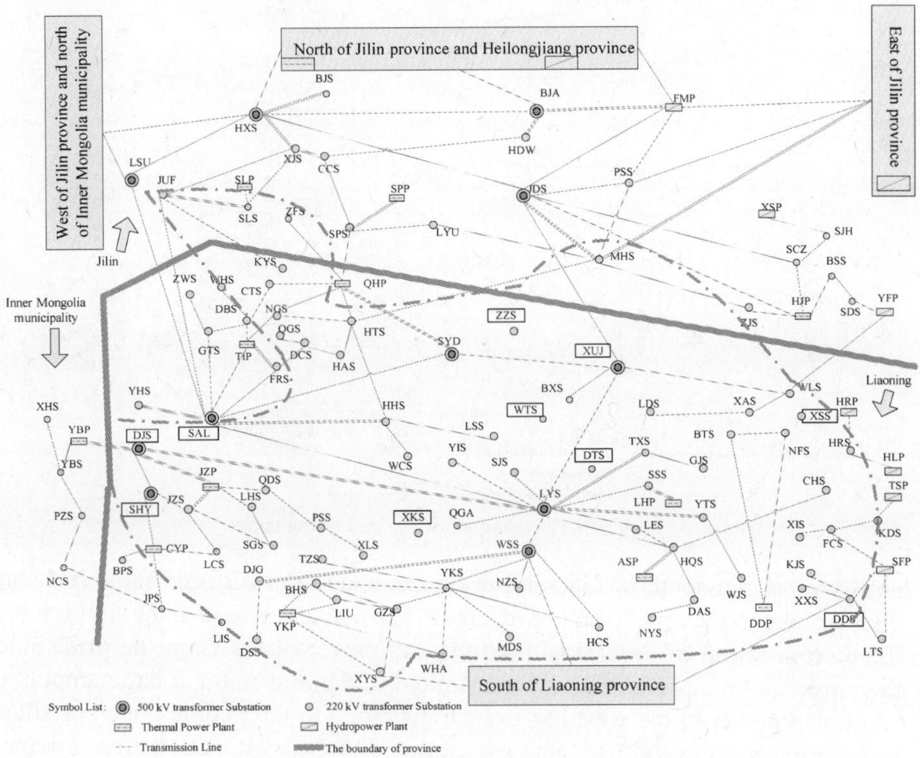

Figure 9.16 Stage 3 of the cascading failure

Stage 4 Now 24 lines and transformers are overloaded and 15 of them are tripped. The dispatch center begins to take actions, as a result of which the generations are increased and 1623 MW loads at 72 locations in the middle and the west of Liaoning province are shed. Finally the system restores its stability. At the end, the cascading blackout affects most of Liaoning province, the south of Jilin province and the east of the Inner Mongolia municipality. As shown in Fig. 9.17, several areas lose their power supply and some substations and power plants have to be disconnected from the grid.

We summarize the above four stages of the blackout as follows. The simulated large-scale blackout is provoked by the outage of a single transformer in the middle of Liaoning province, which first affects the power delivery from the

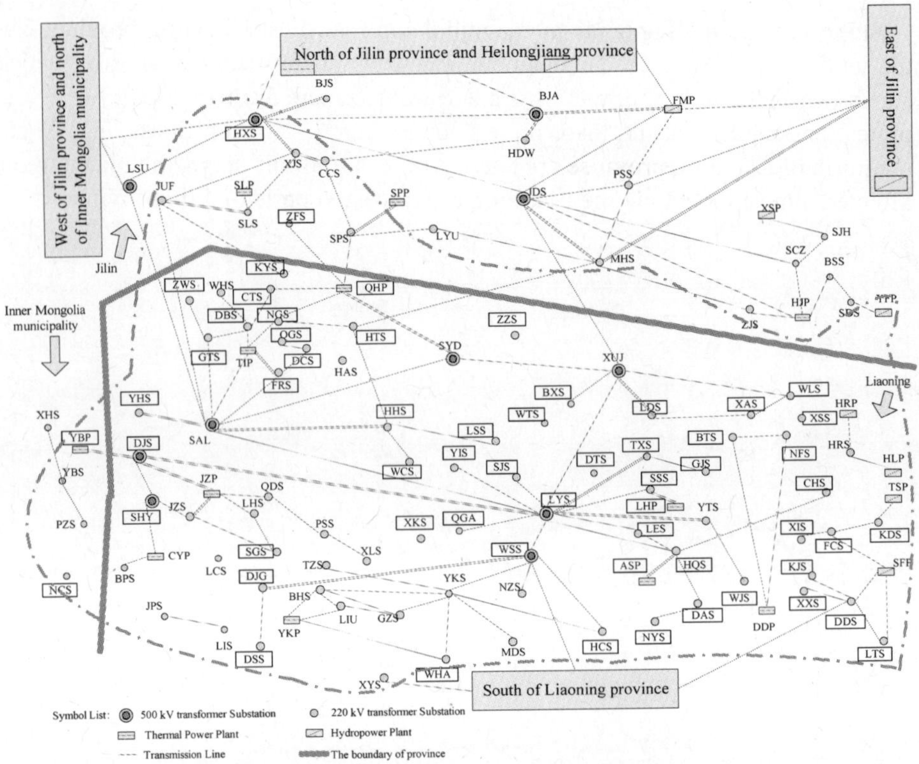

Figure 9.17 Stage 4 of the cascading failure

north to the mid-south in Liaoning province, and then overloads some lines and transformers in the middle and north areas. The associated cascading failure forces the disconnection of several substations and power plants from the grid. Since the power in the north cannot be transmitted to the mid-south, a large amount of additional power in the west has to be transmitted to the middle which produces the overloads at several lines and transformers in the west of Liaoning province. The system is finally stabilized only after regulating extensively the related generations and loads. The amount of the lost load caused by line outage is 6101 MW and the amount of shed load caused by the actions of the dispatch center to eliminate overloads is 1623 MW. So altogether 7724 MW loads have been lost in the whole blackout, which amounts to 22.8% of all the loads. The tripped lines and transformers in the cascading failure are shown in Table 9.5.

From Table 9.5 one can see the simulated whole process of large-scale line overload and load lost in the power grid caused by the combination of line outage, relay actions and communication failures at the dispatch center. In this simulated cascading failure, the Northeast Power Grid of China has been safe because of the joint efforts of the engineers in the plan, design and operation departments and the accurate actions of the automatic relay protections. Hence, the simulated cascading failure nevertheless indicates a new approach for the study of how to

Table 9.5 Outage lines, transformers and the shed loads at each stage

Stage	Outage lines	Transformers in outage	Loads shed
Triggering event		LYS	
1	BTS→GJS, BTS→XAS	LYS, WSS	
2	LYS→WCS, SAL→WCS, BXS→LDS, CHS→HRS, DTS→LSS, DTS→SSS, HHS→YIS	XJS, SAL	DTS
3	XUJ→BTS, SYD→LSS, CHS→SSS, DAS→WJS, DDP→XXS, DDS→FCS, HCS→NYS, HRS→XSS, JZP→SGS, LSS→WTS, MHS→ZZS, QDS→XKS, QDS→TZS, SGS→XLS, SPS→ZFS, XLS→YKS, YKS→NZS, SYD→ZZS, BXS→WTS, CHS→FCS, CTS→JUF, DAS→WSS, DCS→HTS, DSS→XYS, HCS→NZS, HQS→XKS, KYS→QHP, LSS→ZZS, MDS→WHA, NFS→WLS, QDS→SGS, WLS→XSS	DJS, SHY	LDS, LSS, WLS, XAS, ASP, CHS, DAS, GJS, HQS, LHP, NYS, QGA, SJS, SSS, TXS, WJS, YIS, LES, CTS, DBS, DCS, FRS, GTS, HHS, HTS, KYS, NGS, QGS, TIP, WCS, WHA, YHS, ZFS, ZWS, BTS, DDS, FCS, KDS, KJS, LTS, NFS, XIS, XXS, BXS, DJG, DSS, HCS, NZS, SGS, WTS, XKS, ZZS, XSS
4	YFP→WLS, CYP→LCS, CYP→JPS, DCS→HAS, GZS→XYS, HAS→HTS, HTS→QHP, JPS→NCS, JZS→LCS, NCS→PZS, XYS→YKP	HXS, YBP, ASP, QHP	CHS, DAS, DBS, DCS, FCS, FRS, GJS, GTS, GZS, HAS, HHS, HQS, HTS, KDS, KJS, KYS, LTS, NFS, NGS, NYS, PSS, QGA, QGS, SJS, XLS, XXS, YHS, ZFS, ZWS, YIS, LIU

prevent disasters in power grids, especially considering the similarity between the simulated cascading blackout and the real one that took place in 2003 in North America. In fact, the North American blackout took place because there was no central dispatch center that could coordinate different regional networks to deal with the overloads and thus faults could spread out until developing into system-level disasters. So we are really looking at a main potential reason for large disasters in huge power grids.

Also from the above cascading process, one can see that the small disturbance of the outage of a transmission line can lead to vital cascading failures and a significant amount of loads have to be shed. Using the terms from the SOC theory discussed in Chapter 2, the system has entered a critical state and the influence of disturbance might not be restricted in a small region, but propagate over long distances to a much larger area, and even result in the collapse of the whole system. In what follows, we investigate the influence of some system parameters on the probability distribution of blackouts.

3) The influence of maloperation and misoperation of relay protections

In Fig. 9.18, two curves show the probability distributions of the blackouts with the same initialization values of the system parameters except for the different values, 0.01 and 0.001, for the maloperation probability.

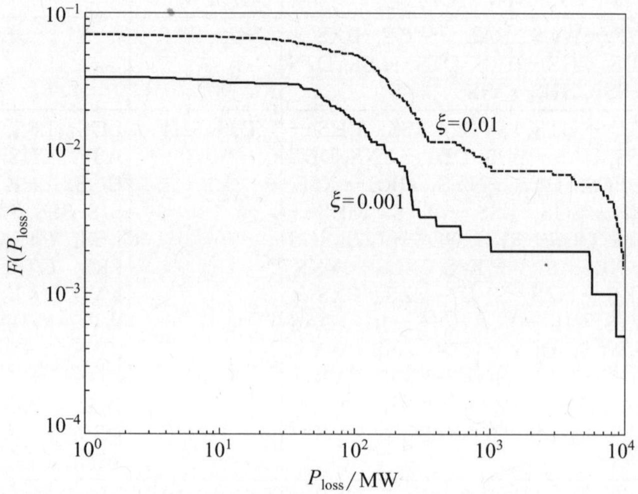

Figure 9.18 Blackout probability distributions with different probabilities of maloperation

Similarly, in Fig. 9.19, two curves are shown to compare the different probability distributions of blackouts, the initialization values of which only differ in the misoperation probability.

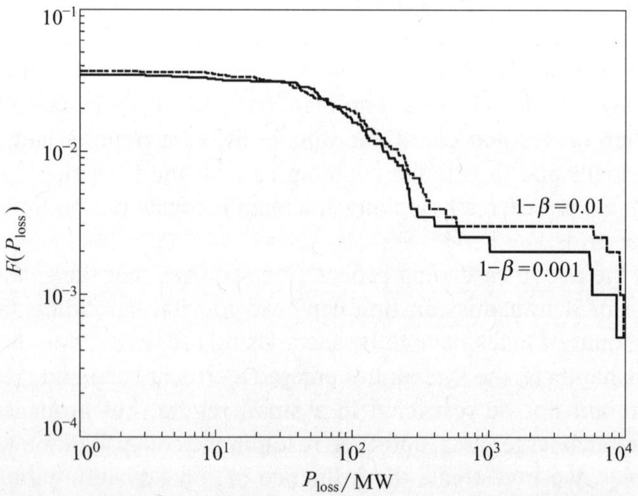

Figure 9.19 Probability distributions with different misoperation probabilities

Figure 9.19 shows that the probability for blackouts is only slightly higher when the probability for misoperation is larger.

Let $\sigma = 0.99$, then the computation results of the VaR and the CVaR are listed in Table 9.6.

Table 9.6 VaR and CVaR with different probabilities for mal and misoperations

ξ	$1-\beta$	VaR/MW	CVaR/MW
0.001	0.001	167.74	17.180
0.01	0.001	708.90	60.183
0.001	0.01	216.92	26.862

Table 9.6 shows that when ξ or $1-\beta$ becomes larger, the cascading failure becomes more serious and the blackout risk is bigger. Notice that the improved model may underestimate the consequence of misoperation of relay protections because the model cannot simulate the short-circuit fault after the overload caused by misoperations. So the real consequence for misoperation might be more serious than the simulation results using the improved OPA model.

4) The influence of the growth rate for transmission capacity

In Fig. 9.20 two curves are used to illustrate the probability distribution for blackouts with different growth rates for transmission capacities.

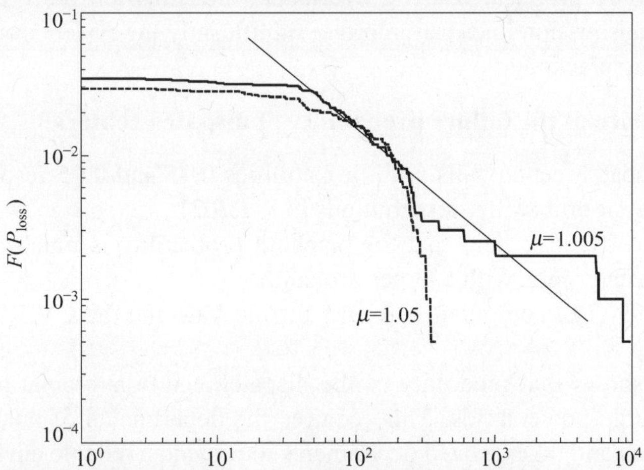

Figure 9.20 Probability distributions with different growth rates for transmission capacities

It is clear from Fig. 9.20 that increasing μ can reduce the number of blackouts, especially large-scale blackouts. Note that the probability distribution curve agrees approximately with the power law with μ being 1.005 and complies with an exponential distribution when μ is 1.05. From the viewpoint of SOC theory, the system is not in a critical state when $\mu = 1.05$ and thus the system operating in this state is comparatively safer. Of course to have a greater μ requires invenstment into the upgrades on lines and transformers. Also from Fig. 9.8, one can see that the probability distributions are almost the same when μ takes the value of 1.008 and 1.005 respectively. Hence, only when μ becomes larger than some threshold value, can the system get rid of the SOC characteristics.

Let $\sigma = 0.99$, the computation results for the VaR and the CVaR are listed in Table 9.7.

Table 9.7 VaR and CVaR with different growth rates for transmission capacity

μ	VaR/MW	CVaR/MW
1.005	167.7443	17.1796
1.05	181.3895	2.4741

After increasing μ to 1.05, the VaR increases slightly and the CVaR decreases greatly. This indicates that, increasing μ has no effect in small-scale blackouts, but can reduce large-scale ones. This can be explained as follows. The reason of small-scale faults is the acidental line outages, and the rebuild of transmission lines can hardly prevent them. On the other hand, the reason for large-scale faults is the cascading failure propagated along overloaded transmission lines, and thus the upgrade of transmission lines can improve significantly the system's ability to deal with cascading blackouts.

5) The influence of the failure probability of dispatch centers

When the dispatch center fails with probabilities 0.15 and 0.05 respectively, we plot the blackout probability distributions in Fig. 9.21.

From Fig. 9.21, it is clear that the blackout probability is much higher when the dispatch center fails with a larger probability.

Let $\sigma = 0.99$. The computation results for the VaR and the CVaR are listed in Table 9.8.

Table 9.8 shows that reliability of the dispatch centre is crucial to assure the security of large power grids. This requires the departments of automation and communication and other related departments to provide a reliable environment for dispatchers.

6) The influence of the load distribution

We consider the following two different load distribution schemes.

Scheme 1 The real load distribution in the Northeast Power Grid of China;

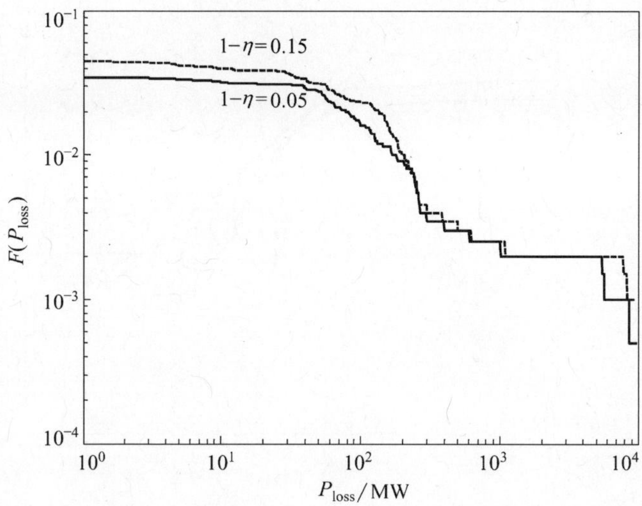

Figure 9.21 Probability distributions with different failure probabilities at dispatch centers

Table 9.8 VaR and CVaR with different failure probabilities at the dispatch centre

$1-\eta$	VaR/MW	CVaR/MW
0.05	167.74	17.180
0.15	197.80	20.575

Scheme 2 To avoid long distance transmission, we reduce the loads in Liaoning province, and increase the loads in Jilin province and Heilongjiang province. So the active power are balanced onsite. Hence, the load distribution in Scheme 1 is not as uniform as that in Scheme 2.

Figure 9.22 shows that, the blackout probability for load distribution scheme 1 is larger than that for load distribution scheme 2. Note that the blackout probability distribution for load distribution scheme 1 agrees approximately with the power law with the exponent being about -0.8288, while the corresponding exponent is about -1.3545 for the load distribution scheme 2. Interpreting using SOC theory, scheme 2 is safer than scheme 1 because smaller slopes without "heavy tails" correspond to smaller likelihoods for large-scale blackouts.

Let $\sigma = 0.99$. The computation results for the VaR and the CVaR are listed in Table 9.9.

Table 9.9 shows that the number and risk of blackouts decrease when the active power are balanced onsite. This can be explained as follows. Interconnected power systems make it possible for different regions to support each other when faults take place. However, if large amounts of powers are transmitted over long distances, the outage of tielines can become fatal. If the powers are balanced onsite, the tielines are light-loaded and different regions can support each other in real contingencies, in which case the system becomes safer.

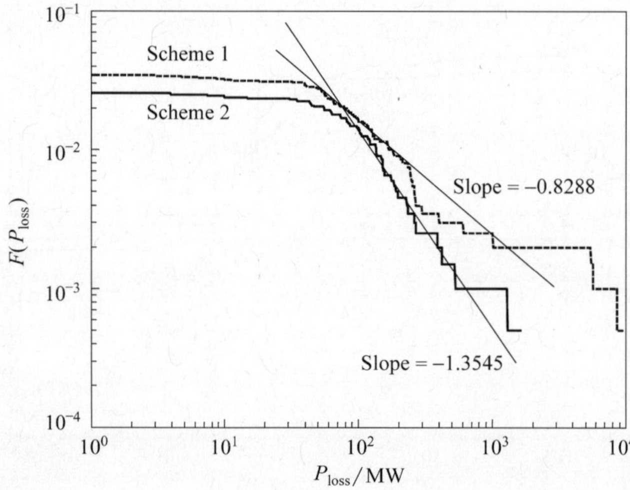

Figure 9.22 Probability distributions with different load distributions

Table 9.9 VaR and CVaR with different load distributions

Load distribution	VaR/MW	CVaR/MW
Scheme 1	167.7443	17.1796
Scheme 2	135.0066	3.4283

7) The influence of load rates of weak lines

Two curves are shown in Fig. 9.23 indicating different probability distributions with different load rates of weak lines.

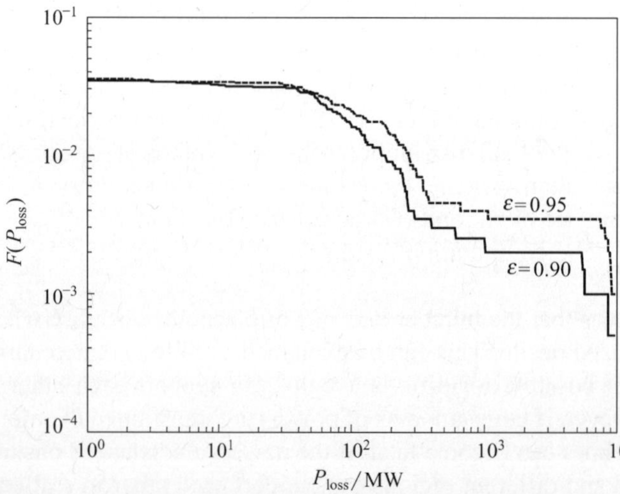

Figure 9.23 Probability distributions with different load rates of weak lines

From Fig. 9.23 we know that increasing the load rates of weak lines increases the probability for large-scale blackouts.

Let $\sigma = 0.99$. The computation results for the VaR and the CVaR are listed in Table 9.10.

Table 9.10 VaR and CVaR with different maximum load rates of weak lines

ε	VaR/MW	CVaR/MW
0.90	167.7443	17.1796
0.95	268.0006	33.0701

Table 9.10 shows that the growth rate ε affects the occurance of large-scale blackouts. This is because if ε is large, it must be true that the planning and construction in the power grid lag behind the urgent need in the system. So small-scale faults may develop into large-scale blackouts, which again emphasizes the importance of the planning department in the operation of power grid. This is especially true when the constuction cycle of a real system is long.

9.4 Discussion

Two blackout models based on DC power flows are discussed in this chapter, which are the OPA model and the improved OPA model. From the simulations using the OPA model and the improved OPA model, one can see that the improved OPA model is more effective in the study of cascading failures, blackout distribution in long-term evolutions, and the influence of different system parameters. Since the improved OPA model reflects the engineering practice more faithfully and its simulation results agree with the operation experiences gained from real systems, it is advantageous to exploit its quantitative analysis results, such as the blackout probability distribution and blackout risks, for tasks including the planning of generation, the upgrade of grids, secondary system construction and operation, and others.

Both the OPA model and the improved OPA model use linear programming techniques from the viewpoint of optimization theory. So the blackout model based on the DC power flows are preferable for long-term system planning and fast analysis of large-scale cascading failures because of its rapid convergence speed. However, such models also have some shortcomings. In particular, the blackout model based on DC power flows cannot be used to analyze the faults that are related to voltages and reactive powers. So it cannot provide guidence for the planing and operation management of reactive powers in power systems. Since it also neglects the dynamic characteristics of some power system components, it cannot simulate the large-scale blackouts that are caused by transient power-angle instability.

References

[1] Wang X F, Song Y H, Irving M. Modern Power System Analysis. New York: Springer, 2008

[2] Dobson I, Carreras B A, Lynch V E, et al. An initial model for complex dynamics in electric power system blackouts. Proceedings of the 34th Hawaii International Conference on System Sciences, Maui, Hawaii, Jan. 2001: 710 – 718

[3] Carsreras B A, Lynch V E, Newman D E, et al. Dynamics, criticality and self-organization in a model for blackouts in power transmission systems. Proceedings of the 35th Annual Hawaii International Conference on System Sciences, Jan. 2002

[4] Carreras B A, Lynch V E, Dobson I, et al. Critical points and transitions in an electric power transmission model for cascading failure blackouts. Chaos, 2002, 12(4): 985 – 994

[5] The operation mode of Northeast Power Grid of China in 2006. Northeast China Power Grid Co. Ltd (in Chinese)

[6] The operation mode of Northeast Power Grid of China in 2007. Northeast China Power Grid Co. Ltd) (in Chinese)

[7] The operation mode of Northeast Power Grid of China in 2004. Northeast China Power Grid Co. Ltd) (in Chinese)

[8] Mei S W, He F, Zhang X M, et al. An improved OPA model and blackout risk assessment. IEEE Transactions on Power Systems, 2009, 24(2): 814 – 823

[9] Yu Q, Guo J B. Statistics and self-organized criticality characters of blackouts in China electric power systems. Automation of Electric Power Systems, 2006, 30(2): 16 – 21 (in Chinese)

[10] Weng X, Hong Y, Xue A, et al. Failure analysis on China power grid based on power law. Journal of Control Theory and Application, 2006, 4(3): 1 – 4

[11] Bak P. How Nature Works: the Science of Self-Organized Criticality. New York: Copernicus Press, 1996

Chapter 10 Blackout Model Based on AC Power Flow

Cascading blackout models based on optimal AC power flows are constructed in this chapter. Since these models are developed using SOC theory and incorporate the Manchester model, they can be utilized to describe the fast dynamics in cascading failures as well as the slow dynamics during the evolution of power grids. The advantage of this approach is that at both the macroscopic and microscopic levels, one can reveal and then study in depth the SOC characteristics in power grids.

From the practical engineering point of view, blackout models based on DC power flows have their own limitations because the computations for DC power flows cannot take into account the influence of reactive powers and voltages, while sometimes blackouts in real systems are caused exactly by voltage collapse following the shortage of reactive powers[1−4]. To deal with this limitation, the Manchester model[5] has been proposed to simulate cascading failures and discuss several criticality criteria based on AC power flows. However, if we examine the Manchester model using SOC theory, it is clear that the model only simulates fast dynamic processes, but without the slow ones. Hence, the model is only related to criticality, but has few connections to the macroscopic self-organization. To overcome this, we develop cascading blackout models based on optimal AC power flows to disclose and then study in depth the SOC characteristics in power grids at both the macroscopic and microscopic levels in this chapter.

10.1 Mathematical Description of Optimal AC Power Flow

The key objective for the dispatch and management of power systems is to lower operation costs and promote social and economic benefits while maintaining the systems' stability. To achieve this objective in a real power grid, the main tool is the computation of optimal power flows (OPF)[6].

Mathematically speaking, the optimal power flow problem is a nonlinear programming problem, which can be formulated as follows. Consider a power system consisting of n buses and m lines. At time k, the objective for the computation of the optimal power flows is to minimize the power loss in the system:

$$\min F(P,Q)=\sum_{i\in G}P_{gi,k}-\sum_{j\in L}P_{dj,k} \tag{10.1}$$

where G is the set of generator buses, L is the set of load buses, $P_{gi,k}$ and $P_{dj,k}$ are the active powers at generator bus i and load bus j respectively, and the vectors P and Q are the decision variables to be determined that correspond to the active and reactive power outputs of all the generators.

The optimization constraints include the power flow constraint, the generation output constraint, the bus voltage constraint and the line capacity constraint. Now we describe these constraints in detail.

1) Power flow constraint

$$P_{i,k} = V_{i,k} \sum_{j=1}^{n} V_{j,k} (G_{ij,k} \cos\theta_{ij,k} + B_{ij,k} \sin\theta_{ij,k}) \tag{10.2}$$

$$Q_{i,k} = V_{i,k} \sum_{j=1}^{n} V_{j,k} (G_{ij,k} \sin\theta_{ij,k} - B_{ij,k} \cos\theta_{ij,k}) \tag{10.3}$$

where $P_{i,k}$ and $Q_{i,k}$ are the injected active and reactive powers at bus i, $V_{i,k}$ and $\theta_{i,k}$ are the amplitude and phase of the voltage at bus i, $\theta_{ij,k} = \theta_{i,k} - \theta_{j,k}$ is the difference in phase-angles of the voltages at bus i and bus j, and $Y_k = (G_{ij,k} + jB_{ij,k})_{n\times n}$ is the admittance matrix of the system.

2) Generator output constraint

$$P_{gi,k}^{\min} \leqslant P_{gi,k} \leqslant P_{gi,k}^{\max}, \quad i \in G \tag{10.4}$$

$$Q_{gi,k}^{\min} \leqslant Q_{gi,k} \leqslant Q_{gi,k}^{\max}, \quad i \in G \tag{10.5}$$

where $P_{gi,k}$ and $Q_{gi,k}$ are the active and reactive powers outputs of generator i, $P_{gi,k}^{\max}$ and $Q_{gi,k}^{\max}$ are their corresponding upper limits respectively, and $P_{gi,k}^{\min}$ and $Q_{gi,k}^{\min}$ are their corresponding lower limits respectively.

3) Bus voltage constraint

$$V_{i,k}^{\min} \leqslant V_{i,k} \leqslant V_{i,k}^{\max} \tag{10.6}$$

where $V_{i,k}$ is the amplitude of the voltage at bus i, and $V_{i,k}^{\max}$ and $V_{i,k}^{\min}$ are its upper and lower limits respectively.

4) Line capacity constraint

$$-F_{l,k}^{\max} \leqslant F_{l,k} \leqslant F_{l,k}^{\max}, \quad l = 1, 2, \cdots, m \tag{10.7}$$

where $F_{l,k}$ is the power transferred through line l, and $F_{l,k}^{\max}$ is the capacity of line l.

10.2 Model Design

10.2.1 Model Structure

Figure 10.1 illustrates the simulation model for cascading blackouts based on AC optimal power flow. The model contains the inner and outer loops. The inner loop is the fast dynamic process simulating the cascading failure while the outer loop is the slow dynamic process simulating the upgrades and developments in power systems, such as the growths in the generation and load levels and the improvements in transmission capacities.

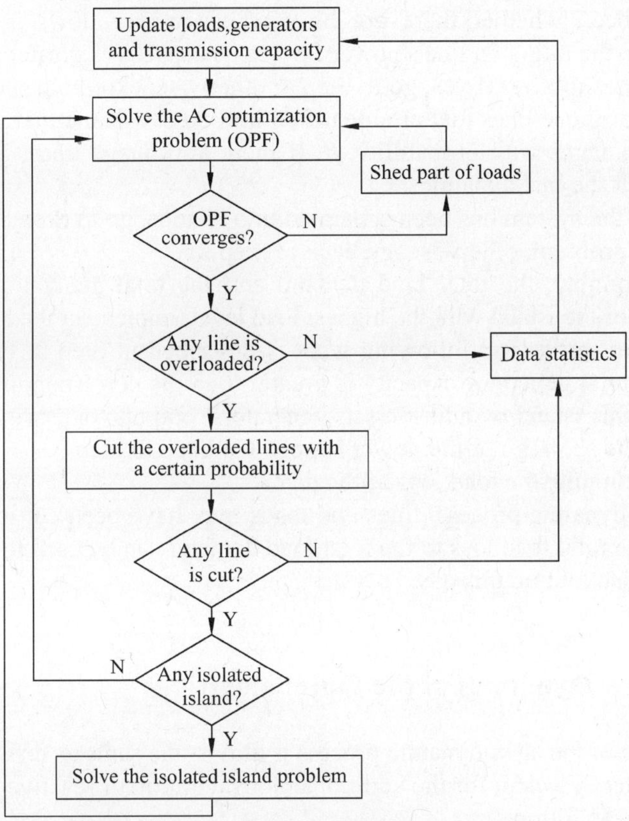

Figure 10.1 Blackout model based on AC OPF

10.2.2 Fast Dynamics of the Inner Loop

The fast dynamic process is considered as the development in one day of the slow dynamic process that uses days as its time unit. In the fast dynamics, typical loads, usually the ones corresponding to the peak load level, are chosen to simulate the operation in that day with possible cascading failures and blackouts. The main idea in the inner loop is as follows: First, determine the system's operation mode by calculating its optimal AC power flow and shedding loads; furthermore, if faults take place in overloaded lines, then recalculate the optimal AC power flow, shed loads and determine the new operation mode; such process continues until no fault takes place any more. The steps in the inner loop are as follows.

Step 1 For day k, the initial load levels and the outputs of the generators are determined by the slow dynamics of the outer loop.

Step 2 Calculate the optimal AC power flow. If the computation converges, go to Step 3; otherwise, continue to shed loads until the optimal power flow converges, and then go to Step 3.

Step 3 Check whether there are lines whose power flows are near their capacities, e.g. the ratio of a line's power flow to its capacity is greater than or equal to a pre-set threshold α. If yes, go to Step 4; otherwise, exit the fast dynamics.

Step 4 For those lines that are identified in Step 3, assume that each of them experiences outages with probability β. If there are outage lines, go to Step 5; otherwise, exit the fast dynamics.

Step 5 If the system has been separated into islands, go to Step 6 to deal with the islanding problem; otherwise, go back to Step 2.

Step 6 Compute the total load demand and the total generator capacity in each island. For the island with the highest load level, implement the load shedding strategy within this island following what is described in Step 2. For any other island, if the total generator capacity is greater than the corresponding total load, then balance this island by adjusting its generators' outputs; otherwise, shed loads to eliminate the shortage of the power supply. Go back to Step 2.

Step 7 Compute the load loss of the day.

In the fast dynamic process, lines and loads may have been cut in which case the ratio of the total load loss to the total load demand can be used to describe the scale of the blackout on this day.

10.2.3 Slow Dynamics of the Outer Loop

The main idea of the slow dynamic process is almost the same as that in the model used in Chapter 9, except for the additional consideration of reactive powers. The main steps are as follows:

Step 1 Increase the daily load levels in a continuous and uniform way to approximate the discrete annual increase in load levels. For each load bus, we have

$$P_{di,k+1} = \lambda P_{di,k}, \quad i \in L \tag{10.8}$$

$$Q_{di,k+1} = \lambda Q_{di,k}, \quad i \in L \tag{10.9}$$

where L is the set of load buses, $P_{di,k}$ and $Q_{di,k}$ are the active and reactive powers at bus i on day k, and λ is the growth rate of the load in the system, which is also set to be the growth rate of the generation output.

Step 2 Increase the generator capacities in a similar way, namely for each generator we have

$$P_{gi,k+1}^{\max} = \lambda P_{gi,k}^{\max}, \quad i \in G \tag{10.10}$$

$$Q_{gi,k+1}^{\max} = \lambda Q_{gi,k}^{\max}, \quad i \in G \tag{10.11}$$

where G is the set of generator buses, and P and Q are the upper limits for the active and reactive power outputs at bus i on day k.

Step 3 For the outage lines in the fast dynamic process, we use the average improvement effect to replicate the improvement of the line capacities, namely

$$F_{j,k+1}^{\max} = \mu F_{j,k}^{\max} \tag{10.12}$$

where $F_{j,k}^{\max}$ is the transmission capacity for line j on day k and μ is the growth rate of the line capacity.

For the lines that are not tripped in the fast dynamic process, their capacities are considered to be sufficient, namely

$$F_{j,k+1}^{\max} = F_{j,k}^{\max} \tag{10.13}$$

Step 4 Go to the fast dynamics.

From the above implementations of the fast and slow dynamic processes, it is clear that the fast dynamics can be used to simulate the daily operation of the systems where possible cascading failures may take place, while the slow dynamics can be used to simulate the long-term development of the power grid. If we further analyze statistically the data during faults in a power grid, such as power shortage, load loss, load level, etc., then it is possible to identify several macroscopic criteria to investigate and evaluate more precisely the SOC characteristics of power grids.

10.2.4 Characteristics of the Model

Within the framework of SOC theory, we have constructed the cascading failure and blackout model based on the optimal AC power flows. The model has the following characteristics:

(1) It takes into account both the active and reactive power flows, and thus can simulate the blackouts that are caused by the shortage of active power and/or reactive power.

(2) It considers the loss in power grids and uses the minimization of the active power loss as the objective function, which is close to the objective function used

in the economic operation of real power systems. Hence, the calculated optimal power flows are close to the actual power flows.

(3) It can simulate the voltage collapse in a system when the computation for optimal power flows fails to converge. At the same time, within the model, actions, such as load shedding, can be taken to prevent voltage instability.

(4) In the process of optimizing power flows, it also considers the adjustments in generators' outputs.

It must be pointed out that the blackout model based on optimal AC power flows correspond to a class of nonlinear programming problems. Compared with its linear programming counterpart based on optimal DC power flows, it needs much more computational time. However, since the AC power flow computation takes into consideration more physical actions, the AC power flow based model can simulate more faithfully the long-term evolution and short-term operation in a real system, and thus its simulation results are more convincing.

10.3 Simulations of Cascading Failures

We have developed a simulation platform for the evolution of power grids based on the model presented in Section 10.2. In this platform, we use the interior point method[7] to carry out the computation for the optimal AC power flows. In this section, we use this platform to simulate and analyze the cascading failures in the IEEE 30-bus system and the 500 kV Northeast Power Grid of China.

10.3.1 Choosing Parameters

As shown in Fig. 10.2, we have chosen four parameters to study power grid cascading failures under different operation modes.

(1) Line overload threshold α

The parameter α is used as a criterion to check whether a line is overloaded. According to the computation results of the optimal power flows, for a given line, if the ratio of its power flow to its capacity is greater than or equal to α, then this line is considered to be overloaded.

(2) Outage probability β for overloaded lines

The parameter β is the probability that an overloaded line is cut.

(3) Growth rate λ of the system load level and the generation output

The parameter λ is the daily growth rate for the total load level and generation output in a system. For example, if the load level in the system grows 20% annually, then the average daily growth rate is set to be $\lambda = 1.0005$.

(4) Growth rate μ of line transmission capacity

If there is a line outage on day k, then the transmission capacity for that line on day $k+1$ will increase to μ times that on day k.

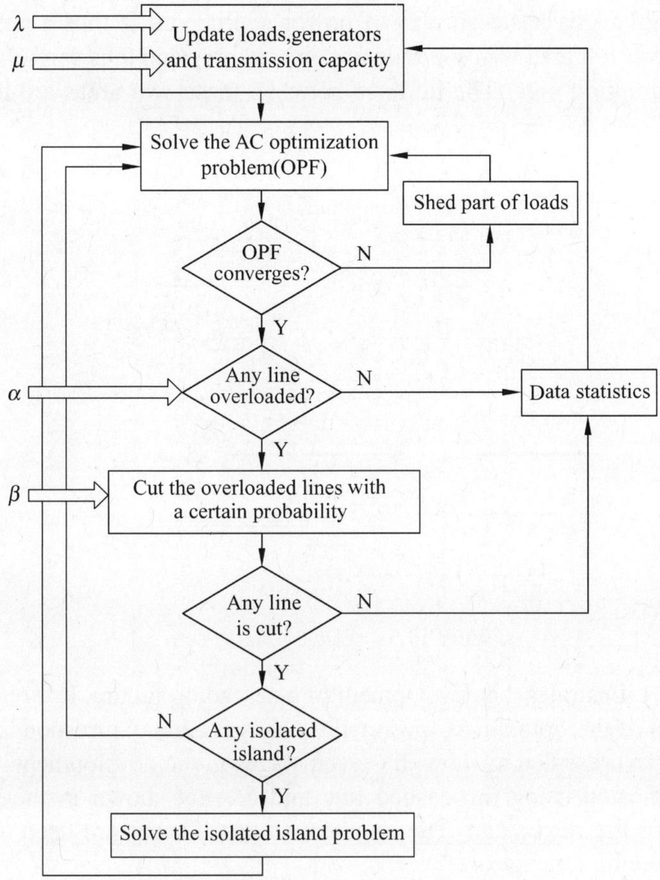

Figure 10.2 Cascading failure model

From SOC theory, α and β determine the strength of the internal forces during the fast dynamics in the power grid, while λ and μ determine the intensity of the external drives during the slow dynamics. In the models in this chapters, unless otherwise specified, we take $\alpha = 0.7$, $\beta = 0.3$, $\lambda = 1.0005$, and $\mu = 1.005$.

10.3.2 IEEE 30-Bus System

The topology of the IEEE 30-bus system is shown in Fig. 10.3. There are six generator buses in the system, two of which have their own loads; and there are twenty-four load buses, six of which are floating buses having zero loads. As indicated by the dashed line, the topology of the system can be divided into two areas: the upper right area containing three generator buses and five load buses and the lower left area containing three generator buses and nineteen load buses.

The upper right area has a surplus of power source and is thus a power supply area. The lower left area is in shortage of power supply and thus can be viewed as a power consumption area. The tie lines between these two areas are lines 15-23, 21-22, 10-22 and 27-28.

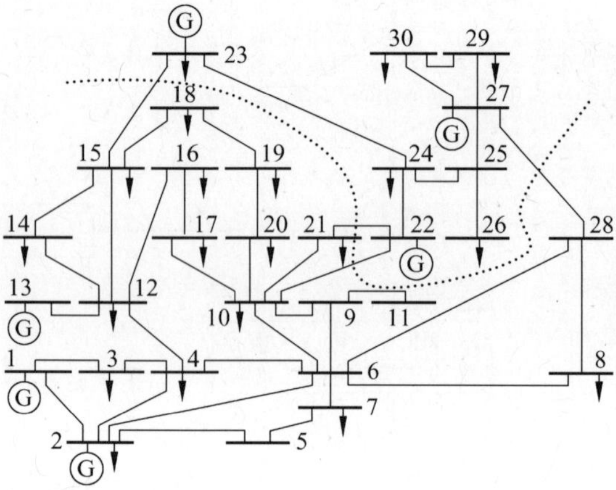

Figure 10.3 IEEE 30-bus system

Figure 10.4 illustrates the development of a cascading failure. It is obtained after seven rounds of the optimal AC power flow computation. Correspondingly, seven sub-figures are presented to show the seven stages in the development. The outage lines are indicated using the dashed line and are not shown in the subsequent sub-figure for the next stage. The numbers in parentheses are used to show the ordering of the line outages.

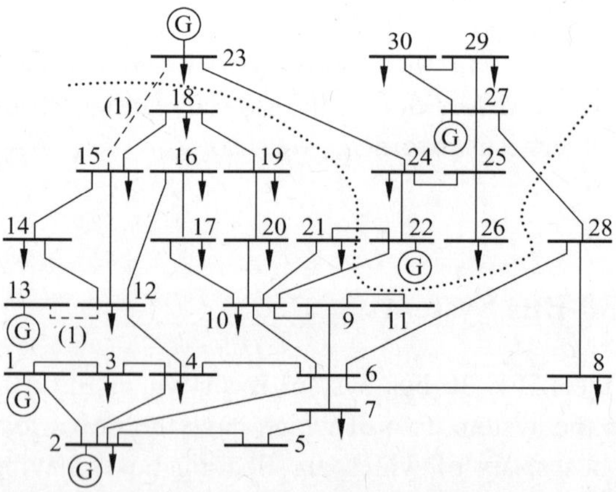

(a) Stage 1 of the cascading failure

Figure 10.4 Cascading failure in the IEEE 30-bus system

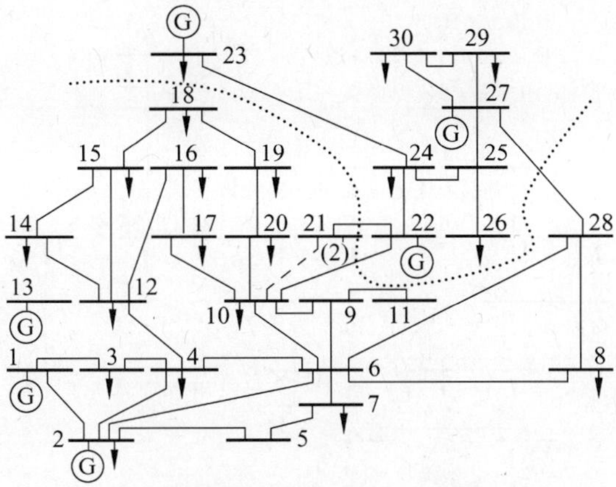

(b) Stage 2 of the cascading failure

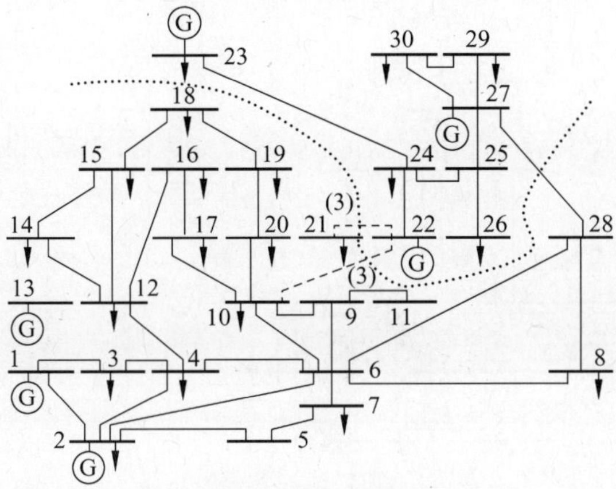

(c) Stage 3 of the cascading failure

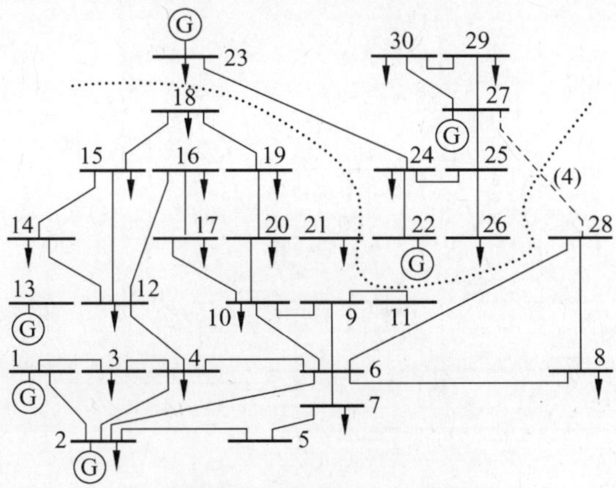

(d) Stage 4 of the cascading failure

Figure 10.4(Continued)

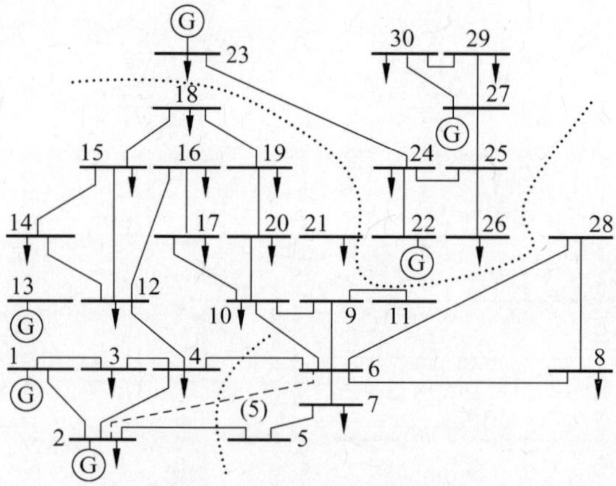

(e) Stage 5 of the cascading failure

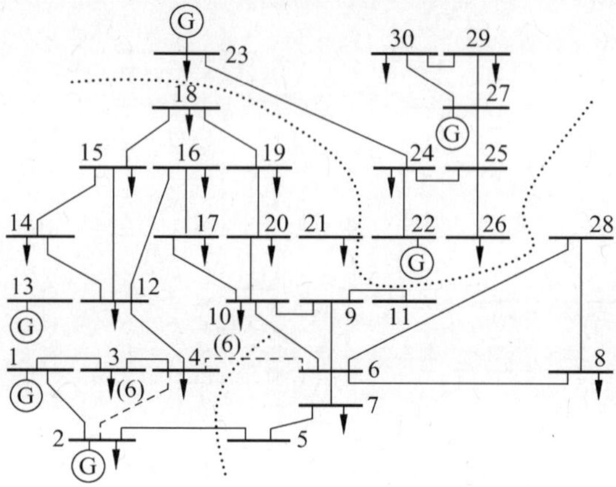

(f) Stage 6 of the cascading failure

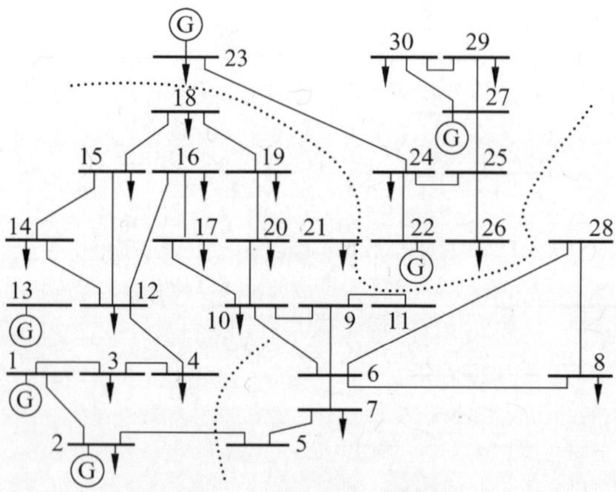

(g) Stage 7 of the cascading failure

Figure 10.4(Continued)

The lines labeled (1) to (6) are cut sequentially in the seven-stage process as shown in Fig. 10.4 and Table 10.1.

Table 10.1 Outage lines of the cascading failure in the IEEE 30-bus system

No.	Overloaded lines	Outage lines
1	12-13,21-22,15-23	12-13,15-23
2	10-21,21-22	10-21
3	10-22,21-22	10-22,21-22
4	27-28,6-28	27-28
5	2-6	2-6
6	2-4，4-6	2-4,4-6

Stage 1 Lines 12-13, 21-22 and 15-23 are overloaded and consequently lines 12-13 and 15-23 are cut. Note that the three lines are all connected to generator buses. The overloads might be caused by the growth in the load levels during the slow dynamics. There are two direct consequences for the cutting of lines 12-13 and 15-23. First, the generator bus 13 is lost and the power supply in the lower left corner is in severe shortage. Second, the power supply from the upper right area to the lower left area cannot be transmitted through bus 15, but is transferred via buses 21, 22 and 28 instead, which ultimately leads to the overloads and outages in the lines connected to buses 21, 22 and 28 in the following rounds of the OPF computation.

Stage 2 Lines 10-21 and 21-22 are overloaded, in which line 10-21 is cut. After the generator bus 13 is lost, the power demands at load buses 14, 15, 16, 17, 18, 19 and 20 can only be met by the supplies from the generator buses 1 and 2 in the lower left corner and the generator buses 22, 23 and 24 in the upper right corner. In particular, the power from the latter must be transmitted through lines 10-21 and 21-22, which causes the overloads in these two lines.

Stage 3 Lines 10-22 and 21-22 are overloaded and cut. After the optimal power flow computation in the previous round, line 10-21 is cut which makes line 10-22 responsible to deliver the power supply from the upper right area to the lower left area. As a result, line 10-22 is overloaded and cut.

Stage 4 Lines 27-28 and 6-28 are overloaded, and line 27-28 is cut. After line 10-22 is cut in the previous round of the optimal power flow computation, line 27-28 becomes the only line that supports the power transmission from the upper right area to the lower left area. It is unavoidable that line 27-28 becomes overloaded, which further causes line 6-28 to be overloaded. Up to now, the system has been divided into two isolated islands in the upper right and lower left areas.

Stage 5 Line 2-6 is overloaded and cut. The outage of line 27-28 results in the lost of power supply at the load buses 6, 8 and 28 from generator bus 27 in the upper right corner. Hence, all the power is supported by generator buses 1 and 2 in the lower left corner, which causes the rapid increase of power flow in line 2-6 and thus line outage takes place.

Stage 6 Lines 2-4 and 4-6 are overloaded and cut. The result of the outage of line 2-6 is the redistribution of its power flow into lines 2-4 and 4-6. These two lines are overloaded consequently.

Stage 7 After shedding most of the loads, the optimal AC power flow still cannot converge. This indicates that voltage collapse may have happened in the system.

The above simulation demonstrates the development of a cascading failure in the IEEE 30-bus system. The system can be divided into the lower left and upper right two areas according to its topological and electrical connections. There are several tie lines to interconnect these two areas to balance the mutual generation capabilities and load demands. The triggering event of the lost of one generator in the lower left area causes the power supply shortage in this area. Correspondingly, large amount of power needs to be delivered from the upper right area to the lower left area and the tie lines between the two areas quickly become overloaded and have to be cut. The cascading events divide the system into two isolated islands and subsequently the large-scale blackout finally takes place.

10.3.3 The 500 kV Northeast Power Grid of China

In order to investigate further the blackout model based on AC power flows, we simulate the operation of the real power system in the northeast of China. Fig.10.5 shows the topology of its 500 kV main grid. There are 32 buses in the system, including 8 generator buses and 23 load buses. All the grids that are below the voltage level of 500 kV are simplified as constant loads. In the figure, the loads at buses 13, 16, 17, 21 and 22 are negative, which means that there are generators of lower voltage levels connected in the nearby areas whose outputs are greater than their surrounding loads.

From Fig. 10.5, it is clear that the lines connecting to buses 1, 2, 3, 5, and 39 are all single lines, which have relatively small capacities and are easy to become overloaded and thus to be cut off. In the mean time, the lines connecting to buses 4, 6, 8 and 9 are double lines, which have relatively large capacities and thus less likely to get overloaded. The generator buses can be divided into three areas according to the capacities of the lines connecting to them. The first area is in the upper left corner containing two generators and the single output line 11-14. The grid structure in this area is vulnerable. The middle area contains three generators, each of which is connected to the system by a single line, and thus its grid structure is hardly resilient. The lower left area has four generators that are all connected to the system by double lines, and its structure is more robust.

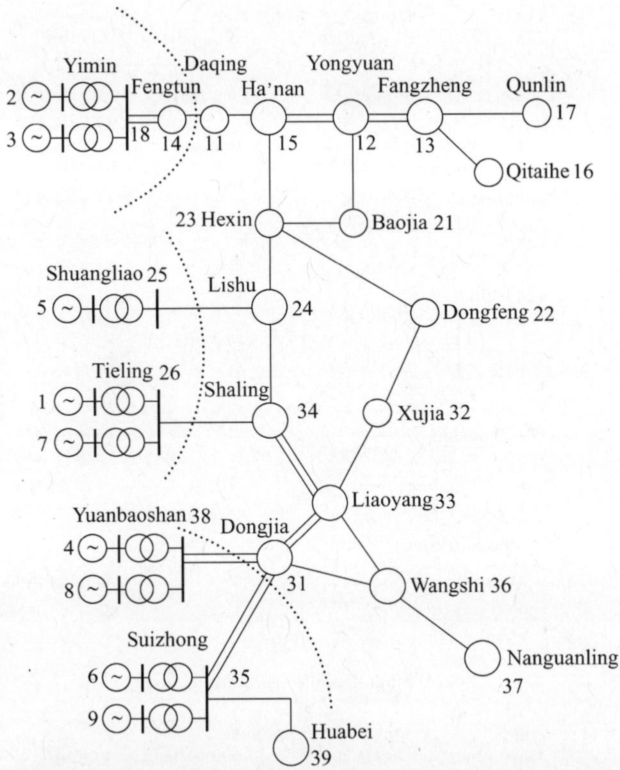

Figure 10.5 The 500 kV Northeast Power Grid of China

Figure 10.6 illustrates the development of a cascading failure in the 500 kV Northeast Power Grid of China that is obtained after seven rounds of computations for the optimal AC power flows. In Table 10.2 and Fig.10.6, the lines labeled (1) to (6) are cut in sequence in different computation rounds.

Stage 1 Lines 1-26, 7-26 and 24-25 are overloaded due to some initial fault in the system and line 1-26 is cut. Therefore, generator 1 is lost in the system. Because the lines connecting to the generator buses 1, 5 and 7 are single lines with smaller capacities, they are likely to become overloaded first.

Table 10.2 The cascading failure in the Northeast Power Grid of China

No.	Overloaded lines	Outage lines
1	24-25,1-26,7-26	1-26
2	24-25,31-36,5-25,6-35,9-35	24-25,9-35
3	35-39,11-14,31-38,14-18,6-35,7-26,2-18,3-18,4-38,8-38	35-39,31-38,7-26
4	11-14,31-36,14-18,6-35,2-18,3-18	3-18
5	22-23,31-33,31-36,31-38,6-35,2-18,4-38	31-33
6	11-14,31-33,31-36,6-35,2-18	11-14

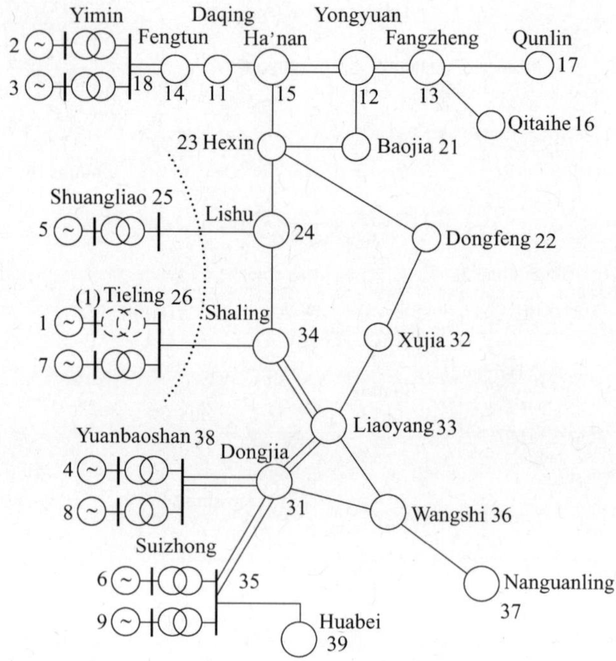

(a) Stage 1 of the cascading failure

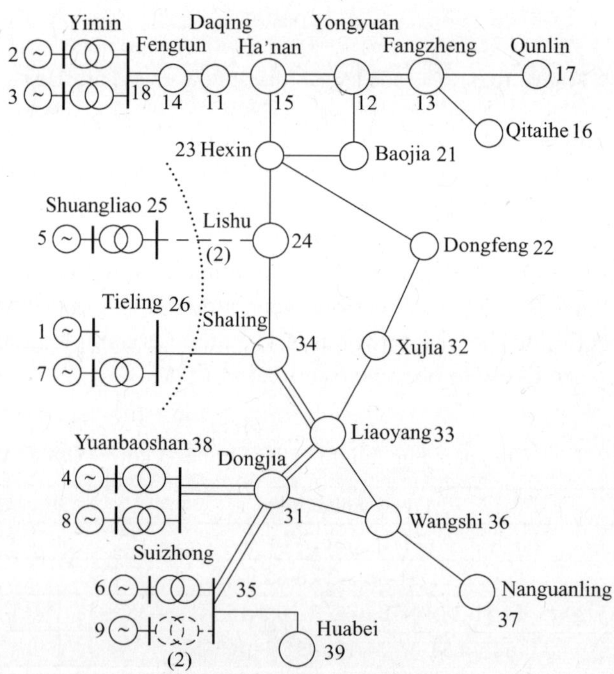

(b) Stage 2 of the cascading failure

Figure 10.6 The cascading process in the 500 kV Northeast Power Grid of China

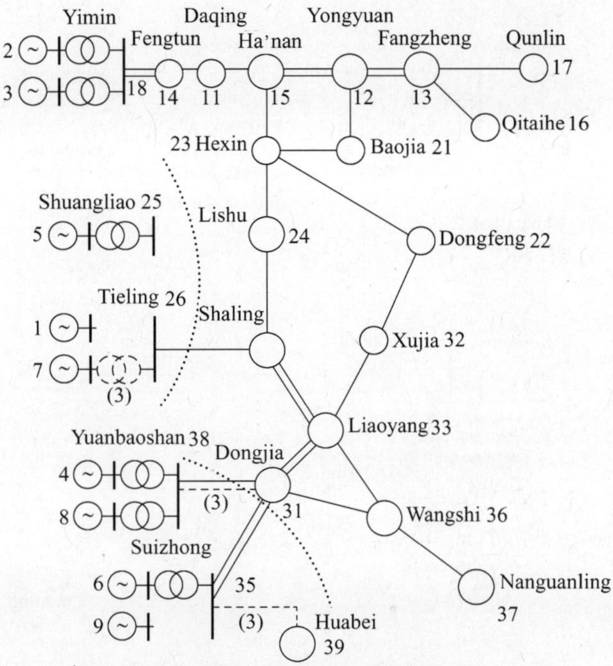

(c) Stage 3 of the cascading failure

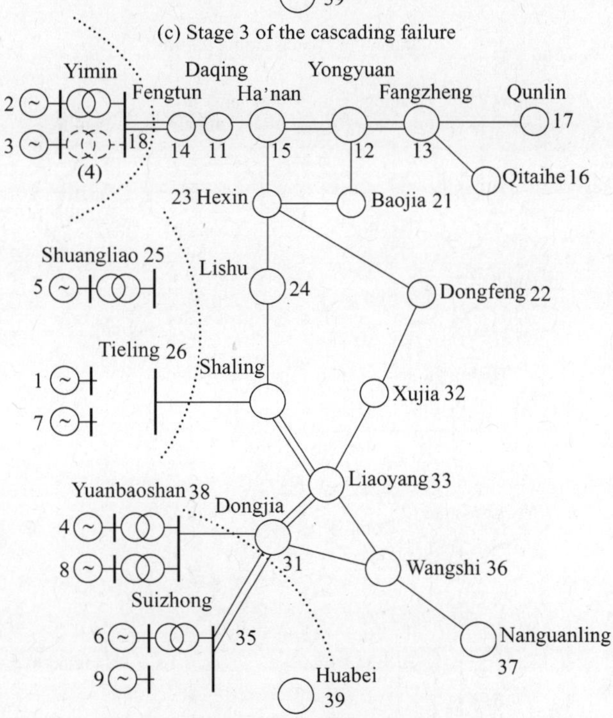

(d) Stage 4 of the cascading failure

Figure 10.6(Continued)

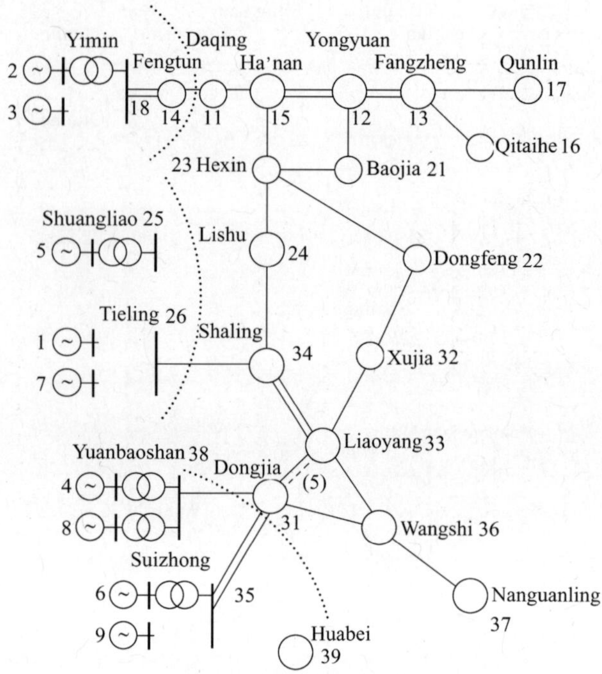

(e) Stage 5 of the cascading failure

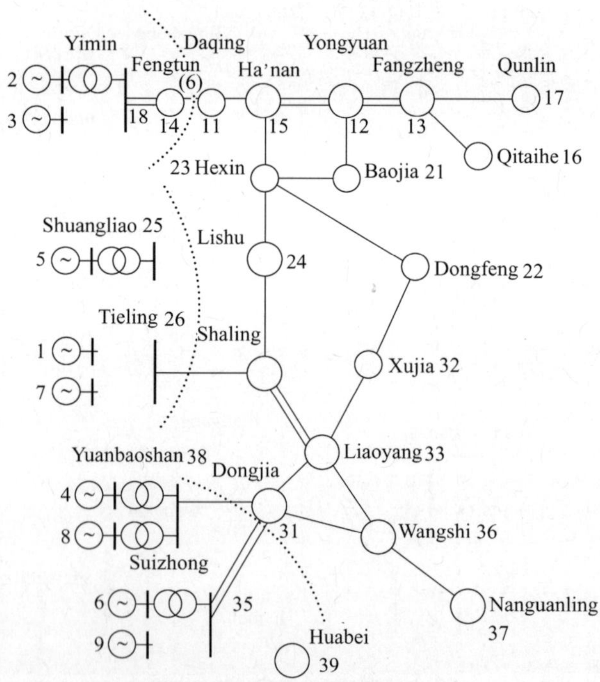

(f) Stage 6 of the cascading failure

Figure 10.6(Continued)

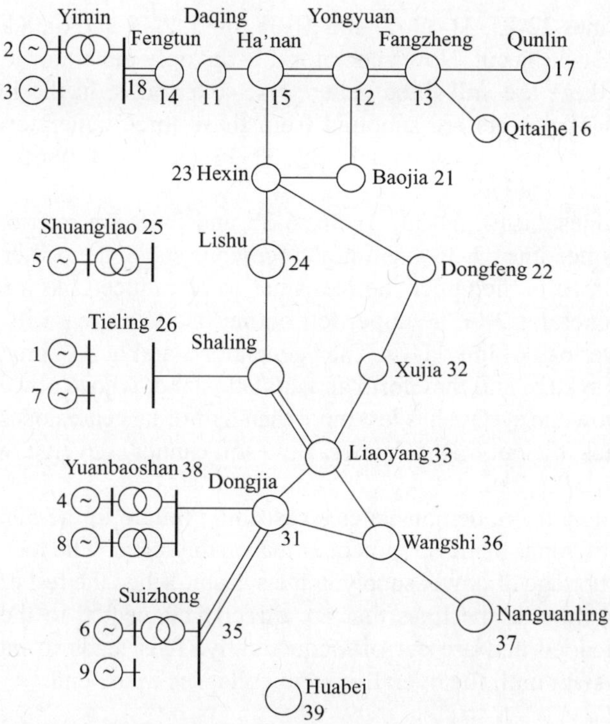

(g) Stage 7 of the cascading failure

Figure 10.6(Continued)

Stage 2 Lines 24-25, 31-36, 5-25, 6-35 and 9-35 are overloaded, among which lines 24-25 and 9-35 are cut. Since line 1-26 is cut and generator 1 is lost in the previous stage, the power in shortage has to be supported by other generators in the system. Consequently, several lines that are directly connected to generators become overloaded which forces generators 5 and 9 to be isolated from the system.

Stage 3 Lines 35-39, 11-14, 31-38, 14-18, 6-35, 7-26, 2-18, 3-18, 6-35 and 8-15 are overloaded, among which lines 7-26, 31-38 and 35-39 are cut. Because of the cutting of line 9-35 in the second stage, the power supply for load bus 35 comes from bus 39 (the inter-connection to the north China grid) instead of originally from generator bus 9. This power redistribution causes line 35-39 to become overloaded. Because of the outage of line 1-26 in the first stage, the power output at generator bus 7 has increased, which leads to the overload in line 7-26. Now generator bus 7 and the bus 39 representing the north China power grid are forced to leave the system.

Stage 4 Lines 11-14, 31-36, 14-18, 6-35, 2-18 and 3-18 are overloaded, and line 3-18 is cut. With more generator buses leaving the system, the output of generator 3 has to keep increasing. Then line 7-26 is overloaded and cut. Now generator 3 also leaves the system.

Stage 5 Lines 22-23, 31-33, 31-36, 31-38, 6-35, 2-18 and 6-38 are overloaded, in which line 31-33 is cut. Now generator 2 is the only generator left in the upper left area and there are still three generators, 4, 6 and 8, in the lower left area. Large amount of powers are supplied from these three generators to the upper right area through bus 31. So lines 31-33, 31-36 and 31-38, which are connected to bus 31, are overloaded.

Stage 6 Lines 11-14, 31-33, 31-36, 6-35 and 2-18 are overloaded, and line 11-14 is cut. Since line 31-33 is cut in the previous stage, the power delivery from the lower left area to the upper right area has to be reduced. As a result, the only functioning generator 2 in the upper left corner has to increase its output, which leads to the overload of line 11-14. Then generator 2 and bus 14 are separated from the rest of the system and they form an isolated island, whose power is balanced.

Stage 7 Now the system has lost more than half of its generators. After shedding most of its loads, the optimal AC power flow still cannot converge, which indicates the system has collapsed.

The simulation above demonstrates a cascading failure in the 500 kV Northeast Power Grid of China. Its development is reasonably clear. The lost of a generator results in the shortage of power supply in the system. When the rest of the generators increase their outputs, the lines that are directly connected to these generators, become overloaded and are cut off sequentially. This leads to more generators leaving the system until the overall system collapses in the end.

10.3.4 Influence of Network Topology on Cascading Failures

Figure 10.7(a) and (b) present the process of cascading blackouts in the IEEE 30-bus system and the 500 kV Northeast Power Grid of China, where the numbers in parentheses indicate the ordering for the line outages. As shown in Fig. 10.7(a), the IEEE 30-bus system has several loops in its topology. According to the concentration of generator buses and the distribution of electrical components, the system can be roughly divided into lower left and upper right two power supply areas, which are interconnected by several tie lines to achieve power balance. Simulation results indicate that the blackout is mainly caused by the outages of the tie lines, which separate the system into two islands. On the other hand, there are few loops in the tree-like topology of the 500 kV Northeast Power Grid of China. Therefore, its blackout process is relatively simple, which is mainly the result of the outage lines that are directly connected to the generators. The system collapses because the overall power supply cannot meet the overall load demand.

In the consideration of system stability, the lost of generators does far more damage than the outage of tie lines. Hence, during power grid planning, more emphasis should be given to establish more loops and divide the system into several self-balanced sub-regions. Such network topologies are more robust to cascading failures and each sub-region is more capable to handle cascading blackouts. In this way, the overall system can effectively prevent the occurrence of large-scale blackouts.

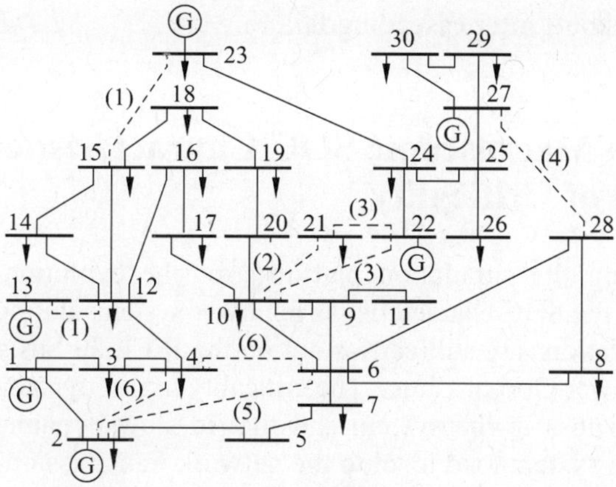

(a) Cascading failure and topology of the IEEE 30-bus system

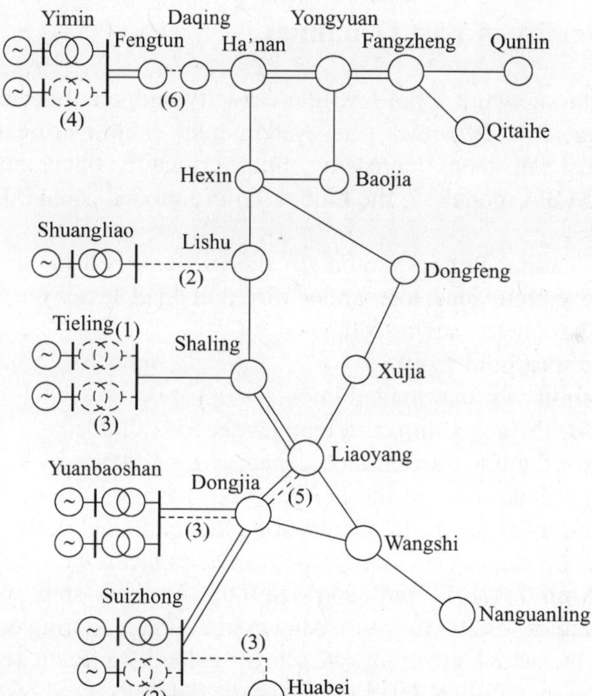

(b) Cascading failure and topology of the 500 kV Northeast Power Grid of China

Figure 10.7 Cascading failures and topologies of the IEEE 30-bus system and the 500 kV Northeast Power Grid of China

The study on the propagation of the cascading failure in grids with different network topologies reveal that multi-loop, multi-region power grids are able to avoid extensive blackouts after cascading failure.

10.4 Study on Macroscopic SOC Characteristics and Analysis of Criticality

In this section, using the simulation platform for the evolution of cascading failures, we disclose the SOC characteristics in power systems during fast and slow dynamics. The simulations are still carries out on the IEEE 30-bus system and the 500 kV Northeast Power Grid of China. The criticality in fast dynamics is described by the average load loss in the system and that in slow dynamics is analyzed using the ratio of the system load level to the network transmission capacity.

10.4.1 Criticality of Fast Dynamics

When keeping the network topology, line capacity and generator capacity fixed, namely ignoring slow dynamics, the system may exhibit critical behaviors if the system load level keeps increasing. Such criticality phenomena have been discussed in the OPA model[8,9], the hidden failure model[10], and the Manchester model[5].

Using the cascading failure simulation model described in Section 10.2, we characterize the system load loss under different load levels in fast dynamics. The simulation parameters are as follows:

Line overload threshold α =0.7,

Outage probability for overloaded lines β =0.3,

Growth rate for the generation and load levels λ =1.0005,

Growth rate for the line transmission capacity μ =1.005.

Set the original load level in the IEEE 30-bus system to be the base value 1. Let k_L be the load level factor; in other words, the load level in the system can be obtained by multiplying all the loads by the load level factor k_L. Let the system load level increase from 0.6 to 1.7 with stepsize 0.05. At every step, we simulate the system fast dynamics for 200 times. As shown in Fig. 10.8, during each simulation, to approximate the actual situations, we allow the loads to fluctuate stochastically provided that the overall load level is consistent. The statistical results are shown in Fig. 10.9. The horizontal axis indicates the system load level factor k_L, and the vertical axis indicates the ratio of the average relative load loss in the system:

$$\overline{P}_{\text{loss}} = \frac{P_{\text{loss}}}{P_{\text{demand}}} \tag{10.14}$$

where P_{loss} is the total shed load and P_{demand} is the total load demand.

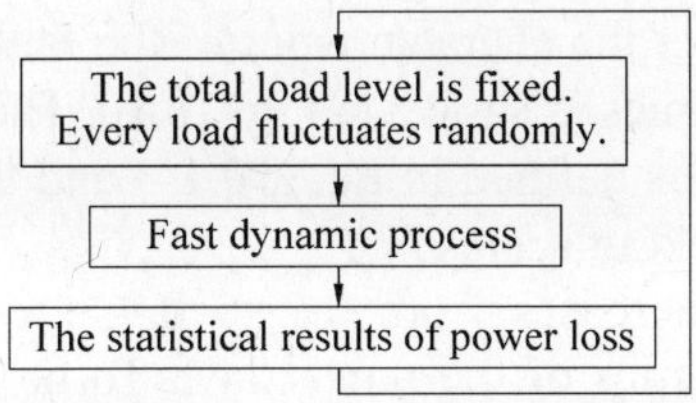

Figure 10.8 Flowchart of statistical analysis for the criticality of fast dynamics

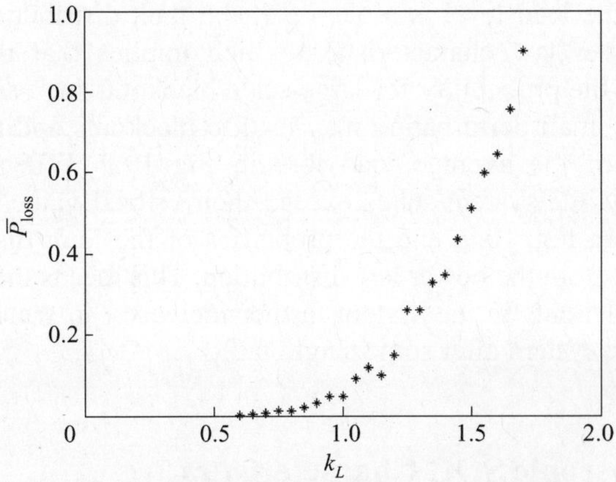

Figure 10.9 Criticality of the fast dynamics in the IEEE 30-bus system

The following conclusions can be drawn from Fig. 10.9. When the system load level k_L is low, the average load lost $\overline{P}_{\text{loss}}$ is small. After the load level $k_L>0.9$, $\overline{P}_{\text{loss}}$ increases dramatically. Especially after $k_L>1.2$, $\overline{P}_{\text{loss}}$ is much larger than that in the low load level situation. This implies that after the load level surpasses its critical value 0.9, any small perturbation may result in large-scale blackouts.

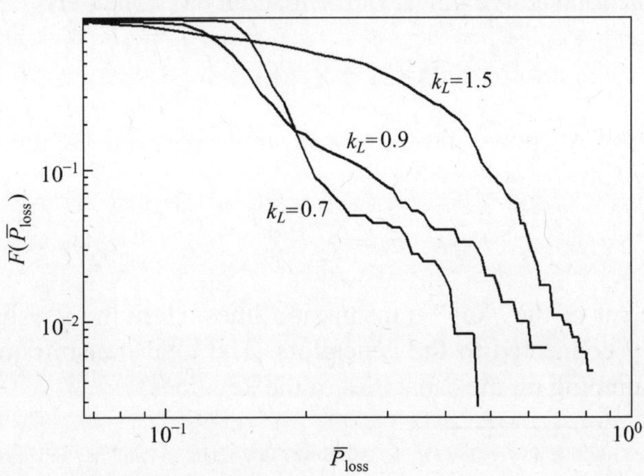

Figure 10.10 Blackout distributions at different load levels in fast dynamics of the IEEE 30-bus system

Furthermore, we consider the characteristics of the fault distribution at different load levels in the fast dynamics as shown in Fig. 10.10. The horizontal axis indicates the relative load loss $\overline{P}_{\text{loss}}$ as defined in Eq. (10.14) and the vertical axis indicates the distribution of the blackout scales $F(\overline{P}_{\text{loss}})$.

When the load level increases from 0.7 to 0.9, the fault distribution curve exhibits certain characteristics of the power law. To be more specific, when the load level is below 0.7, the system is in non-critical states, and small disturbances can hardly affect the system. This corresponds to the small average load loss in Fig. 10.9. When the load level is around 0.9, the fault distribution curve reveals precisely the power law characteristics, which implies that the system is in critical states and the probability for large-scale blackouts increases significantly. At this time, any small perturbation may lead to blackouts and this explains the sudden increase of the average load loss in Fig. 10.9. When the load level continues to grow, the system states exceed their critical values, e.g. load level goes beyond 1.5 in Fig. 10.9, and the properties of the fault distribution at this time differ greatly from the power law distribution. This means that the probability for large-scale blackouts in the system further increases, in which case disasters may happen in the system after some single fault.

10.4.2 Macroscopic SOC Characteristics

Following a different approach than the one used in Section 9.2.2, in this section we study the ratio of the overall load level to the total transmission capacity using AC power flow[1]

$$I(k) = \frac{P_{d,k}}{F_k} \tag{10.15}$$

where $P_{d,k}$ is the total active power of the load on day k, namely

$$P_{d,k} = \sum_{i \in L} P_{di,k} \tag{10.16}$$

In the context of AC power flows, we can similarly define the total transmission capacity on day k as

$$F_k = \sum_{j \in C} F_{j,k}^{\max} \tag{10.17}$$

where C is the set of the "key" transmission lines. Here the key lines are those that are directly connected to the generators. The total transmission capacity is estimated by summing up the capacities of the key lines.

[1] Although there are reactive power flows in the AC power flows, one can still use active powers to construct approximate indexes. The reasons are that reactive powers cannot be transferred over long distances in power systems and thus have only local influences. Of course, one can construct indexes that are more sophisticated and this will be discussed in Chapter 11.

In what follows, using the blackout model based on AC power flows and the just described indices $I(k)$ and $\overline{P}_{\text{loss}}$, we study the evolution of 5000 days (about 14 years) for the dynamics of the IEEE 30-bus system. The simulation results are shown in Fig. 10.11, where sub-figure (a) shows how $I(k)$ changes with time and sub-figure (b) shows how $\overline{P}_{\text{loss}}$ correspondingly changes with time.

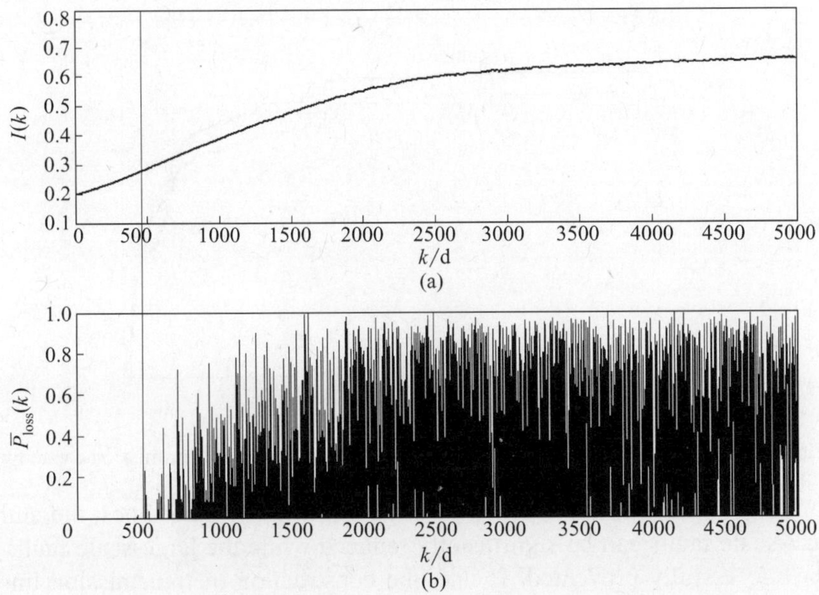

Figure 10.11 The IEEE 30-bus system approaching critical states

As shown in Fig. 10.11, the generation capacity and the load level keep increasing in the process of 5000 days. During days 1 – 480, there are few faults in the system and $I(k)$ grows with time. At the time around the 480th day, faults start to take place. After this, the scale of faults rises with the growth of $I(k)$. On day 480, $I(k)$ is approximately 0.29, which can be taken as a critical value to assess the reliability of the system operation.

We can summarize using SOC theory. When $I(k)$ is small, the system is in normal states and the probability for blackouts is very small. As time goes on, since the upgrades in the system cannot keep up with the demand from the increasing loads, the amount of the transmitted power reaches the limit of the total transmission capacity of the system. This is the moment at which $I(k)$ attains its critical value. Since then, large-scale blackouts get to take form and the system is subject to the danger of disasters after entering critical states.

Now we study the impact of upgrading transmission lines on system blackouts. In the blackout model constructed in Section 10.2, upgrades are performed if there are outages in order to improve transmission capacities. To be more specific, all those lines that were in outage in the fast dynamics in the previous day are upgraded so that the new capacities are now μ times their original capacities. As such, we analyze the fault distribution when μ equals 1.005 and 1.008 respectively during a process of 5000 days in the IEEE 30-bus system. The simulation results are shown in Fig. 10.12 where the horizontal axis indicates the relative load loss $\overline{P}_{loss}$, which suggests the scale of the fault, and the vertical axis indicates the number N of those faults that are greater than a preset scale.

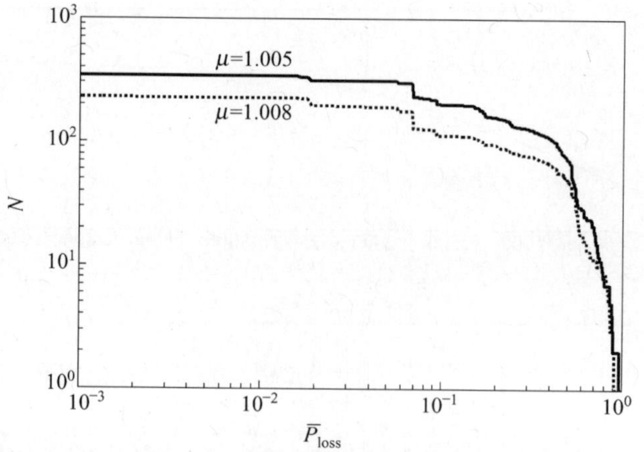

Figure 10.12 Fault distributions under different growth rates of transmission capacity

From Fig. 10.12, one can see that by strengthening the weak lines, the number of small-scale faults can be significantly reduced while the large-scale faults can also be successfully prevented. Hence, the construction of transmission lines is an effective means to prevent various blackouts and disasters.

10.4.3 Criticality Analysis for the 500 kV Northeast Power Grid of China

In order to check the SOC phenomena in a real system, we use the 500 kV main grid in the Northeast Power Grid of China to carry out simulation study. The parameters used are the same as in Section 10.4.1.

The simulation results for the criticality in fast dynamics in the 500 kV Northeast Power Grid of China are shown in Fig. 10.13. The overall trend of the curve agrees with the simulation results for the IEEE 30-bus system. In the figure, the horizontal axis indicates the load level k_L, and the vertical axis indicates the relative load loss $\overline{P}_{loss}$. When the load level is below 0.5, there is no fault. When the load level exceeds 0.65, $\overline{P}_{loss}$ suddenly increases. Especially after the load level goes beyond 1.15, $\overline{P}_{loss}$ becomes much larger than that when the load level is low. This implies that when the load level is greater than some threshold value, any small disturbances may lead to large-scale blackouts. When the load level is around 1.4, the average load loss is about a quarter of all the loads in the system. Note that when the load level equals 1, the ratio of the overall load over the overall generation capacity is 0.7139. When the load level reaches 1.4, the load level is about equal to the overall generation capacity. If taking into consideration the network loss, at this time the system generation output has become smaller than the load demands, and thus the optimal AC power flows cannot converge and the system collapses.

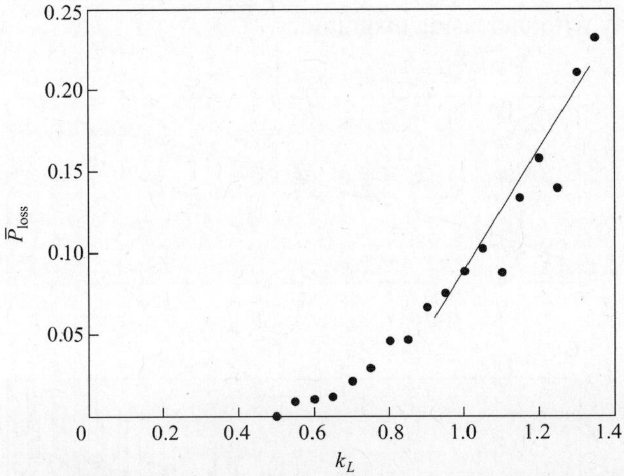

Figure 10.13 Criticality in fast dynamics in the 500 kV Northeast Power Grid of China

Now we further study the distribution of the relative load loss $\overline{P}_{\text{loss}}$ at different load levels, e.g. $k_L = 1.0$, 1.1, 1.2 respectively, in the fast dynamics of the 500 kV Northeast Power Grid of China. As shown in Fig. 10.14, all the three curves exhibit the power-law tail behavior, which implies that the system load levels are all around critical values and the systems are subject to cascading failures.

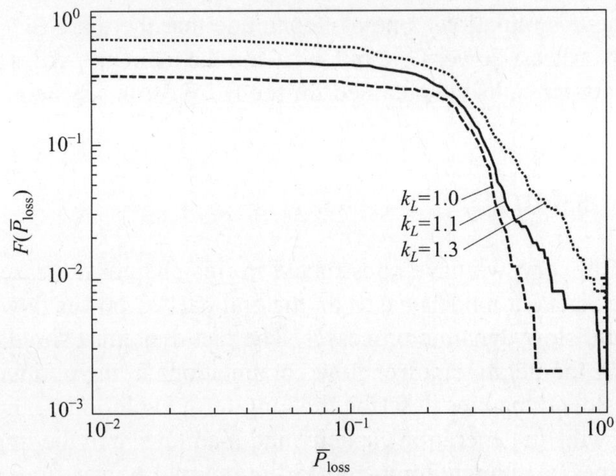

Figure 10.14 Blackout distributions at different load levels in the fast dynamics of the 500 kV Northeast Power Grid of China

Finally, we study the changes in $I(k)$ and $\overline{P}_{\text{loss}}$ over a period of 2000 days in the 500 kV Northeast Power Grid of China. The simulation results are shown in Fig. 10.15. During the period from day 1 to day 250, there are few blackouts. As the system load level keeps increasing up to $I(k)>0.74$, namely during the interval from day 250 to day 570, small-scale faults begin to develop in the system. When $I(k)>0.82$, namely between day 570 and day 2000, large-scale blackouts take place constantly with increasing magnitudes.

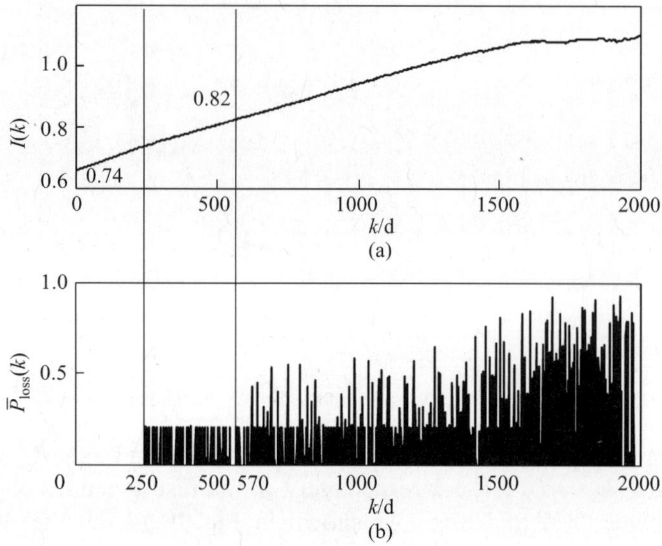

Figure 10.15 The process of approaching criticality in the 500 kV Northeast Power Grid of China

From the above simulations, one can conclude that there are SOC phenomena in the 500 kV Northeast Power Grid of China and the main properties are consistent with the simulation results we obtained for the IEEE 30-bus system.

10.5 Discussion

Using the SOC theory, we have constructed in this chapter the cascading failure and large-scale blackout models based on the optimal AC power flows. The model contains fast and slow dynamic processes. The fast dynamics simulate cascading failures utilizing the iterative power flow computations in the optimal power flow program. The slow dynamics describe the long-term evolution of the power grid, such as the growths in generation capacity and load level and the improvement in transmission lines. Although compared to the optimal power flow model in the previous chapter, the models in this chapter can simulate voltage collapse by exploiting the divergence of the computation of AC power flows, they cannot evaluate the voltage instability quantitatively. To deal with this shortcoming, we will study in depth in the next chapter the modeling techniques to consider the characteristics of the interplay between reactive power and voltages in blackouts.

At the first glance, the Manchester model resembles the fast dynamic process in the model studied in this chapter. However, the model in this chapter takes the generators' outputs as decision variables in the fast dynamics and thus takes care of the inherent limitation of the Manchester model, which assumes that the generators' outputs can be determined before the power flow computation. Note that in real practice, the system's operation mode is set according to the optimal power flow computation that minimizes the network loss, and thus generators' outputs are adjustable. As pointed out in Chapter 1, the Manchester model has not taken full consideration of the adjustment in generators' outputs and may lead to misjudgments on the occurence of blackouts.

References

[1] Taylor C W. Power System Voltage Stability. New York: McGraw-Hill, 1994

[2] IEEE System Dynamic Performance Subcommittee of the Power System Engineering Committee of the PES. Voltage stability of power system: concepts, analytical tools, and industry experience. Technical Report 90 TH 0358-2PWRS, IEEE, 1990

[3] Kundur P. Power System Stability and Control. New York: McGraw-Hill, 1994

[4] Pai M A, O'Grady M G. Voltage collapse analysis with reactive generation and voltage dependent load constraints. Electric Machines and Power Systems, 1989, 17(6): 379 – 390

[5] Nedic D P, Dobson I, Kirschend S, et al. Criticality in a cascading failure blackout model. Proceedings of the 15th Power Systems Computation Conference, Liege, Belgium, Aug. 2005

[6] Wang X F, Song Y H, Irving M. Modern Power System Analysis. New York: Springer, 2008

[7] Wei H, Sasaki H, Kubokawa J, et al. An interior point nonlinear programming for optimal power flow problems with a novel data structure. IEEE Transactions on Power Systems, 1998, 13(3): 870 – 877

[8] Carreras B A, Lynch V E, Dobson I, et al. Modeling blackout dynamics in power transmission networks with simple structure. Proceedings of the 34th Annual Hawaii International Conference on System Sciences, Jan. 2001: 719 – 727

[9] Mei S W, He F, Zhang X M, et al. An improved OPA model and blackout risk assessment. IEEE Transactions on Power Systems, 2009, 24(2): 814 – 823

[10] Chen J, Thorp J S, Dobson I. Cascading dynamics and mitigation assessment in power system disturbances via a hidden failure model. International Journal Electrical Power and Energy Systems, 2005, 27(4): 318 – 326

Chapter 11 Blackout Model Considering Reactive Power/Voltage Characteristics

This chapter embeds two types of voltage stability analysis modules based on the blackout model developed in the previous chapters. The aim is to construct blackout models that take into account the reactive power/voltage characteristics. Improvements can then be made to simulate more faithfully the fast and slow dynamics in power systems. In the meanwhile, the change of voltage stability levels can be tracked more precisely, and the SOC characteristics become more obvious.

In the previous chapter, we have constructed blackout models based on optimal AC power flows, which take into account both slow and fast dynamics in the system and thus describe more precisely the fast evolution process in power system operation and the slow evolution process in power grid development. Furthermore, the evolution mechanism and the SOC characteristics have been revealed by the examination of the pair of active power and load level. However, it needs to be pointed out that blackouts are usually caused not just by cascading failures as a result of overloading transmission lines, but by voltage collapse as well[1–3]. The model used in the previous chapter does not include the module to analyze voltage stability and thus cannot present completely various classes of blackouts and SOC phenomena. Hence, in this chapter we construct blackout models that consider the reactive power/voltage characteristics to track more precisely the trends in the change of voltage stability levels, and reveal more insightfully the SOC characteristics.

Because of the complexity of the voltage stability problem, various indices have been proposed to assess the voltage stability level. Each of these indices has its own advantages and disadvantages. The blackout model constructed in this chapter utilizes two classes of voltage stability indices. One is to use those that are determined by present or optimal power flows, such as the critical voltage and the minimum eigenvalue of the system matrix. This type of indices is simple and easy to use. The other is to use the load margin in the direction of load growth. This is also called the continuation power flow method, which is comparatively complicated in computation, but clearer in physical concepts and consistent with the engineering practice.

11.1 Blackout Model Ⅰ

We first discuss blackout model Ⅰ with the reactive power/voltage characteristics[4] shown in Fig. 11.1. It is obtained by embedding the static voltage stability analysis module into the blackout model discussed in the previous chapter. This model can be directly used to analyze the trend in the voltage stability level and assess the voltage security level. Its structure can be described as follows.

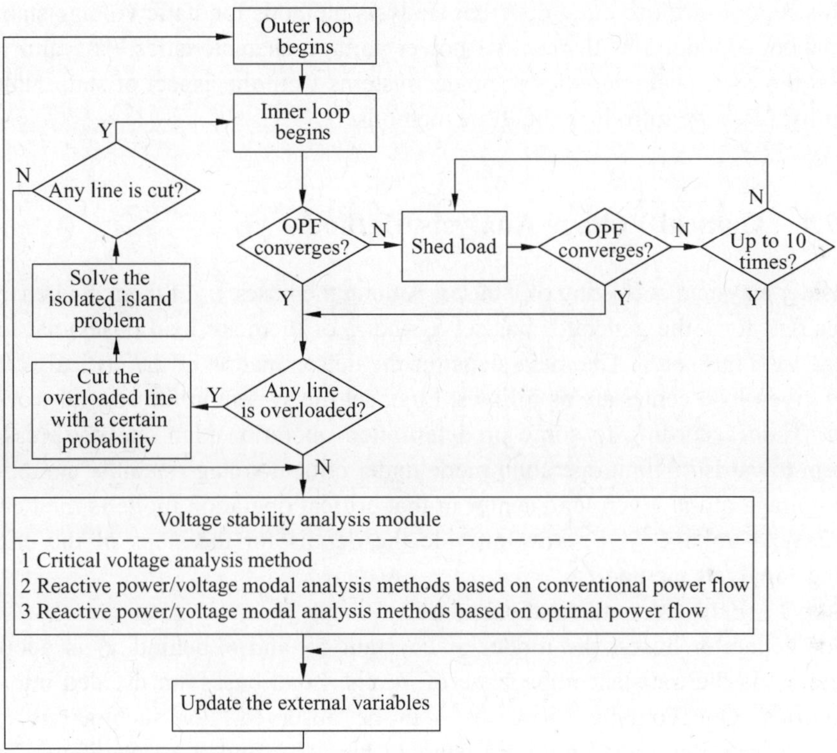

Figure 11.1 Blackout model Ⅰ with reactive power/voltage characteristics

The inner loop simulates the fast dynamics for possible blackouts in daily operations. The main idea is to first determine the system's operation state through the optimization of power flows and load shedding. If some overloaded lines break down, we trip those lines and recalculate the operation state of the system. If there are islands, procedures need to be taken to deal with islanding[4]. After the fast dynamics are over, the analysis for static voltage stability is carried out.

The outer loop simulates the slow dynamics. It uses the average growth rates for the generation output and load level to describe the system's developments.

The lines that have been tripped in the fast dynamics are upgraded according to the growth rate for the line transmission capacity; the transmission capacities of all the other lines remain the same.

11.2 Reactive Power/Voltage Analysis Methods in Blackout Model Ⅰ

In this section, we use three different analysis methods for static voltage stability in blackout model I with reactive power/voltage characteristics. The aim is to reveal the SOC characteristics in power systems from the aspect of static voltage stability. Now we introduce the three methods.

11.2.1 Critical Voltage Analysis Method

Consider a system consisting of n buses. Among the buses, n_g of them are generator buses that form the generator bus set S_g and n_l of them are load buses that make up the load bus set S_l. The main steps for the determination of the critical voltage for a given load center are as follows. First, set the system in the normal working state. Then according to some predetermined operation plan, slowly press the system towards its limit operation mode under critical voltage stability conditions. The voltage at the given load center in that critical operation mode is then called the critical voltage[5]. We now provide the details for the steps of the critical voltage analysis method.

Step 1 Establish mathematical models.

For generator buses, the model of constant E' and θ behind x_d' is adopted, where x_d' is the transient impedance of d-axis. Load buses are divided into two categories. One contains those that can be observed. For such a bus i, its voltage's amplitude and angle are denoted by $|U_i|$ and θ_i respectively. These buses are set to be the $V\theta$ type. Constant impedance model is applied to the rest of the load buses.

Step 2 Construct the admittance matrix corresponding to the load bus i and the generator buses.

We first enlarge the system admittance matrix by adding generators' x_d' and the loads. Then we reduce this augmented matrix to that with only generators' internal buses and load bus i. So we have

$$\begin{pmatrix} Y_{11} & Y_{12} \\ Y_{21} & Y_{22} \end{pmatrix} \begin{pmatrix} E_g \\ U_i \end{pmatrix} = \begin{pmatrix} I_g \\ I_i \end{pmatrix} \tag{11.1}$$

where U_i and I_i represent the voltage and injected current at load bus i respectively, and E_g and I_g are the voltage and injected current vectors of the generators' internal buses respectively. The admittance matrix can be divided into four block submatrices Y_{11}, Y_{12}, Y_{21} and Y_{22}, whose dimensions are determined by those of U_i, I_i, E_g and I_g respectively.

Step 3 From Eq. (11.1), it is easy to compute the equivalent reactive power transimitted from the network to load bus i

$$Q_{gi} = \mathrm{Im}(Y_{22})|U_i|^2 + \mathrm{Im}(Y_{21}E_g)|U_i| \tag{11.2}$$

where Im denotes the operation of taking the imaginery part.

Step 4 From Eq. (11.2), we can compute the critical voltage for load bus i

$$|U_{icr}| = -\frac{\mathrm{Im}(Y_{21}E_g)}{2\mathrm{Im}(Y_{22})} \tag{11.3}$$

As implied in Eq. (11.2), Q_{gi} is quadratic in $|U_i|$, and thus the critical voltage $|U_{icr}|$ is in fact the value at which Q_{gi} attains its maximum. The physical meaning of a maximum Q_{gi} is the maximum reactive power that can be delivered from the network to load bus i in the present opertion mode, i.e.

$$Q_{gi}^{\max} = -\frac{(\mathrm{Im}(Y_{21}E_g\angle -\theta_i))^2}{4\mathrm{Im}(Y_{22})}$$

Said differently, when the voltage of bus i is lower than $|U_{icr}|$, the system losses its voltage stability[6].

Step 5 Calculate the system voltage stability margin

$$\min_{i\in S_l}\left(1-\frac{|U_{icr}|}{|U_i|}\right) \tag{11.4}$$

Obviously, the above voltage stability margin can only be obtained after computing the critical voltages $|U_{icr}|$ for all load buses $i \in S_l$.

Remark 11.1

Sometimes we call

$$U_{i0} = \frac{|U_i| - |U_{icr}|}{|U_i|} \tag{R1}$$

the static voltage stability reserve coefficient for load bus i. In normal operation modes, it is usually true that

$$10\% < U_{i0} < 15\%, \quad \forall i \in S_l \tag{R2}$$

The range considered for the system enduring a fault is

$$U_{i0} > 8\%, \quad \forall i \in S_l \tag{R3}$$

11.2.2 Reactive Power/Voltage Modal Analysis Method Based on Conventional Power Flows

The critical voltage analysis method discussed in the previous section only considers the growth of the reactive power at the given load bus while assuming that the other loads in the system remain fixed. Such a simplification can only determine approximately the voltage stability level in the system. The reactive power/voltage modal analysis method to be discussed in this section, on the other hand, does not assume any particular growth pattern for the loads, and thus can describe the voltage stability in the overall system with higher accuracy[3].

Denote the Jacobian matrix of the system power flow equations by J, whose elements are the partial derivatives of the active and reactive power balance equations with respect to the amplitudes and angles of the voltages of all buses. Since all the angles in the power flow equations are relative, the matrix J is singular (please refer to Remark 11.2 for details).

Because of the singularity of matrix J, the conventional power flow has to be used for the reactive power/voltage modal analysis. Set all those generators whose reactive powers reach their limits to be PQ buses. For the remaining generators within their reactive power limits, pick the one with the maximum active power output to be the slack bus denoted by i_0, and set the rest to be PV buses in the set S_{PV}.

Further, eliminate the $i_0{}^{\text{th}}$ and $(j+n)^{\text{th}}$ rows and columns in the Jacobian matrix J, where $j \in S_{PV}$ or $j = i_0$. Then we obtain a new Jacobian matrix $\hat{J}$ for the power flow iteration equations. We write $\hat{J}$ into its block matrix form

$$\hat{J} = \begin{pmatrix} \hat{J}_{P\theta} & \hat{J}_{PV} \\ \hat{J}_{Q\theta} & \hat{J}_{QV} \end{pmatrix} \tag{11.5}$$

Since voltages are closely related to reactive powers, we can investigate the relationship between voltages and reactive powers while assuming the active powers at all buses are fixed. We first reduce $\hat{J}$ into

$$\hat{J}_R = \hat{J}_{QV} - \hat{J}_{Q\theta}\hat{J}_{P\theta}^{-1}\hat{J}_{PV} \tag{11.6}$$

Obviously, different choices of the slack generator bus lead to different $\hat{J}_R$. However, since the dominant term in Eq. (11.6) is $\hat{J}_{QV}$ which is always the same, the difference is not substantial.

Then rearrange the rows and columns of $\hat{J}_R$ in such an ordering that the floating buses (these buses are associated with neither generators nor loads and thus have zero injected powers) and generator buses, which are treated as PQ bus, are in the front followed by the load buses. The matrix that we obtained is denoted by $\tilde{J}$. Now the sensitivity relationship between voltages and reactive powers are

$$\begin{pmatrix} \tilde{J}_{11} & \tilde{J}_{12} \\ \tilde{J}_{21} & \tilde{J}_{22} \end{pmatrix} \begin{pmatrix} \Delta V_1 \\ \Delta V_2 \end{pmatrix} = \begin{pmatrix} 0 \\ \Delta Q_2 \end{pmatrix} \tag{11.7}$$

where ΔV_1 is the increase in the voltages of the floating buses and ΔV_2 and ΔQ_2 are the incease in the voltages and reactive powers of the load buses respectively

We further reduce $\tilde{J}$ into

$$\tilde{J}_R = \tilde{J}_{22} - \tilde{J}_{21}\tilde{J}_{11}^{-1}\tilde{J}_{12} \tag{11.8}$$

and thus

$$\Delta Q_2 = \tilde{J}_R \Delta V_2 \tag{11.9}$$

where the matrix $\tilde{J}_R$ has an obvious physical meaning: when the reactive power of a load bus increases, the voltage amplitude of this bus also increases proportionally, and $\tilde{J}_R^{-1}$ represents exactly the proportion between the voltage and reactive power increments.

Generally speaking, the reduced Jacobian matrix is only approximately symmetric, and thus may have complex eigenvalues. In this case, the voltage stability index can be chosen to be the smallest of the norms of the eigenvalues. The larger this value is, the more stable the system is. If the resistances in the network are neglected, the admittance matrix then becomes symmetric, so does the reduced Jacobian matrix, whose eigenvalues are consequently real numbers associated with real eigenvectors. Then the minimum eigenvalue of $\tilde{J}_R$ can be used to represent the voltage stability margin. When this eigenvalue is positive, the system has static voltage stability, and the larger this value is, the more stable the system is. Correspondingly, a negative value indicates voltage instability, and when the values is 0, the system is critically stable[3].

Remark 11.2 The proof of the singularity of matrix J

If we only consider the partial derivates with respect to the phase angles of the voltages in J, then the active and reactive power balance equations are functions of differences in phase angles only, namely

$$P_i(\theta_i - \theta_1, \cdots, \theta_i - \theta_{i\text{-}1}, \theta_i - \theta_{i+1}, \cdots, \theta_i - \theta_n) = 0, \quad i = 1, 2, \cdots, n \tag{R1}$$

$$Q_i(\theta_i - \theta_1, \cdots, \theta_i - \theta_{i\text{-}1}, \theta_i - \theta_{i+1}, \cdots, \theta_i - \theta_n) = 0, \quad i = 1, 2, \cdots, n \tag{R2}$$

J can then be divided into blocks

$$\begin{pmatrix} \Delta P \\ \Delta Q \end{pmatrix} = \begin{pmatrix} J_{P\theta} & J_{PV} \\ J_{Q\theta} & J_{QV} \end{pmatrix} \begin{pmatrix} \Delta \theta \\ \Delta V \end{pmatrix} \tag{R3}$$

where the elements in $J_{P\theta}$ and $J_{Q\theta}$ can be computed by

$$(J_{P\theta})_{ii} = \sum_{j=1, j\neq i}^{n-1} (P_i)'_j, \quad i = 1, 2, \cdots, n \tag{R4}$$

$$(J_{P\theta})_{ij} = -(P_i)'_j, \quad i, j = 1, 2, \cdots, n, \ i \neq j \tag{R5}$$

$$(J_{Q\theta})_{ii} = \sum_{j=1, j\neq i}^{n-1} (Q_i)'_j, \quad i = 1, 2, \cdots, n \tag{R6}$$

$$(J_{Q\theta})_{ij} = -(Q_i)'_j, \quad i, j = 1, 2, \cdots, n, \ i \neq j \tag{R7}$$

Here, $(P_i)'_j$ and $(Q_i)'_j$ are partial derivatives of functions P_i and Q_i with respect to the jth element respectively.

Denote the columns of J by $L_1, L_2, \cdots, L_{2n}$, and define their linear combination to be

$$\sum_{k=1}^{n} 1 \cdot L_k + \sum_{k=n+1}^{2n} 0 \cdot L_k \tag{R8}$$

From (R4) and (R5), we know that the first n elements are 0. In view of (R6) and (R7), we conclude that the last n elements are also 0. Therefore

$$\sum_{k=1}^{n} 1 \cdot L_k + \sum_{k=n+1}^{2n} 0 \cdot L_k = 0 \tag{R9}$$

which implies that J is singular.

11.2.3 Reactive Power/Voltage Modal Analysis Method Based on Optimal Power Flows

In the previous sections, we discussed the reactive power/voltage modal analysis method based on conventional power flows. It has assumed that the generators, whose reactive power outputs are within their limits, can be taken as *PV* buses with constant active power outputs and terminal voltages. With this assumption, the sensitivity relationships between reactive powers and voltages at load buses can then be analyzed. However, in real practice, the voltages of generators are adjustable responding to the changes of the loads. This is similar to the fact that in the computation for optimal power flows, the voltages and active powers of generator are all variables and are not constants as in conventional power flows. Motivated by the above observation, we propose in this section the reactive power/voltage modal analysis method which utilizes the Jacobian matrix for optimal power flows.

Using this method, we determine the system's operation mode by computing the optimal power flows, where the objective function is set to be the active power loss in the system, namely

$$\min F(P,Q)=\sum_{i\in S_g} P_{gi}-\sum_{j\in S_l} P_{dj} \tag{11.10}$$

where P_{gi} is the active power output of generator i and P_{dj} is the active power load of load j.

The constraints include the power flow equation constraints, the voltage amplitude constraints, the active power and reactive power output constraints of the generators, and the transmission line capacity constraints. We describe these constraints in detail as follows.

(1) Power flow equation constraints

$$f_i=V_i\sum_{j=1}^{n}V_j(G_{ij}\cos\theta_{ij}+B_{ij}\sin\theta_{ij}-P_{gi}+P_{di})=0,\quad i=1,2,\cdots,n \tag{11.11}$$

$$f_{i+n}=V_i\sum_{j=1}^{n}V_j(G_{ij}\sin\theta_{ij}-B_{ij}\cos\theta_{ij}-Q_{gi}+Q_{di})=0,\quad i=1,2,\cdots,n \tag{11.12}$$

where V_i is the voltage amplitude of bus i, Q_{gi} is the reactive power output of generator i, and Q_{di} is the reactive power load at bus i.

(2) Generator output constraints

$$P_{gi}^{\min}\leqslant P_{gi}\leqslant P_{gi}^{\max} \tag{11.13}$$

$$Q_{gi}^{\min}\leqslant Q_{gi}\leqslant Q_{gi}^{\max} \tag{11.14}$$

where $P_{gi}^{\max}$ and $Q_{gi}^{\max}$ are the upper limits of active and reactive power outputs of generator i respectively, and $P_{gi}^{\min}$ and $Q_{gi}^{\min}$ are the corresponding lower limits respectively.

(3) Bus voltage constraints

$$V_i^{\min}\leqslant V_i\leqslant V_i^{\max} \tag{11.15}$$

where $V_i^{\max}$ and $V_i^{\min}$ are the upper and lower limits of the voltage amplitude of bus i.

(4) Transmission line capacity constraints

$$-F_i^{\max} \leqslant F_i \leqslant F_i^{\max} \tag{11.16}$$

where F_i is the transmission power of line i, $F_i^{\max}$ is the capacity of line i.

We can write the Lagrangian for the above optimization problem as

$$L = \sum_{i \in S_g} P_{gi} - \sum_{j \in S_l} P_{dj} + \sum_{j=1}^{m} \mu_j f_j \tag{11.17}$$

where μ_j is the Lagrangian multiplier, and m is the number of equality and inequality constraints. Note that the first $2n$ items of f_j correspond to equality constraints and the last $m-2n$ items correspond to inequality constraints.

Then the state x for the optimization problem consists of the voltage angle vector θ, the voltage amplitude vector V, the vectors of active and reactive power outputs of generators P_g and Q_g, and the Lagrangian multiplier vector μ, namely

$$x = [\theta, V, P_g, Q_g, \mu]^{\mathrm{T}}$$

One necessary condition for the solution being optimal is that

$$\left[\frac{\partial L}{\partial \theta}, \frac{\partial L}{\partial V}, \frac{\partial L}{\partial P_g}, \frac{\partial L}{\partial Q_g}, f\right]^{\mathrm{T}} = 0 \tag{11.18}$$

where $f = [f_1, f_2, \cdots, f_m]$, and $\frac{\partial L}{\partial \theta}, \frac{\partial L}{\partial V}, \frac{\partial L}{\partial P_g}, \frac{\partial L}{\partial Q_g}$ are all row vectors.

Let

$$y = \left[\frac{\partial L}{\partial \theta}, \frac{\partial L}{\partial V}, \frac{\partial L}{\partial P_g}, \frac{\partial L}{\partial Q_g}, f\right]^{\mathrm{T}}$$

then the iteration equation for the optimal power flow can be written as

$$\begin{pmatrix} H & J_0^{\mathrm{T}} \\ J_0 & 0 \end{pmatrix} \Delta x = y \tag{11.19}$$

$$H = \sum_{j=1}^{n} \nabla^2 P_{gj} + \sum_{j=1}^{m} \mu_j \nabla^2 f_j \tag{11.20}$$

where $\begin{pmatrix} H & J_0^{\mathrm{T}} \\ J_0 & 0 \end{pmatrix}$ is the Jacobian matrix of the iteration equation, ∇^2 is the Hessian operator, J_0 is the Jacobian matrix of the power flow equations together with the equations of active power constraints. Since the angles in the power flow equations are all relative, the matrix $\begin{pmatrix} H & J_0^{\mathrm{T}} \\ J_0 & 0 \end{pmatrix}$ is singular (Please refer to Remark 11.3 for its proof) .

Without loss of generality, we take bus 1 as the reference bus. Then θ_1 can be deleted from x, and the equation $\dfrac{\partial L}{\partial \theta_1}=0$ can be removed from Eq. (11.18). Correspondingly, the first row and the first column are removed from $\begin{pmatrix} H & J_0^{\mathrm{T}} \\ J_0 & 0 \end{pmatrix}$ to get $\begin{pmatrix} H_r & J_r^{\mathrm{T}} \\ J_r & 0 \end{pmatrix}$, which is the Jacobian matrix for the new set of iteration equations.

The operation model determined by the optimal power flows can be computed by either the Newton method or the interior point method. If the optimal power flows are obtained by the successive linear or quadratic programming methods, or if the bus voltages are expressed using orthogonal coordinates, then μ_j can be determined by the Kuhn-Tucker optimality condition[7]. We propose to take the following approach to compute μ_j.

Substituting the solution for optimal power flows into the equations below

$$\frac{\partial L}{\partial \theta}=0,\quad \frac{\partial L}{\partial V}=0,\quad \frac{\partial L}{\partial P_g}=0,\quad \frac{\partial L}{\partial Q_g}=0$$

then we get the linear equations about μ

$$J_r^{\mathrm{T}}\mu = B \tag{11.21}$$

There are $2n+2n_g-1$ equations with m unknowns. Since the number of equations is less than the number of unknowns, there might be multiple solutions. So we look for the solution to the following least square problem to determine μ:

$$\min_{\mu} \quad (J_r^{\mathrm{T}}\mu - B)^{\mathrm{T}}(J_r^{\mathrm{T}}\mu - B) \tag{11.22}$$

And its least square solution is

$$\mu = (J_r J_r^{\mathrm{T}})^{-1} J_r B \tag{11.23}$$

According to the Kuhn-Tucker condition, if the solution obtained is optimal, then the corresponding value of the objective function (11.22) must be the minimum.

Now we calculate $\begin{pmatrix} H_r & J_r^{\mathrm{T}} \\ J_r & 0 \end{pmatrix}$ using the μ just obtained. Denote the inverse of the matrix that we are interested in by W. It follows that

$$W\left[\frac{\partial L}{\partial \theta},\frac{\partial L}{\partial V},\frac{\partial L}{\partial P_g},\frac{\partial L}{\partial Q_g},\Delta P_B,\Delta Q_B,\Delta h\right]^{\mathrm{T}} = [\Delta\theta,\Delta V,\Delta P_g,\Delta Q_g,\Delta\mu]^{\mathrm{T}} \tag{11.24}$$

where $\Delta P_B,\Delta Q_B$ are the increments in the active and reactive power at the load buses respectively and Δh is the increment in the active constraints.

Consider the following set of equations

$$\frac{\partial L}{\partial \theta}=0,\frac{\partial L}{\partial V}=0,\frac{\partial L}{\partial P_g}=0,\frac{\partial L}{\partial Q_g}=0$$
$$\Delta P_B=0,\Delta Q_b=0,\Delta h=0$$

where ΔQ_b is the increment of the reactive power of the floating buses. Then the change in ΔQ_l (reactive power increment of load buses), which is caused by the change of ΔV_l (voltage increment of load buses), can be described by

$$\Delta V_l = W_l \Delta Q_l \tag{11.25a}$$

where W_l consists of the rows corresponding to ΔV_l and columns corresponding to ΔQ_l in W. For optimal power flows, the relationship indicates the sensitivity between the increments in voltages and reactive powers at the load buses when we fix the active powers at these buses and the reactive powers at the floating buses.

In order to better compare the results with those in Section 11.2.2, we write the above equation as

$$W_l^{-1}\Delta V_l = \Delta Q_l \tag{11.25b}$$

When W_l^{-1} has complex eigenvalues, the minimum of the norms of the eigenvalues is considered to be the voltage stability index. The larger this value is, the more stable the system is. If eigenvalues are all real numbers, the smallest eigenvalue represents the voltage stability margin. If the smallest eigenvalue is positive, then the system can maintain its static voltage stability; and the larger this value is, the more stable the system is. Correspondingly, a negative value indicates voltage instability and the zero value implies the system is critically stable.

Remark 11.3 The proof for the singularity of $\begin{pmatrix} H & J_0^{\mathrm{T}} \\ J_0 & 0 \end{pmatrix}$

Denote the row vectors of $\begin{pmatrix} H & J_0^{\mathrm{T}} \\ J_0 & 0 \end{pmatrix}$ by H_1, $\cdots$, H_{2n+2n_g+m}. Let the ith component of the row vector

$$\sum_{k=1}^{n} 1 \cdot H_k + \sum_{k=n+1}^{2n+2n_g+m} 0 \cdot H_k \tag{R1}$$

be h_i. Since the terms in the first summation of H in Eq. (11.20) are Hessian matrices of linear functions, they must be 0.

For $h_i\,(i = 1, 2, \cdots, n)$, the contributions from the items in the second summation of H are $\mu_j \dfrac{\partial}{\partial \theta_i}\left(\sum_{k=1}^{n} \dfrac{\partial f_j}{\partial \theta_k}\right)$. From Remark 11.2, we know they also have to be zero.

For $h_i(i = n+1, \cdots, 2n)$, the contributions from the items in the second summation of H are $\mu_j \dfrac{\partial}{\partial V_i}\left(\sum_{k=1}^{n} \dfrac{\partial f_j}{\partial \theta_k}\right)$. Also from remark 11.2, we know that they must be zero.

For h_i $(i = 2n+1, \cdots, 2n+2n_g)$, the contributions from the items in the second summation of H are the second-order partial derivatives of the power flow equations and the active inequality constraints with respect to angles/active power outputs and angles/reactive power outputs. Since there are no couplings between angles and power outputs of the generators, these partial derivatives must be 0. Therefore, $h_i = 0\,(i = 2n+1, \cdots, 2n+2n_g)$.

For h_i $(i = 2n+2n_g+1, \cdots, 2n+2n_g+m)$, $h_i = \sum_{k=1}^{n} \dfrac{\partial f_{i-2n-2n_g}}{\partial \theta_k}$. From the above discussion, we know that their values are zero.

Summarizing the cases above, we have proven that

$$\sum_{k=1}^{n} 1 \cdot H_k + \sum_{k=n+1}^{2n+2n_g+m} 0 \cdot H_k = 0 \tag{R2}$$

which implies that $\begin{pmatrix} H & J_0^{\mathrm{T}} \\ J_0 & 0 \end{pmatrix}$ is singular.

11.3 SOC Analysis for Blackout Model Ⅰ

11.3.1 Three Indices for Critical Characteristics

In this section, we propose three macroscopic indices to investigate the SOC phenomena and reveal critical characteristics[4].

(1) Index I_s is the ratio of the total load demand over the network effective transmission capability:

$$I_s = \frac{\sum_{i \in L} \sqrt{P_{di}^2 + Q_{di}^2}}{\sum_{j \in C} F_j^{\max}} \quad (11.26a)$$

where L is the set of load buses, P_{di} and Q_{di} are the active and reactive power of load i respectively, $F_j^{\max}$ is the transmission capacity of line j, $\sum_{j \in C} F_j^{\max}$ represents the effective transmission capability of the network, and C is the set of key lines. Here, the key lines include those that are directly connected to generators and those with smallest transmission capabilities in the network.

(2) Index I_p is the ratio of the total active power demand over the effective transmission capability of the network:

$$I_p = \frac{\sum_{i \in L} P_{di}}{\sum_{j \in C} F_j^{\max}} \quad (11.26b)$$

(3) Index I_q is the ratio of the total reactive power demand over the effective transmission capability of the network:

$$I_q = \frac{\sum_{i \in L} Q_{di}}{\sum_{j \in C} F_j^{\max}} \quad (11.26c)$$

11.3.2 Simulation

In order to check whether the SOC characteristics are presented in the voltage stability behaviors, and also to verify the blackout model I , we carry out simulation studies in this section on the IEEE 30-bus system and the IEEE 118-bus system respectively.

1) IEEE 30-bus system

We compute and analyze the blackout processes in the following two scenarios.

Case 1:

(1) The line overload threshold $\alpha = 0.7$,

(2) The tripping probability of overloaded lines $\beta = 0.3$,

(3) The growth rate of load level and generator capacity $\lambda = 1.0005$,

(4) The growth rate of line transmission capacity $\mu = 1.005$,

(5) The simulated period is 5000 days.

Case 2:

All the simulation parameters are the same as those in case 1 except for the growth rate of the reactive powers of loads, which is set to be 1.0007.

Figures 11.2(a) and (b) show the trends in the changes of indices I_s and $\overline{P}_{\text{loss}}$ in Cases 1 and 2 respectively, where $\overline{P}_{\text{loss}}$ defined by Eq. (10.14) indicates the scale of blackout failures. Figures 11.3(a) and (b) show the trajectories of the evolution of indices I_p and I_q respectively.

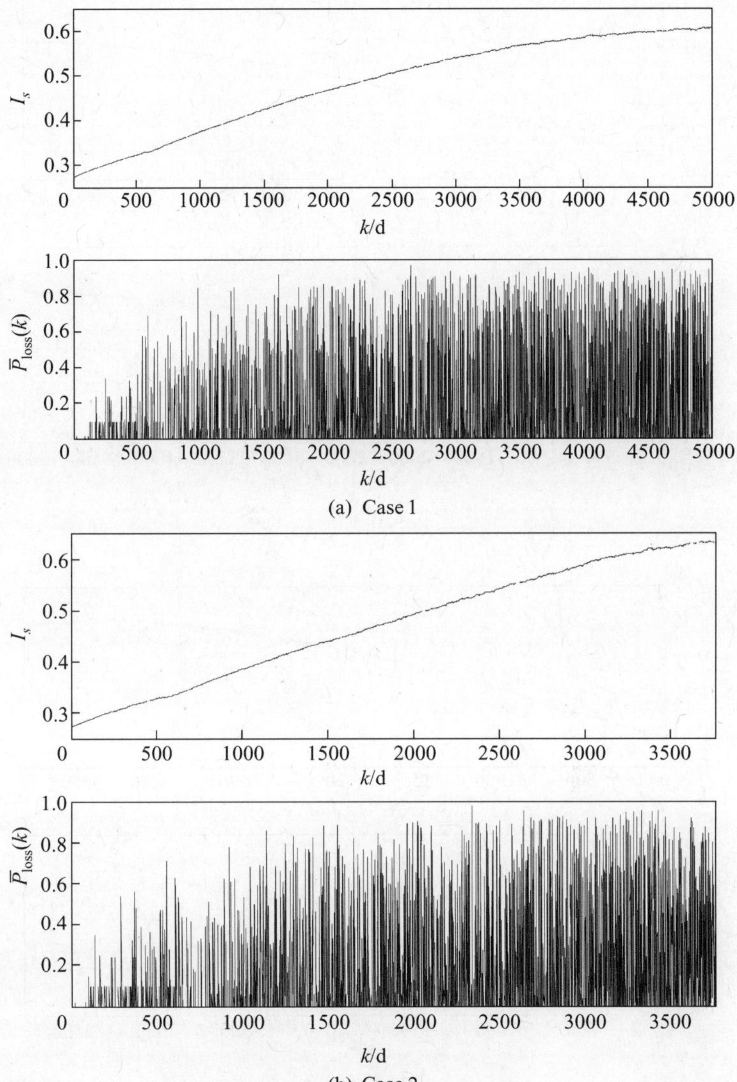

Figure 11.2 Changes in I_s and $\overline{P}_{\text{loss}}$

From Fig. 11.2(a) it can be seen that in case 1, large-scale blackouts take place, indicating the system is at critical states, on around day 540 with $I_s = 0.3259$. Around the same time in Fig. 11.3(a), we get $I_p = 0.2778$ and $I_q = 0.1580$. From Fig. 11.2(b), it is clear that in case 2, the system enters a critical state on around day 525 with $I_s = 0.3300$. Then from Fig. 11.3(b), we find that $I_p = 0.2742$ and $I_q = 0.1725$. In the end, after day 3767, as a result of exceedingly increasing reactive power loads, the optimal power flows fail to converge and the system's voltages

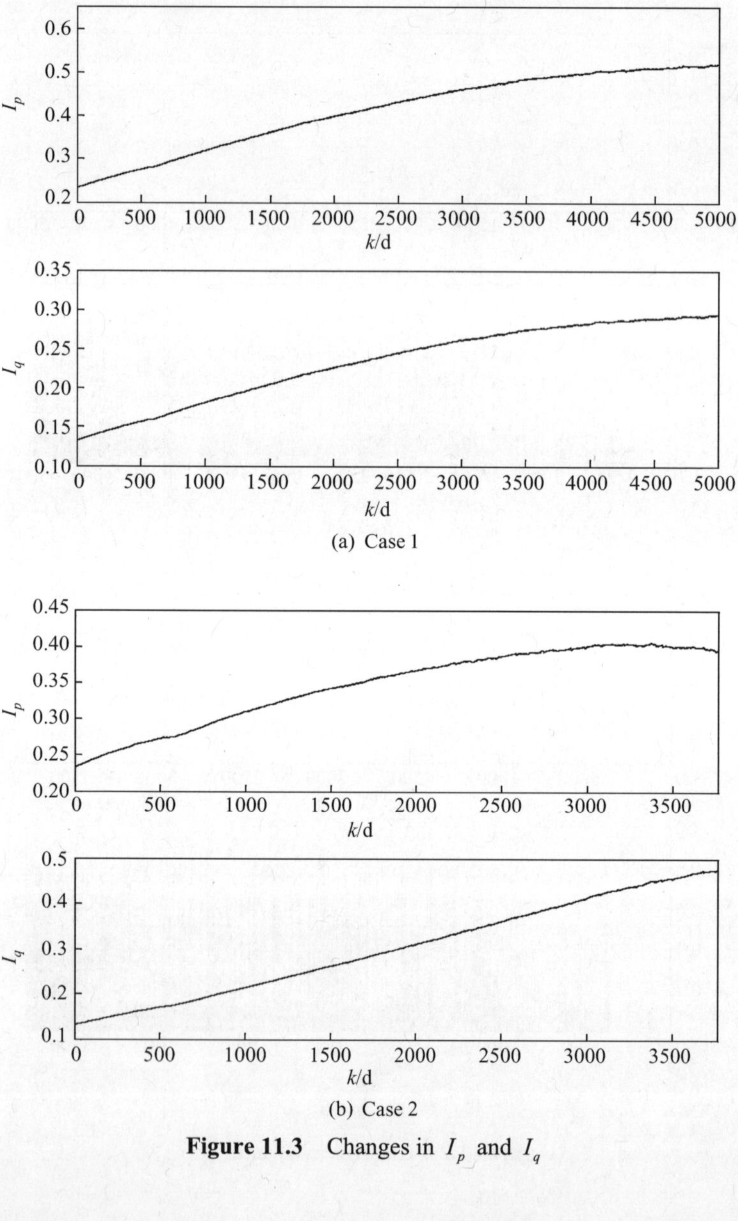

Figure 11.3 Changes in I_p and I_q

collapse. Comparing the two cases, the system enters a critical state earlier in Case 2 than in Case 1 because the reactive power loads grow faster in Case 2. This phenomenon is consistent with the real practice in power systems.

Figures 11.4(a) and (b) show the trends in the change of the minimum static voltage reserve coefficient $\overline{U}_0 = \min_i \overline{U}_{i0}$ using the critical voltage analysis method. What is also shown in the figures are the distributions of $F(\overline{U}_0)$ with respect

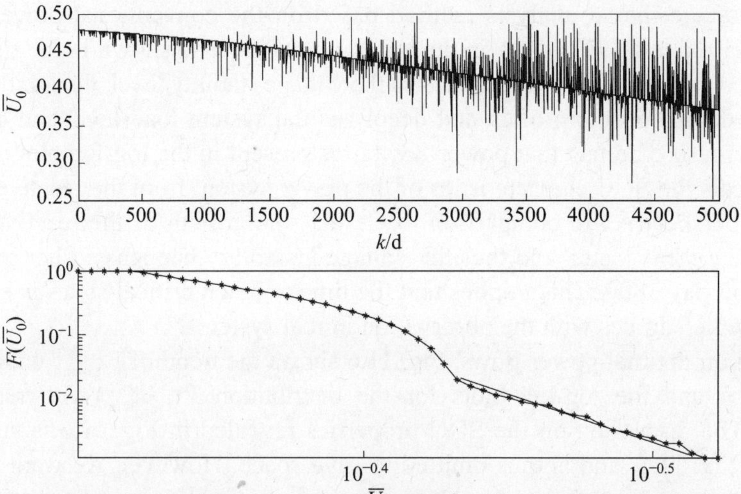

(a) Case 1

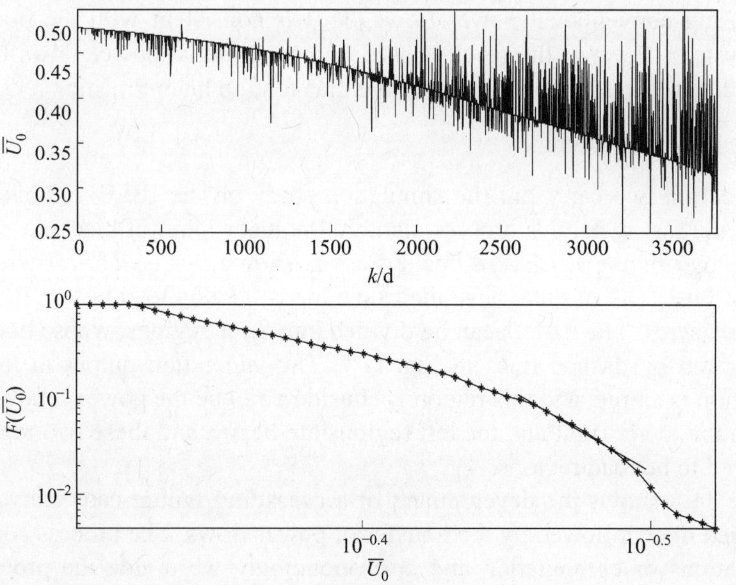

(b) Case 2

Figure 11.4 Trends of the static voltage reserve in the system

to $\bar{U}_0$ in the log-log coordinates. From these figures, we can see that $\bar{U}_0$, which indicates the voltage stability level, decreases gradually as the system load level and the blackout scale increase. Significant changes of $\bar{U}_0$ happen when large-scale blackouts take place. The distribution of $\bar{U}_0$ in the log-log plot exhibits the power law properties which reveals the SOC characteristics that are related to reactive powers and voltages in power systems.

Figures 11.5(a) and (b) show the trend of changes in $|\lambda|_{\min}$ using the reactive power/voltage modal analysis method based on the conventional power flows. The distribution of $F(|\lambda|_{\min})$ in the log-log plots are also shown in the figures. It can be seen that $|\lambda|_{\min}$, which indicates the voltage stability level, though fluctuates sharply, does drop more often and deeply as the system load level and blackout scale increase. The fact that power-law tail is present in the log-log plot of $|\lambda|_{\min}$ also reveals the SOC characteristics of the power system from the reactive power/voltage perspective. In comparison to Case 1, the growth of the reactive power load in Case 2 is faster, and thus the voltage instability happens earlier in Case 2 on around day 3800. This implies that the time to reach critical states is earlier in Case 2 which agrees with the observation in real systems.

Based on optimal power flows, Fig. 11.6 shows the trend of $|\lambda|_{\min}$ with respect to time k and the log-log plots for the distribution $F(|\lambda|_{\min})$ with respect to $|\lambda|_{\min}$. The discussion on the SOC properties revealed in Fig. 11.6 is similar to that for Fig. 11.5, and is thus omitted to save space. However, we want to point out that comparing the modal analysis results in Figs. 11.5 and 11.6, the stability margin estimated in the latter using optimal power flows is lower than that in the former using conventional power flows. This is consistent with the practice in power systems because the results based on conventional power flows have not been further adjusted by optimization and thus tend to be optimistic.

2) IEEE 118-bus system

In this section, we carry out the simulation study on the IEEE 118-bus system utilizing the fast dynamics process, namely the inner loop, of blackout model I. The topology of the IEEE 118-bus system is shown in Fig. 11.7. There are 54 generator buses (37 of them have their own loads) and 64 load buses (10 of them are always zero). The system can be divided into three regions, whose boundaries are indicated by dashed lines in Fig. 11.7. The generation output in the upper right region is ample, and this region is considered to be the power supply region. Loads in the lower right and the left regions are heavy, and these two regions are considered to be load regions.

Figure 11.8 shows the development of a cascading failure caused by tripping overloaded lines followed by the transfer of power flows. The process consists of two iterations of computation and correspondingly we divide the process into two stages. In the first figure, the dashed lines indicate the tripped lines which do not appear in the second figure.

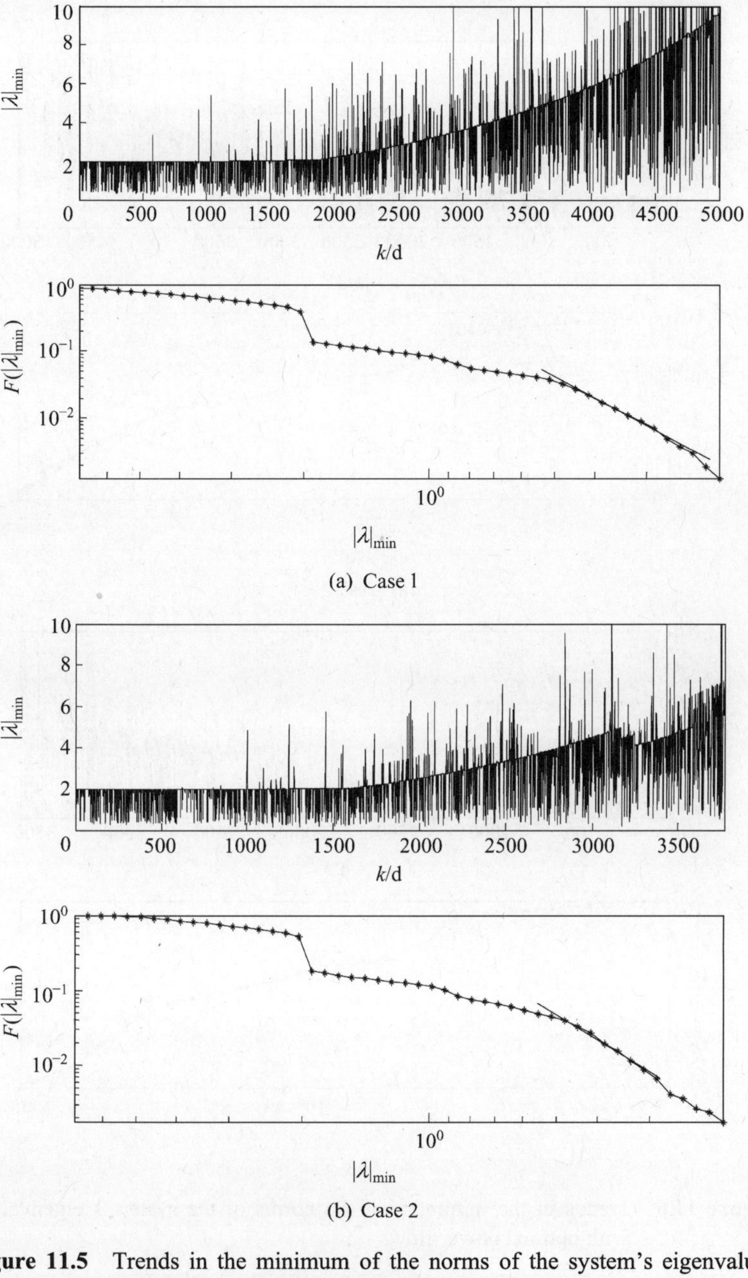

Figure 11.5 Trends in the minimum of the norms of the system's eigenvalues with conventional power flows

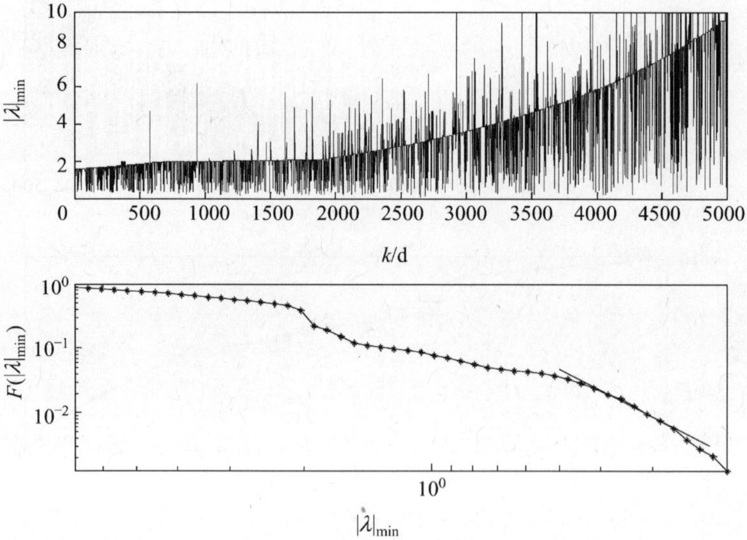

(a) Case 1

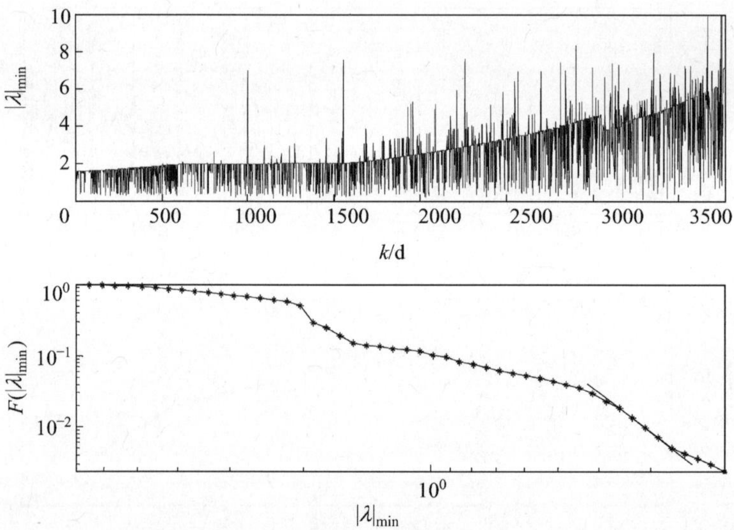

(b) Case 2

Figure 11.6 Trends in the minimum of the norms of the system's eigenvalues with optimal power flows

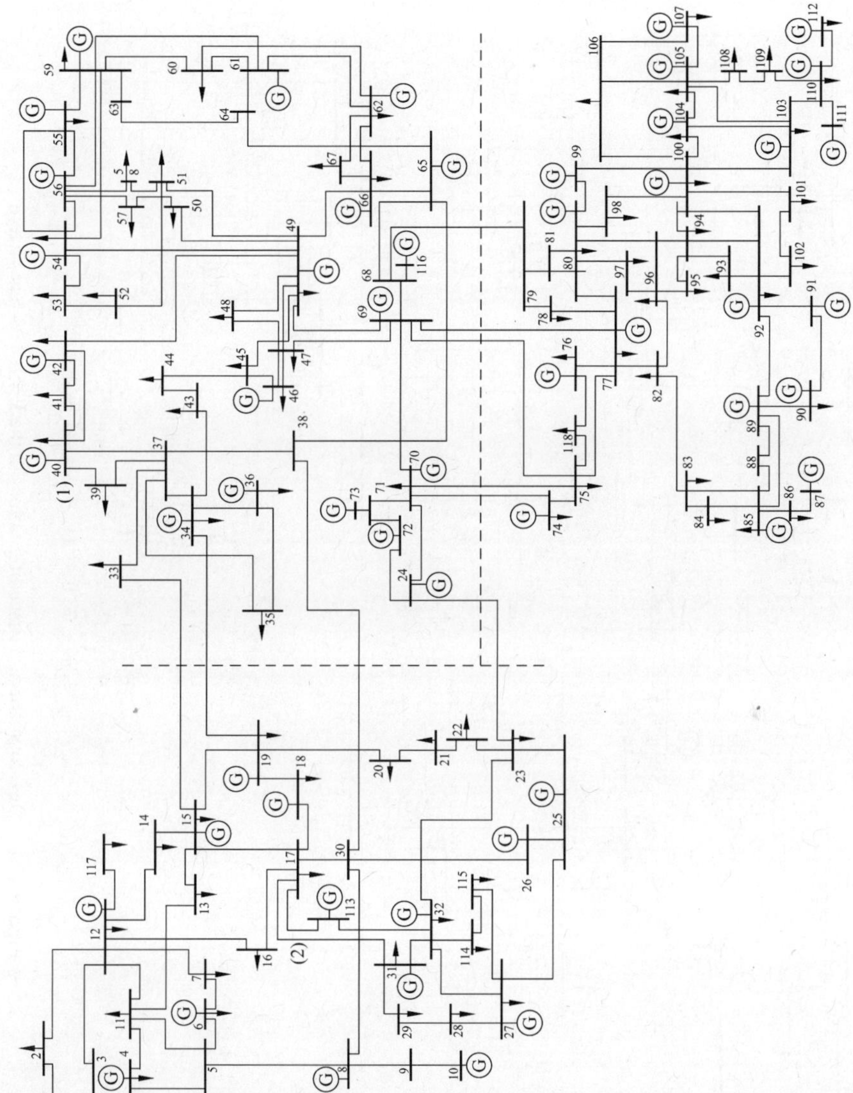

Figure 11.7 Topology of the IEEE 118-bus system

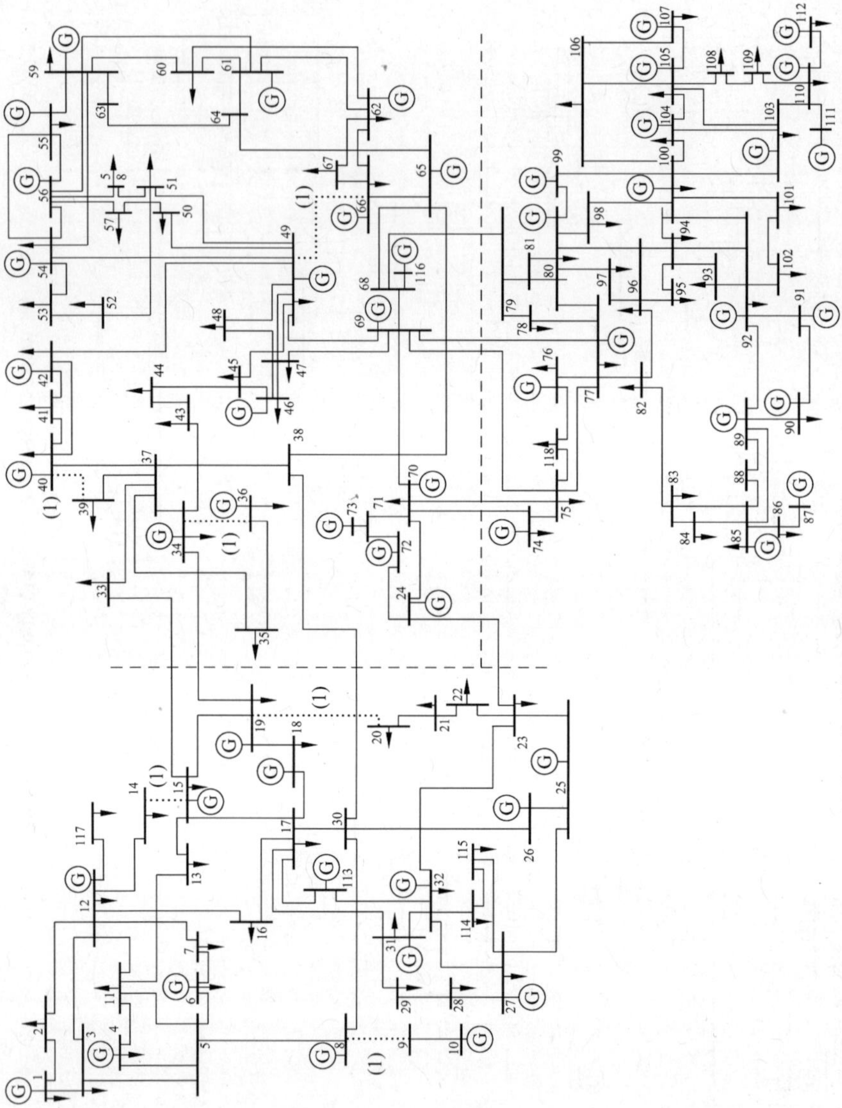

(a) Cascading failure in the IEEE 118-bus system, stage 1

Figure 11.8 Development of a cascading failure in the IEEE 118-bus system

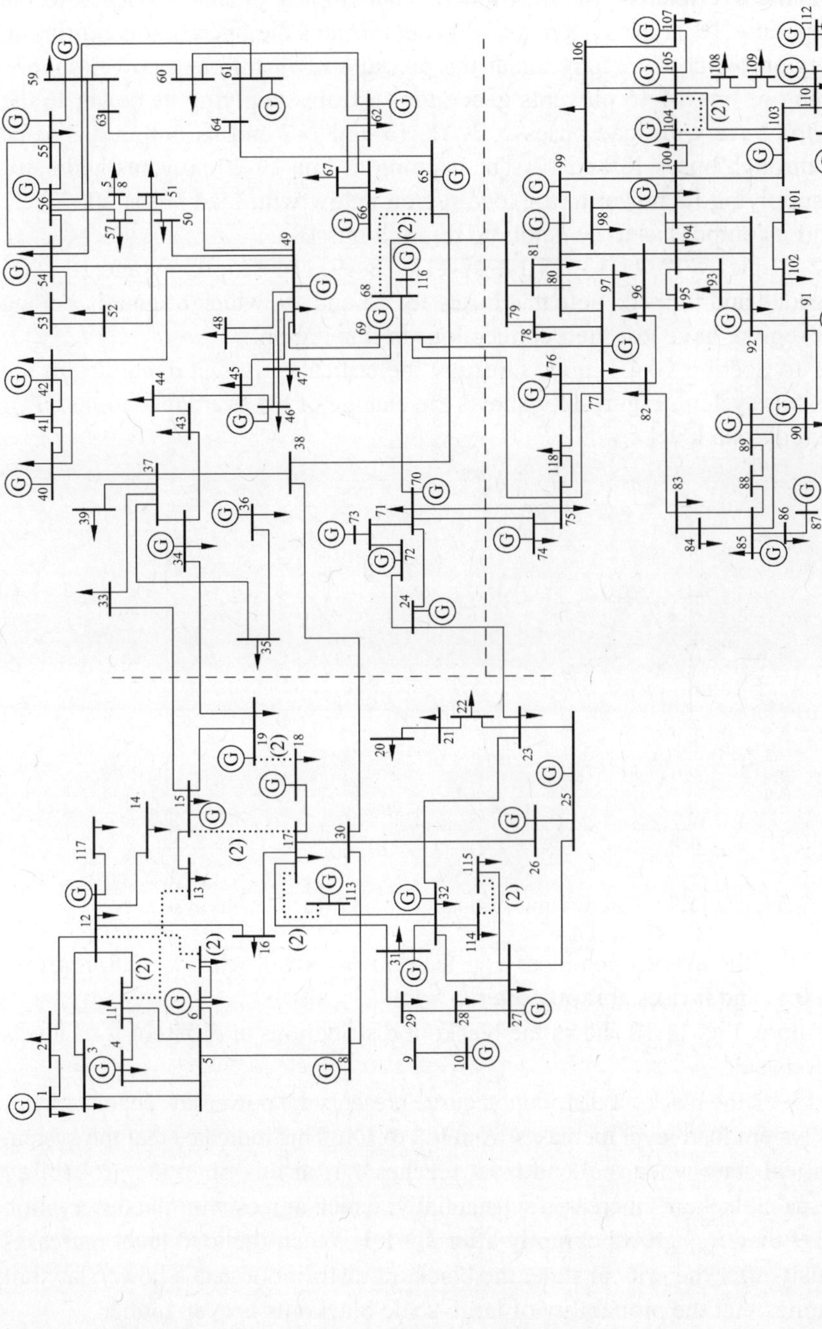

(b) Cascading failure in the IEEE 118-bus system, stage 2

Figure 11.8(Continued)

The cascading failure process, as described in Fig. 11.8, can be summarized as follows.

Stage 1 Because of some initial fault, lines 8-9, 14-15, 19-20, 34-36, 39-40 and 49-66 become overloaded and are tripped. The tripping of line 8-9 leads to the loss of generator 10 in the system and this deteriorates the operation condition in the left region which is already under the pressure of insufficient power supply. The tripping of line 14-15 prevents generator 15 from supplying its power to the load region above with load buses 2, 3, 11, 14 and 117 and its output has to be detoured through buses 13 and 17. The tripping of line 19-20 prevents generator 19 from supplying its power to the load region below with load buses 20, 21, 22 and 23 and its output has to be detoured through bus 18.

Stage 2 Lines 7-12, 11-13, 15-17, 17-113, 18-19, 114-115, 65-69 and 104-105 are overloaded and tripped. Note that buses 13, 15 and 19, which originally belong to the left region, have lost their connections to this region.

Similar to Section 10.4.1, now we study the criticality in fast dynamics in the IEEE 118-bus system. Figure 11.9 shows the change of the average load loss $\overline{P}_{\text{loss}}$ with different load level k_L.

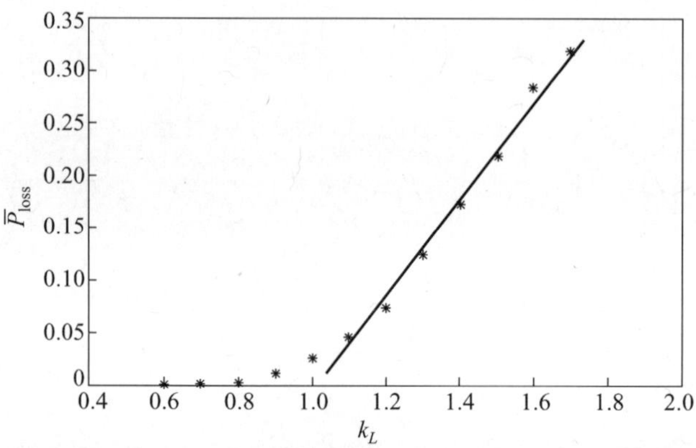

Figure 11.9 Fast dynamic criticality in the IEEE 118-bus system

In Fig. 11.9, the average load loss $\overline{P}_{\text{loss}}$ is relatively small when k_L<0.8, it grows when k_L>0.8, and it rises abruptly after k_L>1.1.

Furthermore, Fig. 11.10 shows the blackout distributions at different load levels on a log-log scale.

In Fig. 11.10, the blackout distribution curve presents the power law characteristics when the system load level increases from 0.8 to 1.1. This indicates that the system is in a critical state when the load level reaches 1.1; at this time, the probability for large-scale blackouts increases substantially, which agrees with the observation in Fig. 11.9 that $\overline{P}_{\text{loss}}$ grows abruptly after k_L>1.1. When the load level increases continuously after the critical state, the blackout distribution has a power-law tail, which implies that the probability of large-scale blackouts is even higher.

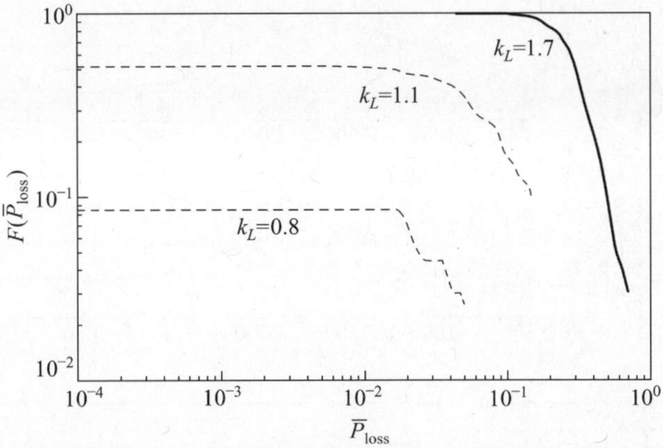

Figure 11.10 Blackout distributions at different load levels in the IEEE 118-bus system

We simulate IEEE 118-bus system for a longer period of 3500 days. The change in blackout scales is shown in Fig. 11.11. Here, we use the relative load shedding amount $\overline{P}_{\text{loss}}$ to represent the blackout scale.

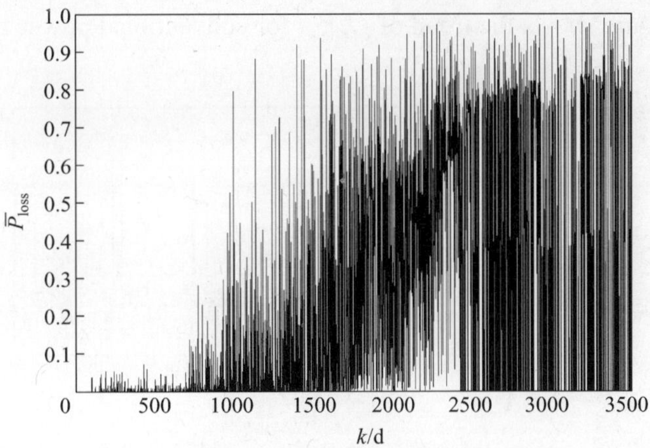

Figure 11.11 Trend of $\overline{P}_{\text{loss}}$

The evolutions of the macroscopic criticality indices I_s, I_p and I_q in these 3500 days are shown in Fig. 11.12. In the first 971 days, there is almost no failure. After day 972, large-scale blackouts begin to happen and these three indices increase. On day 972, $I_s = 0.6494$, $I_p = 0.6071$ and $I_q = 0.2058$.

After day 2500, the indices I_s, I_p and I_q are in steady states, which implies that the system has evolved into a quasi-stable critical state after self-organization. Hence, from the evolutions of I_s, I_p and I_q, and the analysis based on them, we conclude that power systems show typical SOC properties.

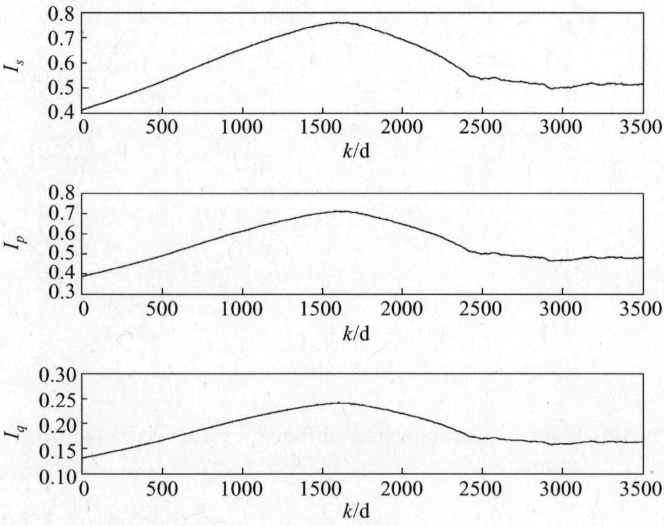

Figure 11.12 Trends of I_s, I_p and I_q

Moreover, as shown in Fig. 11.13, we can also study the trend of the voltage stability margin using the voltage stability analysis module in the blackout model I. The upper figure shows the trend of $|\lambda|_{\min}$ for conventional power flows and the

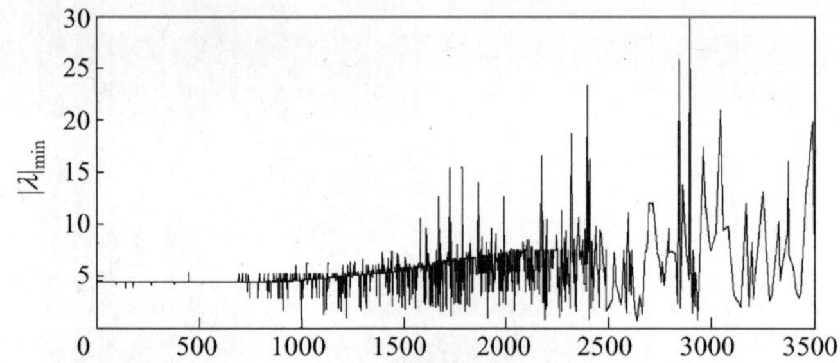

(a)

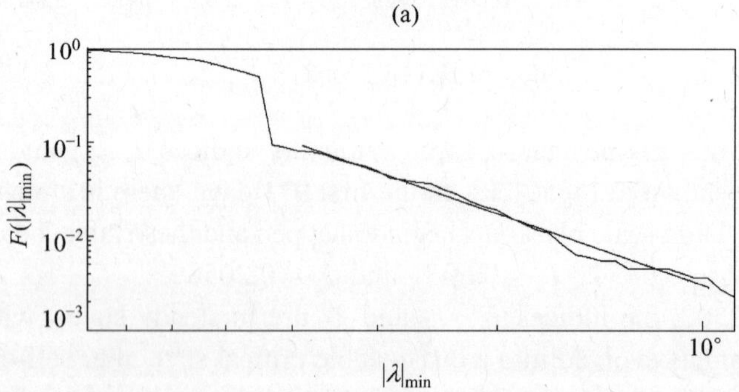

(b)

Figure 11.13 Trend of the minimum of the norms of eigenvalues

lower figure shows the log-log plot of the distribution $F(|\lambda|_{\min})$. The power-law tail in the latter reveals the SOC properties of power systems from the reactive power/voltage perspective. Under normal operation conditions, $|\lambda|_{\min}$ is relatively big while it decreases fast and approaches critical values for voltage stability when blackouts take place. In other words, the voltage stability margin decreases progressively as the system evolves.

11.3.3 Effects of Typical Operation Parameters on Blackout Distributions

Using the blackout model I, in this section we study the effect of three parameters on the blackout distributions. These parameters include the overload threshold α, the tripping probability β for overloaded lines and the growth factor μ for transmission line capacities. We also assess and compare the risks and scales for blackouts using the indices VaR and CVaR that have been discussed in Chapter 3.

The setting of the parameters is as follows:

(1) The overload threshold $\alpha = 0.7$,

(2) The tripping probability for overloaded lines $\beta = 0.3$,

(3) The growth rate for load level and generator capacity $\lambda = 1.0005$,

(4) The growth rate for transmission line capacity $\mu = 1.005$.

The above parameter values are called the initial values.

1) The effect of the overload threshold α on blackout distributions

Figure 11.14 shows the blackout distribution under different overload thresholds while the other parameters remain their initial values.

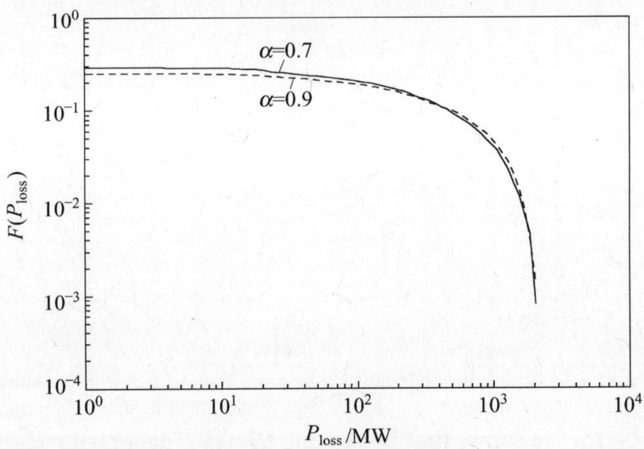

Figure 11.14 Blackout distributions with different overload thresholds

Remark 11.4 For convenience, we use load loss P_{loss} to represent the blackout scale and $F(P_{\text{loss}})$ to denote its distribution.

From Fig. 11.14 we know that to increase α (i.e., to increase the set value of line protection) reduces the probability for small-scale blackouts, but, on the other hand, increases the probability for large-scale blackouts. This is because as the threshold value increases, the number of overloaded lines decreases and in turn the probability of small-scale blackouts decreases. But this also reduces the actions for line upgrade and reconstruction, and consequently the probability for large-scale blackouts increases.

We further set the confidence level $\sigma = 0.99$ and calculate the VaR and CVaR. The results are shown in Table 11.1.

Table 11.1 The risks under different overload thresholds (MW)

Overload threshold	VaR	CVaR
0.7	1598	18.30
0.9	1652	18.58

Table 11.1 shows that the increase of the threshold value puts the system under higher risk, which is to the disadvantage for the secure operation of power systems.

2) The effect of the tripping probability β on blackout distributions

Figure 11.15 shows the blackout distribution under different tripping probability β for overloaded lines while the other parameters remain their initial values.

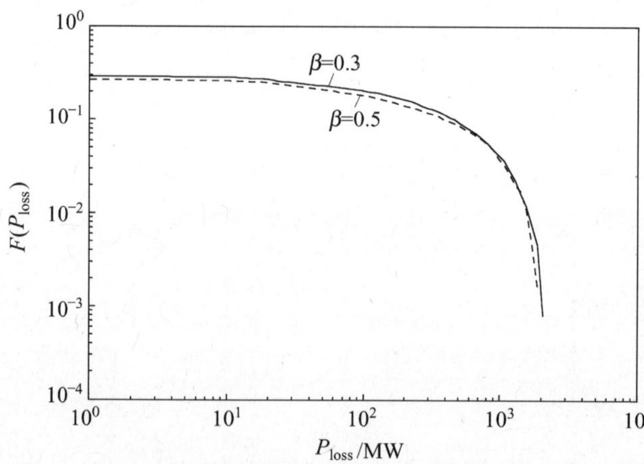

Figure 11.15 Blackout distributions under different tripping probabilities

From Fig. 11.15, we know that increasing β (i.e., decreasing the capability of recovering overloaded lines) reduces the probability for blackouts. This is because as the probability for tripping overloaded lines increases, the actions for line upgrades and reconstructions are more likely and as a result, the probability for large-scale blackouts becomes lower.

We further set the confidence level $\sigma = 0.99$ and calculate the VaR and CVaR. The results are presented in Table 11.2.

Table 11.2 Risks under different tripping probabilities (MW)

Tripping probabilities	VaR	CVaR
0.3	1598	18.30
0.5	1583	17.37

Table 11.2 implies that the increase of β slightly reduces the risk for blackouts.

3) The effect of the growth rate μ for line capacity on blackout distributions

Figure 11.16 shows the blackout distribution under different growth rate μ while the other parameters remain their initial values.

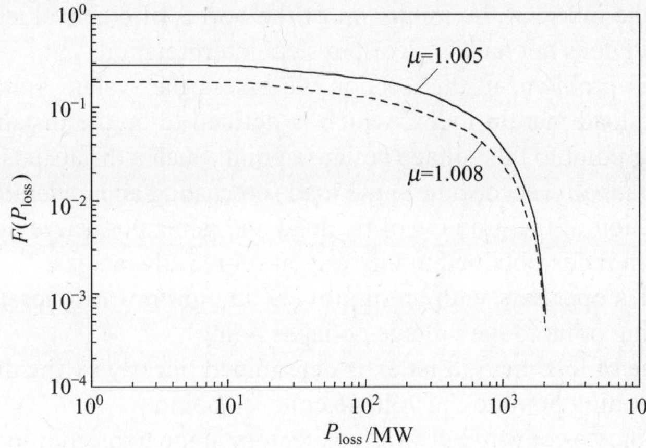

Figure 11.16 Blackout distributions under different growth rates for line capacities

As shown in Fig. 11.16, the increase of μ reduces the number of blackouts significantly, which is especially true for large-scale blackouts.

We set the confidence level $\sigma = 0.99$ and calculate the VaR and CVaR. The results are listed in Table 11.3.

Table 11.3 Risks under different growth rates for line capacities (MW)

Transmission capacity growth rate	VaR	CVaR
1.005	1598	18.30
1.008	1446	16.60

Table 11.3 shows that the increase of μ significantly reduces the system risk. This can be explained by the fact that large-scale blackouts are usually results of cascading failures, and the increase of μ makes the system more robust to cascading failures. Hence, increasing μ is critical to improve the system's capability to deal with disastrous events.

11.4 Blackout Model Ⅱ

Two aspects of the blackout model Ⅰ can be further improved.

1) The assessment using the index of the static voltage stability margin can be more accurate

It is obvious in the blackout model Ⅰ that the indices including the static voltage reserve coefficient and the critical voltage can only assess the system voltage stability level approximately because only the most recent information about the power flow solution has been utilized. In particular, in conventional and optimal power flows, the index of the minimum of the norms of eigenvalues is far from being linear and does not have a clear physical interpretation.

To solve this problem, in this section we assess the system voltage stability level using the load margin index, which is defined to be the distance from the given operating point to the voltage collapse point. Such a distance is obtained by approaching gradually, according to the load forecasting and generation schedule, along the direction of the growths of the load and generator active power output. The load margin index obtained in this way has three advantages:

(1) It provides operators with an intuitively straightforward measure from the present operating point to the voltage collapse point;

(2) The value of load margin index is determined linearly by the distance from the current operating point to the voltage collapse point;

(3) It incorporates conveniently various factors in the transient process, such as the influence of constraints, the active power output allocation among generators and the growth pattern of loads.

2) The control actions for dealing with voltage instability are overly simplified

If the computation for the optimal power flows does not convergence after tripping overloaded lines, it is an indication for the appearance of voltage instability. Load shedding is an effective strategy that has been applied widely in practice. In blackout model I, the load shedding strategy is to proportionally trip loads at both ends of overloaded lines in a continuous manner. However, in real power system operations, operators usually prefer tripping loads at the buses where voltage instability is more prominent in order to guarantee the secure and stable operation. Since the buses that are weak in voltage stability are not always those at the ends of overloaded lines determined by the optimal power flow computation, the load shedding strategy in blackout model I is too simplified and the results obtained accordingly hold only with a certain probability. Because of this, in this section we propose an accurate modeling method in the load shedding strategy. The main steps are as follows.

Step 1 Identify the weak buses using the tangent plane equation for the boundary of the voltage stability region, and construct a set of buses that are both weak and adjustable by operators. Shed the loads in this set according to a preset proportion.

Step 2 Determine the system's operation mode by computing the optimal power flows. If the computation does not converge, increase the proportion for shedding the load in the set constructed in Step 1. If the number of shedding load computations exceeds a preset constant and the optimal power flow computation still diverges, then the system is considered to have experienced voltage collapse.

Hence, using the load margin as the voltage stability index can assess and describe more accurately the voltage stability level for an evolving system. To model more accurately the load shedding strategy can help to simulate more faithfully the real practice in power systems. We thus propose the blackout model II considering the reactive power/voltage characteristics which is summarized in Fig. 11.17.

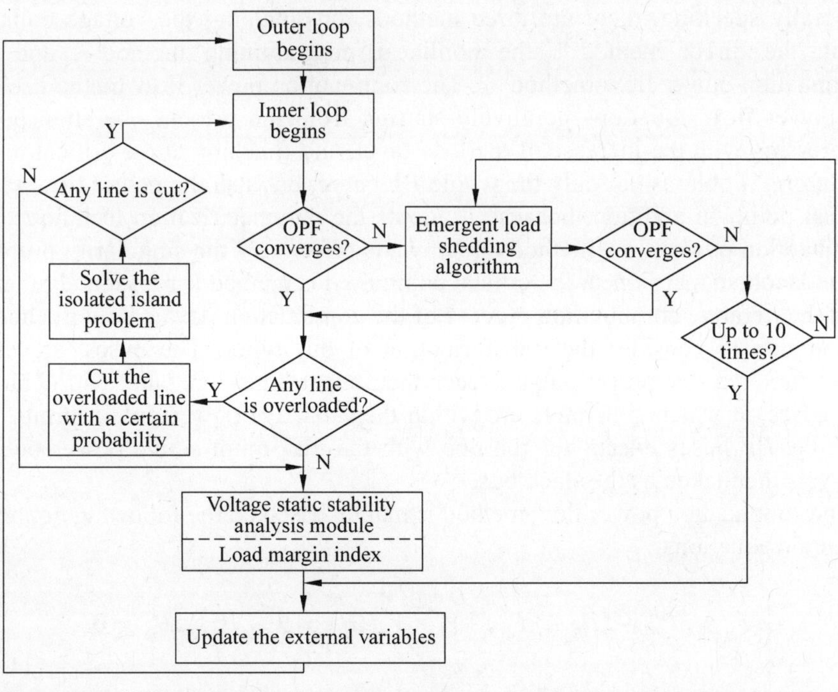

Figure 11.17 Blackout model II considering the reactive power/voltage characteristics

Similar to blackout model I, this model also consists of the inner and outer loops, with the former simulating fast dynamics and the latter simulating slow dynamics. The load shedding module in this model is based on the precise model for emergent load shedding that we have just described, and the static voltage stability analysis module utilizes the load margin index. The other modules are the same as those in blackout model I. As analyzed in the previous discussion, the blackout model II overcomes the shortcomings of the blackout model I that we have discussed in the beginning of this section.

11.5 Reactive Power/Voltage Analysis Methods in Blackout Model II

11.5.1 Load Margin Algorithm

The critical step in the calculation of the load margin index is to determine the voltage collapse point. The Jacobian matrix of the power flow equations becomes singular at this point, and for this reason the computation usually diverges when conventional power flow methods are used to approach the voltage collapse point. Generally speaking, there are three methods for obtaining the voltage collapse point: the direct method[8], the nonlinear programming method[9], and the continuation power flow method[10]. The continuation power flow method solves the power flow equations iteratively starting from the present operation point, progressing with the increase of the load level, and finishing at the critical point. Therefore, it obtains not only the whole *PV* curve, but also the power flow at the critical point. In addition, because it adopts the parameterization technique, the continuation power flow method circumvents effectively the singularity problem of the Jacobian matrix near the critical point. So this method is robust and reliable.

In the iterative computation process of the continuation power flow method, it is common to consider the transformation of bus types. The buses, at which generators' reactive power outputs reach their upper limits, are taken as *PQ* buses. The buses, at which generators are within their reactive power output limits, are taken as *PV* buses except for the one with the maximum active power output, which is then taken as the slack bus.

The continuation power flow method is mainly based on the following nonlinear algebraic equations:

$$\begin{cases} (P_{gi} + \gamma P_{gi}^L) - (P_{di} + \gamma P_{di}^L) - V_i \sum_{j=1}^{n} V_j (G_{ij} \cos\theta_{ij} + B_{ij} \sin\theta_{ij}) = 0 \\ -(Q_{di} + \gamma Q_{di}^L) - V_i \sum_{j=1}^{n} V_j (G_{ij} \sin\theta_{ij} - B_{ij} \cos\theta_{ij}) = 0 \end{cases} \tag{11.27}$$

where P_{gi}, P_{di} and Q_{di} are the active power outputs of generators, active power and reactive power demands of loads at the present operating point respectively, P_{gi}^L, P_{di}^L and Q_{di}^L are the corresponding increasing rates of the above three variables, γ is the parameter of the growth factor, and V denotes the voltage magnitude.

We can write Eq. (11.27) into its matrix form:

$$H(x,\gamma)=0 \tag{11.28}$$

where H denotes the functions on the left-hand side of Eq. (11.27), and x is the state variable, including the phase angles of the PQ and the PV buses and the voltage magnitudes of the PQ buses. Equation (11.28) is also called the extended power flow equations with growth factor.

In each iteration of the continuation power flow algorithm, there are mainly two steps: prediction and correction, which are summarized as follows.

1) Prediction

The variable in the extended power flow Eq. (11.28) is $[x,\gamma]^{\mathrm{T}}$. To estimate its value in the next iteration, it is assumed that the loads increase in the direction of the linearization.

Differentiating Eq. (11.28), we have

$$[H_x \quad H_\gamma]\begin{bmatrix}\mathrm{d}x\\ \mathrm{d}\gamma\end{bmatrix}=0 \tag{11.29}$$

where H_x is the Jacobian matrix for the conventional power flow equations, and $\begin{bmatrix}\mathrm{d}x\\ \mathrm{d}\gamma\end{bmatrix}$ is the tangent vector for approximation. Since the growth factor γ is the new independent variable introduced to the conventional power flow equations, we need another equation to solve the new power flow Eq. (11.28) in order to determinate the tangent vector. Thus, set the kth component of the tangent vector to be

$$\mathrm{d}x_k=\pm 1 \tag{11.30a}$$

which is also called the continuation parameter. Then Eq. (11.29) after augmentation is

$$\begin{bmatrix}H_x & H_\gamma\\ & e_k\end{bmatrix}\begin{bmatrix}\mathrm{d}x\\ \mathrm{d}\gamma\end{bmatrix}=\begin{bmatrix}0\\ \pm 1\end{bmatrix} \tag{11.30b}$$

where e_k is the unit row vector whose k^{th} element is one. In the beginning of the calculation, set the tangent vector component of the continuation parameter to be one. Since the tangent is known, the continuation parameter can then be taken as a state variable, which attains its maximum rate of change around the given solution point, and the sign of its slope specifies the sign of the corresponding component in the tangent vector. Note that when load level reaches its maximum, the voltage at the corresponding bus has the maximum rate of change.

Once the tangent vector is obtained, the prediction is given by

$$\begin{bmatrix} x^1 \\ \gamma^1 \end{bmatrix} = \begin{bmatrix} x^0 \\ \gamma^0 \end{bmatrix} + \sigma \begin{bmatrix} \mathrm{d}x \\ \mathrm{d}\gamma \end{bmatrix} \tag{11.31}$$

where $\begin{bmatrix} x^1 \\ \gamma^1 \end{bmatrix}$ is the predicted value, $\begin{bmatrix} x^0 \\ \gamma^0 \end{bmatrix}$ is the given power flow solution, namely the initial value for the prediction, and σ denotes the step size, which should guarantee the predicted value falls in the convergence region to assure the existence of the power flow solution with the given continuation parameter. In other words, for a specific step size σ, if the solution does not converge, then it needs to be reduced.

2) Correction

Because the number of the unknowns is greater than the number of equations in Eq. (11.28) by one, we set the kth element of x_k to be the predicted value, and get the augmented equation

$$\begin{bmatrix} H(x,\gamma) \\ x_k - \xi \end{bmatrix} = 0 \tag{11.32}$$

where x_k is the kth component of the continuation parameter $\begin{bmatrix} x \\ \gamma \end{bmatrix}$, and ξ is the kth component of $\begin{bmatrix} x^1 \\ \gamma^1 \end{bmatrix}$. Substituting the prediction $\begin{bmatrix} x^1 \\ \gamma^1 \end{bmatrix}$ as the initial value into the augmented power flow Eq. (11.28), we get a new power flow solution. It is easy to see that the introduction of the additional equation of x_k makes the Jacobian matrix at the critical operating point nonsingular. Thus, by solving iteratively the power flow Eq. (11.32), we can obtain the complete *PV* curve, and in the end get the load margin index L_{rsv}.

The load margin index computed by the continuation power flow method has several advantages such as the desired linearity and the clear physical meaning, and the more accurate description about the trend of the voltage stability level in the system.

11.5.2 Accurate Modeling of the Load Shedding Module in Blackout Model Ⅱ

We first give the definition of the voltage security region. Voltage security region refers to the set of parameters with which the power flow equations are solvable. In this section, we assume that loads are constant power ones. If the load shedding strategies are implemented according to power factors, then the loads are assumed to be of the constant power factor type.

As the load level increases, if the Jacobian matrix of the power flow equations becomes singular, then the system encounters the saddle-node bifurcation at the critical point for voltage stability, which implies that the system has reached the boundary of the voltage security region. At the boundary, it holds that

$$\begin{cases} g(x,\mu)=0 \\ \eta g_x = 0 \\ \eta\eta^{\mathrm{T}} = 1 \end{cases} \tag{11.33}$$

where $g(x,\mu)=0$ are the power flow equations, g_x is the corresponding Jacobian matrix, η is the left eigenvector associated with the zero eigenvalue of g_x, x denotes the state, including the real and imaginary parts of the voltage at the *PQ* and *PV* buses, and μ incorporates all the parameters, including the voltage magnitude V_g of the *PV* buses and slack bus, the active power P_g at the *PV* buses, and active power P_d at all the load buses.

From the discussion above we know that the boundary of the voltage security region is composed of all the critical points for voltage stability. Here, the critical points are usually obtained by the following method: starting from the present operating point (x_0,μ_0), let the loads increase in the direction that is determined by the load forecasting and the generation plan. Then use the continuation power flow method to obtain the maximum load for which the power flow equations still have a solution. This is a critical point denoted by (x_c,μ_c).

Equation (11.33) gives an equation about μ, which defines the boundary surface of the security region. Linearizing around the critical point (x_c,μ_c), we have the following relationship:

$$g_x|_c\,\Delta x + g_\mu'|_c\,\Delta\mu = 0 \tag{11.34}$$

where $g_x|_c$ and $g_\mu|_c$ represent the values of g_x and g_μ at the point (x_c,μ_c) respectively, and $\Delta\mu=\mu-\mu_c$.

Multiplying from the left both sides of Eq. (11.34) by η, we obtain the tangent plane equations for the boundary of the voltage security region

$$\eta g_\mu|_c\,\Delta\mu = 0 \tag{11.35}$$

namely

$$k_v^0 V_g + k_{P_g}^0 P_g + k_{Pd}^0 P_d = 1 \tag{11.36}$$

where V_g represents the voltages of generators, P_g represents the active power of generators, P_d represents the active power of loads, and k_v^0, $k_{P_g}^0$ and k_{Pd}^0 are their coefficients respectively (please refer to Remark 11.5 for the derivation of Eq. (11.36)).

Besides the saddle-node bifurcation caused by the singularity of the Jacobian matrix, the operation constraints can also lead the system to the critical points for voltage stability. As the loads increase, the reactive power outputs of generators increase correspondingly until reaching their limits. Once the reactive power output of the last generator reaches its limits, the system encounters limit-induced bifurcation at the critical point for voltage stability. The boundary of the voltage security region at such points satisfies

$$\begin{cases} g(x,\mu) = 0 \\ r(x,\mu) = 0 \end{cases} \tag{11.37}$$

where $r(x,\mu) = 0$ is the equation stating the fact that the reactive power outputs of generators equal their limits. Let

$$\mu = [\mu_1, \mu_2, \cdots, \mu_n]^{\mathrm{T}}$$

Then Eq. (11.37) determines an implicit equation of parameters $\mu_i (i = 1, 2, \cdots, n)$

$$z(\mu) = 0$$

whose tangent plane equation at the critical point (x_c, μ_c) is

$$\sum_{i=1}^{n} z_{\mu_i} |_c \Delta\mu_i = 0 \tag{11.38}$$

where $z_{\mu_i} |_c$ is the partial derivate of z with respect to μ_i at the critical point and in the sequel we denote it by z_{μ_i}. Assume $z_{\mu_1} \neq 0$, then

$$\Delta\mu_1 + \sum_{i=2}^{n} \frac{z_{\mu_i}}{z_{\mu_1}} \Delta\mu_i = 0 \tag{11.39}$$

From the implicit function theorem, the tangent plane equation for the boundary of the voltage security region can be written as

$$\Delta\mu_1 - \sum_{i=2}^{n} \frac{\partial \mu_1}{\partial \mu_i} \Delta\mu_i = 0 \tag{11.40}$$

Taking partial derivates on both sides of Eq. (11.37) with respect to every

$\mu_i (i = 1, 2, \cdots, n)$, we have

$$\begin{pmatrix} g_x & g_{\mu_1} \\ r_x & r_{\mu_1} \end{pmatrix} \begin{pmatrix} \dfrac{\partial x}{\partial \mu_i} \\ \dfrac{\partial \mu_1}{\partial \mu_i} \end{pmatrix} = \begin{pmatrix} -g_{\mu_i} \\ -r_{\mu_i} \end{pmatrix} \tag{11.41}$$

Solving the above set of equations and substituting the derivative of μ_1 with respect to μ_i into Eq. (11.40), we obtain the tangent plane equation for the boundary of the voltage security region as follows:

$$\Delta \mu_1 + \sum_{i=2}^{n} \left(r_{\mu_1}^{-1} r_x \frac{\partial x}{\partial \mu_i} + r_{\mu_1}^{-1} r_{\mu_i} \right) \Delta \mu_i = 0 \tag{11.42}$$

The form of this equation is similar to that of Eq. (11.36). Note that for different μ_i, the coefficient matrices for the set of linear Eq. (11.41) are the same, and the only difference lies in the terms on the right-hand side. So for the coefficient matrix, we only need to compute its inverse or triangular decomposition once.

In fact, emergency load shedding strategies try to recover the solvability of the power flow equations by taking the loads as controllable variables[5]. Since the number of equations is less than the number of unknowns by one, the procedure in essence is to solve an optimization problem. In order to simplify the emergency load shedding problem, one selects some buses to construct a set of weak buses to reduce the dimension of the solution space. Compared with load shedding strategies for other sets, it is more likely to get optimal or suboptimal solutions in the constructed set.

The construction of the set of weak buses is a critical step in searching for solutions to the emergency load shedding problem. However, the set of weak buses differs as the objective function and the fault location change, and it cannot be easily identified using the information about present operation point. Usually emergency load shedding strategies involve the minimum amount of shed load, which changes only slightly the operating point in the parameter space before and after faults. Thus, one may assume that the coefficients of the tangent plane equations for the boundary of the voltage security region almost remain the same. With this assumption, in this section we have adopted the tangent equation before faults to identify weak buses in voltage stability, namely we have used the set of the load buses with the greatest $k_{P_d}^0$ in Eq. (11.36) as the candidate set of weak buses. And then we choose from this candidate set those load buses that are adjustable by operators to construct the set of weak buses. Since most of the faults in power systems are small, it is reasonable to identify weak buses using the boundary of the voltage security region before the fault and this also agrees with the physical characteristics of power systems. When the computation for optimal power flows does not converge, we shed loads at the buses in the set according to a preset proportion, and then determine the system state by computing again the optimal power flow. If the computation still diverges, then we shed more loads. If the number of computation has reached the upper limit and the computation has not converged, then the system is considered to have encountered voltage collapse.

Compared with the blackout model I, the blackout model Ⅱ deals with the load shedding procedure more carefully and can guarantee a certain voltage stability margin in the system, and is thus more useful for engineering practice.

Remark 11.5 The derivation of Eq. (11.36)

$\Delta\mu$ in Eq. (11.35) can be written into the following form

$$\Delta\mu = \mu - \mu_c \tag{R1}$$

Substituting Eq. (R1) into Eq. (11.35) and rearranging terms, we have

$$\eta g_\mu |_c \, \mu = \eta g_\mu |_c \, \mu_c \tag{R2}$$

Without loss of generality, assume $\eta g_\mu |_c \, \mu_c = const \neq 0$. Then Eq. (R2) becomes

$$\frac{\eta g_\mu |_c}{\eta g_\mu |_c \, \mu_c} \mu = 1 \tag{R3}$$

Finally, substituting $\mu = [V_g, P_g, P_d]^{\mathrm{T}}$ into Eq. (R3), we get the coefficients in Eq. (11.36):

$$k_v^0 = \frac{\eta \left.\dfrac{\partial g}{\partial V_g}\right|_c}{\eta g_\mu |_c \, \mu_c}, \quad k_{P_g}^0 = \frac{\eta \left.\dfrac{\partial g}{\partial P_g}\right|_c}{\eta g_\mu |_c \, \mu_c}, \quad k_{P_d}^0 = \frac{\eta \left.\dfrac{\partial g}{\partial P_d}\right|_c}{\eta g_\mu |_c \, \mu_c} \tag{R4}$$

11.6 SOC Analysis Based on Blackout Model Ⅱ

11.6.1 Trend of Critical Characteristic Indices in Blackout Processes

In this section, we carry out the simulation study on the IEEE 30-bus system using the blackout model Ⅱ. The setting of the parameters is as follows.

(1) The line overload threshold $\alpha = 0.7$,

(2) The probability of tripping overloaded lines $\beta = 0.3$,

(3) The growth rate of load level and generator capacity $\lambda = 1.0005$,

(4) The growth rate of the transmission line capacity $\mu = 1.005$,

(5) The simulated duration is 5000 days.

The above parameter values are called the initial values.

The trend of the relative load loss $\overline{P}_{\text{loss}}$ is shown in Fig. 11.18.

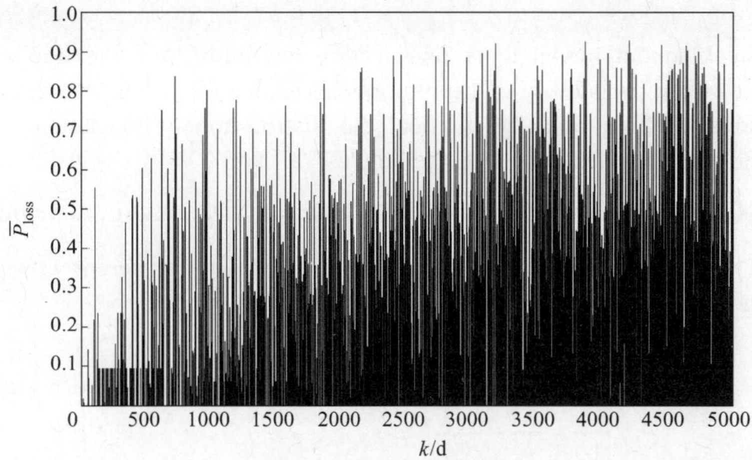

Figure 11.18 Trend of $\overline{P}_{\text{loss}}$

From the figure, it is clear that the scales of the blackouts are small in the first 540 days. Starting from around day 972, large-scale blackouts begin to take place and around this time the system can be considered to be in a critical state.

The trend of the load margin L_{rsv} for assessing voltage stability is shown in Fig. 11.19.

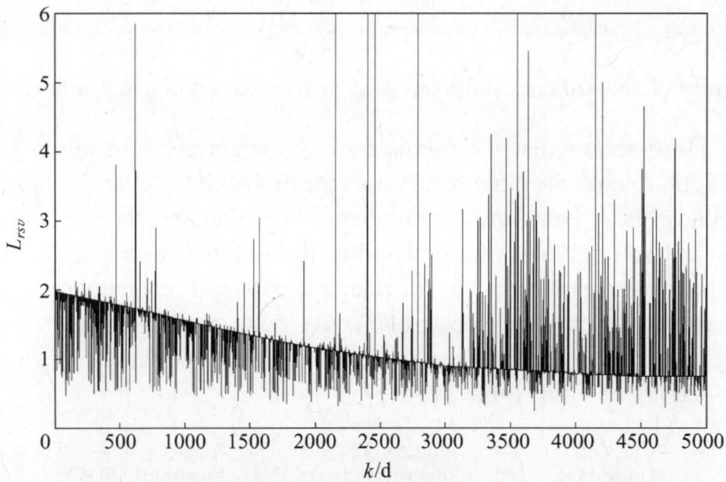

Figure 11.19 Trend of the load margin for voltage stability

It can be seen that the load margin index L_{rsv} decreases gradually in a statistical sense as the loads grow and the blackout scale increases. When the blackout happens, its value varies dramatically.

11.6.2 Effects of Typical Operation Parameters on Blackout Distributions

Using the blackout model II, in this section we study how the line overload threshold α, the probability of tripping overloaded lines β and the growth rate of transmission line capacity μ affect the distribution of blackouts. We also assess the risk for blackouts using the indices VaR and CVaR.

1) The effect of the line overload threshold α on the blackout distribution

Figure 11.20 shows the blackout scale distributions under different values of α while the other parameters remain fixed at their initial values.

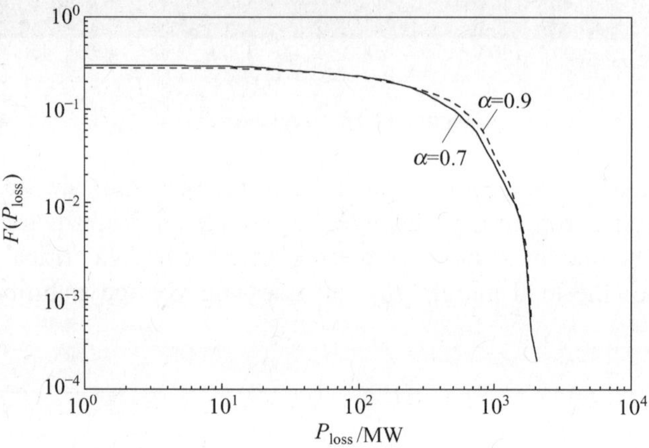

Figure 11.20 Different distributions of the blackout scales under different α

Figure 11.20 shows that the increase of α lowers the probability for small-scale blackouts, but on the other hand raises the probability for large-scale blackouts. This is because as α increases, the number of lines that are taken to be overloaded decreases and consequently the probability for small-scale blackouts decreases. However, this reduces the actions for line upgrades and reconstructions, and as a result the probability for large-scale blackouts increases.

Setting the confidence level $\sigma = 0.99$, we calculate the values of the VaR and CVaR, which are listed in Table 11.4.

Table 11.4 Risks under different overload thresholds (MW)

Overload threshold	VaR	CVaR
0.7	1369	16.25
0.9	1421	16.55

Table 11.4 indicates that the increase of the threshold value increases the system risk slightly, which is a disadvantage for the secure operation of power systems.

2) The effect of the probability β of tripping overloaded lines on blackout distributions

Figure 11.21 shows the blackout scale distributions under different probabilities of tripping overloaded lines, while the other parameters remain fixed at their initial values.

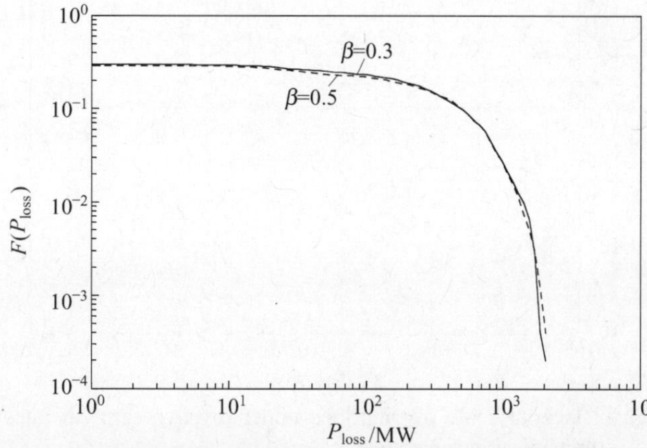

Figure 11.21 Blackout scale distributions under different probabilities of tripping overloaded lines

Figure 11.21 shows that the increase of β, namely decreasing the capability of recovering the overloaded lines, reduces the probability of blackouts. This is because as the probability of tripping overloaded lines increases, more lines are more likely to be tripped and as a result the planning department might strengthen their actions to upgrade transmission lines, and thus decrease the probability for large-scale blackouts.

Setting the confidence level $\sigma = 0.99$, we calculate the values of the VaR and CVaR and the results are listed in Table 11.5.

Table 11.5 Risks under different line tripping probabilities (MW)

Tripping probability	VaR	CVaR
0.3	1369	16.25
0.5	1322	15.98

Table 11.5 indicates that the increase of β reduces the system risk slightly.

3) The effect of the growth rate μ of transmission line capacities on blackout distributions

Figure 11.22 shows the blackout scale distributions under different growth rates of transmission line capacities when the other parameters remain fixed at their initial values. It can be seen that the increase of μ reduces the number of blackouts significantly, which is especially true for large-scale blackouts.

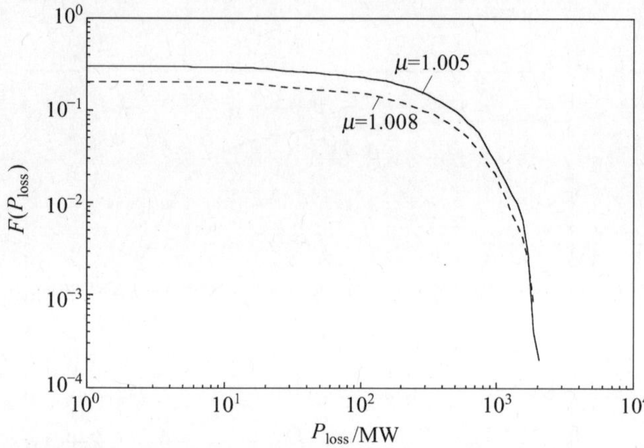

Figure 11.22 Blackout scale distributions under different growth rates for line transmission capacities

Setting the confidence level $\sigma = 0.99$, we calculate the values of the VaR and CVaR and list the results in Table 11.6.

Table 11.6 Risks under different growth rates for the transmission capacities (MW)

Transmission capacity growth rate	VaR	CVaR
1.005	1369	16.25
1.008	1235	15.04

Table 11.6 indicates that the increase of μ significantly reduces the system risk. The reason for this is that large-scale blackouts are mainly caused by cascading failures, which can be effectively prevented by upgrading transmission lines. The systems after upgrade are more robust to random faults. Hence, to strengthen the construction of transmission lines is one of the most important ways to prevent disastrous event in power systems.

Comparing Tables 11.1 – 11.3 and 11.4 – 11.6, one can see that the load shedding strategy used in blackout model II greatly reduces the risk for blackouts. This also proves the importance of advanced control strategies whose effects, with a much smaller cost, can be comparable to line constructions.

11.7 Notes

(1) In this chapter we have constructed two blackout models taking into account the reactive power/voltage characteristics, both making use of optimal power flows to simulate the fast dynamics and at the same time considering the changes in voltage stability levels. The two models can also simulate the slow dynamics in system constructions with great precision. The two models consist of four different, yet complementary, voltage stability analysis modules, which are able to reveal the system evolution mechanisms and the SOC phenomena not only from the active power/load level perspectives, but also from the reactive power/voltage perspectives.

(2) The blackout model Ⅱ has incorporated the load margin index using the continuation power flow method and integrated a more accurate module for emergency load shedding. Thus, the results obtained accordingly are more reliable. However, we need to point out that the above conclusion has been drawn based on the assumption that the operation point in state space only differs slightly before and after a fault, and consequently the coefficients in the tangent plane equation for the boundary of the voltage security region can be taken as fixed. If this assumption does not hold anymore, the emergency load shedding module might not be able to provide a practical control strategy to recover the solvability of the power flow equations. Of course, there is no stability control system that works absolutely correctly in all situations in reality, and if the optimal power flow computation in the blackout model Ⅱ does not converge after shedding loads, it is possible to use this model to approximately simulate the system collapse after the failure of stability control.

(3) There is still room for improving both the blackout models Ⅰ and Ⅱ. For example, we can consider the dynamics of generators and loads. In addition, up to now, the research has been confined to be carried out at the macroscopic level to reveal the SOC characteristics from the reactive power/voltage perspectives. These issues will be further discussed in the next chapter.

References

[1] U S-Canada Power System Outage Task Force. Interim report: causes of the August 14th blackout in the United States and Canada. Nov. 2003

[2] IEEE System Dynamic Performance Subcommittee of the Power System Engineering Committee of the PES. Voltage stability of power system: concepts, analytical tools, and industry experience. Technical Report 90 TH 0358-2PWRS, IEEE, 1990

[3] Kundur P. Power system stability and control. New York: McGraw-Hill, 1994

[4] Mei S W, Ni Y X, Wang G, Wu S Y. A study of self-organized criticality of power system under cascading failures based on AC-OPF with voltage stability margin. IEEE Transactions on Power Systems, 2008, 23(4): 1719 – 1726

[5] Taylor C W. Power System Voltage Stability. New York: McGraw-Hill, 1994

[6] Ni Y X, Chen S S, Zhang B L. Theory and Analysis of Dynamic Power System. Beijing: Tsinghua University Press, 2002 (in Chinese)

[7] Kuhn H W, Tucker A W. Linear inequalities and related system. Annals of Mathematics Studies, No.38, Priceton, N J: Princeton University Press, 1956

[8] Alvarado F L, Jung T H. Direct detection of voltage collapse conditions. Proceedings: Bulk power system voltage phenomena-voltage stability and security, EPRI EL26183, Project 2473-21 Proceedings, 1989: 5.23 – 5.38

[9] Cutsem T V. A method to compute reactive power margins with respect to voltage collapse. IEEE Transactions on Power Systems, 1991, 6(1): 145 – 156

[10] Canizares C A, Alvarado F L. Point of collapse and continuation methods applied to AC/DC systems. IEEE Transactions on Power Systems, 1992, 7(1): 1 – 8

Chapter 12 Blackout Model Based on the OTS

The blackout models discussed in this chapter present transient dynamics, power flows and power system planning on three time scales. The main tools used in building these models are the index of transient stability margin and the optimal power flow with transient stability constraints (OTS). The advantage of using the models constructed in this chapter is that they are close to the physical realities of the cause and the evolution of cascading failures. Most importantly they incorporate preventive and emergency control actions, so we can use them to evaluate operators' roles during blackouts. Furthermore, weak links in the system can be identified in the mean time, which, for the purpose of preventing large-scale blackouts, can be critical for decision making in secure operations.

Power grids are highly complex systems subject to various random events, and thus it is difficult to study cascading failures and blackouts at a microscopic level. SOC theory, on the other hand, provides a new approach to study blackouts from a macroscopic angle. However, it needs to be pointed out that all the models discussed in the previous chapters are based on the same idea that the tripping of overloaded lines causes load-flow redistribution, which in turn leads to the tripping of new overloaded lines. Studies on blackout data reveal that quite a few cascading failures are prompted by transient instability as a result of large disturbances, such as permanent three-phase short circuit faults and the tripping of corresponding lines. Such transient instability is usually followed by generator tripping, load shedding and power flow redistribution, which finally result in cascading failures. For example, the blackout in North America on August 14, 2003 was triggered by a ground fault that developed into cascading failures causing the outage of more than 100 power stations, including 22 nuclear plants, and tens of high voltage transmission lines.

In order to simulate cascading failures and the developments of blackouts more comprehensively and overcome the shortcoming, as we just described, of the previous models, transient dynamics have to be introduced into the modeling. To deal with the transient stability problem directly is equivalent to solving a class of high-dimensional differential-difference-algebraic equations, for which there is no systematic solution. We follow another promising approach using the transient stability margin index to incorporate the transient processes in the blackout model. With this end in view, in this chapter we construct blackout models upon the index of transient stability margin and the optimal power flow with transient stability constraints (OTS)[1].

12.1 Index of Transient Stability Margin

Secure and economic operation of power systems depends on the system's stability when incidents occur; in other words, the system has to satisfy certain constraints on transient stability. Since such constraints are usually in the form of differential equations that are difficult to solve, conventional optimal power flow computations assume that all the transient stability constraints have been satisfied and thus only consider static constraints. In fact, several key technical issues in the area of secure and economic operations, such as preventive control, the computation of total transmission capabilities, and the dynamic congestion management, are related to both resource optimization and transient stabilization. Such problems can be conveniently described by the optimal power flow problem with transient stability constraints (OTS). Compared with conventional optimal power flow problems, the main difficulty in solving the OTS problem is how to handle the transient stability constraints, for which we refer the interested readers to [1] for detailed discussions.

For the OTS problem, there are generally two approaches to deal with the transient stability constraints. One is the iterative correction method[2]. It first obtains an initial operating point by solving the optimal power flows with only static constraints. Then it checks dynamic constraints at this point. If the system is not stable, it adjusts the operating point until all the security requirements are satisfied. This method is prone to sacrifice optimality and build up computational burdens. The other method is to transform the original problem into a conventional static optimization problem. Different transformation tools include converting differential equations into difference equations[1], special integration[3] and transient stability indices and their sensitivities. The transformation of differential equations into difference equations can be applied to large-scale systems, but usually with poor convergence performances. The special integration method integrates along the unstable swing trajectories to assess the system's transient stability level, which is then used as the constraint for the optimization problems. But it needs to integrate in the given time interval for different faults in order to obtain the transient stability index, and thus requires sufficiently long computational time. The method that utilizes transient stability indices and their sensitivities usually encounters the difficulty of constructing energy functions and other modeling restrictions[4,5].

So a key step in solving the OTS problem is to find an effective method to assess transient stability. A new class of transient stability margin indices has been constructed, which calculates the relative position of the operation point when clearing a fault and that of the stable equilibrium point after the fault with respect to the boundary of the system's stability region[6]. The use of such an index circumvents the difficulty in constructing a Lyapunov energy function while in the mean time replacing those transient stability constraints that are usually in the form of differential equations. Hence, in this chapter we are able to build the OTS model using the transient stability index and its sensitivity information after approximating the boundary of the system's stability region.

12.1.1 Construction of Transient Stability Margin Index

Consider a system with n generators. Assume that the loads are of the constant impedance type. The generator dynamics are described by the classical 2nd order models

$$\begin{cases} \dot{\delta}_k = \omega_0 \omega_k \\ \dot{\omega}_k = \dfrac{1}{2H_k}(P_{mk} - P_{ek} - D_k \omega_k) \end{cases} \tag{12.1}$$

where $k = 1, 2, \cdots, n$, $\omega_0 = 2\pi f_B$, δ_k, ω_k are the kth generator's angle and angular velocity respectively with respect to the synchronous reference frame, D_k and H_k are the damping coefficient and the inertia constant respectively, and P_{mk} and P_{ek} are the mechanical power and the electromagnetic power respectively. Here,

$$P_{ek} = E_k^2 G_{kk} + E_k \left(\sum_{j=1, j \neq k}^{n} E_j (G_{kj} \cos \delta_{kj} + B_{kj} \sin \delta_{kj}) \right)$$

where E_k is the internal voltage of the generator k which is set to be constant; $\delta_{kj} = \delta_k - \delta_j$, and $Y = (Y_{ij})_{n \times n} = (G_{ij} + jB_{ij})_{n \times n}$ is the system admittance matrix after eliminating load buses.

In the case of uniform damping $\lambda = D_k / 2H_k$, we choose the nth generator as the reference machine, and then system (12.1) can be rewritten as

$$\begin{cases} \dot{\delta}_{kn} = \omega_0 \omega_{kn} \\ \dot{\omega}_{kn} = \dfrac{P_{mk} - P_{ek}}{2H_k} - \dfrac{P_{mn} - P_{en}}{2H_k} - \lambda \omega_{kn} \end{cases} \tag{12.2}$$

where $k = 1, 2, \cdots, n-1$ and $\omega_{kn} = \omega_k - \omega_n$.

Correspondingly, in the case of non-uniform damping, system (12.1) can be rewritten as

$$\begin{cases} \dot{\delta}_{kn} = \omega_0 \omega_{kn}, & k = 1, 2, \cdots, n-1 \\ \dot{\omega}_k = \dfrac{P_{mk} - P_{ek} - D_k \omega_k}{2H_k}, & k = 1, 2, \cdots, n \end{cases} \tag{12.3}$$

For the case of uniform damping, let

$$u = [P_{m1}, P_{m2}, \cdots, P_{mn}]^{\mathrm{T}}$$
$$x = [\delta_{1n}, \delta_{2n}, \cdots, \delta_{(n-1)n}, \omega_{1n}, \omega_{2n}, \cdots, \omega_{(n-1)n}]^{\mathrm{T}}$$
$$N = 2n - 2$$

and for the case of non-uniform damping, let

$$u = [P_{m1}, P_{m2}, \cdots, P_{mn}]^{\mathrm{T}}$$
$$x = [\delta_{1n}, \delta_{2n}, \cdots, \delta_{(n-1)n}, \omega_1, \omega_2, \cdots, \omega_n]^{\mathrm{T}}$$
$$N = 2n - 1$$

Hence, system (12.1) after network reduction can be described by

$$\dot{x} = f(x,u) \tag{12.4}$$

where $f \in \mathrm{C}^{\infty}, x \in \mathbb{R}^N$.

Assume $u = u_0$ before and after a fault, namely the active power outputs of the generators are fixed, then the dynamics of system (12.4) when encountering a fault can be divided into three stages that correspond to the evolutions before, during and after the fault:

$$\begin{cases} \dot{x} = F_1(x,u), & t < 0 \\ \dot{x} = F_2(x,u), & 0 \leqslant t < t_F \\ \dot{x} = f(x,u), & t \geqslant t_F \end{cases} \tag{12.5}$$

where t_F is the fault clearing time. Using the approximate expression for the boundary of the stability region[7−14], the following index can be used to assess the system's transient stability

$$I(x(t_F),u) = \frac{h(x(t_F),u)}{h(x_S,u)} \tag{12.6}$$

where $x(t_F) = \Phi(u) = \Phi(t_F, x_0, u)$ is the initial state of the post-fault system, x_S denotes the stable equilibrium of the post-fault system, $h(x,u) = 0$ is the approximate expression for the boundary of the stability region near the controlling unstable equilibrium point[15] satisfying

$$f^{\mathrm{T}} \cdot D_x h = \mu \cdot h,\ h(x_e,u) = 0 \tag{12.7}$$

where μ is the unstable eigenvalue of the Jacobian matrix $J = D_x f(x)|_{x=x_e}$ at the controlling unstable equilibrium point x_e.

The index (12.6) can be used as follows:

(1) If $I(x(t_F)) > 0$ (resp. < 0), the initial state of the post-fault system $x(t_F)$ is

inside (resp. outside) the stability region, and the system is stable (resp. unstable);

(2) If $I(x(t_F))=0$, the initial state of the post-fault system $x(t_F)$ is on the boundary of the stability region, and the system is critically stable.

In power systems, the stable operation point x_0 before a fault is usually close to the stable operation point x_S after the fault, so

$$I(x(0),u)=\frac{h(x_0,u)}{h(x_S,u)}\approx 1 \tag{12.8}$$

From the above analysis, we know when $I(x(t_F),u)$ is close to 1, the system is quite stable; if $I(x(t_F),u)$ is close to 0, the system approaches being unstable; if $I(x(t_F),u)$ is less than 0, the system is unstable. Therefore, the transient stability index (12.6) normalizes the system's stability level. For a given fault, the transient stability index $I(x(t_F),u)$ gradually decreases from around 1 towards a negative value as the system loses its stability when the fault duration time increases. We call the index (12.6) the transient stability margin index.

12.1.2 Calculation of Transient Stability Margin Index

It is usually difficult to obtain analytical solutions to the partial differential equations like Eq. (12.7). In [6], using semi-tensor products, an approximate expression for the transient stability margin index has been defined as

$$I_Q(x(t_F),u)=\frac{h_Q(x(t_F),u)}{h_Q(x_S,u)} \tag{12.9}$$

where

$$h_Q(x,u)=[x-x_e(u)]^{\mathrm{T}}\eta(u)+\frac{[x-x_e(u)]^{\mathrm{T}}Q(u)[x-x_e(u)]}{2} \tag{12.10}$$

Here, the first-order term coefficient vector $\eta(u)=(\eta_1,\eta_2,\cdots,\eta_N)^{\mathrm{T}}$ is the left eigenvector associated with the unstable eigenvalue μ of the Jacobian matrix $J=D_x f(x,u)|_{x=x_e}$, that is

$$\begin{cases} J^{\mathrm{T}}\eta=\mu\eta \\ \eta^{\mathrm{T}}\eta=1 \end{cases} \tag{12.11}$$

The quadratic term coefficient matrix Q satisfies the Lyapunov matrix equation

$$CQ+QC^{\mathrm{T}}=H \tag{12.12}$$

where $C=\left(\frac{\mu}{2}I_N - J^{\mathrm{T}}\right)$, I_N is an $N\times N$ identity matrix, $H=\sum_{i=1}^{N}\eta_i H_i$, and H_i is the Hessian matrix of f_i at x_e, i.e., $H_i=\frac{\partial^2 f_i}{\partial x^2}$.

We rewrite Eq. (12.12) into linear equations that have to be satisfied by the matrix elements, then we get the symmetric coefficient matrix for the quadratic term[9, 16]

$$Q=V_c^{-1}[(C\otimes I_N + I_N\otimes C)^{-1}V_c(H)] \tag{12.13}$$

where V_c is the operator that obtains a column vector by stacking all the elements of a given matrix[13], i.e. for matrix $B=(b_{ij})_{N\times N}$,

$$V_c(B)=[b_{11},b_{21},b_{31},\cdots,b_{N1},b_{12},b_{22},\cdots,b_{N2},\cdots,b_{1N},b_{2N},\cdots,b_{NN}]^{\mathrm{T}}.$$

The operator $\otimes$ in Eq. (12.13) denotes the Kronecker tensor product. For any arbitrary $s\times t$ matrix $A=(a_{ij})_{s\times t}$ and $k\times l$ matrix $B=(b_{ij})_{k\times l}$, the Kronecker tensor product is defined to be

$$A\otimes B=\begin{pmatrix} a_{11}B & \cdots & a_{1t}B \\ \vdots & \ddots & \vdots \\ a_{s1}B & \cdots & a_{st}B \end{pmatrix}$$

We summarize the steps to compute the transient stability margin index (12.9) as follows.

Step 1 Specify the location and the type of a given fault. Look for the controlling unstable equilibrium point using the BCU-based method[12], and the stable equilibrium point in the post-fault system.

Step 2 Calculate the system state $(\delta(t_F),\omega(t_F))$ at the fault clearing moment.

Step 3 Compute the Jacobian matrix J and the Hessian matrix H_k at the controlling unstable equilibrium point.

Step 4 Find J's unstable eigenvalue μ and its associated left eigenvector η.

Step 5 Get Q according to (12.13) using the semi-tensor method.

Step 6 Substitute the system state at the fault clearing moment into (12.9), and calculate the transient stability index at that moment.

12.1.3 Calculate the Sensitivity of the Transient Stability Margin Index

In power system operations, the stable equilibrium point $x_0(u)$ of the pre-fault system, the system's state $\Phi(u)$ at the fault clearing moment, the controlling unstable equilibrium point $x_e(u)$ of the post-fault system and its unstable eigenvalues (eigenvectors) and the stable manifold $\{x \mid h(x,u)=0\}$, all change continuously as the parameter u changes within a certain range. In fact, the functions describing such relationships are continuously differentiable[11].

When the scale of the system is small, the sensitivity of the transient stability margin index with respect to u can be determined analytically. We provide the details as follows.

The sensitivity of the stable equilibrium point $x_0 = X(u)$ of the pre-fault system is

$$A_1(u) = D_u X(u) = -[D_x F_1(x,u)]^{-1} D_u F_1(x,u) \tag{12.14}$$

The system's state at the fault clearing moment can be written as

$$\Phi(u) = x(t_F) = \Phi(t_F, x_0(u), u) = x_0 + \int_0^{t_F} F_2(x,u)\mathrm{d}t \tag{12.15}$$

The sensitivity $A_2(u)$ of the system's state at the fault clearing moment can be derived directly from the differential equations that have to be satisfied by the trajectories of the system[17, 18], i.e.

$$A_2(u) = D_u x(t_F) = A_1(u) + \int_0^{t_F} [D_x F_2(x(t),u) D_u x(t) + D_u F_2(x(t),u)]\mathrm{d}t \tag{12.16}$$

Furthermore, the sensitivity of the controlling unstable equilibrium $x_e(u)$ of the post-fault system can be computed by

$$A_3(u) = -[D_x f(x_e,u)]^{-1}[D_u f(x_e,u)] \tag{12.17}$$

The sensitivity of the unstable eigenvalue μ is denoted by

$$E_1 = [D_{u_1}\mu, D_{u_2}\mu, \cdots, D_{u_n}\mu],$$

and the sensitivity of the associated left eigenvector η is denoted by

$$E_1 = [D_{u_1}\eta, D_{u_2}\eta, \cdots, D_{u_n}\eta].$$

They can be determined by

$$\begin{pmatrix} D_{u_i}\eta \\ D_{u_i}\mu \end{pmatrix} = \begin{pmatrix} J^{\mathrm{T}} - \mu I & -\eta \\ \eta^T & 0 \end{pmatrix}^{-1} \begin{pmatrix} -(D_{u_i}J^{\mathrm{T}})\eta \\ 0 \end{pmatrix}, \quad i = 1, 2, \cdots, n \tag{12.18}$$

From Eq. (12.13), the approximate sensitivity of the quadratic coefficient matrix Q can be obtained by

$$Q_{u_i} = D_{u_i} Q = V_c^{-1}\{(C \otimes I_N + I_N \otimes C)^{-1} V_c [D_{u_i} H - (D_{u_i} C)Q - Q(D_{u_i} C^{\mathrm{T}})]\} \tag{12.19}$$

where

$$D_{u_i}H = \sum_{j=1}^{m}[(D_{u_i}\eta_j)H_j + \eta_j D_{u_i}H_j], \quad i=1,2,\cdots,n$$

$$D_{u_i}C = \left(\frac{1}{2}D_{u_i}\mu I - D_{u_i}J^{\mathrm{T}}\right), \quad i=1,2,\cdots,n$$

So the analytical form of the sensitivity of $h_Q(x,u)$ [14] is

$$\frac{\partial h_Q(x,u)}{\partial u} \approx \eta^{\mathrm{T}} A_0 + a^{\mathrm{T}} E_2 + a^{\mathrm{T}} Q A_0 + K_1 \tag{12.20}$$

where

$$a = \Phi(u_0) - x_e(u_0)$$
$$A_0 = A_2 - A_3$$
$$K_1 = [a^{\mathrm{T}} Q_{u_1} a, a^{\mathrm{T}} Q_{u_2} a, \cdots, a^{\mathrm{T}} Q_{u_n} a]$$

As to Eq. (12.6), for simplicity we usually assume that $h(x_S,u)$ almost remains the same when u changes slightly.

Finally, the approximate analytical expression for the sensitivity of $I(x(t_F),u)$ is

$$\frac{\partial I(x(t_F),u)}{\partial u} \approx \frac{\partial h_Q(x,u)}{\partial u} \frac{1}{h(x_S,u)} \tag{12.21}$$

When the scale of the system is large, it becomes difficult to compute analytically the sensitivity of the transient stability margin index. In this case, we usually use numerical methods, e.g. the difference equations

$$\frac{\partial I(x(t_F),u)}{\partial u_i} = \frac{I(x(t_F),u_0 + \Delta u_i) - I(x(t_F),u_0)}{\Delta u_i}, \quad i=1,2,\cdots,n \tag{12.22}$$

Then the sensitivity of $I(x(t_F),u)$ in its vector form can be computed by

$$\frac{\partial I(x(t_F),u)}{\partial u} = \left[\frac{\partial I(x(t_F),u)}{\partial u_1},\cdots,\frac{\partial I(x(t_F),u)}{\partial u_n}\right]^{\mathrm{T}} \tag{12.23}$$

12.2 Mathematical Model of the OTS and Its Algorithmic Implementation

In this section we first give mathematical descriptions about the conventional optimal power flow model with transient stability constraints, and then we establish the mathematical model of the OTS.

12.2.1 Conventional Optimal Power Flow (OPF)

The conventional OPF problem can be formulated as an optimization problem

$$\begin{aligned} \min \quad & F(u) \\ \text{s.t.} \quad & G(u,x)=0 \\ & H(u,x)\leqslant 0 \end{aligned} \tag{12.24}$$

where $F(u)$, $G(u,x)$, and $H(u,x)$ are the objective function, equality constraints, and static inequality constraints respectively; x represents the system state, including voltages, active powers and reactive powers, etc.; u denotes the adjustable variables, which, without further specification, are chosen to be generators' active power outputs $P=[P_{g1},P_{g2},\cdots,P_{gn}]$. For convenience, we sometimes write $F(u)$, $G(u,x)$ and $H(u,x)$ as $F(P)$, $G(P,x)$ and $H(P,x)$. The objective function and the constraints can be defined as follows.

1) Objective function

The objective function of the optimization problem (12.24) is to minimize the generation fuel cost

$$\min \quad F(P)=\sum_{i\in S_G}\phi_i(P_{gi}) \tag{12.25}$$

where

$$\phi_i(P_{gi})=a_i+b_iP_{gi}+c_iP_{gi}^2$$

is the quadratic function for the ith generator's fuel cost with coefficients a_i, b_i and c_i, and S_G is the set of adjustable generators.

2) Equality constraints

The equality constraints are the power flow equations

$$\begin{cases} P_g-P_L-P(V,\theta)=0 \\ Q_g-Q_L-Q(V,\theta)=0 \end{cases} \tag{12.26a}$$

where P_g and Q_g are the injected active and reactive power vectors respectively, and P_L and Q_L are the active and reactive powers of the load respectively. Eq. (12.26a) can be further written as

$$\begin{cases} P_{gi} - P_{Li} = V_i \sum_{j=1}^{n} V_j (G_{ij} \cos\theta_{ij} + B_{ij} \sin\theta_{ij}) \\ Q_{gi} - Q_{Li} = V_i \sum_{j=1}^{n} V_j (G_{ij} \sin\theta_{ij} - B_{ij} \cos\theta_{ij}) \end{cases} \tag{12.26b}$$

where P_{gi} and Q_{gi} are the active and reactive power outputs at generator bus i respectively, P_{Li} and Q_{Li} are the active and reactive power loads at bus i respectively, $\theta_{ij} = \theta_i - \theta_j$, V_i and θ_i are bus i's voltage amplitude and angle, and $Y = (G_{ij} + \mathrm{j}B_{ij})_{n \times n}$ is the system admittance matrix and the total number of buses is n.

3) Static inequality constraints

$$\begin{cases} P_{gi}^{\min} < P_{gi} < P_{gi}^{\max}, & i \in S_G \\ Q_{gi}^{\min} < Q_{gi} < Q_{gi}^{\max}, & i \in S_R \\ V_i^{\min} < V_i < V_i^{\max}, & i \in S_N \\ S_{ij}^{\min} < S_{ij} < S_{ij}^{\max}, & (i,j) \in S_{CL} \end{cases} \tag{12.27}$$

The above inequality constraints include the upper and lower limits of generators' active and reactive power outputs, and the bus voltage amplitudes and the transmission lines' thermal stability. Here, S_G, S_R, S_N and S_{CL} represent respectively the set of generators whose active power outputs are adjustable, the set of generators whose reactive power are adjustable, the set of buses and the set of transmission lines.

12.2.2 Treatment of Transient Stability Constraints

Usually the fault clearing time t_F is determined by the time setting of the relay protection devices. In this section, we assume t_F is fixed, and as we have explained the generators' active power outputs $P = [P_{g1}, P_{g2}, \cdots, P_{gn}]$ are considered to be adjustable. For a given fault, the transient stability margin index I changes only with P, namely I depends only on P, and thus can be denoted by $I(P)$.

When $I(P) > 0$, the system satisfies the transient stability constraints, for which case

$$I(P) \geqslant \varepsilon, \quad \varepsilon > 0 \tag{12.28}$$

where $\varepsilon > 0$ is used here to guarantee a certain stability margin.

Since it is very difficult to deal with the differential equation constraints as transient stability constraints directly, we follow the approach in [14] where a quadratic approximation of the transient stability margin index is obtained utilizing the boundary of the stability region. In addition, the sensitivity of the transient stability margin index with respect to the system parameter u can also be determined analytically. Hence, the transient stability constraints in the OTS can be replaced by the transient stability margin index and its sensitivity. Thus, we have

$$I(P)+[\nabla I(P)]^{\mathrm{T}}\Delta P>\varepsilon',\ \varepsilon'>0 \tag{12.29}$$

where

$$\Delta P=[P_{g1}-P_{g1}^{0},P_{g2}-P_{g2}^{0},\cdots,P_{gn}-P_{gn}^{0}]^{\mathrm{T}}$$

$P_{g1}^{0},P_{g2}^{0},\cdots,P_{gn}^{0}$ are the generators' active power outputs before the optimization, $P_{g1},P_{g2},\cdots,P_{gn}$ are the generators' active power outputs after the optimization, and $\nabla I(P)$ denotes the sensitivities of the index to the adjustable parameters P, i.e.,

$$\nabla I(P)=\left[\frac{\partial I(P)}{\partial P_{g1}},\frac{\partial I(P)}{\partial P_{g2}},\cdots,\frac{\partial I(P)}{\partial P_{gn}}\right]^{\mathrm{T}}.$$

In the case of multiple faults, e.g. k faults, for each fault, there is a set of transient stability margin indices with associated sensitivities. The $I(P)$ and ε' in Eq. (12.29) can be generalized to their vector forms respectively

$$I(P)=[I_1(P),I_2(P),\cdots,I_k(P)]^{\mathrm{T}}$$

$$\varepsilon'=[\varepsilon_1',\varepsilon_2',\cdots,\varepsilon_k']^{\mathrm{T}}.$$

So for the multiple-fault case, the expression for the inequality constraints of the transient stability remains the same, and Eq. (12.28) can be generalized to its vector form, where $\varepsilon=[\varepsilon_1,\varepsilon_2,\cdots,\varepsilon_k]^{\mathrm{T}}$.

12.2.3 OPF with Transient Stability Constraints (OTS)

The conventional OPF only takes into account static constraints. If we consider in addition the transient stability constraints (12.28), we get the general mathematical description for the OTS problem

$$\begin{aligned}&\min\ \ F(P)\\&\text{s.t.}\ \ G(P,x)=0\\&\qquad H(P,x)\leqslant 0\\&\qquad I(P)\geqslant\varepsilon,\ \ \varepsilon>0\end{aligned} \tag{12.30}$$

Inequality (12.28) can be replaced by inequality (12.29) when solving the optimization problem, and thus the OTS problem can be further written into a different form that is easier to solve

$$\begin{aligned} \min\ & F(P) \\ \text{s.t.}\ & G(P,x)=0 \\ & H(P,x)\leqslant 0 \\ & I(P)+[\nabla I(P)]^{\mathrm{T}}\Delta P>\varepsilon',\varepsilon'>0 \end{aligned} \tag{12.31}$$

The above optimization problem can be solved using the standard nonlinear programming methods, whose computation complexity is about the same as the conventional OPF.

12.2.4 Implementation of the OTS Algorithms

In order to accelerate the convergence of the OTS algorithm, we can take the solution of the conventional OPF as the initial solution to the OTS problem. For a specified fault, we calculate the transient stability margin index and their associated sensitivities to decide whether the transient stability index satisfies $I(P)\geqslant\varepsilon$. If the inequality is satisfied, then we stop the computational routine and the result obtained by the OPF is the solution to the OTS problem. If on the other hand, it is not satisfied, then we add transient stability inequality constraints (12.29) into the optimization problem and try to solve the OTS problem (12.31). After solving the optimization problem, the index needs to be checked by recalculating the transient stability margin index to see whether it satisfies Eq. (12.28). New rounds of the optimization process (12.31) will be carried out until the constraint is satisfied.

We summarize the steps as follows.

Step 1 Solve the conventional OPF problem.

Step 2 For a specified fault, calculate the transient stability margin index and its associated sensitivities to assess the system stability.

Step 3 Check whether inequality (12.28) is satisfied. If so, save the result and exit the routine. Otherwise go to the next step.

Step 4 Solve the OTS optimization problem (12.31), and return to Step 2.

To speed up the convergence and reduce the number of iterations, the ε' in problem (12.31) can be set to be a bit larger than ε in problem (12.30) or can be increased gradually using intuitions gained from practical experiences.

12.2.5 Simulations of the New England 39-Bus System

In this section we carry out the simulation study on the New England 39-Bus system to verify the effectiveness of the OTS algorithm. The system is shown in Fig. 12.1.

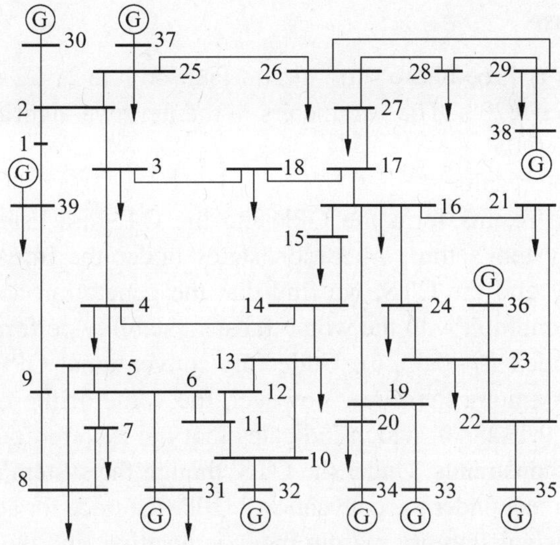

Figure 12.1 The New England 10-generator 39-bus system

The generators' parameters are listed in Table 12.1.

Table 12.1 The dynamic parameters of the generators in the New England 10-generator 39-bus system

Generator	Bus	Rotary inertia /(H/s)	Synchronous reactance/ (p.u.)	Transient reactance/ (p.u.)	Active power limit /MW	
					minimum	maximum
1	30	42.00	1	0.31	200	350
2	31	30.30	2.95	0.697	300	750
3	32	35.80	2.495	0.531	300	750
4	33	28.60	2.62	0.436	300	732
5	34	26.00	6.7	1.32	250	608
6	35	34.80	2.54	0.5	300	750
7	36	26.40	2.95	0.49	250	660
8	37	24.30	2.9	0.57	250	640
9	38	34.50	2.106	0.57	400	930
10	39	500.0	0.2	0.06	400	1100

The 10th generator is set to be the reference machine. The generators are described by the classical second order generator models, and the loads are assumed to be of the constant impedance type. The generation cost coefficients of generators 1 to 8 are $a_i = 0.01, b_i = 0.3, c_i = 0.2$, and those for generators 9 and 10 are $a_i = 0.006$, $b_i = 0.3, c_i = 0.2$. The upper and lower voltage limits of all the buses in the OTS model are set to be 1.1p.u and 0.94p.u. respectively.

1) Single fault case

The given fault is a three-phase short circuit fault on line 21-22 near bus 21. The line is tripped after 0.23 s. The parameters in the iterative algorithm are set to be $\varepsilon = 0.03$ and $\varepsilon' = 0.05$.

(1) Optimization results

The calculation results form the OPF and the OTS are listed in Table 12.2. Comparing the system's three operation states under the initial condition, the conventional OPF and the OTS, we find that the generation cost is the highest under the initial condition with the worst transient stability performance (the value of the stability index is –0.5326). Under the conventional OPF, the generation cost is the lowest among the three, however, the value of the transient stability margin index is –0.1328<0, which indicates that the system does not satisfy the transient stability constraints. Under the OTS, though the system's generation cost is a bit higher than that under the conventional OPF (the price for achieving transient stability), the transient stability margin index is positive, and thus the system has a certain stability margin.

Table 12.2 Results for the single-fault case (MW)

Generator	Initial state	OPF	OTS
1	250.00	350.00	350.00
2	572.93	577.65	603.95
3	650.00	574.20	594.54
4	632.00	562.94	555.02
5	508.00	562.66	539.07
6	650.00	567.28	537.17
7	560.00	564.46	533.39
8	540.00	554.44	567.69
9	830.00	909.22	873.06
10	1005.37	968.05	1033.74
Generation cost/($/h)	37,044	34,441	36,170
Transient stability index	–0.5326	–0.1328	0.0395
Stability assessment	unstable	unstable	stable

(2) Validation of optimization results

In order to check the effectiveness of the OTS algorithm and the reliability of the transient stability margin index, we simulate for three different situations in which the generators' outputs are under the initial condition, the OPF and the OTS. A three-phase short circuit fault occurs on line 21-22 near bus 21 at $t_0=0.1$ s. After a delay of t_F, the line is tripped off. The power angle (denoted by δ) curves in these three situations are shown in Fig. 12.2 – Fig. 12.4. The 10th generator is set to be the reference machine, and we show the power angle curves of the other nine generators.

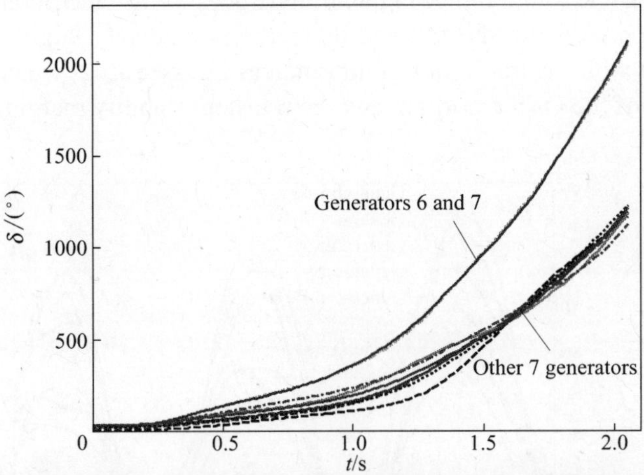

Figure 12.2 Power angle curves under the initial condition

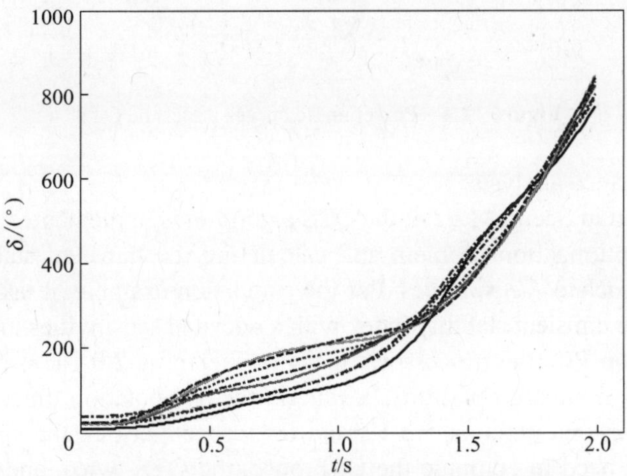

Figure 12.3 Power angle curves under the OPF

Figure 12.2 shows that generators 1 to 9 evolve into two groups both of which lose synchronization with the reference generator 10 under the initial condition. The power angle curves of generators 6 and 7 diverge the farthest because their electric distances to the reference generator are the longest (Please refer to Fig. 12.1). Figure 12.3 shows that after the OPF adjustment, all the nine generators are synchronized, but their power angle curves still diverge from the reference although improvements have been made compared with to those in Fig. 12.2. Figure 12.4

shows that after the OTS adjustment, the system is transiently stable. In addition, the swing amplitudes of generators 6 and 7 are the largest because of the long electric distances between these two generators and the reference. In the meantime, generator 1 has the minimum swing amplitude because its electric distance to the reference machine is the shortest and its inertia is relatively large. These results are consistent with the transient stability indices in Table 12.2, which verifies the effectiveness of the OTS algorithm and the transient stability margin index.

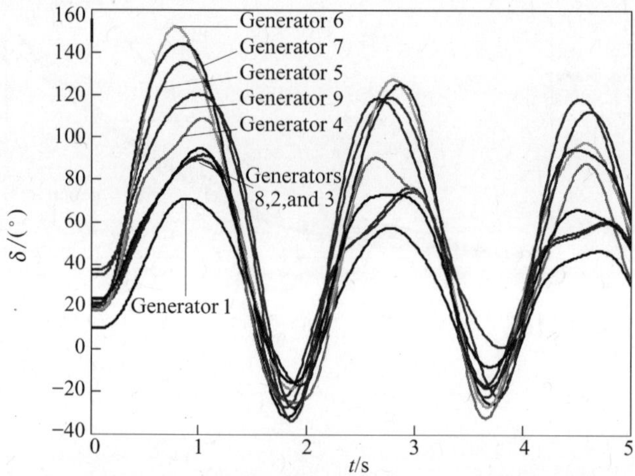

Figure 12.4 Power angle curves under the OTS

(3) Performance analysis

As explained in Section 12.2.4, the OTS algorithm is implemented by iteratively solving the optimization problem and calculating the transient stability margin index with associated sensitivities. For the simulation example, it takes about 20 s to compute the transient stability index with associated sensitivities in one iteration using a desktop PC (the processor's main frequency is 2.0 GHz) installed with the MATLAB simulation platform. Similarly the computation times for the OPF and the OTS in one iteration are 5 s and 6 s respectively. In the simulation just described, we need to compute the OPF once, the OTS twice, and the transient stability margin indices and their sensitivities for three times. So the simulation takes a total time of $5+20\times 3+6\times 2=77$ s.

The OTS algorithm does not incorporate the angle constraints for transient stability directly into the OPF computation, but incorporates the transient stability margin indices and their sensitivities as transient stability constraints. So the inequality constraints of adjustable variables can be dealt with directly in the OTS optimization computation. Therefore, the computation complexity of the OTS algorithm is about the same as that of the conventional OPF.

2) Multiple fault case

We choose two faults from the contingency set:

Fault 1: A three-phase short circuit fault occurs on line 21-22 near bus 21 at $t_0 = 0.1$ s. The faulted line is tripped off after t_F s.

Fault 2: A three-phase short circuit fault occurs on line 21-22 near bus 22 at $t_0 = 0.1$ s. The faulted line is tripped off after t_F s.

In the simulation, we let $t_F = 0.23$ s. The parameters in the iterative algorithms are set to be $\varepsilon = [0.03 \quad 0.03]^{\mathrm{T}}$ and $\varepsilon' = [0.05 \quad 0.05]^{\mathrm{T}}$.

(1) Optimization results

The simulation results under the initial condition, the OPF and OTS are listed in Table 12.3. Comparing the simulation results under the three conditions, we find that the generation cost under the initial condition is the highest with the worst transient stability level (the stability index value is $[-0.4681 \quad -0.6190]^{\mathrm{T}}$). The generation cost for the system under the conventional OPF is the lowest, however, the system does not satisfy the transient stability constraints and the values of the transient stability margin indices $[-0.0711 \quad -0.0853]^{\mathrm{T}}$ are negative. The OTS considers the transient stability constraints and thus the system's generation cost under the OTS is a bit higher than that under the conventional OPF. The transient stability margin indices $[0.0434 \quad 0.0453]^{\mathrm{T}}$ are both positive, and the system is stable with a certain stability margin.

Table 12.3 Simulation results in the multiple fault case (MW)

Generator	Initial state	OPF	OTS
1	250.00	350.00	350.00
2	572.93	577.65	595.24
3	650.00	574.20	588.12
4	632.00	562.94	558.29
5	508.00	562.66	547.43
6	650.00	567.28	547.43
7	560.00	564.46	544.01
8	540.00	554.44	563.62
9	830.00	909.22	886.07
10	1005.37	968.05	1008.56
Generation cost/($/h)	37,044	34,441	34,473

(2) Validation of optimization results

To verify the OTS algorithm and the transient stability margin index, numerical simulations are carried out for the system under the two given faults using the OPF and the OTS adjustments. The results are listed in Table 12.3.

The power angle curves under fault 1 are shown in Fig. 12.5 and Fig. 12.6 where the 10th generator is taken as the reference machine.

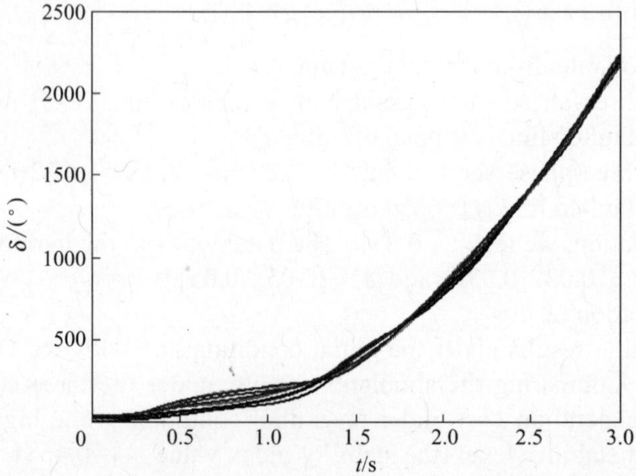

Figure 12.5 Power angle curves under fault 1 using the OPF adjustments

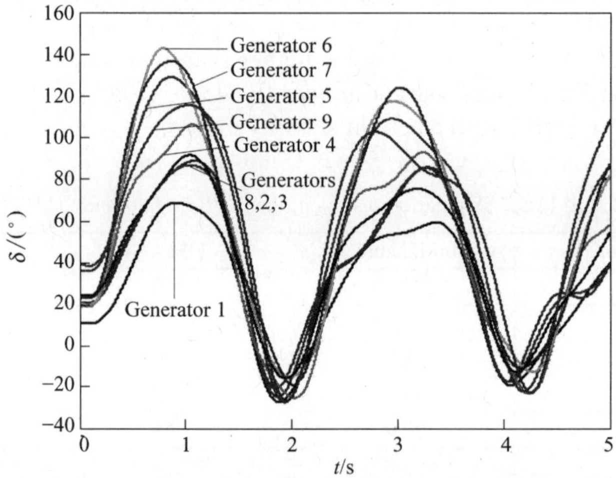

Figure 12.6 Power angle curves under fault 1 using the OTS adjustments

Figure 12.5 and Fig. 12.6 show that under fault 1, even using the OPF adjustments, the power angle curves still diverge, which implies that the system is unstable. After the OTS adjustment, the system is transiently stable.

The power angle curves under fault 2 are shown in Fig. 12.7 and Fig. 12.8 where the 10th generator is taken as the reference machine.

Figure 12.7 and Fig. 12.8 show that under fault 2, even using the OPF adjustments, the power angle curves still diverge, which implies that the system is unstable. After the OTS adjustments, the system is transiently stable. In addition, similar to the single fault case, in the multiple fault case, the swing amplitudes of generator 6 and 7 are the largest because of the long electric distances from generators 6

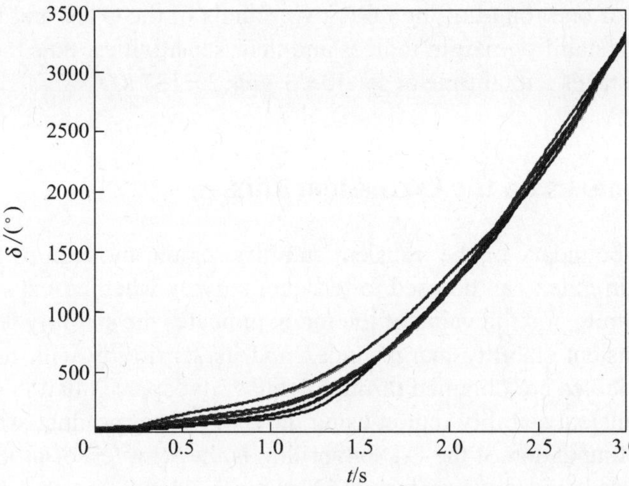

Figure 12.7 Power angle curves under fault 2 using the OPF adjustments

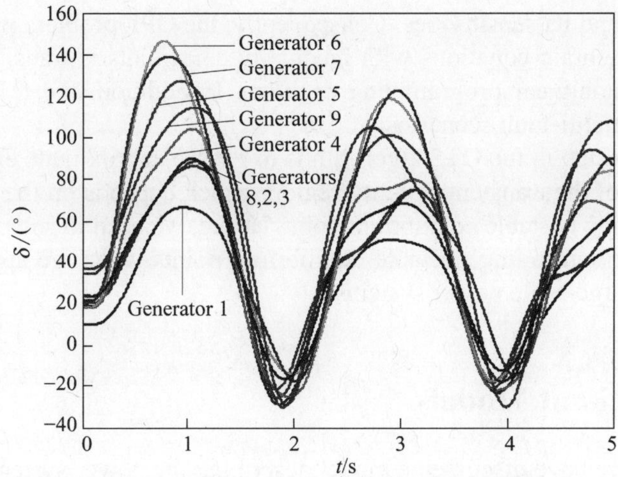

Figure 12.8 Power angle curves under fault 2 using the OTS adjustments

and 7 to the reference machine. In the mean time, generator 1 has the minimum swing amplitude since its electric distance to the reference machine is the smallest.

The above simulation results have validated the effectiveness of the OTS algorithm and the transient stability margin index.

(3) Performance analysis

In the previous computational example, it takes about 40 s for each iteration to compute the transient stability indices and their sensitivities under the two faults on a desktop PC whose main frequency is 2.0 GHz (20 s for the one fault case, 40 s for the two fault case). The computation time for one round of the OPF is 5 s, and that for one round of the OTS is 6 s. In the previous example, there are computations of one round of the OPF, two rounds of the OTS, and three rounds of the transient stability margin indices and their sensitivities, thus the running of this algorithm takes a total time of $5+40\times 3+6\times 2=137$ s.

12.2.6 Remarks on the OTS Algorithm

(1) Using the boundary of the transient stability region, the sign of the transient stability margin index can be used to tell qualitatively whether a system remains stable after a fault, and the value of the index indicates the stability level.

(2) The transient stability margin index and its sensitivity with respect to the adjustable variables are obtained through a quadratic approximation of the boundary of the transient stability region using the semi-tensor product. Therefore, the mathematical foundation of the OTS algorithm is the semi-tensor product method, and the interested reader can refer to [16] for more details.

(3) The feature of the OTS algorithm is to replace the transient stability constraints by the inequality constraints in the form of the transient stability margin index and its sensitivities. Consequently, the OPF problem in the form of differential-algebraic equations with inequality constraints is transformed into a conventional nonlinear programming problem. In addition, the OTS algorithm can deal with multi-fault scenarios.

(4) One key step in the OTS algorithm is to get the approximate expression for the boundary of the transient stability region, which depends on the computation of the controlling unstable equilibrium point. Hence, we need to solve the problem of looking for controlling unstable equilibrium points when we apply the OTS algorithm to large-scale power systems.

12.3 Blackout Models

In Chapter 3 we have discussed a model describing the power system's evolution mechanism on three different time scales. In this section, we give the mathematical description[19] for the cascading failures and blackouts using the OTS method. The model to be established consists of the inner, the middle and the outer loops that are shown in Fig. 12.9.

The inner loop describes the transient dynamics, which simulates generator and protection dynamics after large random disturbances. The middle loop implements the fast dynamics, which simulates the post-transient dynamics, including the undervoltage and underfrequency load shedding, generator output adjustment, load flow transfer, and line overload protection. The outer loop corresponds to the slow dynamics, which simulates the development of power systems including the continuous growth of the generation, load and transmission capacity.

In what follows, we discuss the three loops in detail in three sub-sections.

12.3.1 Transient Dynamics (Inner Loop)

The transient dynamics describe how a system evolves under large random disturbances. The tripping of faulted lines, and the actions of the protection and emergency control are some of the main driving forces for large-scale cascading failures. The inner loop is concerned mainly with the transient dynamics. The cascading failure that is caused by a sudden large disturbance in the transient dynamics can be classified into self-organized procedures triggered by impulsive changes (Please refer to Section 3.3).

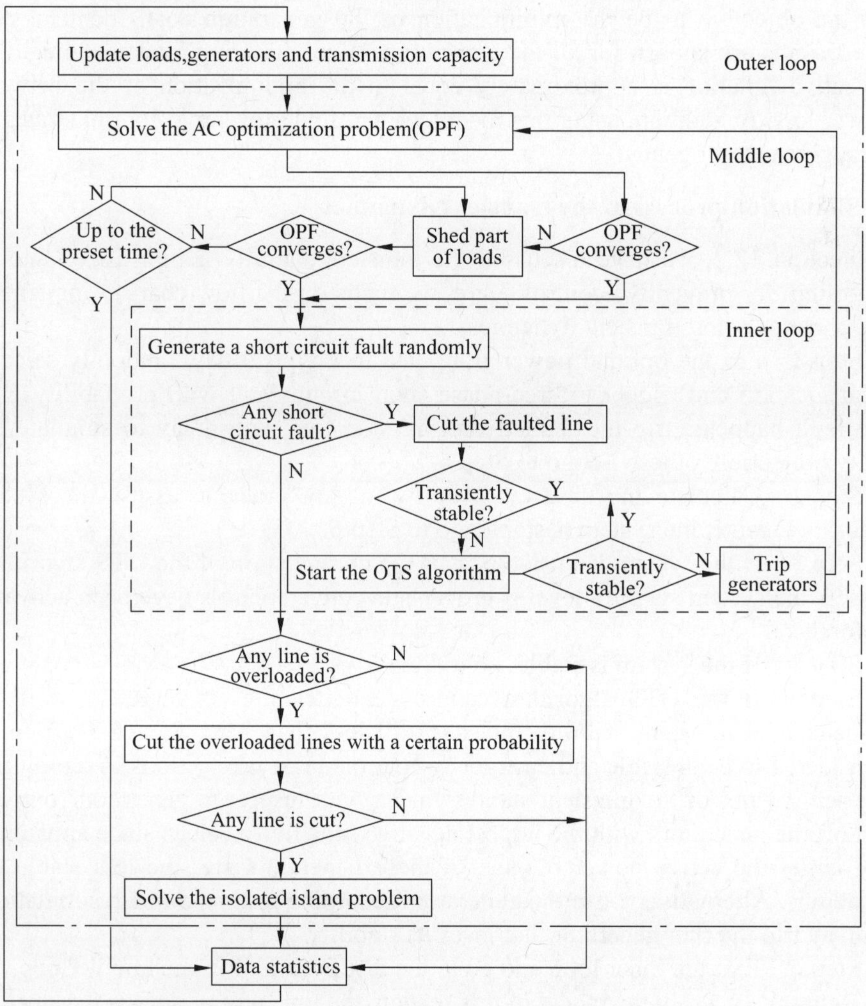

Figure 12.9 Power system blackout model using the OTS

1) Transient stability index and its OTS algorithm

The secure and economic operation of power systems requires that the system remains stable under common faults; in other words, the system must satisfy the transient stability constraints. If we directly take into consideration the transient stability conditions into the blackout model, then we have to face high dimensional differential equations and the problem becomes intimidatingly complicated. To simplify the problem, we utilize the transient stability margin index $I_Q(x(t_F),u)$ defined in (12.9) to estimate the transient stability level of the system (Please refer to [14] for more detailed discussions). The OTS algorithm based on the transient stability margin index has been demonstrated in Eq. (12.31). In this section we set the objective to be the minimization of the generation cost. Identical cost coefficients are chosen for all the generators. When ignoring load shedding, this objective function is approximately equal to the minimization of the network power loss. In addition, using this model we can find preventive control strategies for a given contingency.

2) Simulation process of the transient dynamics

In Section 12.2, we have discussed the transient stability margin index and its algorithm for preventive control. Here, we focus on the flow chart to implement the corresponding transient dynamics.

Step 1 For the optimal power flow data in a typical day, randomly select a fault location and trigger a three-phase short circuit fault with probability ξ. If the fault happens, trip the faulted transmission line with delay to simulate the relay protection; otherwise, go to Step 6.

Step 2 Calculate the transient stability margin index to assess the system stability level. If the system is stable, go to Step 6.

Step 3 If the system is unstable, start the computation of the OTS algorithm. Assess the system stability level if proper preventive controls have been activated beforehand.

Step 4 If the system is stable, go to Step 6.

Step 5 If the OTS algorithm cannot guarantee the convergence or if the re-calculated transient stability index is still less than zero, then the system is considered to be unstable and emergency controls must take actions: According to the sensitivities of the transient stability index with respect to generators' outputs, cut off the generators with the largest negative sensitivity values since apparently the larger the active power outputs of these generators are, the less stable the system is. Alternatively, trip the generators according to time domain simulations, namely trip the first generator that loses its stability.

Step 6 Exit the inner loop and go to the fast dynamics in the middle loop.

Remark 12.1 In the model of this section, the time unit of days is the smallest time unit in the simulation and may not correspond directly to days in the physical sense. Note that generator tripping and load shedding belong to different time scales in the physical process after encountering faults. Generator tripping is mainly activated by transient dynamics, while load shedding usually happens after some delay as a result of undervoltage and underfrequency load shedding, and line protection of overloaded lines. Therefore, in the blackout model with the OTS discussed in this section, generator tripping is classified into the transient dynamics, i.e., the inner loop, while load shedding is classified into the fast dynamics, i.e., the middle loop.

3) The treatment of stability control

As discussed in Chapter 3, power systems have exploited several strategies to guarantee the security and stability of the system operation. The goal is to keep the system stable in the first place and in the meanwhile maintain the frequencies and the voltages within their required ranges. We have simplified the actions of security and stability control in our blackout model in this chapter, but we still keep their main influences on the system. This can be explained as follows:

(1) Emergency control

Emergency control strategies for tripping generators and shedding loads are usually determined beforehand to deal with large disturbances, such as three-phase short circuit faults, which endanger the system stability. When a serious fault that can be dealt with by emergency control happens in the system, such control strategies can be activated with minimum time delay to assure the power angel (synchronous) stability. In this chapter, the model only incorporates approximately the generator tripping actions which are the mostly commonly used emergency control strategies to maintain the synchronous stability. Since load shedding in emergency control is difficult to simulate and has only limited effects on the control performance, we have not considered it in the blackout model.

(2) Frequency control

After the emergency control, the system may be synchronized but with an unacceptable frequency. Then the AGC of the corresponding generators are activated for the secondary frequency control. This process is relatively slow because the power sources of the generators have to be adjusted. The blackout model with the OTS does not contain a particular module to simulate the frequency control actions, however the resulting changes in the generators' active power outputs can be taken as a control variable that is adjustable in the OPF computation.

(3) Underfrequency load shedding

If the system's frequency keeps below the required value, the underfrequency load shedding equipments cut off part of the loads according to the preset strategy to recover the system's frequency to the desirable range. The failure of maintaining the system's frequency implies that the adjustments in the generators' outputs cannot satisfy the load demands, which is shown in the computation as the divergence of the OPF because the total generator outputs is less than the total load demands. In the blackout model in this chapter, we use load shedding to simulate this underfrequency load shedding behavior.

(4) Undervoltage load shedding

If the voltage of some buses keeps below the required value, the undervoltage load shedding equipments will cut off part of the loads to recover the voltage. In the blackout model in this chapter, when the OPF does not converge because some of the bus voltages cannot satisfy the constraints, load shedding strategies take actions to simulate the undervoltage load shedding behavior.

(5) Congestion control

When the transmission capacity of the network is not sufficient and thus the power flow exceeds the constraints, loads will be shed manually or automatically. Since this happens when the OPF diverges because of the fact that the power flow constraints are not satisfied, we can use the shedding load module to manage network congestion.

As to the transmission lines in the network, except for those faulted lines that are tripped after the response time of the relay protections in the transient dynamics, it takes some time to trip the overloaded lines and implement the underfrequency or undervoltage load shedding strategies. Thus these are considered in the fast dynamics in the next subsection. In that subsection, the phenomenon of islanding as a result of tripping generators or lines is also considered.

Hence, the blackout model discussed here incorporates generator tripping, load shedding and other important system behaviors, and at the same time avoids complicated computations to stay away from the overly detailed discussions in transient analysis. We have done so since the main idea of the book is to study the macroscopic collective behavior of large-scale power grids.

12.3.2 Fast Dynamics (Middle Loop)

The fast dynamics simulate "one day" in the slow dynamics. As in previous chapters, within one day, we choose the peak load level to represent the daily load level in the fast dynamics process in which cascading failures and blackouts may happen. The fast dynamics process is determined as follows. The operation state is first determined by the OPF and load shedding. Once an overloaded line is tripped, the system's operation state is recalculated by the OPF. This process repeats until there are no more overloaded lines.

1) Process of the fast dynamics

Step 1 For the kth day, the generators' maximum outputs and the load demands are determined by the slow dynamics.

Step 2 Solve the AC OPF. If it converges, go to Step 3. Otherwise, to simulate underfrequency load shedding and undervoltage load shedding, trip the loads on both ends of the overloaded lines and the loads at those buses whose voltages are below the lower limits until the OPF converges.

Step 3 If it is the first time for the OPF to convergence within that day, trigger a randomly generated three-phase short circuit fault with probability ξ and go into the inner loop of the transient dynamics. Otherwise, go to Step 4.

Step 4 If there are lines for which the ratio of its power flow to its capacity exceeds a given threshold α, go to Step 5. Otherwise, exit the fast dynamics.

Step 5 For the overloaded lines, trip them with a given probability β. If no line is tripped, go to Step 8. Otherwise, go to Step 6.

Step 6 If there are islands, go to Step 7 to deal with the islanding. Otherwise, go back to Step 2.

Step 7 For each island, compute the total load demand and the total generation capacity. For the island with the largest load, compute the amount of load to shed within this island according to the undervoltage and underfrequency load shedding strategies. For other smaller islands, if the total generation capacity of the island is greater than its total load demand, local balance can be achieved; otherwise, take the difference between the load and the generation capacity to be the amount of load shedding. Return to Step 2.

Step 8 Compute the load loss of the day.

To improve the computational efficiency, only one three-phase short circuit fault is generated with probability ξ after the first convergence of the OPF in a single day, which initiates the inner loop. The OPF adjusts generation outputs and loads after tripping the faulted lines and some of the generators. After the fault, the system evolves according to its inherent characteristics, but no more faults are added. This assumption is reasonable in practice. Take the North American blackout on August 14, 2003 for example. In the first hour after the first outage event, the five 345 kV lines were tripped off sequentially and the corresponding time intervals were 22 minutes, 9 minutes, 5 minutes and 29 minutes respectively. During any of these time intervals, the system had entered fast dynamics from the transient dynamics.

Remark 12.2 The main component of the fast dynamics is the calculation of the OPF. We use the OPF instead of the OTS for the following reasons:

(1) The model has divided the system dynamics into the transient dynamics, the fast dynamics and the slow dynamics. So it is not necessary to consider the transient stability problem in the fast dynamics, and thus the OTS needs not to be included.

(2) By using the OPF, we save quite some computational time (please refer to Section 12.2.1 for the mathematical model of the OPF).

2) Probability for tripping overloaded lines

To simulate the influence of uncertain factors, such as temperature, wind, etc, on the upper limit of line transmission capacity, in the fast dynamics, we can use different schemes to describe the tripping probability of overloaded lines. We list three approaches here.

(1) Scheme A

We use parameters α and β to describe the tripping probability of overloaded lines. Referring to the discussion in Step 5 in Section 12.3.2, if the load-capacity ratio (the ratio of the power flow to the line capacity) is greater than or equal to the threshold α, then we trip this line with probability β. Denote the ratio by x, and then the probability function $f(x)$ for tripping overloaded lines is

$$f(x)=\begin{cases}0, & x<\alpha\\ \beta, & x\geqslant\alpha\end{cases} \tag{12.32}$$

Although scheme A is easy to implement, the effect of different overload levels on the tripping probability cannot be taken into account in this scheme. Scheme B is thus proposed which is obtained by making improvements on scheme A.

(2) Scheme B

We use parameters α and γ to describe the tripping probability of overloaded lines:

$$f(x)=\begin{cases}0, & x<\alpha\\ \gamma(x-\alpha), & \alpha\leqslant x<\alpha+\dfrac{1}{\gamma}\\ 1, & x>\alpha+\dfrac{1}{\gamma}\end{cases} \tag{12.33}$$

Figure 12.10 compares the probability functions of tripping overloaded lines in schemes A and B. We can see that when the load-capacity ratio is less than α, the tripping probabilities in both of the two schemes are 0. In scheme A, the tripping probability is identically β once the load-capacity ratio is greater than α. In scheme B, once the load-capacity ratio is greater than α, the tripping probability increases with the rate γ until its value reaches one.

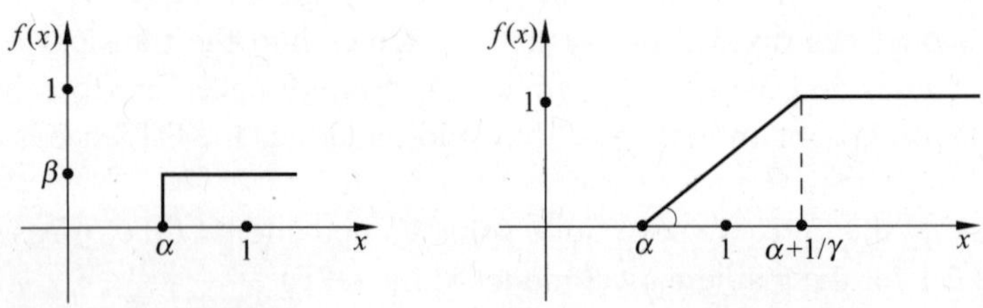

(a) Scheme A (b) Scheme B

Figure 12.10 Probability function for tripping lines

In fact, from the OPF model we know that the load-capacity ratio must be less than or equal to one when the OPF converges due to static stability constraints. When the load-capacity ratio is greater than one, load shedding strategies take actions. Then why is it needed to trip overloaded lines with a certain probability when the load-capacity ratio is less than or equal to 1? There are two reasons. One is to simulate the influence of uncertainties such as temperature and wind on the upper limits of line transmission capacities. The other is to simulate the inaccurate setting values and the misoperation of relays in practice. Take the European blackout on November 4, 2006 for example. The UCTE investigation committee concluded that one of the causes was that the dispatcher had mistakenly taken the setting value of relays at the Wehrendorf substation in Landesbergen-Wehrendorf as 2000A (which should be 1900A) and consequently believed according to his experience that the system would satisfy the $N-1$ criteria if the line was tripped (which was contrary to the reality). In this example, should we take the maximum line capacity as 2000A, the simulated behavior would be inconsistent with reality. In fact, the relay acts when the current reaches 1900A. The load-capacity ratio is 0.95 if 2000A is taken as the line capacity. This example sheds light on how we should choose parameters for the probability of tripping overloaded lines. In general, the ranges of the parameters are $0.7 \leqslant \alpha \leqslant 0.95$. $0 < \beta \leqslant 1$ and $0 < \gamma \leqslant 1/(1-\alpha)$ (To assure that the tripping probability is less than one when the load-capacity ratio is one).

It needs to be pointed out that the probability function used for tripping overloaded lines in the model needs not necessarily to be the exact characteristics of a relay protection. Rather, it is a probabilistic approximation of the behaviors in a real system with considerable simplification for the real devices. Similar treatments have been used in other models as well[20–23].

(3) Scheme C

We can also consider the probability function for tripping overloaded lines following the idea proposed in [24]. That paper considers the relationship between the line flow and the probability of line outage. It concludes that the increase of line flows can lead to the increase of the probability of line outage for mainly two reasons:

① As the power flow increases, the transmission line heats up, which in turn decreases its mechanical strength. The line may sag and lower its distance to the ground. When the line flow continues to increase and the line operates above its thermal limit for an extended period, the line is going to fuse.

② The overload and overcurrent protections act when the line flow is above the line's thermal limit. Their action time depends on the value of the line flow, which obeys an inverse relationship. The lager the line flow is, the smaller the tripping time of the relay protection is set to be. So in this case, it is more difficult to recover the line flow to the normal range under the control actions, and the line is more likely to be cut off.

For the above reasons, in [24] a probability function is proposed as follows:

In Fig. 12.11, the horizontal axis L denotes the line flow. Correspondingly, $L_{\max}^{\text{normal}}$ and $L_{\min}^{\text{normal}}$ are the upper and lower bounds for the normal line flows respectively

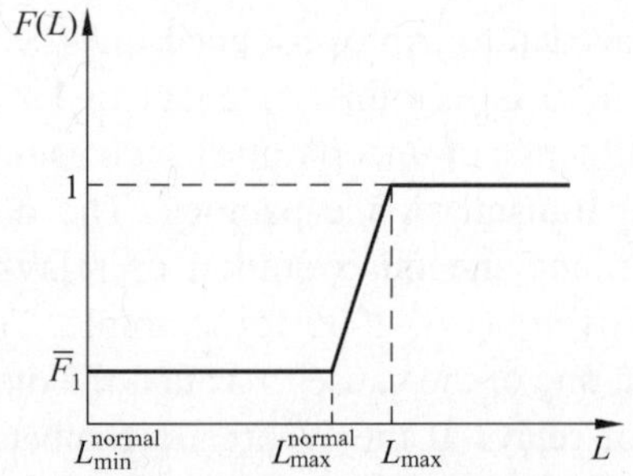

Figure 12.11 Probability function for tripping overloaded lines

and L_{max} is the limit value of the line flows. The vertical axis $F(L)$ denotes the line outage probability, which is a piecewise linear function of the line flow. When the value of the line flow is within the normal range, the outage probability is $\overline{F}_1$, when the value of the line flow is above the limit, the probability is 1, and when the value of the line flow is in between, the probability function is linear in the power flow.

In the model of this section, such a method can also be adopted as the probability function for tripping overloaded lines, which is thus called scheme C. In scheme C, even if the line flow is below the threshold, the probability of line outage is still not zero, but a small positive number. Hence, this model can be used to simulate hidden failures.

For simplicity, without further specification, scheme A is used in this chapter.

12.3.3 Slow Dynamics (Outer Loop)

The slow dynamics are mainly the same as that in Chapter 10. We only need to consider in addition the parameters for generator dynamics.

Step 1 To simulate the annual load growth, we assume that each load increases uniformly each day, namely

$$\begin{aligned} P_{di,k+1} &= \lambda P_{di,k}, \quad i \in L \\ Q_{di,k+1} &= \lambda Q_{di,k}, \quad i \in L \end{aligned} \tag{12.34}$$

where L denotes the set of load buses, $P_{di,k}$ and $Q_{di,k}$ denote the active and reactive power load at bus i on day k, and λ is the growth rate.

Step 2 Similarly, we assume the generation capacity also increases uniformly each day, namely

$$\begin{aligned} P_{gi,k+1}^{\max} &= \lambda P_{gi,k}^{\max}, \quad i \in G \\ Q_{gi,k+1}^{\max} &= \lambda Q_{gi,k}^{\max}, \quad i \in G \end{aligned} \tag{12.35}$$

where G denotes the set of generator buses, $P_{gi,k}^{\max}$ and $Q_{gi,k}^{\max}$ denote the upper limits of active and reactive power outputs of the generator at bus i on day k, and λ is the growth rate of generation, which is the same as the growth rate of the load level.

As discussed in Chapter 3, when increasing the generation capacity, dynamic parameters of the generators need to be adjusted accordingly. So we have

$$x_{d,gi,k+1} = \frac{1}{\lambda} x_{d,gi,k}, \quad i \in G$$
$$x'_{d,gi,k+1} = \frac{1}{\lambda} x'_{d,gi,k}, \quad i \in G \tag{12.36}$$

where $x_{d,gi,k}$ and $x'_{d,gi,k}$ are the d-axis synchronous reactance and transient reactance of the generator at bus i on day k respectively;

$$x_{q,gi,k+1} = \frac{1}{\lambda} x_{q,gi,k}, \quad i \in G$$
$$x'_{q,gi,k+1} = \frac{1}{\lambda} x'_{q,gi,k}, \quad i \in G \tag{12.37}$$

where $x_{q,gi,k}$ and $x'_{q,gi,k}$ are the q-axis synchronous reactance and transient reactance of the generator at bus i on day k respectively;

$$H_{gi,k+1} = \lambda H_{gi,k}, \quad i \in G \tag{12.38}$$

where $H_{gi,k}$ is the inertia constant of the generator at bus i on day k; and finally

$$D_{gi,k+1} = \lambda D_{gi,k}, \quad i \in G \tag{12.39}$$

where $D_{gi,k}$ is the damping coefficient of the generator at bus i on day k.

Step 3 For those lines that have been tripped in fast dynamics, use the average improvement effect to simulate the line capacity improvement, namely

$$F_{j,k+1}^{\max} = \mu F_{j,k}^{\max} \tag{12.40}$$

where $F_{j,k}^{\max}$ denotes the transmission capacity of line j on day k, and μ denotes the growth rate of the transmission capacity.

When updating the transmission capacity, line parameters need to be updated correspondingly

$$\begin{cases} r_{j,k+1} = \dfrac{1}{\mu} r_{j,k} \\ x_{j,k+1} = \dfrac{1}{\mu} x_{j,k} \\ b_{j,k+1} = \mu b_{j,k} \end{cases} \tag{12.41}$$

where $r_{j,k}$, $x_{j,k}$, $b_{j,k}$ are the resistance, reactance and admittance of line j on day k respectively.

For those lines that are operating in the fast dynamics, their capacities are considered to be sufficient and no upgrade is necessary, namely

$$F_{j,k+1}^{\max} = F_{j,k}^{\max} \tag{12.42}$$

And similarly, the line parameters need not to be further adjusted

$$\begin{cases} r_{j,k+1} = r_{j,k} \\ x_{j,k+1} = x_{j,k} \\ b_{j,k+1} = b_{j,k} \end{cases} \tag{12.43}$$

Step 4 Go to fast dynamics.

12.3.4 Control Parameters

From the previous three subsections, we know that six control parameters have been added in the inner, middle and outer loops in the blackout model with the OTS.

1) Overload threshold α

It is the criterion to tell whether a line is overloaded. If the ratio of the line flow to the line capacity is greater than or equal to this threshold, the line is considered to be overloaded.

2) The probability β for tripping overloaded lines

The overloaded lines are tripped with probability β. The larger β is, the more likely the overloaded line is tripped.

3) Growth rate λ of the load level and the generation capacity

The total load level and generation capacity of a given day is λ times those of the previous day. If the annual increase of the load level is 20%, then the corresponding daily growth rate is $\lambda = 1.2^{1/365} \approx 1.0005$. The generation capacities grow with the same rate while their parameters are adjusted proportionally.

4) Growth rate μ of the line capacity

For the tripped overloaded lines in fast dynamics, their capacities increase by μ folds in the next day after the upgrade, and the line parameters are adjusted accordingly.

5) Probability for large disturbance ξ

At the beginning of the fast dynamics of the inner loop, the probability ξ determines how likely for a random large disturbance (mainly three-phase short circuit fault) to take place. If $\xi = 0$, the system will not enter the transient dynamics and it evolves with only fast and slow dynamics, which can be modeled by the blackout model based on the OPF discussed in Chapter 10. If $\xi = 1$, the system enters transient dynamics for sure everyday. In a real system, the value of ξ should be within the interval (0, 1), and can be identified using the statistical data. In this chapter, for the sake of simplicity and computational efficiency, we take $\xi = 1$ as default. In other words, transient dynamics are triggered every time to reveal potential problems and weak parts in the system as soon as possible.

6) Fault clearing time t_F

Relays take actions to trip the faulted lines after a delay of t_F seconds, which determines the size and strength of a fault once its location and type are specified.

From the SOC point of view, α and β appear in the fast dynamics of the middle loop, and determine the strength of the system's internal forces; λ and μ appear in the slow dynamics of the outer loop, and determine the strength of the system's external forces; ξ and t_F appear in the transient dynamics of the inner loop, and determine the frequency and strength of random disturbances.

It needs to be pointed out that in our simulation we choose the fault location completely randomly, i.e. the probabilities at the two ends of any line are equal. The type of the fault that we use is the three-phase short circuit default. Of course, to simulate a real system, the location and the type of a fault can be determined using the statistics of historical data and engineering experiences, and so one can adjust the weighted probabilities for different locations and types.

In short, the system is influenced jointly by external and internal driven forces and random disturbances during its evolution. The system evolves gradually to its critical state under the constant impact of the external forces. When the internal and external factors satisfy certain conditions, a random disturbance can initiate an abrupt and dramatic change in the system, such as cascading failures and blackouts.

12.4 Simulations of Cascading Failures

In this section we use the New England 39-bus system to simulate the development of cascading failures caused by three-phase short circuit faults. As shown in Fig. 12.12, we pick the day with the most severe security conditions during the simulated 2000 days. The development includes seven stages. In the figure, the dashed lines indicate the faulted or outage devices that will not appear in the following diagram. The numbers in parentheses represent the ordering of the cutting of the lines. Without further specification, the control parameters in the simulation are set to be

(1) The overload threshold $\alpha = 0.7$,
(2) The probability for tripping overloaded lines $\beta = 0.3$,
(3) The growth rate of load level and generation capacity $\lambda = 1.0005$,
(4) The growth rate of line capacity $\mu = 1.005$,
(5) The probability for large disturbances $\xi = 1$,
(6) The fault clearing time $t_F = 0.16$ s.

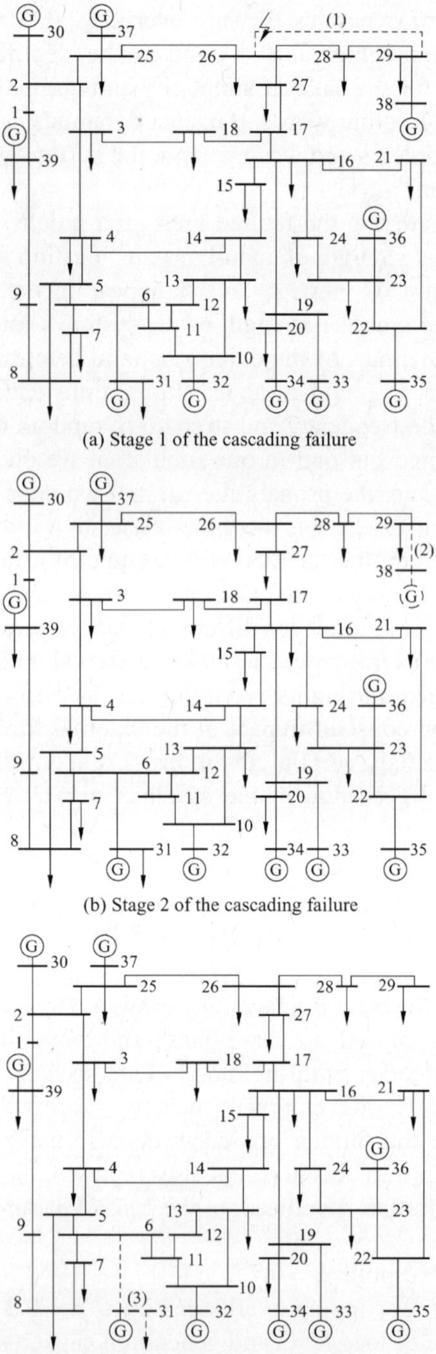

(a) Stage 1 of the cascading failure

(b) Stage 2 of the cascading failure

(c) Stage 3 of the cascading failure

Figure 12.12 The development of a cascading failure in the New England 10-generator 39-bus system

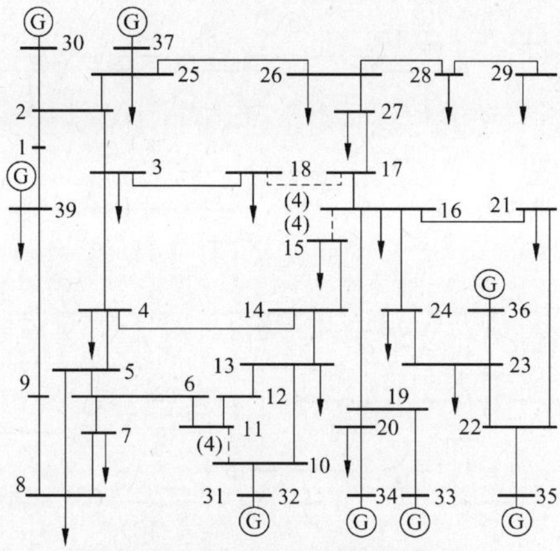

(d) Stage 4 of the cascading failure

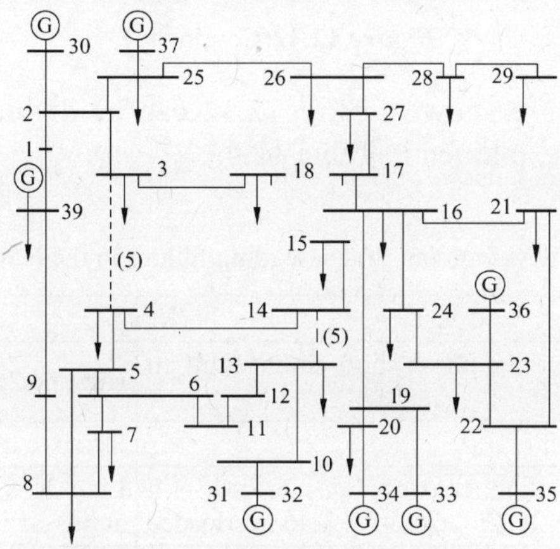

(e) Stage 5 of the cascading failure

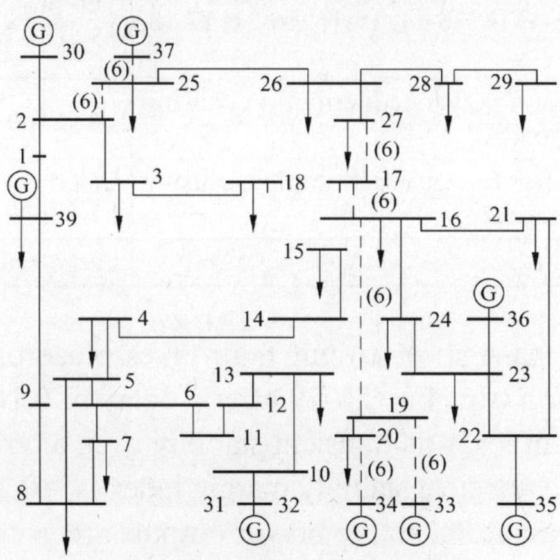

(f) Stage 6 of the cascading failure

Figure 12.12(Continued)

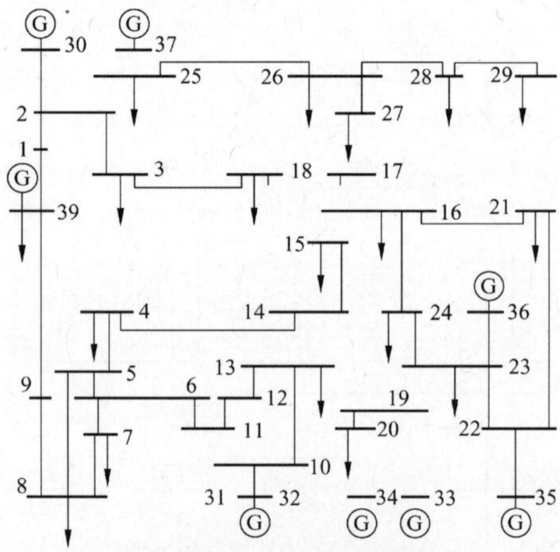

(g) Stage 7 of the cascading failure

Figure 12.12(Continued)

The cascading failure shown in Fig. 12.12 can be divided into seven stages. Each stage is further explained in Table 12.4.

Table 12.4 The development of a cascading failure in the New England system

Stage	Cause	Events
1	Line 26-29, the three phase short circuit fault on the bus 26	Cut line 26-29
2	System is unstable	Remove the generator at bus 38
3	Power redistribution causing lines 6-5, 13-14, 17-27, 21-22, 22-35, 25-26, 26-28, 31-6 overloaded	Line 31-6 out of work; generator at bus 31 is cut
4	Underfrequency load shedding resulting in the overloads in lines 6-11, 10-11, 15-16, 16-21, 17-18, 21-22	The protection of the lines 10-11, 15-16, 17-18 cuts these lines
5	Load shedding and power redistribution causing more lines overloaded	Lines 3-4 and 13-14 are cut off
6	Load shedding and power redistribution causing more lines overloaded	Lines 2-25, 16-17, 16-19, 17-27, 19-33, 20-34, 25-37 are cut off
7	—	Large-scale blackout

Stage 1 A three-phase short circuit fault takes place on line 26-29 near bus 26. The relay protection trips line 26-29 after a delay of 0.16 s.

Stage 2 The system loses its transient stability even after the preventive control (OTS) is activated. The transient stability margin index is $-0.2407<0$. Its sensitivities with respect to the generators' active power outputs are

$\nabla I(P) = [0.0109,\ 0.0147,\ 0.0143,\ 0.0144,\ 0.0165,\ 0.0143,\ 0.0143,\ 0.0035,\ -0.2568,\ 0.0113]^{\mathrm{T}}$

The generator at bus 38 has the largest negative sensitivity –0.2568 in absolute value, which implies that reducing its active power output is the most effective way to stabilize the system. Moreover, this generator is the nearest to the fault and thus the most likely one to lose stability. Therefore, the emergency control is implemented to trip this generator.

Stage 3 Transient dynamics ends after tripping the generator at bus 38. The system is now synchronously stable and enters a quasi steady state. However, the loss of the generator forces the rapid increase of active power outputs of other generators, which in turn results in the growth of the power flows in several lines over the threshold α (lines 6-5, 13-14, 17-27, 21-22, 22-35, 25-26, 26-28, and 31-6). Among the overloaded lines, line 31-6 is tripped, and the associated generator 31 is then disconnected from the grid.

Stage 4 The power source is severely inadequate after losing generators 31 and 38. Underfrequency load shedding is performed to shed the loads at both ends of the overloaded lines until the OPF finally converges. A new set of lines are overloaded (lines 6-11, 10-11, 15-16, 16-21, 17-18, and 21-22), among which lines 10-11, 15-16 and 17-18 are tripped by relay protections.

Stage 5 The OPF converges after shedding a small amount of load, but leads to the overload in some lines. Lines 3-4 and 13-14 are tripped.

Stage 6 After shedding loads, the OPF now converges, but lines 2-25, 16-17, 16-19, 17-27, 19-33, 20-34 and 25-37 are tripped. There are altogether 18 buses that are cut off from the grid now.

Stage 7 The blackout takes place and the system collapses.

Remark 12.3 In the above description of the cascading failure, the overloaded lines refer to those whose load-capacity ratios are above the threshold α.

The cascading failure process consists of the transient dynamics and the fast dynamics. It reveals that when a power system satisfies a certain condition, an instantaneous disturbance (such as a three-phase short circuit fault) can trigger SOC behaviors. In the above cascading failure simulation, the three-phase short circuit fault is a random disturbance, under which the simulated system starts to evolve step by step according to the rules set by the platform of the blackout model, until develops into a large-scale blackout in the end.

12.5 Macroscopic SOC Characteristics and Criticality Analysis

In this section, we study the power system SOC phenomena by the examination of the macroscopic variables such as the total load level, the total transmission capacity and the total shed load. Similar to Section 9.2.2, we first give some primitive explanations of the SOC characteristics through the ratio $I(k)$ of the total load demand over the network total transmission capability.

Figure 12.13 (a) shows the ratio I of the total active power load demand to the network effective transmission capability changing in time. The way in which I converges confirms the self-organization behavior. The lower figure shows the corresponding percentage of load shedding which represents the scales of the blackouts. The generation and load grow gradually in this process of 2000 days.

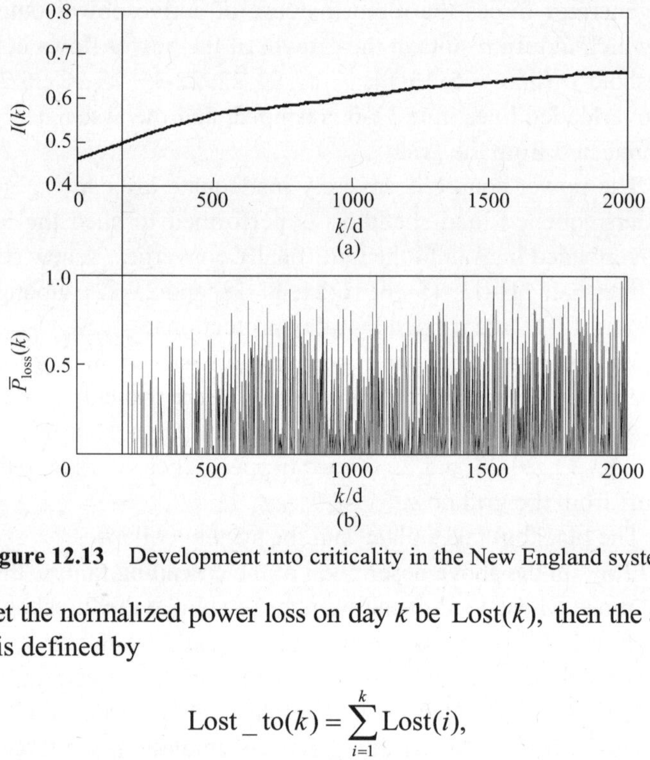

Figure 12.13 Development into criticality in the New England system

Further, let the normalized power loss on day k be $\mathrm{Lost}(k)$, then the accumulated power loss is defined by

$$\mathrm{Lost\_to}(k) = \sum_{i=1}^{k} \mathrm{Lost}(i), \tag{12.44}$$

which is the sum of the power loss from the first day to the kth day during the evolution.

Remark 12.4 Computation time. The calculation of transient constraints is performed through an alternate iteration of solving the optimization problem, calculating the transient stability margin index and associated sensitivities, and validating the obtained results. In one round, it takes about 20 seconds to assess the transient stability index and its associated sensitivities by a desktop PC with a 2.0 GHz processor using the Matlab simulation platform. The computation time for the OPF in one round is 5 seconds, and that for the validation of the transient constraints is 6 seconds. For example, in the simulation of transient dynamics, there is one round of the OPF, two rounds of the validation of the transient constraints, and three rounds of the assessment of the transient stability margin indices and their sensitivities. So the running of the algorithm takes altogether

$5+6\times 2+20\times 3=77$ seconds. In addition, another one second is needed to simulate the fast dynamics. So a SOC simulation for a long-term evolution, e.g. 2000 days, takes about 16 hours.

Remark 12.5 Compared with the test system in [25], the number of the buses in the 39-bus system might not seem to be large enough. However, the system complexity of the proposed blackout model with the OTS for the 39-bus system is much greater than those used in [25]. A set of 20 differential equations is used to simulate the 39-bus system, and this is the main source of complexity. Although the proposed model takes a computation time that is 4 times longer than that for the improved OPA model, it can deal with the transient stability by obtaining the CUEP and the approximate boundary of the stability region. To simulate the SOC behavior in a longer process, e.g. 2000 days, the needed computation time is 15 times longer than that for the improved OPA model. Such costs are justifiable considering the capability of simulating cascading failures that are caused by the transient instability. We are currently working on some further development for the proposed model to reduce the computation time in order to make it applicable to systems with larger scales.

From Fig. 12.14, one can see that after day 170, the growth rate of the accumulated power loss increases. This also indicates that the probability for large-scale blackouts grows rapidly[19].

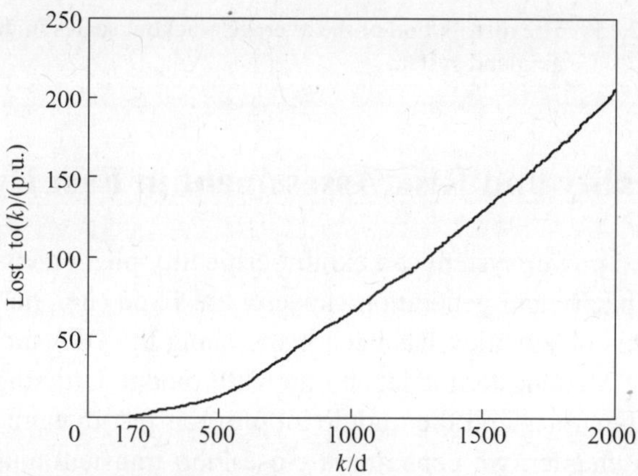

Figure 12.14 The accumulated power loss during the evolution of the New England system

Now we define the average power loss on day k by

$$\text{Lost\_av}(k)=\frac{\sum_{i=k+1-N}^{k}\text{Lost}(i)}{N} \tag{12.45}$$

The main reason for defining the average power loss is to identify the scales of the blackouts when the system is evolving towards criticality as shown in Fig. 12.13. By averaging every N points in the sequence of the blackout scales, which is one way of filtering, one can find the trend of the scale of the blackout. Figure 12.15 shows that the average power loss increases rapidly after the 170th day. Here, we have chosen $N=50$. On day 700, the average power loss reaches a certain level and from that day on, this level is maintained while a slowly rising trend emerges with oscillations in subintervals.

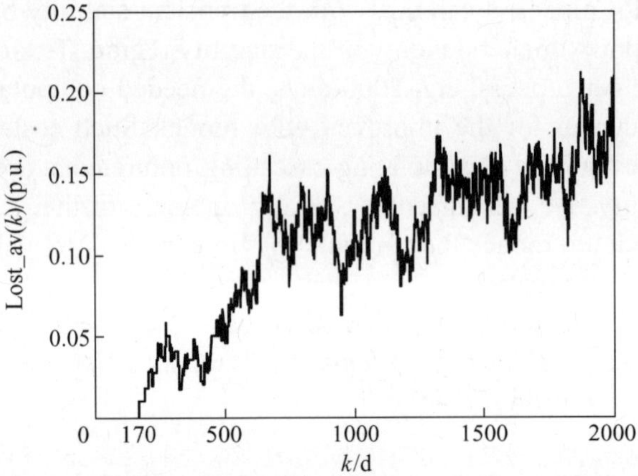

Figure 12.15 The distribution of the average blackout scales in the New England system

12.6 Criticality and Risk Assessment in Fast Dynamics

An interconnected power system can exhibit criticality phenomena even when its topology, line capacity and generation capacity are fixed (i.e., no slow dynamics are involved). This observation has been made using the OPA model, the hidden failure model, the Manchester model and the OPF model. Utilizing the inner loop of the transient dynamics and the middle loop of the fast dynamics in the model discussed in this chapter, we construct a close-loop transient and fast dynamics model as shown in Fig. 12.16. With this model, we can analyze statistically the load loss in blackouts at different load levels and further study the criticality in fast and transient dynamics of power systems. The setting of the parameters is as follows:

The overload threshold $\alpha = 0.7$;
The probability for tripping overloaded lines $\beta = 0.3$;
The probability for large disturbances $\xi = 1$;
The fault clearing time $t_F = 0.16$ s.

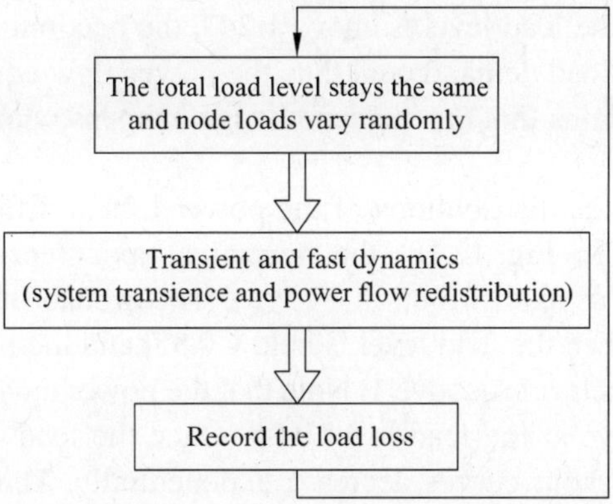

Figure 12.16 Diagram for the statistical analysis of the criticality in fast dynamics

We consider the New England system shown in Fig. 12.1 and set the initial load level to be the base value 1. The load level k_L is defined to be the factor that is multiplied to all the loads in the system. The load level is chosen to grow from 0.6 to 1.3 with a step size of 0.05. For each load level, we run 500 times simulations on fast dynamics. In such simulations, to be closer to the real scenario, we allow each load to fluctuate randomly as long as the total load level is maintained.

Figure 12.17 shows how the average load loss $\overline{P}_{\text{loss}} = P_{\text{loss}} / P_{\text{demand}}$ changes with the load level k_L. $\overline{P}_{\text{loss}}$ is relatively small when the load level is below 0.85, and rises abruptly when the load level is between 0.85 and 1.2. Once the load level goes beyond 1.2, the system collapses and the power flow cannot converge. The maximum total generation output is 7670 MW. When the load level is 1, i.e. the total load is 6151 MW, the ratio of the maximum generation output to the total

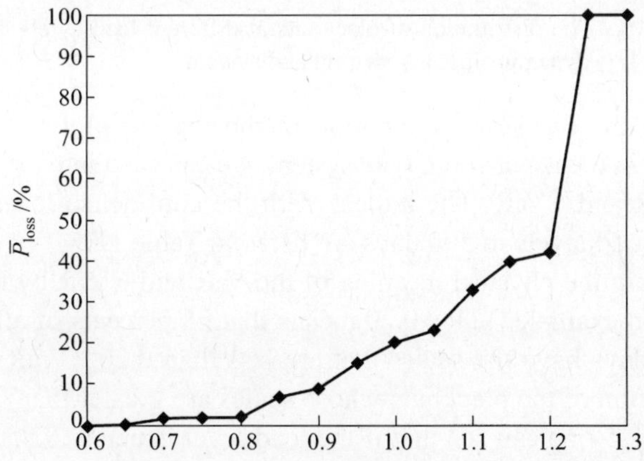

Figure 12.17 The criticality in the fast dynamics in the New England system

load is 1.247. Once the load level is above 1.247, the maximum generation output is less than the total load demand, and thus the power flow equations do not have a solution. This explains that the sudden change happens when the load levels is between 1.2 and 1.25.

We further study the distributions of the power loss at different load levels in the fast dynamics. In Fig. 12.18, the horizontal axis represents $\overline{P}_{\text{loss}}$ and the vertical axis represents its distribution $F(\overline{P}_{\text{loss}})$. The probabilities for blackouts of all sizes are small when the load level is below 0.85, and the probability increases rapidly when the load level is above 1. Note that the power law tail (long tail) starts to appear in the curve at the load level 0.85. Once the load level goes between 1.0 and 1.2, the blackout curves decrease exponentially. This indicates that the load level of 0.85 is a critical characteristic value for system catastrophes.

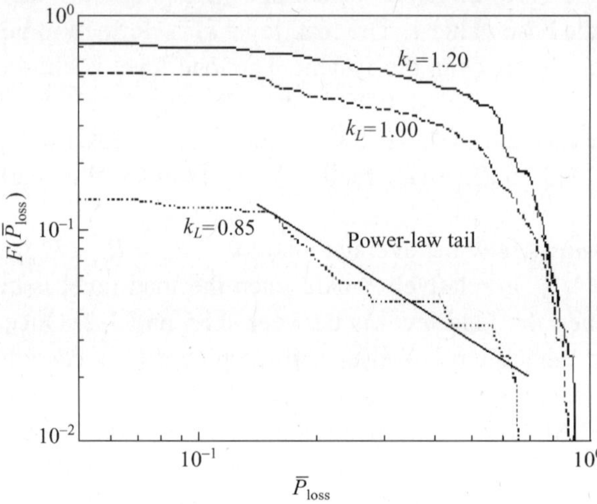

Figure 12.18 The distribution of blackouts at different load levels in the fast dynamics in the New England system

Transforming the blackout distribution in the log-log plot to the probability density function in the usual coordinate system, we can calculate the corresponding risk indices VaR and CVaR. The indices with the confidence level chosen to be 0.95 at various load levels in 500 days are listed in Table 12.5.

We now explain the physical meaning of the VaR and CVaR by taking the load level of 0.6 as an example. VaR = 0.20 means that 95 percents of all the blackouts during the 500 days have the scales that are less than 0.20. CVaR = 0.27 implies that the expectation of the blackouts whose scales are greater than 0.20 is 0.27.

From Table 12.5 one can see that if the load level is below 0.85, both the VaR and CVaR are small. Once the load level exceeds 0.85, the two risk assessment indices jump, as shown in Fig. 12.19.

Table 12.5 The risk indices at different load levels

Load level	VaR ($\sigma = 0.95$)	CVaR ($\sigma = 0.95$)
0.60	0.20	0.27
0.65	0.19	0.272
0.70	0.21	0.244
0.75	0.25	0.32
0.80	0.23	0.302
0.85	0.59	0.714
0.90	0.55	0.694
0.95	0.63	0.77
1.00	0.65	0.782
1.05	0.64	0.732
1.10	0.69	0.826
1.15	0.67	0.764
1.20	0.66	0.738

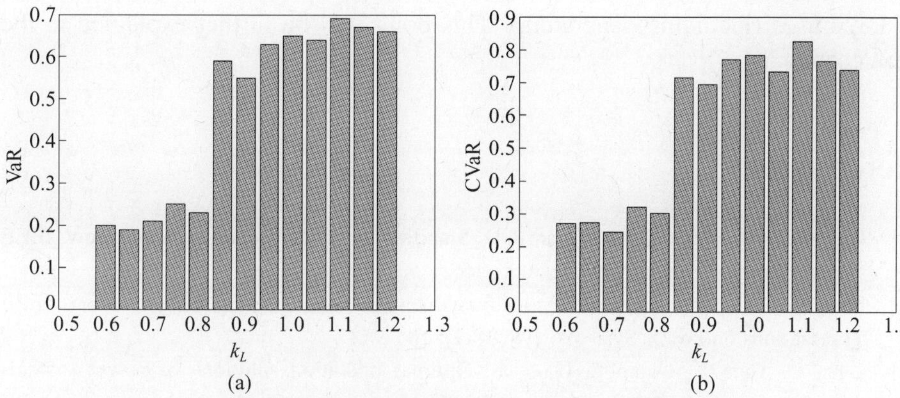

Figure 12.19 Risk indices at different load levels

Therefore, we conclude that the load level 0.85 is a critical characteristic value, which agrees with the conclusions drawn from Fig. 12.18.

Remark 12.6 In this chapter, we use the following equation, not Eq. (3.35) to calculate CVaR

$$\text{CVaR} = \frac{1}{1-\sigma}\int_{\text{VaR}_\sigma}^{\infty} xp(x)\mathrm{d}x$$

12.7 Notes

(1) Similar to the blackout models based on DC or AC power flows in the previous chapters, in this chapter we have constructed a blackout model using the OTS.

This model has been applied to simulate cascading failure processes, and analyze the criticality in the fast dynamics and the macroscopic SOC characteristics. This model is also suitable to simulate more comprehensively and accurately the triggering events and the physical evolution processes of cascading failures. It also takes into consideration the preventive and emergency control actions, which makes it possible to evaluate the effects of human operators' control actions during blackouts.

(2) There are two main shortcomings for the blackout model using the OTS. One is that the analysis, simulation and computation based on it can only incorporate predefined contingencies. As a result, this model is difficult to be generalized for the online analysis of power systems' security levels. The other is that in order to obtain the transient stability index, a couple of complex problems have to be solved first, which include the CUEP, and the quadratic approximation of the boundary of the transient stability region. So the model involves too much computation to be applied to large-scale power systems.

(3) One advantage of the blackout model constructed in this chapter lies in the planning of the generation and grid development, in particular, the identification of key buses (including generators). This point will be further explained in the next chapter.

References

[1] Gan D, Thomas R J, Zimmerman R D. Stability-constrained optimal power flow. IEEE Transactions on Power Systems, 2000, 15(2): 535 – 540

[2] Fouad A A, Tong J. Stability constrained optimal rescheduling of generation. IEEE Transactions on Power Systems, 1993, 8(1): 105 – 112

[3] Chen L, Tada Y, Okamoto H, et al. Optimal operation solutions of power systems with transient stability constraints. IEEE Transactions on Circuits and systems, Part I: Fundamental Theory and Applications, 2001, 48(3): 327 – 339

[4] David A K, Lin X. Dynamic security enhancement in power-market systems. IEEE Transactions on Power Systems, 2002, 17(2): 431 – 438

[5] Scala M L, Trovato M, Antonelli C. On-line dynamic preventive control: an algorithm for transient security dispatch. IEEE Transaction on Power Systems, 1998, 13(2): 601 – 610

[6] Xue A C, Mei S W. A new transient stability margin based on dynamic security region and its applications. Science in China (Series E), 2008, 51 (6): 751 – 760

[7] Chiang H D, Hirsch M W, Wu F F. Stability regions of nonlinear autonomous dynamical systems. IEEE Transactions on Automatic Control, 1988, 33(1): 16 – 27

[8] Chiang H D, Wu F F, Varaiya P P. Foundation of direct methods for power systems transient stability analysis. IEEE Transactions on Circuits and Systems, 1987, 34(2): 160 – 173

[9] Cheng D, Ma J. Calculation of stability region. Proceedings of the 2003 (42nd) IEEE Conference on Decision and Control, Maui (Hawaii, USA): 2003(6): 5615 – 5620

[10] Xue A, Wu F F, Ni Y X, et al. Power system transient stability assessment based on the quadratic approximation for the stability region. Proceedings of the IEEE/PES Transmission and Distribution Conference, Asian Pacific, Dalian (China), 2005: 1 – 6

[11] Yu Y X, Wang C S. Power System Stability Theory and Methods. Beijing: Science Press, 1999 (in Chinese)

[12] Chiang H D, Chu C C. Theoretical foundation of the BCU method for direct stability analysis of network-reduction power system models with small transfer conductances. IEEE Transactions on Circuits and Systems, Part I: Fundamental Theory and Application, 1995, 42(5): 252 – 265

[13] Cheng D. Matrix and Polynomial Approach to Dynamic Control System. Beijing: Science Press, 2002

[14] Xue A C, Hu W, Mei S W, Ni Y X, et al. Comparison of the linear approximations for the dynamic security region of network-reduction power systems. 2006 IEEE Power Engineering Society General Meeting, Jun. 2006, Montreal, Canada

[15] Athay T, Podmore R,Virmani S. A practical method for direct analysis of transient stability. IEEE Transaction on Apparatus and Systems, 1979, 98(2): 573 – 584

[16] Cheng D Z, Ma J, Lu Q, et al. Quadratic form of stable sub-manifold for power systems. International Journal of Robust and Nonlinear Control, 2004, 14 (9 – 10): 773 – 788

[17] Laufenberg M J, Pai M A. A new approach to dynamic security assessment using trajectory sensitivities. IEEE Transactions on Power Systems, 1998, 13(3): 953 – 958

[18] Hiskens I A, Pai M A. Trajectory sensitivity analysis of Hybrid System. IEEE Transaction on Circuits and Systems, Part I: Fundamental theory and Applications, 2000, 47(2): 204 – 220

[19] Mei S W, Wu S Y, Zhang X M, et al. Power system blackout model with transient constraints and its criticality. European Transactions on Electrical Power, 2011, 21(1): 59 – 69

[20] Sarshar N, Roychowdhury V P. Multiple power-law structures in heterogeneous complex networks. Physical Review E, 2005, 72(2): 026114

[21] Carreras B A, Lynch V E, Dobson I, et al. Modeling blackout dynamics in power transmission networks with simple structure. Proceedings of the 34th Annual Hawaii International Conference on System Sciences, Jan. 2001: 719 – 727

[22] Phadeke A G, Thorp J S. Expose hidden failures to prevent cascading outages. IEEE Computer Application in Power, 1996, 9(3): 20 – 23

[23] Nedic D P, Dobson I, Kirschend S, et al. Criticality in a cascading failure blackout model. Proceedings of the 15th Power Systems Computation Conference, Liege, Belgium, Aug. 2005

[24] Sun Y Z, Cheng L, Liu H T. Power system operating reliability evaluation based on real-time operating state. Power System Technology, 2005, 29(15): 6 – 12 (in Chinese)

[25] Mei S W, He F, Zhang X M, et al. An improved OPA model and blackout risk assessment. IEEE Transactions on Power Systems, 2009, 24(2): 814 – 823

Chapter 13 Applications to Generation Expansion Planning and Power Network Planning

This chapter utilizes the OTS model to construct a general method to identify critical lines and buses in power grids. This leads to plannings for generation expansion and power network development. Since parameters of the OTS model may affect cascading failures and large-scale blackouts, we discuss the rules to set the line capacity growth rate as well as the probability for tripping overloaded lines.

We have discussed four different blackout models in the previous chapters, namely those based on DC power flows[1], AC power flows[2], reactive power/voltage characteristics[3] and the OTS. A close look from both the mathematical and the physical (engineering) points of view reveals that the first three models are all special cases of the last one using the OTS. Since the OTS model incorporates the system's transient, fast and slow dynamics on three different time scales, it is advantageous to use this model to analyze the macroscopic and microscopic SOC behaviors in power grids, and then use SOC theory to solve the problems arising in the area of generation expansion planning and power network planning. The existing solutions to generation expansion planning and power network planning have various shortcomings and need to be improved. Towards this end, we present in this chapter a general framework to make decision support for generation expansion planning and power network planning based on the OTS model.

13.1 Shortcomings of Existing Methods

In order to meet the load demand, the generation expansion planning and power network planning are carried out using the information provided by load forecasting. The generation expansion planning and power network planning are in fact inseparable and should be viewed as a single entity[4]. However, since the two have different focuses and cannot be easily solved at the same time following a general approach, they are usually treated independently, and then optimized in coordination when necessary.

The generation expansion planning is an important component of the power system development plannings. With the growths of electricity loads and capacities of generation units, coupled with the diversification of generators with various primary energy resources, the power supply structure has become increasingly complicated. Consequently, how to determine a reasonable power supply structure and how to cooperatively and effectively utilize various types of power resources are becoming urgent problems to solve. The power network planning, on the other hand, is to determine when and where to build which type of transmission lines with the appropriate numbers of loops to satisfy the requirements for upgrading transmission capacities in planning cycles, and at the same time to lower the transmission cost under various performance indices[4].

What follows is the WASP (Wien Automation System Planning Package) model[5] for the generation expansion planning. The objective function is

$$\min \quad \mathrm{PVC}_j = \sum_{t=1}^{T} (\overline{I}_{j,t} - \overline{S}_{j,t} + \overline{F}_{j,t} + \overline{M}_{j,t} + \overline{O}_{j,t}) \tag{13.1}$$

where the meanings of the notations are listed in Table 13.1.

Table 13.1 Notations in the objective function of the WASP model

Notation	Description
PVC_j	Present value of the total cost of scheme j
$\overline{I}_{j,t}$	Investment cost
$\overline{S}_{j,t}$	Salvage value of investment
$\overline{F}_{j,t}$	Fuel cost
$\overline{M}_{j,t}$	Maintenance cost
$\overline{O}_{j,t}$	Blackout cost
Subscriptions j and t	The cost in year t related to scheme j
Bar "—"	The current value of the cost in year t at a given time after conversion using the given discount rate
T	The total number of years in the planning period, also known as the level years

Obviously, the optimal scheme is the one with the minimum present value of the total cost PCV_j.

The constraints in the WASP include the power balance constraints and the reliability constraints. Let $K_{j,t}$ be the number of generators that operate in the planning scheme j in year t. Then it holds that

$$K_{j,t} = K_{j,t-1} + A_{j,t} - R_{j,t} + U_{j,t} \tag{13.2}$$

where $A_{j,t}$ is the number of the expanded generation units in year t according to the mandatory planning in scheme j, $R_{j,t}$ is the number of the generation units that are estimated to ware out in year t in scheme j, and $U_{j,t}$ is the number of the expected new generation units in year t in scheme j satisfying $U_{j,t} \geqslant 0$. Here, $A_{j,t}$ and $R_{j,t}$ are given, and $U_{j,t}$ is the unknown, which is called the system layout. Reliability constraints require that each system's layout must be implemented in such a way that the probabilities for load losses in different periods of time are always less than a preset reliability value.

From the mathematical description of the above planning problem, it is clear that the generation expansion planning and the power network planning problems are complex combinatorial optimization problems or mixed integer programming problems, which are difficult to solve directly[6–8]. To add one generator or to build one transmission line is always a 0-1 discrete change. However, the OTS model uses difference equations to describe the growth in the generator capacity and load level and uses probability theory to describe the gradual growth in the transmission line capacity, and thus circumvents the difficulties for solving the 0-1 integer programming problems.

In addition, in the traditional power system planning, the system is usually checked by deterministic methods, such as the $N-1$ calibration method and the contingency ranking method with some deterministic load patterns. In reality, power loads change with time frequently and may differ greatly at different times. Therefore, even when the system satisfies the $N-1$ calibration under certain conditions, it may still fail the test when the amount or the distribution of load changes, and this may even lead to network islanding and other serious faults in the system and cause blackouts and load losses. Furthermore, the slow dynamic process is another important cause for cascading failures. The OTS blackout model can, to a certain extent, deal with the above problems because it simulates and analyzes cascading failures and blackout processes when taking the system evolution into consideration. Hence, the OTS model can provide more information for the generation expansion planning and power network planning, and in turn becomes an important supportive tool complementing the traditional means. At the same time, the system-level risk assessment indices VaR and CVaR[9–11] discussed in Chapter 3 are also available to evaluate quantitatively the effects of different planning schemes and thus compare their advantages and disadvantages.

13.2 Identification of Critical Lines

In the slow dynamics of the OTS model, we need to update all the lines that have been tripped off in the previous day in order to improve the transmission capacity. The total outage times $\mathrm{Ncut}_{j,k}$ for line j during the system evolution up to day k can be calculated by dividing line j's transmission capacity $F_{j,k+1}^{\max}$ on day k by its transmission capacity $F_{j,1}^{\max}$ on day one, that is

$$\mathrm{Ncut}_{j,k} = \frac{\ln(F_{j,k+1}(k+1)/F_{j,1}(1))}{\ln(\mu)} \tag{13.3}$$

where μ is the growth rate for the transmission capacity.

By the examination of the total outage times for all the lines in the network, one can tell which area in the network is weak and which lines are prone to induce blackouts. Take the New England 10-generator 39-bus system in Chapter 12 as an example. There are altogether 46 lines, and we rank them according to their outage times in ascending orders over the periods of 365 days (one year), 1095 days (three years) and 2000 days respectively (denoted by j_R). The rankings are shown in Fig. 13.1.

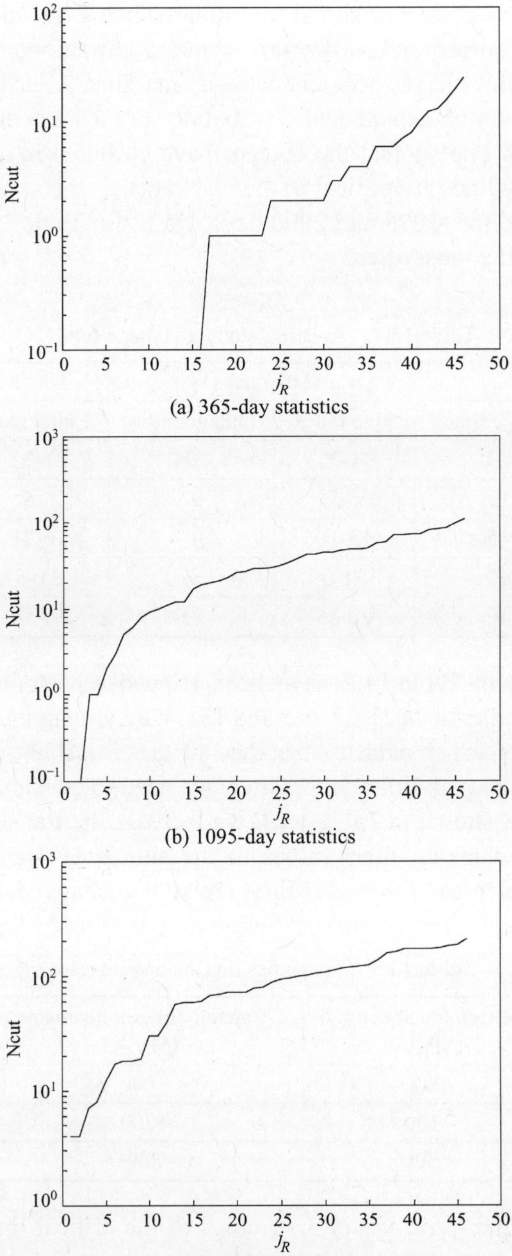

(a) 365-day statistics

(b) 1095-day statistics

(c) 2000-day statistics

Figure 13.1 Line ranking according to outage times

From Fig. 13.1 it can be seen that at the end of the 365th day, the total outage times of the 16 lines are zero, which implies that the line capacities can meet the growing demand in a year. Checking up the system evolution curves for the 1095-day and 2000-day processes, one can find that there is a turning point around the spots with the horizontal coordinates being 17. Before the turning points, the curves correspond to the safe working conditions before the end of day 365, when all lines have sufficient capacity margins. After the turning points, the curves agree with the exponential distribution (Fig. 13.1 is in semi-logarithmic coordinates), which implies that the outages have clustered in a small number of lines and that a few lines are critical in the blackouts.

In the periods of 365, 1095 and 2000 days, we list in Table 13.2 five lines that have been tripped the most often.

Table 13.2 Accumulated line outage times

365 days		1095 days		2000 days	
Line index	Outage times	Line index	Outage times	Line index	Outage times
21-22	24	21-22	108	21-22	209
6-5	19	6-5	97	6-5	185
13-14	15	13-14	84	16-21	180
16-19	13	31-6	82	31-6	173
10-13	10	22-35	81	13-14	172

The lines listed in Table 13.2 have been tripped repeatedly in the network evolution. In particular, lines 21-22, 6-5 and 13-14 are among the top 5 in all of the three chosen periods, which indicates that they are the critical lines causing blackouts and are the system's weak links. In practice, we need to consider upgrading these lines in priority. As shown in Table 13.3, we increase the transmission capacities of these three lines taking their values in the simulated 365-day evolution as reference. The capacities of the other lines are kept unchanged.

Table 13.3 Transmission capacity upgrade

Line index	Original capacity /(MV·A)	Capacity after adjustment /(MV·A)	Growth rate
21-22	900	1000	11.1%
6-5	1100	1200	9.1%
13-14	750	800	6.67%

After adjusting the transmission capacities of the critical lines while keeping the other simulation conditions fixed, we simulate again for a 2000-day process using the OTS model. The simulation results are shown in Figs. 13.2 – 13.4.

Fig. 13.2 shows how $I(k)$ and $\overline{P}_{\text{loss}}(k)$ change with time, where $I(k)$ is the ratio of the total load demand to the total transmission capability and $\overline{P}_{\text{loss}}(k)$ is the relative value of load shedding.

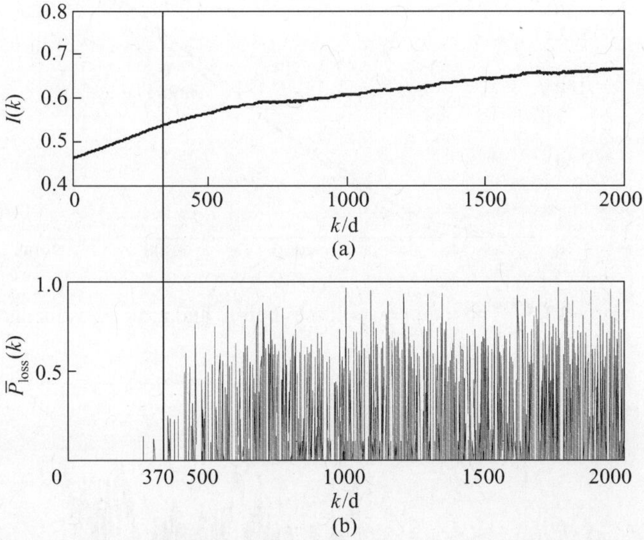

Figure 13.2 The process of approaching criticality after the adjustments of the critical lines

Figure 13.2 shows that $I(k) \approx 0.542$ when the system evolves to day 370, and this value can be taken as a critical characteristic quantity for the system reliability (before day 370, almost no blackout takes place in the system). Comparing to the original system in Fig. 12.13, we conclude that after the capacities of the three critical lines have been improved by about 10%, the system's critical time for blackouts has been postponed by 200 days, and $I(k)$ increases from 0.496 to 0.542. By increasing the capacities of the three critical lines to the level of those of the original system at the end of the first year, we ensure there is no blackout in that year. Therefore, increasing the capacities of the critical lines has a clear effect on preventing blackouts. It also confirms the correctness and effectiveness for using the OTS model to identify critical lines.

Next, we look at the total amount of power loss $\text{Lost\_to}(k)$ before and after the adjustments. The comparison results are shown in Fig. 13.3. It is clear that after the adjustments, the starting time of blackouts is postponed by 200 days, and in the first 1100 days the scales of blackouts after the adjustments are always less than those before the adjustments. However, the growth rate of the total amount of blackouts after the adjustments is higher than that before the adjustments.

Figure 13.4 shows how the average power loss distribution $\text{Lost\_av}(k)$ changes with time k.

Comparing Fig. 13.4 with Fig. 12.15, we know that the blackout time has been

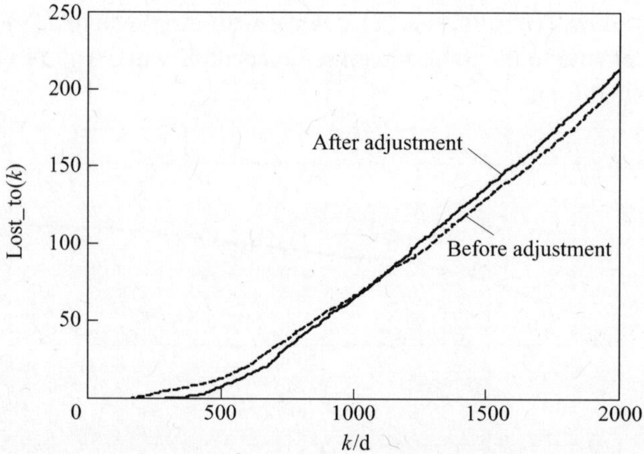

Figure 13.3 The total power losses before and after adjustments

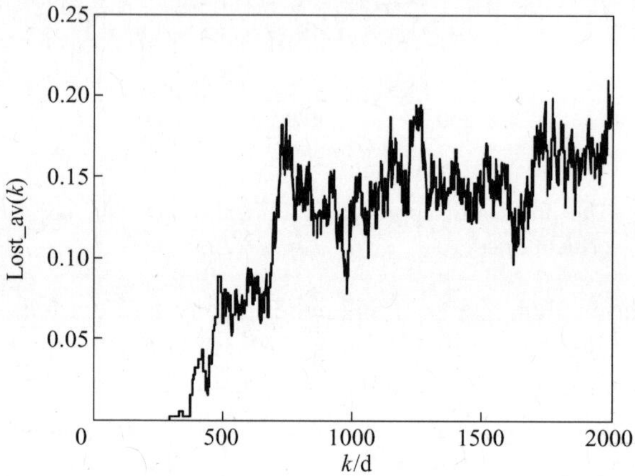

Figure 13.4 The average blackout distribution after adjustment

postponed after the adjustments, and the transition from having almost no blackout to experiencing large-scale blackouts is also postponed. However, the scales of blackouts are always oscillating in the interval [0.10,0.20].

The distributions of the blackout scales, denoted by $F(\overline{P}_{\text{loss}}(k))$, for the system evolution of 2000 days before and after the adjustments are shown in log-log coordinates in Fig. 13.5. From the figure, it can be seen that the frequencies of the occurrence of various blackouts decrease.

The risk indices before and after the adjustments are shown in Table 13.4.

From the examination of the VaR, we know that the system's risk indices are the same before and after the adjustments. But from the examination of the CVaR, it is clear that the average level of the system loss after the adjustments is lower than that before the adjustments. Said differently, the system after the adjustments is of lower risk for blackouts. Here, the reason that the VaR index has not been able to identify this change is that the loss at the tail of the VaR measurements is not sufficient, and this has been discussed in Section 3.6.4.

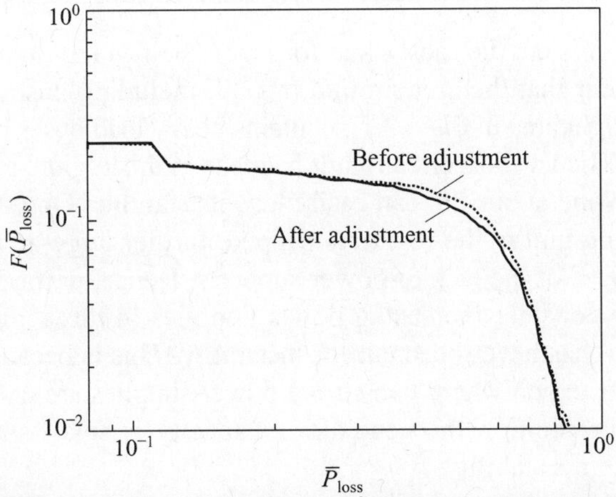

Figure 13.5 Load loss distributions before and after the adjustments

Table 13.4 System risk indices before and after the adjustments

Risk index	VaR ($\sigma = 0.95$)	CVaR ($\sigma = 0.95$)
Before adjustment	0.67	0.0413
After adjustment	0.67	0.0388

13.3 Identification of the Critical Power Supply

The OTS model has considered the system transients, so it can describe the whole system dynamic process that begins with a large disturbance (e.g. a three-phase short circuit fault), and then develops into the loss of the system transient stability, further progresses with the cutting of generators, load-shedding, power flow redistribution and tripping of overload transmission lines, until in the end finishes with cascading failures. In simulations, by counting the times for which a generator has been cut off in the transient process because of the loss of transient stability, and combining with the analysis of the network topology on how generators and loads are distributed, we can check whether the power supplies are appropriately distributed in the network.

Now we take the New England 10-generator 39-bus system discussed in Chapter 12 as an example and count the total cut-off times for each generator in the transient process. The results are listed in Table 13.5.

From Table 13.5, one can see that in the evolution of the 2000 days, there are altogether 113 times for emergent generator tripping control, among which the

Table 13.5 The total cut-off times for each generator

Generators	30	31	32	33	34	35	36	37	38	39
Cut-off times	0	6	5	2	13	5	3	9	55	15

generator at bus 38 is cut the most often for nearly half of the times (48.7%). From Fig. 12.1, it is clear that the area around bus 38, including buses 15 to18 and 26 to 29 (the upper right area in Fig. 12.1), contains heavy load buses but with stringent power supplies. When a short circuit fault happens in this region, the only generator in the region, the one at bus 38, can easily lose its stability. Once this generator as the main supplying unit in the region is tripped, further large-area blackouts may be triggered because of the lack of power supports. Hence, in the system planning, it is necessary to consider increasing power supplies in this region. At the same time, generator 30 has never been cut for instability. This is because this generator is in the upper left region where two strong power supplies are positioned at buses 37 and 39, and in addition the generator's capacity is small and thus has little effect on the system's stability.

According to the above analysis, the upper right region in Fig. 12.1 is an area with heavy loads and in short of power supplies. In order to improve the system's power supply reliability and reduce the risk for blackouts, we should add power supply buses to the region. Keeping the rest of the original system and the parameters fixed, we add generator 40 that is connected to bus 27 through a new line 27-40. The parameters of generator 40 are the same as those of generator 30, and the parameters of line 27-40 are the same as those of line 2-30. The system with the added generator 40 and line 27-40 is shown in Fig. 13.6.

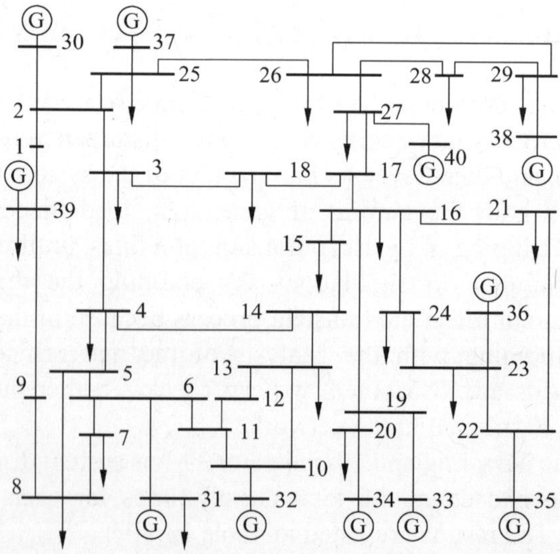

Figure 13.6 The New England system after the power supply adjustment

With the fixed control parameters, we simulate the macroscopic self-organized evolution of the adjusted system for 2000 days. The simulation results are shown in Fig. 13.7.

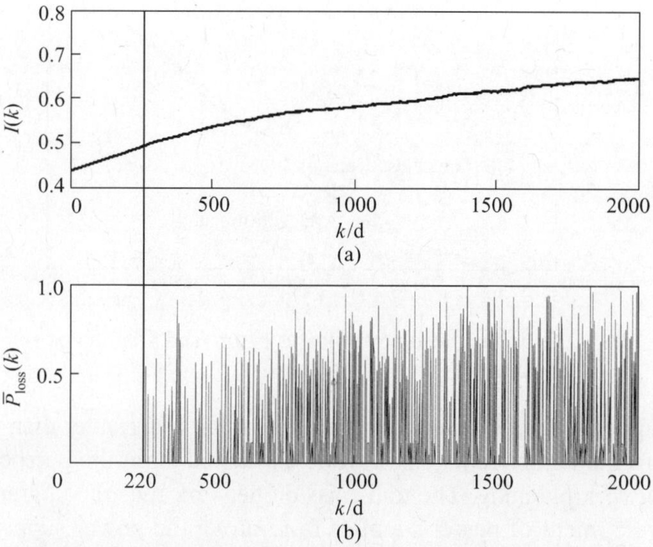

Figure 13.7 The adjusted system develops into the criticality

In Fig. 13.7, there is no fault or blackout during the first 220 days; during the period of days 220 – 2000, the scale of blackouts keeps increasing. The value of $I(k)$ on day 220 is about 0.495 and this can be taken as a critical value for the reliable operation of the system. It can also be seen from the figure that after the power supply adjustment, the time for the occurrence of large-scale blackouts is postponed by about 50 days compared to the system before the adjustment (Fig. 12.13).

We further examine the total amounts of power loss before and after the power supply adjustments, and the results are shown in Fig. 13.8. Within 2000 days, after the power supply adjustment, the amount of the accumulated system power loss is always less than that before the power supply adjustment.

Figures 13.7 and 13.8 imply that the power supply reliability has been greatly improved after adding one more power supply node to the original system. The critical time for the occurrence of large-scale blackouts is postponed by about 50 days and the accumulated blackout amount is also effectively reduced.

We then check quantitatively the system's blackout risk levels before and after the adjustments by looking at the risk assessment indices VaR and CVaR.

As shown in Table 13.6, after the adjustment of the system power supply, the risk assessment indices VaR and CVaR are smaller than those of the system before the adjustments for all the first three years, and thus the system's power supply reliability has been improved significantly. However, in the 4th and 5th

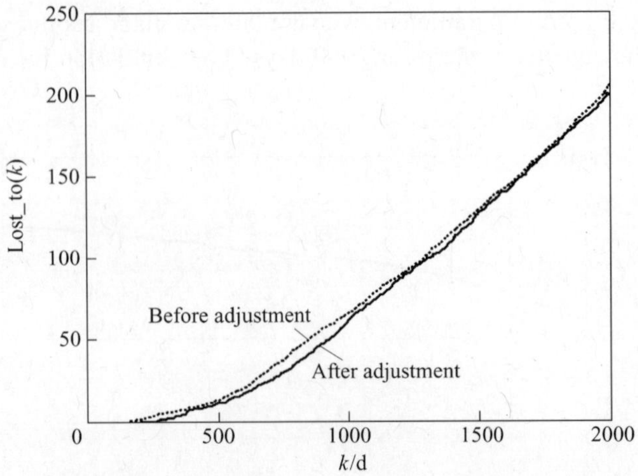

Figure 13.8 The accumulated power losses before and after the power supply adjustment

years, the indices for the system after the adjustment are greater than those of the system before the adjustment. This can be explained from the perspective of the long-term network planning. The transmission network must be upgraded together with the development of power supplies to improve the power supply reliability. In the simulation of this section, only one extra power supply node is added to the system, and the needed transmission grid upgrade has not been considered simultaneously.

Table 13.6 Risk indices before and after the power supply adjustment

Period for analysis	VaR ($\sigma = 0.90$)		CVaR ($\sigma = 0.90$)	
	After adjustment	Before adjustment	After adjustment	Before adjustment
1-365 (the first year)	0.43	0.45	0.0543	0.0543
366-730 (the second year)	0.52	0.52	0.0653	0.0654
731-1095 (the third year)	0.59	0.60	0.0728	0.0738
1096-1460 (the 4th year)	0.62	0.59	0.0743	0.0670
1461-1825 (the 5th year)	0.65	0.60	0.0777	0.0756

13.4 Adjustment of Transmission Capacity Growth Rate

In this section, we discuss the impact of upgrading transmission lines on blackouts in long-term evolutions. In Fig. 13.9, we show the blackout distribution characteristic curves for a period of 2000 days with the growth rate μ being 1.005 and 1.007 respectively.

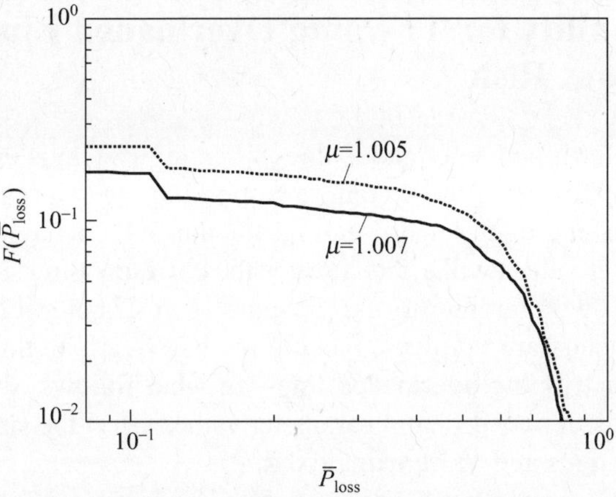

Figure 13.9 Load loss distributions with different growth rates for transmission capacities

As shown in Fig. 13.9, by strengthening the upgrades on faulted lines, namely increasing μ, one can effectively control various blackouts and thus reduce the system loss and prevent catastrophes. The risk assessment indices are calculated under different μ, and the results are listed in Table 13.7.

Table 13.7 Risk assessment indices with different growth rates for transmission capacities

Growth rate for transmission capacities	VaR ($\sigma = 0.95$)	CVaR ($\sigma = 0.95$)
1.005	0.68	0.0413
1.007	0.67	0.0405
1.009	0.65	0.0397

As shown in Table 13.7, the risk indices VaR and CVaR when $\mu = 1.007$ are less than those when $\mu = 1.005$ all the time, and similarly those when $\mu = 1.009$ are still less than those when $\mu = 1.007$. This indicates that increasing μ, namely strengthening the upgrades for faulted lines, can effectively lower the probability for blackouts, and is also of particular importance to reduce the risk for power system catastrophes.

It is worth mentioning that, during the 10 years before the blackout in North America on Aug 14, 2003, the US electric power demand had increased steadily by 30%, while the capacity of high voltage transmission lines had only grown by 15%, which finally forced the power grid to be overly exploited and led to the large-scale blackouts (although in general the US power grid construction requires that the system has to be able to handle the power flow redistribution under multiple faults). This proves that to strengthen the network construction and keep up the generation with respect to the load demand is essential to prevent blackouts.

13.5 Probability for Tripping Overloaded Lines and Blackout Risk

In the OTS model, tripping overloaded lines when the system is under perturbation is an internal mechanism in the system to respond to the external increasing load demand, and also one of the main reasons for the SOC behavior in the system. Different parameter values will affect directly the corresponding simulation results.

In this section, we continue to use Scheme A in Chapter 12 as an example, where two parameters are involved, namely the line overload threshold α and the probability β for tripping overloaded lines. In what follows, we adjust the two parameters to find out how different parameter values affect the simulation results.

Case 1: α changes and β remains fixed.

We take $\beta = 0.3$, the same value as in Chapter 12 and set $\alpha = 0.8$, which is a little bit greater than the value of 0.7 in Chapter 12. Here, we examine the macroscopic long-term self-organized evolution to show the effects of the related parameters on the simulations. We show the simulation results in Fig. 13.10 and Fig. 13.11.

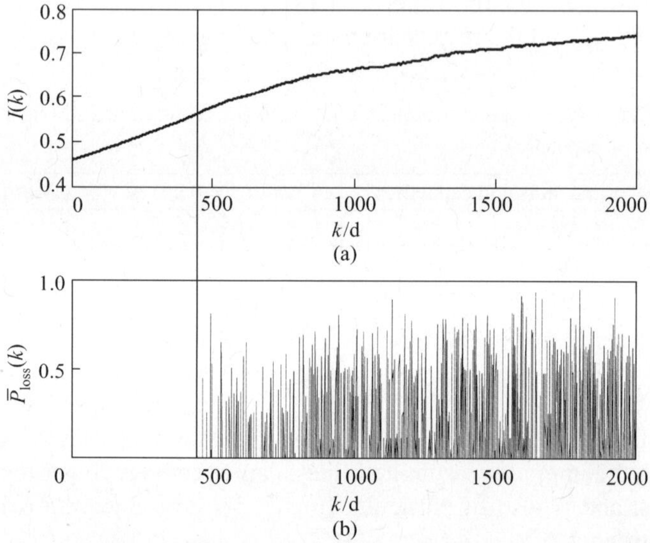

Figure 13.10 System evolutions when $\alpha = 0.8$

From Fig. 13.10, one can see that the critical time for the occurrence of blackouts is on day 447 with the associated critical characteristic $I(k) = 0.565$. In comparison, in Fig. 12.13 ($\alpha = 0.7$), the critical time is on day 170 with the associated $I(k) = 0.496$. So α has increased by $0.8/0.7 \approx 1.14$ times, while the associated critical characteristic has increased by $0.565/0.496 \approx 1.14$ times also.

At the same time, the occurrence of large-scale failures is postponed by 277 days from day 170 to day 447. In this period, the total load has increased by

$1.0005^{277} \approx 1.148$ times. So increasing the value of α is approximately equivalent to increasing the initial capacity of the system with a similar rate. The blackout distributions under different α are shown in Fig. 13.11. From the figure, it is clear that after increasing the value of α, the occurrence of large-scale blackouts is postponed by about 280 days and the accumulated amount of power loss is far more less than that in the original system.

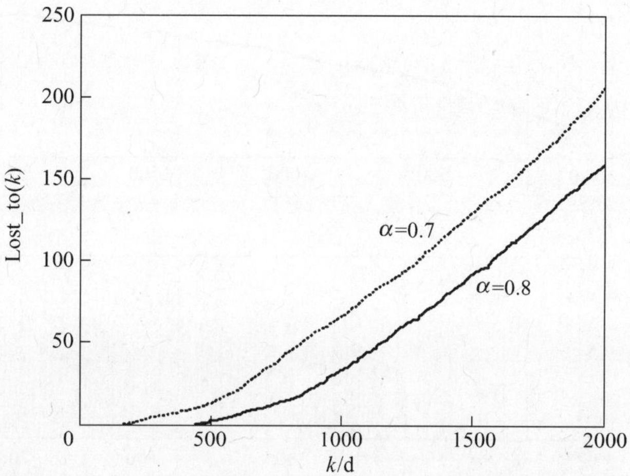

Figure 13.11 The accumulated power loss distributions under different α

Furthermore, we compare the times when the accumulated power loss reaches 0, 50, 100 and 150 p.u. (relative to the total load level of the day). The results are shown in Table 13.8.

Table 13.8 Comparison of the occurrence times of blackouts under different α

Blackout scale (p.u.)	$\alpha = 0.8$	$\alpha = 0.7$	Time difference
0	447	170	277
50	1160	855	305
100	1571	1292	279
150	1932	1659	273

From Table 13.8, one can see that the time differences for the occurrence of blackouts of similar scales are almost the same. The smallest difference is 273 days while the biggest is 305 days, so the variation is less than 10%. This also shows that increasing the value of α is approximately equivalent to increasing the initial capacity of the system with a similar rate. The impact on the simulations can be approximately taken as to equivalently postpone the occurrence times of blackouts of similar scales. Hence, the choice of the value of α needs to be consistent with the system upgrades.

Case 2: α remains fixed and β changes.

We take $\alpha = 0.7$ as in Chapter 12 and set $\beta = 0.15$, which is half of its value in Chapter 12. The simulation results are shown in Fig. 13.12. Comparing Fig. 13.12 with Fig. 12.13, one can see that after the parameter adjustment, the occurrence time for blackouts is postponed slightly by 18 days from day 170 to day 188.

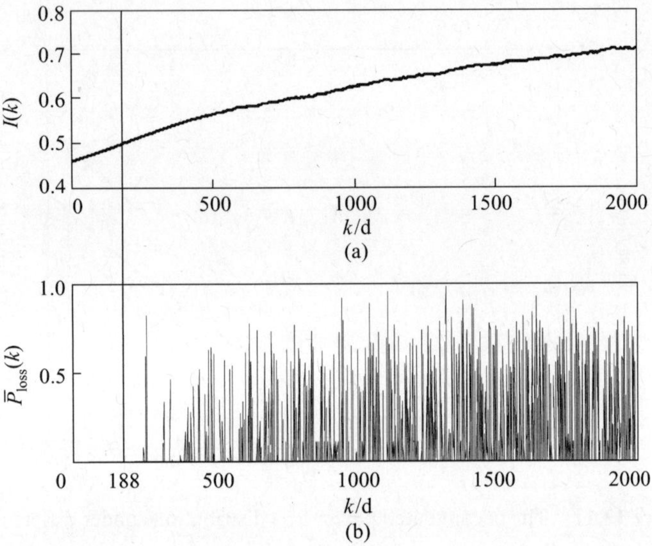

Figure 13.12 System evolutions when $\alpha = 0.7$ and $\beta = 0.15$

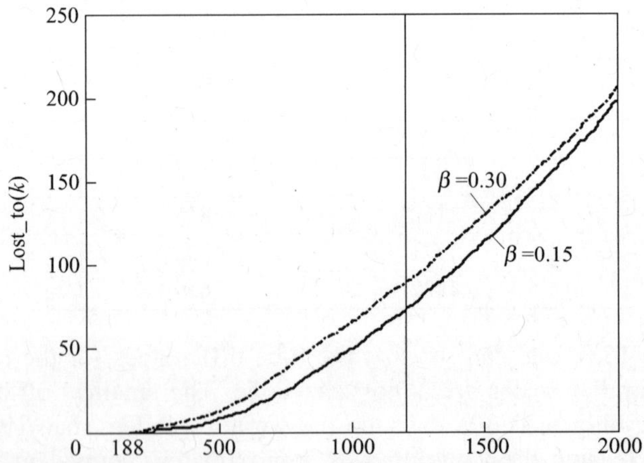

Figure 13.13 The accumulated power loss distributions with different β

The blackout distributions under different values of β are shown in Fig. 13.13. It can be seen that within days 188 – 1200, the gap between the blackout scales for $\beta = 0.15$ and $\beta = 0.30$ increases continuously, which implies that within this period, the decrease of the value of β can effectively prevent blackouts which are usually of small to medium scales. However, during days 1200 – 2000, the gap decreases gradually, which indicates that the decrease of β has disadvantages of preventing long-term threats in forms of large-scale blackouts.

References

[1] Mei S W, He F, Zhang X M, et al. An improved OPA model and blackout risk assessment. IEEE Transactions on Power Systems, 2009, 24(2): 814 – 823

[2] Nedic D P, Dobson I, Kirschend S, et al. Criticality in a cascading failure blackout model. Proceedings of the 15th Power Systems Computation Conference, Liege, Belgium, Aug. 2005

[3] Mei S W, Ni Y X, Wang G, Wu S Y. A Study of self-organized criticality of power system under cascading failures based on AC-OPF with voltage stability margin. IEEE Transactions on Power Systems, 2008, 23(4): 1719 – 1726

[4] Wang X, Mcdonald J. Modern Power System Planning. London. UK: McGraw-Hill, 1994

[5] WASP-Ⅲ Guidebook. International Atomic Energy Agency, 1984

[6] Garver L L. Power generation scheduling by integer programming: development of theory. AIEE Transactions, 1962, 81: 730 – 835

[7] Cote G, Laughton M. Decomposition techniques in power system planning: the benders partitioning method. Electric Power and Energy Systems, 1979, 1(1): 57 – 64

[8] William D D, Thomas R B. Planning for new electric generation technologies: a stochastic dynamic programming approach. IEEE Transactions on Power Apparatus and Systems, 1984, PAS-103(6)

[9] Hull J C. Options, Futures and Other Derivatives (5th Edition). Upper Saddle River, New Jersey: Prentice Hall, Inc, 2003

[10] Morgan J P. Risk metrics: technical document (4th Edition). 1996. http://www.riskmetrics.com

[11] Rockafeller T, Uryasev S. Optimization of conditional value-at-risk. Journal of Risk, 2000, 2(3): 21 – 42

Chapter 14 Applications in Electric Power Emergency Management Platform

In this chapter, to discuss the electric power emergency management platform, we begin with a brief introduction about its background information and the overall structure. Correspondingly, the fundamental principles of the decision support system are also discussed. We use the improved OPA algorithm to design catastrophe evolution models in this chapter since it converges fast and is applicable to large-scale systems. Then the decision support system is constructed for the disaster assessment and disaster prevention evaluation. As case studies, we evaluate the operational risk of the Northeast Power Grid of China under several simulated disasters. The analysis of the affected areas are carried out, based on which disaster prevention plans are made. A power grid disaster forcasting and early-warning model based on AC power flow is also included in this chapter.

Human society has made great achievements in science and technology but is still at risk, sometimes even vulnerable, when facing large-scale emergencies like natural disasters. Therefore, many countries have established public emergency management systems to respond to various disasters due to human or natural causes to reduce casualties and property losses as much as possible. The electric power emergency management system, which copes with power grid large-scale blackouts, is one of the most indispensable and critical components in such public emergency management systems.

The construction of the electric power emergency management platform not only deals with complex modern power systems, but also has to consider factors such as weather conditions, geographical features, transportation limits, available human resources and so on. Hence, without exaggeration, it is an extremely challenging task. On the other hand, we have established different blackout models in the previous chapters and consequently we have been able to describe the dynamic behaviors of large-scale power grids on the fast and slow two different time scales. This has enabled us to analyze quantitatively the probability for potential accidents and their possible scales, which in turn has provided us with powerful decision-making support for the establishment of the eclectic power emergency management platform, especially for the analysis of power grid evolution mechanism and the forecast and early warning for catastrophes.

14.1 Brief Introduction on Electric Power Emergency Management Platform

With the interconnection of regional power grids and the parallel operation of large-capacity generation units, a high-voltage, long-distance and trans-regional transmission pattern has taken shape in China. However, in the mean time, China's power grid construction has been lagging behind its load growth, especially in large- and medium-scale cities because of the difficulty in choosing construction sites, strong resistence during land acquisition and the high cost for removal, resettlement and compensation. As a result, China's power grid structure is still relatively weak and not capable enough to handle all the transmission and distribution demands. In addition, those load centers of large- and medium-scale cities in China are in general in short of dynamic reactive power supports and thus have an uneven reactive power resource distribution. Due to the above reasons, the power system stability problem is becoming increasingly prominent and even considered to be a threat to the secure and stable operation of large-scale power grids. In recent years, catastrophic blackouts and large generation unit accidents have taken place in China and elsewhere[1–5], which indicates that there is still a long way to go to develop the power system safety and control theory to keep up with the increasing scale and complexity of modern power systems. One can safely say that the risk for disasterous events in the Chinese power grids as well as other main power grids in the world cannot be ignored.

Besides faults within a power system, accidents have also been induced by external destructions for China's power grids. Electric power facilities are often damaged or stolen, and unapproved buildings that pose threat againt the power grid safty can often be seen in cities. According to incomplete statistics from the Chinese State Grid Corporation, there are altogether 22 major transmission faults in 2006 (41 faults in 2005) that are caused by building construction projects, car accidents damaging electric poles, theft, deforestation and other external destructions, which accounts for 50% of the power transmission accidents (43.2% in 2005) and 15% of all the accidents. External forces are in fact a serious threat to power grid safety. What needs to be pointed out in particular is that in the past few years, China's power grids have suffered from typhoons, tornadoes, mudslides, icing and other natrual disasters for several times. For example, the 9·26 large-scale blackout in Hainan Province in 2005[1] and the icing accident in China Southern Power Grid in the spring of 2008[2] are both sudden catastrophic incidents in power systems.

Large-scale power grid blackouts have taken place around the world. Some happened in international metropolises in developed countries, and some in cities in developing countries. It seems that the more developed a country or region is, the bigger the losses and impacts of the power grid catastrophe can bring. In fact, no matter whether the region is developed or not and whether the power grid is strong or not, there is always a possibility that large-area blackouts may occur. So it is necessary all the time to take precautious steps before it is too late. For developed international metropolises and central cities, it is quite necessary and urgent to carry out emergence management plannings for large-scale blackouts that take into account the economic losses and social impacts associated with blackouts.

It is a demanding and complex task to deal with sudden large-scale blaclouts, which requires the integration and constant update of various of information for the emergency pre-planning. In addition, the implementation of the plans needs the full cooperation from all the realted departments. To accomplish such tasks, we need to build an electric power emergency management platform as the basis.

14.1.1 Overall Structure of Electric Power Emergency management Platform

The electric power emergency management platform is a security system that exploits tools from the modern computer technology, communication technology, and power system analysis and control technology. It serves to deal with major emergencies and public security accidents, such as severe power production accidents, equipment damage, power supply crisis and natural disasters affecting power production. Its functions include emergency information collection and management, emergency guard, forecast and early warning, dispatching and commanding, decision support, electronic pre-planning, resource management, exercise evaluation and information release, etc[6].

As to system structures, the electric power emergency management platform consists of five systems: the commanding and dispatching system, information management system, resource management system, decision support system and emergency disposition system. The diagram illustrating the relationships between these systems is shown in Fig. 14.1.

The main functions of these systems are explained as follows.

1) Commanding and dispatching system

This system is the highest decision-making component of the whole electric power emergency management platform. Its functions include guarding and training in normal states, activating the emergency plan in alert states and issuing emergency

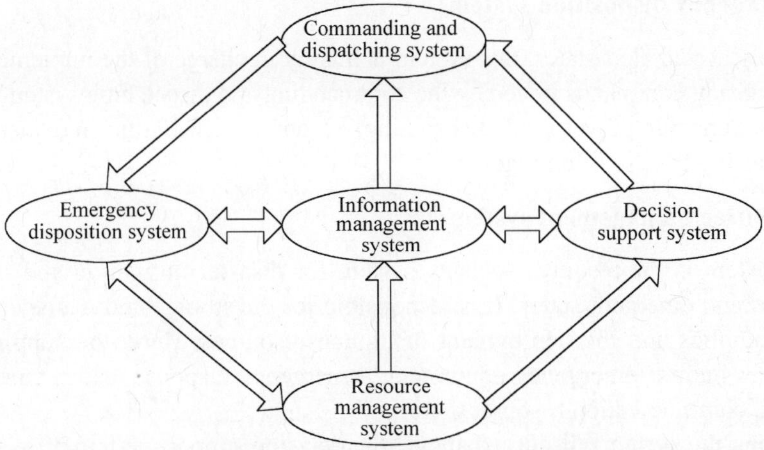

Figure 14.1 The overall frame of the electric power emergency management platform

commands in emergent states. The information of this system comes from the information management system and the decision support system. Through this system, a supervisor can obtain comprehensive information about the disaster, the available resources, the early-warning and the prevention and mitigation actions. He or she can then give specific instructions through the emergency disposition system to dispatch various resources to prevent or minimize the impact of disasters.

2) Information management system

This system is the support system that guarantees the normal operation of the electric power emergency management platform. It is in charge of the communication between various systems and the data management, such as emergency information acquisition and processing, pre-planning management and information broadcasting.

3) Decision support system

This system is an intelligent system that provides decision support for the commanding and dispatching system. It provides forecasting and early warning, and it also supports intelligent decision-making. For forecasting and early warning, it analyzes the risk for crucial equipments to be damaged, assesses the power grid security, forecasts and warns in early stages the scales and locations of possible blackouts, and in the end broadcasts in a timely manner the early-warning signals through the information management system. For supporting intelligent decision-making, it analyzes the overall impact of power grid accidents and natural disasters and correspondingly makes disaster prevention plans and large-scale blackout rescue plans.

4) Emergency disposition system

In emergent and alert states, this system is mainly in charge of the implementation of emergency commands issued by the commanding and dispatching system. It also takes actions for power grid dispatching, resource distributing and instruction broadcasting in case of emergency.

5) Resource management system

This system is the resource security system for disaster mitigation and relief in the alert and emergent states. It is responsible for the storage and management of relief supplies and the deployment of human resources. Here, the supplies and resources include emergency equipments, emergency response teams, emergency experts, communication resources, etc.

Among the systems discussed above, the decision support system provides fast supportive plans for decision makers and is thus a critical component. We explain the basic principles for the decision support system in the next section.

14.1.2 Basic Principles for the Decision-Support System

In the electric power emergency management platform, there exists a variety of information. If these massive data are collected and analyzed mannualy for decision making, the system will suffer from low efficiency, and is prone to errors. Such situation is by no means acceptable when facing disasters. The decision support system is exactly the powerful tool that helps the emergency supervisor to process data and make decisions. Its principles are shown in Fig. 14.2.

1) Power grid equipment damage forecast

This module is responsible for the forecast of the damages that are caused by formidable natural conditions and public security accidents. It provides early warning to possible severe equipment losses. Firstly, this module evaluates how likely an equipment breaks down according to the forecast of natural disasters such as typhoons, heavy fogs and forest fire. It considers the factors such as the location of the disaster, the geographical interconnection of the electric power network and the endurance of power grid equipments against disasters. Then it informs relevant departments the predicted damages so that preparation for equipment emergent maintenance can be made beforehand. In addition, it also provides the information about the equipment's failure probabilities under disasterous conditions for the next power grid security evaluation.

2) Disaster assessment and power grid security evaluation

This module evaluates the security level of the power grid according to its current

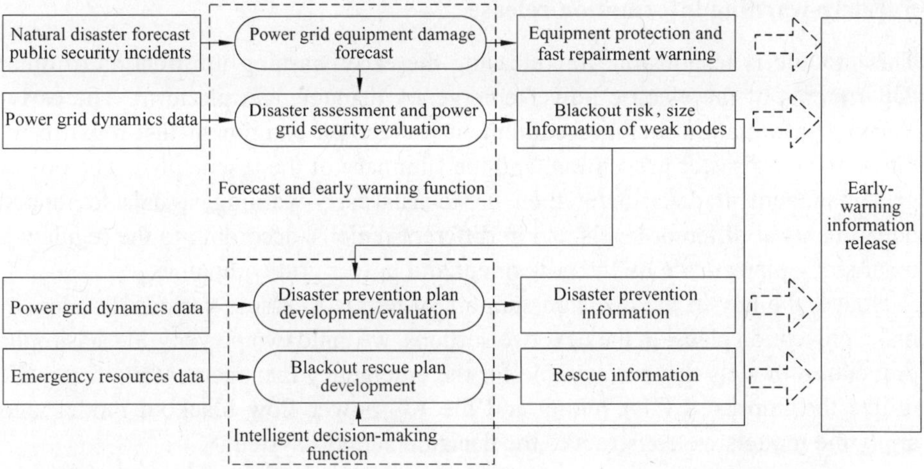

Figure 14.2 The schematic diagram for the functions of the decision support system

operation state and the equipment failure probability under disasterous conditions. It provides the related information, such as the scale, location and probability of the load loss and the weak buses. From this, the severity of disasters can be quantitatively evaluated.

3) Disaster prevention plan development/evaluation

For some low-intensity disasters, it is highly possible to prevent load loss through appropriate dispatching. Even for serious disasters, adjusting operation modes in advance can still reduce load loss. Therefore, the main function of this module is to exploit the dynamics data of the power grid and the information about the disaster assessment either to make disaster-prevention plans in order to reduce the impact of disasters, or to screen different plans made beforehand in order to pick out the most favorable plan for the given disaster.

4) Blackout rescue plan development

Because of the ample uncertainties in disasters, load fluctuations and equipment defects, in combination with the limited disaster prevention actions compared to the many possible types and scales of disasters, it is impossible to completely avoid the occurance of blackouts. Once large-area load loss occurs, it will become important to restore power supply. Therefore, the main task of this module is to make blackout rescue plans in advance for different blackout scenarios, in response to the predicted scope and probability of the blackout and the information about the emergency resources. This is especially effective to reduce the duration of blackouts and minimize economic losses.

5) Early-warning information release

This module is responsible for releasing the early-warning information to other departments of the electric power emergency management platform. The early-warning information includes equipment protection, equipment fast repairment, blackout and disaster prevention, and the summary of the rescue plan. The power grid management department then broadcasts early-warning signals to related departments at different levels and in different regions, according to the regulating emergency plans made by the goverment and power grid companies.

Hence, the key in the decision support system is to assess various disasters to make prevention plans. In the next two sections, we build two power grid catastrophe prevention models that are suitable for the emergency management platform. We utilize the improved OPA model and the AC power flow blackout model, and apply the models we construct to the decision support system.

14.2 Power Grid Disaster Forecasting and Early-Warning Model Based on DC Power Flow

From the discussion in the previous chapters, we know that the outage of an individual electric component may lead to complicated cascading failures. In fact, any device is at risk for being damaged during the development of a disaster. Therefore, in order to quantitatively analyze the impact of a disaster and further estimate the scales and probabilities of blackouts, an appropriate model has to be constructed first.

14.2.1 Model Design

According to the improved OPA model discussed in Chapter 9, we propose to construct the power system catastrophe model shown in Fig. 14.3. This model consists of seven modules: the system operating condition acquisition module, disaster geographic information collection module, disaster prevention plan acquisition module, system parameter update module, the simulation module for fast dynamics based on the improved OPA blackout model[7], load loss probability calculation module and risk index calculation module.

The steps of running this forecasting and early-warning model can be further described as follows:

(1) Acquire the system operating conditions, i.e. the present power flows.

(2) Load the geographic information of disasters, which includes the information about all the disasters that might cause power system catastrophes in the main disaster chains.

(3) Implement the disaster prevention plan to reduce the impact of disasters by adjusting the operation modes, such as changing the network structure and activating stand-by generator units.

(4) Analyze the preliminary impact of the disasters on the power grid according to the system operating conditions and the disaster's geographic information. In other words, assess the impact of disasters on the power system using the disaster information, such as which areas are seriously affected by the disaster, which equipments have increasing failure probabilities, etc. On top of such analysis, modify the parameters of the improved OPA model correspondingly.

(5) Execute repeatedly the fast dynamics of the improved OPA for N times to simulate the evolution of the catastrophic processes and record the load loss each time.

(6) Identify the statistic distribution characteristics of load losses and compute the risk and probability for blackouts at weak buses during the current disaster.

(7) Calculate the risk indices of VaR, CVaR and Risk for the current disaster using the blackout probability density (see Remark 14.2).

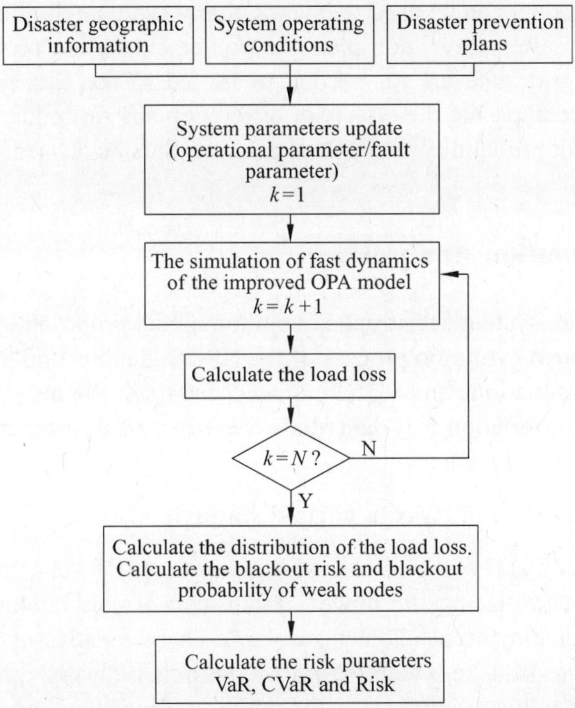

Figure 14.3 The power system catastrophe forecasting and early-warning model

Remark 14.1 Compared with the improved OPA model, this model does not consider the slow dynamic process of system upgrade when simulating the disaster evolution process. Thus it can be computed quickly with the running time at the scale of seconds. When combined with GIS, this model can realize 2D and 3D visualization. In particular, this model can make highly reliable evaluations of the current system operating conditions and the severity of the catastrophe.

Remark 14.2 The system blackout Risk is defined to be the value of the CVaR when σ equals zero, that is

$$\text{Risk} = \text{CVaR}_{\sigma=0} = \int_0^{\infty} xp(x)\text{d}x \quad \text{(R1)}$$

This catastrophe evolution model can be used for the assessment of disasters and the evaluation of disaster prevention plans in the decision support system.

When assessing a disaster, the forecasting and early-warning model can automatically analyze the scale, location and probability for load loss requiring only the information about the equipment failure probability in case of disasters and the current state of the power grid. The severity of different disasters can further be quantitatively compared. In addition, the model can also provide the information about weak buses for the development of disaster prevention plans.

When disaster prevention plans have been set beforehand, the forecasting and early-warning model can be used to quantitatively evaluate their effectiveness. In this case, the disaster prevention plan, the equipment failure probability and the current power grid state are all needed to be fed to the catastrophe evolution program. After comparing the effects of different plans for reducing the blackout risk and blackout probability, an optimal plan can thus be chosen.

14.2.2 Simulation Analysis

We use the power system catastrophe evolution model proposed in Section 14.2.1 to study the system evolution process of the NPGC (see Section 9.2.2 for details) both in normal situations and with simulated disasters. We also provide disaster early-warning information and then study the effect of disaster prevention plans on reducing disaster losses.

1) Power grid operation risks in normal states

We use the Risk, CVaR and VaR to evaluate the security level of the NPGC when there is no disaster. Using the power system catastrophe evolution model just described, we obtain the blackout distribution characteristics of the NPGC that are shown in Fig. 14.4, in which the horizontal axis indicates the load loss P_{loss}, the vertical axis n in subfigure (a) is the number of blackouts when the load loss is P_{loss}, and the vertical axis N in subfigure (b) is the number of blackouts when the scale of load loss is greater than P_{loss}.

The results in Fig. 14.4 show that when there is no disaster, the NPGC can guarantee continuous power supply in general and the maximum blackout loss is

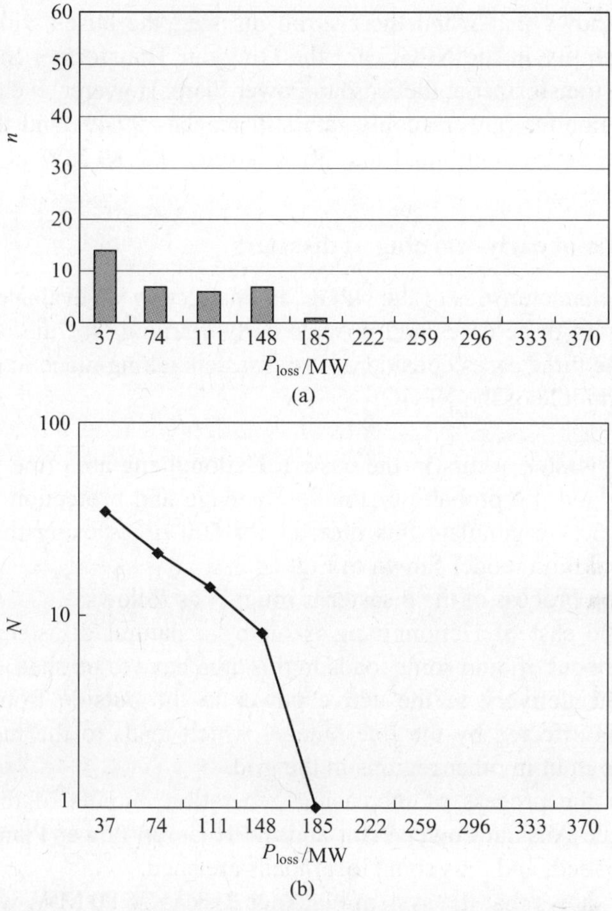

Figure 14.4 Blackout distribution of the NPGC in normal states

less than 200 MW. In view of Remark 14.2, the system blackout Risk is an expected load loss of 5.792 MW. When σ is set to be 99%, the VaR is 137.7 MW (i.e. the scales of 99% of the blackouts are less than 137.7 MW), and the CVaR is 2.004 MW (i.e., the expected extra loss for the blackouts whose scales are greater than 137.7 MW is 2.004 MW). The five buses with the highest Risk in the system are shown in Table 14.1.

Table 14.1 Weak buses in normal states

Power plants or substations	Risk/MW	Blackout probability
Qingyun substation	0.5789	0.006
Anshan power plant(No.1 main transformer)	0.5512	0.006
Daliansan power plant	0.4239	0.002
Yuhong substation	0.3712	0.002
Anshan power plant(No.2 main transformer)	0.3035	0.004

Table 14.1 shows that when there is no disaster, the buses with the highest blackout probability in the NPGC are the Qingyun Transformer Substation and the No. 1 main transformer at the Anshan Power Plant. Howerer, even for the weak buses, the continuous power supply rate still reaches 99.4% and the maximum expected value of the node load loss Risk is only 0.5789 MW at the Qingyun Transformer Substation.

2) Assessment and early-warning of disasters

In view of the characteristics of the NPGC, in this section we evaluate the system's operation risks in three cases and provide early warnings for the corresponding weak links. The three cases considered are disasters taking place in the north, the south and the middle of the NPGC.

Case 1:

Consider a disaster occurs in the eastern Heilongjiang area (the northern part of the NPGC) and the probability for line damage and protection misoperation increases to 0.3. We simulate this disaster for 500 times using the power grid catastrophe evolution model shown in Fig. 14.3.

The evolution process of the disaster is roughly as follows:

Stage 1 The east of Heilongjiang is hit by a natural disaster. As a result, several lines are cut off and some loads in this area have to be shed.

Stage 2 The delivery of the active power to the outside from the east of Heilongjiang is affected by the line outage, which leads to the increase of the power plants' output in other regions in the grid.

Stage 3 In the process of increasing generation outputs of the Liangjiang Power Plant, the Anshan Power Plant and the Huanren Power Plant, some lines become overloaded, and thus some local loads are shed.

Calculation shows that the system blackout Risk is 76.80 MW, which is much larger than 5.792 MW, the value in normal states. When σ is set to be 99%, the VaR is 221.2 MW (i.e. with 99% confidence, the blackout scale is less than 221.2 MW power loss) and the CVaR is 2.385 MW (i.e. the expected extra amount of power loss greater than 221.2 MW is 2.385 MW). Specifically, the distributions for different blackout scales are shown in Fig. 14.5.

Comparing Fig. 14.4 with Fig. 14.5, one can see that after the eastern Heilongjiang area is hit by a disaster, the number of power system blackouts and the blackout scale increase. The Risk and blackout probability for the five most fragile buses in the system are shown in Table 14.2.

The above data imply that the continuous power supply rate of the load buses nodes reduces to as low as 48.6% (at the Liangzihe Power Plant). In the mean time, the load buses with high blackout risks change from the buses at the Qingyun Transformer Substation and the Anshan Power Plant to the buses at the Lulinshan Transformer Substation and the Jinshan Transformer Substation. Among all the buses, the Lulinshan Transformer Substation is of the highest blackout risk of 44.10 MW with a blackout probability of 34%, and the Liangzihe Power Plant has

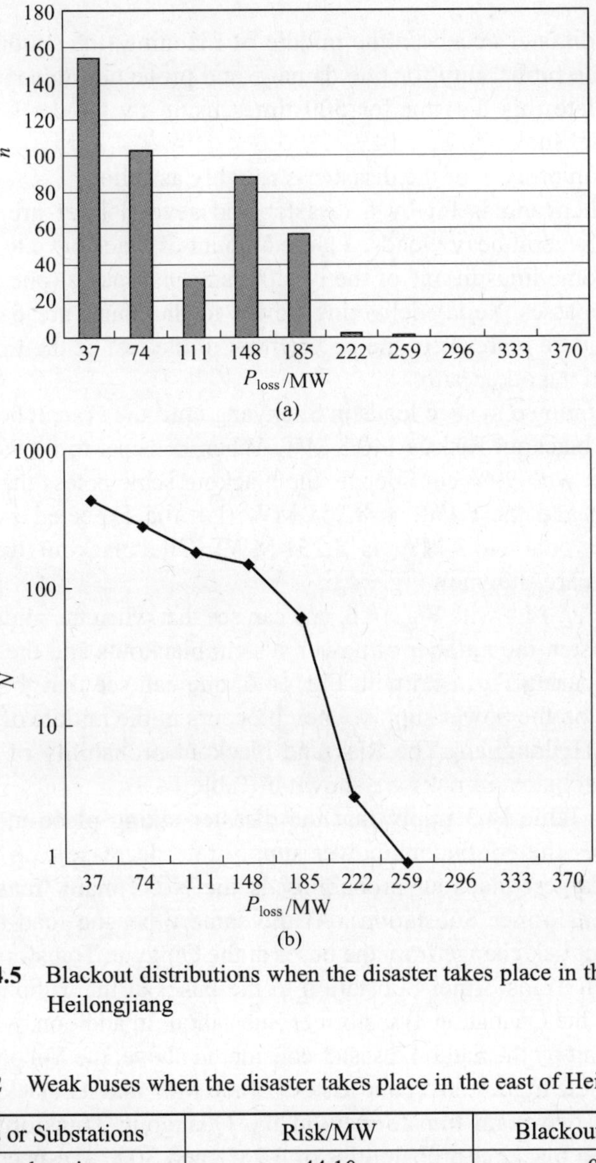

Figure 14.5 Blackout distributions when the disaster takes place in the east of Heilongjiang

Table 14.2 Weak buses when the disaster takes place in the east of Heilongjiang

Power plants or Substations	Risk/MW	Blackout probability
Lulinshan substation	44.10	0.34
Jinshan substation	17.73	0.334
Liangzihe power plant	6.617	0.514
Hedongjiao	2.127	0.308
Yichun substation	2.520	0.334

the highest blackout probability of 51.4% with a blackout Risk of 6.617 MW. Hence, it is necessary to develop blackout restoration plans for such weak buses.

Case 2:

Consider a disaster occurs in the middle of Liaoning (the southern area of the NPGC), and the probability for line damage and protection misoperation rises to 0.3. We simulate this disaster for 500 times using the power grid catastrophe evolution model shown in Fig. 14.3.

The evolution process of the disaster is roughly as follows.

Stage 1 Shenyang is hit by a disaster and several lines are cut off. Since Shenyang has several heavy loads, a large amount of loads have to be shed.

Stage 2 Some lines in one of the two transmission paths (one passes Shaling and the other passes Xujia) delivering power to the south are damaged, which leads to a dramatic increase in the power flow in the rest of the lines in this path and the lines in the other path.

Stage 3 Continue to shed loads in Shenyang until the system becomes stable.

The system blackout Risk is 140.3 MW. When σ is set to be 99%, the VaR is 644.3 MW (i.e. with 99% confidence, the blackout scale is less than a power loss of 644.3 MW) and the CVaR is 8.251 MW (i.e. the expected extra amount of power loss beyond 644.3 MW is 8.251 MW). The blackout distributions for different scales are shown in Fig. 14.6.

Comparing Fig. 14.4 with Fig. 14.6, one can see that when the middle of Liaoning is hit by a disaster, the number of power system blackouts and the blackout scale increase. Comparing Fig. 14.5 with Fig. 14.6, one can see that the disaster has a greater impact on the power supply when it occurs in the middle of Liaoning than in the east of Heilongjiang.The Risk and blackout probability of the five buses with the highest blackout risks are shown in Table 14.3.

The data in Table 14.3 imply that the disaster taking place in the middle of Liaoning makes the continuous power supply rate decrease to 67.2% (the load bus with the largest blackout probability is the No. 2 main transformer of the Quangong Transformer Substation). At the same time, the load buses with the highest blackout risk change from the buses at the Qingyun Transformer Substation and the Anshan Transformer Substation to the buses at the Yuhong Transformer Substation and the Quangong Transformer Substation. In addition, when the middle of Liaoning is hit by the natural disaster considered above, the Yuhong Transformer Substation has the highest blackout Risk of 56.66 MW and its blackout probability is 30.4%. The two main transformers at the Quangong Transformer Substation have the highest blackout probabilities of 32.8% and 30.8% respectively and their Risk are 26.47 MW and 16.78 MW respectivly. Hence, it is very important and urgent to develop blackout restoration plans for these weak buses.

Remark 14.3 The right end of the horizontal axis of the histogram in Fig. 14.6 indicates "greater than 666" and its corresponding vertical cooridinate indicates the number of blackouts, the scales of which are greater than 666 MW. Since there is a blackout whose scale is about 2000 MW in Case 2, if we draw diagrams in the conventional way, there will be unnecessary blank areas. This is why we draw the figure in the chosen fashion.

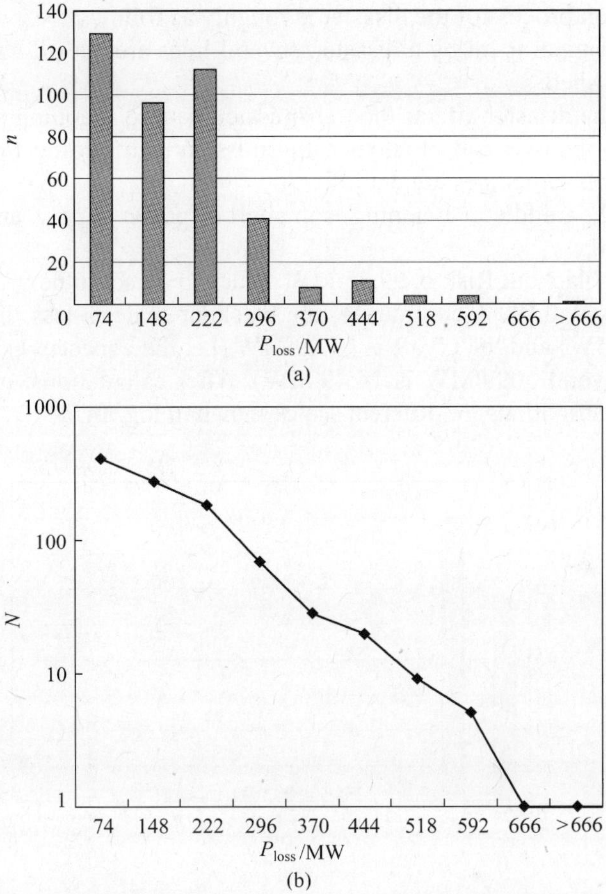

Figure 14.6 Blackout distributions when the disaster occurs in the middle of Liaoning

Table 14.3 Weak buses when the disaster occurs in the middle of Liaoning

Power plants or Substations	Risk/MW	Blackout probability
Yuhong substation	56.66	0.304
Quangong substation (No. 1 main transformer)	26.47	0.308
Quangong substation (No. 2 main transformer)	16.78	0.328
Zhanggong substation	12.53	0.1
Dacheng substation	6.454	0.034

Case 3:

Consider a disaster takes place in Fengbai in the eastern area of Jilin (the middle of the NPGC), and the probability for line damage and protection misoperation rises to 0.3. We simulate the disaster for 500 times using the fast dynamics of the improved OPA model.

The evolution process of the disaster is roughly as follows.

Stage 1 Fengbai is hit by a disaster, several lines are cut off and some loads in Fengbai are shed.

Stage 2 The disaster affects the transmission path to Liaoning passing Xujia, which leads to the overload of another transmission path starting from Lishu and Siping, in which some lines are cut off.

Stage 3 The middle of Liaoning is in short of power supply, and some loads have to be shed.

The system blackout Risk is 99.25 MW. When σ is set to be 99%, the VaR is 1099 MW (i.e. with 99% confidence, the blackout scale is less than the power loss of 1099 MW) and the CVaR is 14.49 MW (i.e. the expected extra amount of power loss beyond 1099 MW is 14.49 MW). After calculations, one can obtain the blackout distributions for different scales shown in Fig. 14.7.

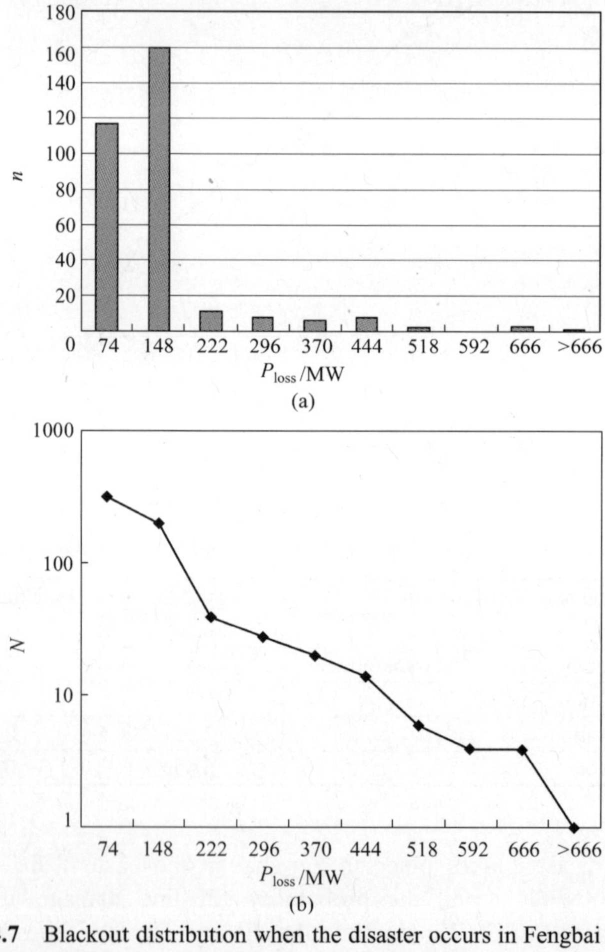

Figure 14.7 Blackout distribution when the disaster occurs in Fengbai in the east of Jilin

Comparing Fig. 14.4 with Fig. 14.7, one can conclude that when the disaster takes place in Fengbai in the east of Jilin, the number of blackouts whose scale is between 100 MW and 300 MW is smaller than that in Case 2, but the probability for large-scale blackouts is greater than that in Case 2. From the VaR index, one can tell clearly that the VaR is 1099 MW when the disaster takes place in Fengbai in the east of Heilongjiang, which is much greater than 644.3 MW in Case 2.

The Risk and blackout probability of the five most vulnerable buses (i.e, the buses with the highest blackout risk) are shown in Table 14.4.

Table 14.4 Weak buses when the disaster takes place in Fengbai in the east of Jilin

Power plants or Substations	Risk/MW	Blackout probability
Miaosheng substation	33.04	0.344
Gaojigang substation	5.118	0.12
Panshi substation	4.795	0.116
Anshan power plant	1.027	0.038
Wuchang substation	0.5152	0.034

The data in Table 14.4 show that the disaster taking place in Fengbai in the east of Jilin makes the continuous power supply rate reduce to 65.6% (the bus with the highest blackout probability is the Miaosheng Transformer Substation). At the same time, the most fragile load buses change from the buses at the Qingyun Transformer Substation and the Anshan Power Plant to the buses at the Miaosheng Transformer Substation and the Gaojigang Transformer Substation. In short, when the disaster occurs in Fengbai in the east of Jilin, the weakest bus is the Miaosheng Transformer Substation, and the expected value of load loss (the Risk) at which is 33.04 MW with blackout probability being 34.4%. The next most fragile buses are the Gaojigang Transformer Substation and the Panshi Transformer Substation. Therefore, it is necessary and urgent to develop blackout restoration plans for these weak buses.

From the above analysis, one can see that the disasters taking place in the middle of Liaoning and in Fengbai in the east of Jilin have greater impact than that in the east of Heilongjiang. The blackout Risks in the former two cases are 140.3 MW and 99.25 MW respectively, which are much greater than that of 76.80 MW in the last case. Also, the blackouts in the former two cases have a larger scale, and in each of the 500 simulations there always exist blackouts whose scales are greater than 666 MW (in fact in both of the two cases, one can find blackouts with power losses of about 2000 MW) with the corresponding CVaR being 8.2511 MW and 14.4934 MW respectively. The largest blackout in the 500 simulations when the disaster happens in the east of Heilongjiang is only 250 MW and the corresponding CVaR is only 2.3845 MW.

3) Assessment of disaster prevention plans

In this section, we consider two types of disaster prevention plans: one is to activate standby generator units and the other is to use backup lines. These two types of disaster prevention plans can both alleviate the pressure on heavy loaded lines, reduce the load shedding initiated by inadequate generation supplies and help to restore the service to interrupted loads.

Considering the operation level and the condition of backup generation units and lines in the NPGC in July 2005, we present several sets of plans below in Table 14.5. We then evaluate these plans using the catastrophe evolution model.

Table 14.5 Emergency plans

Plan number	Starting standby generator units	Using backup lines
Plan 1	No.2 unit of Baishan, No.11 unit of Fengman, No.12 unit of Fengman, No.3 unit of Hongshi, No.4 unit of Hongshi, No.4 unit of Hunjiang	
Plan 2	No.2 unit of Huanren, No.3 unit of Huanren, No.4 unit of Huanren, No.5 unit of Huanren, No.1 unit of Huilong, No.3 unit of Taipingshao, No.4 unit of Taipingshao	
Plan 3	No.4 unit of Jinbohu, No.5 unit of Jingbohu, No.6 unit of Jingbohu	
Plan 4	No.4 unit of Liaoning, No.5 unit of Liaoning, No.7 unit of Liaoning, No.5 unit of Qinghe, No.6 unit of Qinghe, No.2 unit of Jinzhou, No.3 unit of Jinzhou, No.5 unit of Jinzhou, No.1 unit of Tieling	
Plan 5		Shangling substation-Shendong substation, Shendong substation-Zhanggong substation, Shendong substation-Qinghe power plant, Lishizhai substation –Yuanling substation, Yingkou substation-Liushu substation
Plan 6		Baishan power plant-Meihe substation No.1 line and No.2 line, Gaizhou substation-Liushu substation, Niugang substation-Qinghe power plant, Qinghe power plant-Tieling substation, Qinghe power plant-Zhifu substation
Plan 7		Chengxiang substation-Hasan power plant, Hanan substation-asan power plant No.1 line and No.2 line,Hasan power plant-Nangjin substation No.1 line and No.2 line

Next according to different disaster locations, we investigate in three cases the effects of the above plans on the reduction of the probability and loss of disasters.

Case 1:

The disaster takes place in the east of Heilongjiang.

We use the fast dynamics in the improved OPA model to analyze the above plans. The three risk indices after adopting different plans are shown in Table 14.6.

Table 14.6 Comparison of the three risk indices when the disaster happens in the east of Heilongjiang (MW)

	Risk	VaR	CvaR
Without any plan	76.80	221.2	2.385
Plan 1	76.53	258.6	2.987
Plan 2	74.43	232.7	2.828
Plan 3	73.57	217.3	3.121
Plan 4	81.22	281.3	3.086
Plan 5	71.87	226.0	2.819
Plan 6	75.46	275.6	3.026
Plan 7	74.31	224.0	2.923

The results in Table 14.6 indicate that the seven plans shown in Table 14.5 do not have an obvious disaster mitigation effect on the disaster taking place in the east of Heilongjiang. The Risks for plan 3 and plan 5 decrease slightly, but the VaR for plan 5 is greater than that without executing any plan, which implies that the probability for large-scale blackouts has increased rather than decreased. In fact, the number of blackouts with scales greater than 259 MW increases from 0 to 1 in the 500 simulations.

The disaster mitigation effects of plan 3 and plan 5 on the buses with the highest blackout risks are shown in Table 14.7.

Table 14.7 Comparison of disaster mitigation effects on weak buses when the disaster takes place in the east of Heilongjiang

Weak nodes	Risk/MW			Blackout probability		
	Without any plan	Plan 3	Plan 5	Without any plan	Plan 3	Plan 5
Lulinshan substation	44.10	40.26	41.51	0.34	0.31	0.32
Jinshan substation	17.73	17.71	16.23	0.334	0.334	0.302
Liangzihe power plant	6.617	6.529	6.364	0.514	0.512	0.498
Hedongjiao	2.127	2.148	2.088	0.308	0.312	0.304
Yichun substation	2.520	2.563	2.306	0.334	0.334	0.302

Table 14.7 implies that plan 3 and plan 5 do not have an obvious effect on the Risk and blackout probability, so we need to search for other effective plans. We can install in advance self-sustained power sources at the weak buses that are identified using the catastrophe evolution model; in other words, when the disaster happens in the east of Heilongjiang, we activate self-sustained generator sets in the Lulinshan Transformer Substation, the Jinshan Transformer Substation, the Liangzihe Power Plant, the Hedongjiao and the Yichun Transformer Substation assuming that each bus has a 30 MW self-sustained power source. Then the disaster mitigation effects after adopting the self-sustained power source plan are shown in Tables 14.8 and 14.9.

Table 14.8 Comparison of the three indices when the disaster happens in the east of Heilongjiang (MW)

	Risk	VaR	CVaR
Without any plan	76.80	221.2	2.385
Self-sustained power source plan at weak buses	58.35	171.4	2.202

Table 14.8 shows that the self-sustained power source plan focusing on the weak buses has some mitigation effects. The Risk decreases from 76.80 MW to 58.35 MW, the VaR decreases from 221.2 MW to 171.4 MW, and the CVaR decreases from 2.385 MW to 2.202 MW.

Table 14.9 Comparison of disaster mitigation effects on weak buses when the disaster happens in the east of Heilongjiang

Weak buses	Risk/MW		Blackout probability	
	Without any plan	Self-sustained power source plan at weak buses	Without any plan	Self-sustained power source plan at weak buses
Lulinshan Substation	44.10	39.05	0.34	0.3
Jinshan Substation	17.73	17.55	0.334	0.33
Liangzihe Power Plant	6.617	7.485	0.514	0.584
Hedongjiao	2.127	1.910	0.308	0.282
Yichun Substation	2.520	2.497	0.334	0.33

The results in Table 14.9 show that the self-sustained power source plan focusing on weak buses has an obvious effect on the reduction of the blackout risk and blackout probability.

Case 2:

The disaster takes place in the middle of Liaoning.

We analyze in this case the effect of the seven plans in Table 14.5 using the fast dynamics of the improved OPA model. The three risk indices after adopting plans are shown in Table 14.10.

Table 14.10 Comparison of the three risk indices when the disaster takes place in the middle of Liaoning (MW)

	Risk	VaR	CVaR
Without any plan	140.3	644.3	8.251
Plan 1	136.2	659.5	8.541
Plan 2	141.3	626.8	6.992
Plan 3	151.2	612.5	8.486
Plan 4	133.7	525.1	6.342
Plan 5	136.1	541.9	6.196
Plan 6	151.7	537.3	9.683
Plan 7	125.9	508.7	5.617

From Table 14.10, one can see that plan 7 has some disaster mitigation effect on the disaster taking place in the middle of Liaoning. The Risk decreases from 140.3 MW to 125.9 MW, the VaR from 644.3 MW to 508.7 MW and the CVaR from 8.25 MW to 5.617 MW.

The disaster mitigation effects of plan 7 on the buses with high blackout risks are shown in Table 14.11.

Table 14.11 Comparison of disaster mitigation effects on weak buses when the disaster happens in the east of Heilongjiang

Weak buses	Risk/MW		Blackout probability	
	Without any plan	Plan 7	Without any plan	Plan 7
Yuhong Substation	56.66	49.10	0.304	0.262
Quangong Substation (No.1 main transformer)	26.47	28.89	0.308	0.336
Quangong Substation (No.2 main transformer)	16.78	16.73	0.328	0.326
Zhanggong Substation	12.53	7.931	0.1	0.064
Dacheng Substation	6.454	7.352	0.034	0.038

From Table 14.11, it is clear that plan 7 reduces the blackout Risk and the blackout probability of the Yuhong Substaion and the Zhanggong Substation while making the operating conditions at the Quangong Transformer Substation and the Dacheng Transformer Substation worse. In general, plan 7 has some effects on the reduction of the system blackout risk, but it does not have completely satisfactory performance on the improvement of some weak buses. So we need to design related blackout restoration plans to complement this plan.

We consider installing in advance self-sustained power sources at the weak buses that are identified by the catastrophe evolution model. In other words, when the disaster happens in the middle of Liaoning, we activate small-size generator units at the Yuhong transformer Substation, the No. 1 main transformer at the Quangong Transformer Substation, the No. 2 main transformer at the Quangong Transformer Substation, the Zhanggong Transformer Substation and the Dacheng Transformer Substation assuming that each of them is installed with a 30 MW self-sustained power source. Then the disaster mitigation effect after adopting the self-sustained power source plan are shown in Tables 14.12 and 14.13.

Table 14.12 Comparison of the three risk indices when the disaster happens in the middle of Liaoning (MW)

	Risk	VaR	CVaR
Without any plan	140.3	644.3	8.251
Self-sustained power source plan at weak nodes	99.33	436.6	6.068

Table 14.12 shows that the self-sustained power source plan at weak buses has some disaster mitigation effects on the disaster taking place in the middle of Liaoning. The Risk decreases from 140.3 MW to 99.33 MW, the VaR decreases from 644.3 MW to 436.6 MW, and the CVaR decreases from 8.25 MW to 6.068 MW.

Table 14.13 Comparison of disaster mitigation effects on weak buses when the disaster happens in the middle of Liaoning

Weak buses	Risk/MW		Blackout probability	
	Without any plan	Self-sustained power source plan at weak buses	Without any plan	Self-sustained power source plan at weak buses
Yuhong Substation	56.66	45.72	0.304	0.294
Quangong Substation (No.1 main transformer)	26.47	17.07	0.308	0.304
Quangong Substation (No.2 main transformer)	16.78	7.137	0.328	0.334
Zhanggong Substation	12.53	9.189	0.1	0.096
Dacheng Substation	6.454	5.182	0.034	0.032

From Table 14.13, it is clear that the plan focusing at weak buses can significantly reduce the system blackout risk and the blackout probability.

Case 3:

The disaster happens in Fengbai in the east of Jilin.

We analyze the plans described earlier using the fast dynamics in the improved OPA model. The three risk indices after adopting the plans are shown in Tables 14.14 and 14.15.

Table 14.14 Comparison of the three risk indices when the disaster takes place in the east of Jilin (MW)

	Risk	VaR	CVaR
Without any plan	99.25	1099	14.49
Plan 1	69.70	568.0	7.790
Plan 2	57.44	559.0	6.860
Plan 3	86.51	1267	14.26
Plan 4	42.87	251.5	2.689
Plan 5	101.5	1039	16.30
Plan 6	94.18	830.6	11.27
Plan 7	89.47	902.5	13.08

Table 14.15 Comparison of the disaster mitigation effects on weak buses when the disaster happens in the east of Jilin

Weak nodes	Risk /MW			Blackout probability		
	Without any plan	Plan 1	Plan 4	Without any plan	Plan 1	Plan 4
Miaosheng Substation	33.04	25.17	30.78	0.344	0.26	0.308
Gaojigang Substation	5.118	4.504	4.887	0.12	0.098	0.096
Panshi Substation	4.795	3.824	3.425	0.116	0.036	0.07
Anshan Power Plant	1.027	1.250	2.139	0.038	0.016	0.014
Wuchang Substation	0.5152	0.5511	0.4360	0.034	0.018	0.068

From Table 14.14, one can see that plan 4 has the best disaster mitigation performance when the disaster happens in Fengbai in the east of Jilin. The Risk decreases from 99.25 MW to 42.87 MW, the VaR from 1099 MW to 251.5 MW and the CVaR from 14.49 MW to 2.689 MW. Plans 2 and 1 also have some disaster mitigtion effects.

From Table 14.15 it is clear that plan 4 reduces the Risk and blackout probability at the Miaosheng Transformer Substation, the Gaojigang Transformer Substation and the Panshi Transformer Substation, but at the same time increases the blackout risk at the Anshan Power Plant and the Wuchang Transformer Substation. Overall, plans 4 and 1 have an obvious effect on the reduction of the system blackout risk, but do not have satisfactory performance on the improvement of some weak buses. So we need to design related blackout restoration plans to implement this plan.

Similar to what we have done before, we consider installing in advance self-sustained power sources at the weak buses identified by using the catastrophe evolution model. When the disaster happens in Fengbai in the east of Jilin, we activate small-size generator units at the Miaosheng Transformer Substation, the Gaojigang Transformer Substation, the Panshi Transformer Substation, the Anshan Power Plant and the Wuchang Transformer Substation assuming that each of them is installed with a 30 MW self-sustained power source. The disaster mitigation effect after adopting the self-sustained power source plan are shown in Tables 14.16 and 14.17.

Table 14.16 Comparison of the three risk indices when the disaster takes place in the east of Jilin (MW)

	Risk	VaR	CVaR
Without any plan	99.25	1099	14.49
Self-sustained power source plan at weak buses	63.88	689.4	9.788

From Table 14.16, one can see that the self-sustained power source plan at weak buses has an obvious disaster mitigation effects on the disaster taking place in Fengbai in the east of Jilin. The Risk decreases from 99.25 MW to 63.88 MW, the VaR from 1099 MW to 689.4 MW, and the CVaR from 14.49 MW to 9.788 MW.

Table 14.17 Comparison of the disaster mitigation effects on weak buses when the disaster takes place in the east of Jilin

Weak buses	Risk(MW)		Blackout probability	
	Without any plans	Self-sustained power source plan at weak buses	Without any plans	Self-sustained power source plan at weak buses
Miaosheng Substation	33.04	21.65	0.344	0.318
Gaojigang Substation	5.118	2.149	0.12	0.116
Panshi Substation	4.795	4.887	0.116	0.062
Anshan Power Plant	1.027	0.8533	0.038	0.024
Wuchang Substation	0.5152	0.06879	0.034	0.016

From Table 14.17, it is clear that the prevention and control plans at weak buses can significantly reduce the system blackout risk and the blackout probability.

Using the above calculation results, we list in Table 14.18 the quantitative comparison of the impacts of the three simulated disasters on the NPGC, together with the fragile buses and the effect of prevention plans.

Table 14.18 Comparison of the three risk indices (MW)

	Without any plan			Self-sustained power source plan at weak buses			Weakest buses
	Risk	VaR	CVaR	Risk	VaR	CVaR	
The east of Heilongjiang is hit by a disaster	76.80	221.2	2.385	58.35	171.4	2.202	Lulinshan, Jinshan, Liangzihe, Hedongjiao, Yichun

(Continued)

	Without any plan			Self-sustained power source plan at weak buses			Weakest buses
	Risk	VaR	CVaR	Risk	VaR	CVaR	
The middle of Liaoning is hit by a disaster	140.3	644.3	8.251	99.33	436.6	6.068	Yuhong, Quangong, Zhanggong, Dacheng
The east of Jilin is hit by a disaster	99.25	1099	14.49	63.88	689.4	9.788	Miaosheng, Gaojigang, Panshi, Anshan, Wuchang

14.3 Power Grid Disaster Forecasting and Early-Warning Model Based on AC Power Flow

The power grid disaster forecasting and early-warning model discussed in Section 14.2 is based on the DC power flow. This model has a fast computational speed and can be applied large-scale systems. However, it is also useful to consider another situation that the power grid security and warning system has enough time to forecast and analyze cascading failures and blackout disasters caused by the faults of electric equipments, and consequently develop an emergency management plan. In this case, a different power grid disaster prediction and warning model can be constructed utilizing the AC blackout model discussed in Chapter 10. Since the AC power flow is widely used in the practical power grid dispatching and operation, the disaster prediction and warning model based on AC power flows can take into account both system active power flows and reactive powers/voltages as well, and thus monitor the system operational condition in a more comprehensive manner.

14.3.1 Model Design

The disaster prediction and early-warning model based on AC power flows consists of several modules including catastrophe identification, equipment cut-off, system parameters update, island identification, island management, fast dynamic computatin of the AC blackout model, system safety evaluation and emergency dispatching, emergency load shedding, load loss statistics and early-warhing signal broadcasting. The structure of the modules is shown in Fig. 14.8.

This model can jointly consider the impact of different natural disasters, such as typhoons, icing, etc. It also takes into account the electric equipment outage caused by aging and faults. The main idea is as follows. Under normal working conditions, it identifies those electric components that might exit their normal

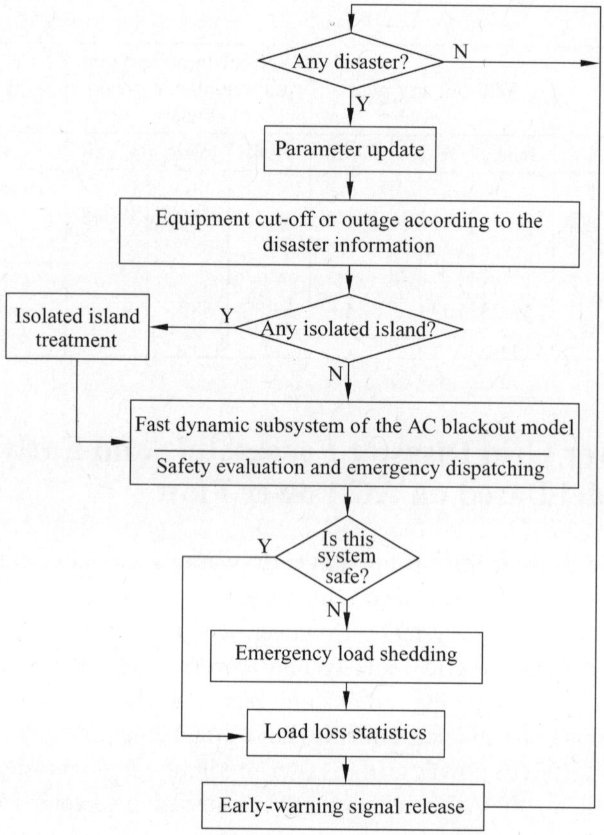

Figure 14.8 The power grid disaster prediction and warning model based on AC power flows

states by analyzing the disaster forecasting information and the network monitoring information. Then it can further provide early warning information for the emergency management platform. Under abnormal working conditions, it provides prevention and emergency control strategies to ensure the system's safety while trying to minimize the power loss.

Since the model is able to carry out the tasks of safety assessment, contingency dispatching and emergency control, it can not only be used to analyze and predict large-scale blackouts, but also play the role of dispatchers in the emergency management system. This makes the model more useful in engineering practice. The model can be triggered by two different mechanisms in the power system emergency management platform: one is to be activated manually, and the other is to be called by the event analysis and processing subsystem[8, 9]. The calculation steps are listed below.

Step 1 Update the power grid model according to the natural disaster information.

Generally speaking, most natural disasters develop much slower than power grid catastrophe. So the impact of these disasters can be lumped into the model as triggering events. For example, if a big fire takes place somewhere in the grid, it will trigger immediately the prediction and early-warning model to modify the fire impact parameter. It needs to be pointed out that the prediction and early-warning model will be called only when some event occurs; in other words, the model does not take part in the daily power grid operation under normal working conditions.

Step 2 Cut off equipments.

Since the selection of electric equipments has considered common natural disasters that might be encountered, these equipments have certain ability to work against natural disasters. Hence, one can define the damage probability for various equipments in different disaster conditions. In this step, electric equipments are cut with a certain probability mainly according to the disaster intensity and the disaster-proof ability of the electric equipment. We develop various power grid accident probability models by the Monte Carlo simulation method[10], simulate the occurrence and development process of accidents by sampling, and predict whether there is any equipment out of service. We can thus predict the faults in power sources, transmission lines and loads, and construct the new grid's topological structures.

Step 3 Check whether there is any island after the cutting of the faulted equipments.

If two or more islands appear, go to Step 4. Otherwise, go to Step 5.

Step 4 Calculate the load demand and generation capacity for each island.

In view of the load level, we determine the load threshold M in such a way that if the load in an island is less than M, then the island can be approximately considered to be stable. If the load of an island is greater than M, go to Step 5. For an island whose load is less than M, if its generation capacity is greater than its load demand, we think there will be no load loss; otherwise, we take the difference between the generation capacity and the load demand to be the estimation of the island's load loss.

Step 5 Security assessment and emergency dispatching.

When the load level is beyond the generation capacity or the network transmission capacity, the power grid is not operating in a secure situation. Then the dispatching department must take control actions to make sure that the system runs within an allowable range in terms of power flow constraints, voltage constraints, generator output constraints, and transmission line capacity constraints. If all the constraints are satisfied, then the system is considered to be in an acceptable state. Obviously, if the system still has converging optimal power flows under certain disasterous conditions, then it is possible to satisfy all the constraints by changing the system's generation pattern. Go to Step 7 in this case. If, on the other hand, the optimal power flow diverges, and the system cannot run in a secure and stable way, then go to Step 6.

Step 6 Emergent load shedding.

When the power system is under attack that threatens the system's safety, emergency load shedding is one possible choice to stabilize the system and avoid further worsening the situation. Emergency load shedding is one of the most important methods of unconventional control actions. It sheds some less important loads to guarantee the stability at the system level, which sacrifies local performance in exchange for the global stability. In current practice, a list of the ordering of loads to be shed is usually determined in advance in view of the system's operation modes and the importance of the loads. As a result, a dispatching operator can implement load shedding strategies according to this list in an emergent situation.

The emergency load shedding problem can be formulated as the following optimization problem

$$\begin{aligned} &\min\ F(u) \\ &\text{s.t.}\ \ G(u,x)=0 \\ &\qquad H(u,x)\leqslant 0 \end{aligned} \tag{14.1}$$

where $F(u)$ is the objective function that usually represents the system's load loss taking into consideration the concerns like how important a load is and whether it has backup power supply; $G(u,x)$ and $H(u,x)$ are constraints, including power flow constraints, the upper and lower limits for generation outputs, line transmission limits, bus voltage constraints, etc., to guarantee the system's stability after shedding the loads; x is the system state, such as voltage and reactive power; u is the decision (adjustable) variable such as the active power loads at load buses.

Step 7 Aggregate the load losses.

By inspecting the cutting of lines and loads, we list all the lost lines and loads at each bus, record all the equipments that have been out of service, and describe the cascading process and the scope affected by the failures.

Step 8 Broadcast early-warning signals.

In this step, the forecast and early-warning information and analysis results are delivered to the emergency management platform. Then usually red, orange, yellow, or blue warning signals are distributed according to the severity, from high to low, of the accident. The analysis results should include the severity of the accident, the possible blackout areas caused by the accident, the outage equipments and the candidate responding plans. Other disasters associated with or initiated by the blackouts are analyzed and evaluated by the dispatching system.

14.3.2 Simulation Analysis

In this section, we study the evolution process of the IEEE 30-bus system in both normal conditions and under simulated disasters using the disaster prediction and warning model based on AC power flows. We then provide disaster early-warning information and verify the effectiveness and correctness of this model. Figure 14.9 shows the two-stage evolution process of the IEEE 30-bus system under the attack of a hypothetical typhoon. We list the details below.

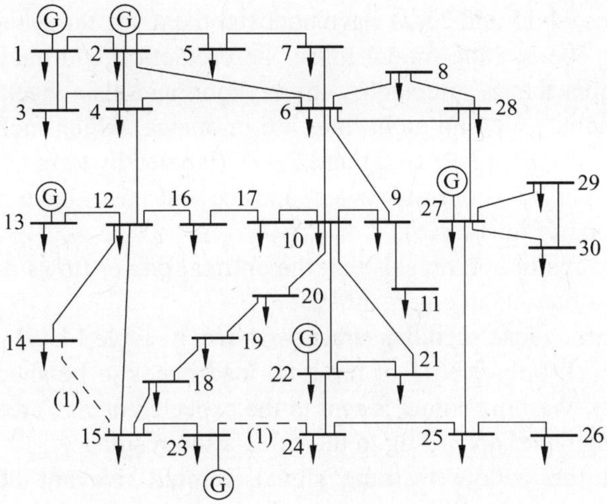

(a) Stage 1 of the catastrophe

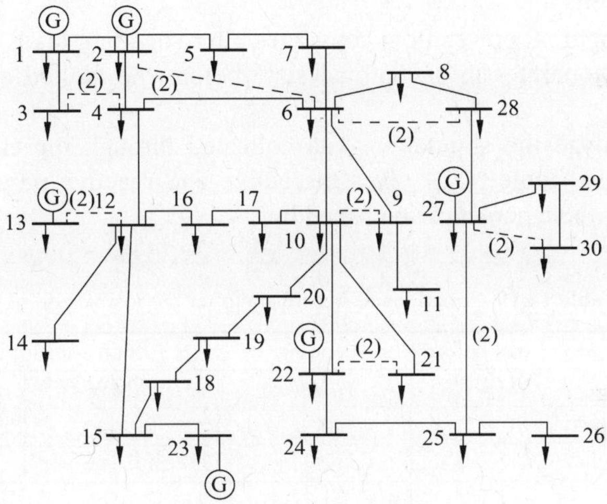

(b) Stage 2 of the catastrophe

Figure 14.9 Catastrophe process of the IEEE 30-bus system under the attack of a hypothetical typhoon

(1) At the initial time t_0, the system is in a normal state. At time t_1, a typhoon lands in the lower left corner of the system, and then the model we have discussed is activated to perform the prediction and analysis. As shown in Fig. 14.9(a), lines 14-15 and 23-24 (labeled by (1)) are in outage.

(2) The system safety assessment shows that the optimal power flow converges, which indicates that the system can still run in a stable state by adjusting its operation mode without shedding any load. At this time, no warning signal needs to be released and it only needs to inform the maintenance department to fix lines 14-15 and 23-24.

(3) At t_2, lines 14-15 and 23-24 have not been fixed yet, the typhoon has moved farther inwards. We use the model to do the prediction and analysis again. By looking at the area that is affected by the typhoon and also checking the related probability model[10], we find eight lines are in outage, which includes lines 3-4, 6-2, 6-28, 27-30, 10-12, 13-9, 25-21 and 22-27 (labeled by (2)); at the same time, the generator at bus 13 is out of service, and the rest buses have not formed any islands, as shown in Fig. 14.9(b).

(4) The safety evaluation reveals that the optimal power flows do not converge and the system is now in an emergent state.

(5) The emergency load shedding strategy (shown in Table 14.19) is implemented, according to which only 9.86% of the total load needs to be shed to ensure the system's stability. Warning signal is sent to the expected outage area so that power supply can be regulated according to the calculation results.

(6) Broadcast the yellow warning signal, remind relevant departments for attention, repair the faulted lines, and prepare resources and rescue workers for the predicted fault.

(7) The AC optimal power flow converges after shedding loads which ensures that the system operates in an optimal state before the faulted equipments are recovered.

Now we analyze the simulation data collected through the emergency load shedding module. Table 14.19 gives the active and reactive power load levels before and after the emergency load shedding.

Table 14.19 Load levels before and after the load shedding

Load bus	Load before load shedding		Load after load shedding	
	P/MW	*Q*/Mvar	*P*/MW	*Q*/Mvar
1	0	0	0	0
2	21.7	12.7	21.7	12.7
3	2.4	1.2	2.1027	1.0514
4	7.6	1.6	7.5977	1.5995
5	0	0	0	0
6	0	0	0	0
7	22.8	10.9	22.7982	10.8991
8	30	30	29.997	29.997
9	0	0	0	0
10	5.8	2	5.7977	1.9992
11	0	0	0	0
12	11.2	7.5	11.1957	7.4971
14	6.2	1.6	6.1968	1.5992
15	8.2	2.5	8.1999	2.5
16	3.5	1.8	3.4628	1.7809

(Continued)

Load bus	Load before load shedding		Load after load shedding	
	P/MW	*Q*/Mvar	*P*/MW	*Q*/Mvar
17	9	5.8	8.9986	5.7991
18	3.2	0.9	3.0866	0.8681
19	9.5	3.4	9.4988	3.3996
20	2.2	0.7	1.8838	0.5994
21	17.5	11.2	0	0
22	0	0	0	0
23	3.2	1.6	3.2	1.6
24	8.7	6.7	8.6979	6.6984
25	0	0	0	0
26	3.5	2.3	3.4356	2.2577
27	0	0	0	0
28	0	0	0	0
29	2.4	0.9	2.1051	0.7894
30	10.6	1.9	10.5983	1.8997

From Table 14.19 it can be seen that the total active power loads before the load shedding is 189.2 MW, and that after the load shedding is 170.55 MW, so 9.86% of the load has been shed to stabilize the system.

We further look at the generator outputs after the load shedding, and the analysis results are shown in Table 14.20.

Table 14.20 Generator outputs after the load shedding

Generator bus	Active power/MW	Reactive power/Mvar	Active power limit/MW		Reactive power limit/Mvar	
			P_{max}	P_{min}	Q_{max}	Q_{min}
1	42.3904	24.6153	80	0	150	−20
2	61.9028	14.6336	80	0	60	−20
22	26.0407	29.629	50	0	62.5	−15
23	16.7553	21.7281	30	0	40	−10
27	31.2472	15.6613	55	0	48.7	−15

Table 14.20, shows that all the generators are within their output limits after the load shedding actions. In other words, each generator has the potential to be further adjusted. Although the output of the generator at bus 2 is close to its upper limit, it is acceptable to keep it at its maximum output for a short period of time in emergency.

The magnitudes of the bus voltages after the emergency load shedding are listed in Table 14.21.

Table 14.21 Bus voltage after the load shedding

Bus number	Voltage magnitude /(p.u.)	voltage upper limit /(p.u.)	Voltage lower limit /(p.u.)
1	1.014	1.1	0.9
2	0.99	1.1	0.9
3	1.012	1.1	0.9
4	0.930	1.1	0.9
5	0.952	1.1	0.9
6	0.920	1.1	0.9
7	0.922	1.1	0.9
8	0.91	1.1	0.9
9	0.920	1.1	0.9
10	0.950	1.1	0.9
11	0.920	1.1	0.9
12	0.918	1.1	0.9
14	0.905	1.1	0.9
15	0.939	1.1	0.9
16	0.925	1.1	0.9
17	0.938	1.1	0.9
18	0.930	1.1	0.9
19	0.928	1.1	0.9
20	0.933	1.1	0.9
21	0.950	1.1	0.9
22	0.996	1.1	0.9
23	0.994	1.1	0.9
24	0.977	1.1	0.9
25	0.982	1.1	0.9
26	0.964	1.1	0.9
27	0.994	1.1	0.9
28	0.940	1.1	0.9
29	0.947	1.1	0.9
30	0.909	1.1	0.9

As shown in Table 14.21, the voltage at each bus is still strictly within the allowed range after the optimal load shedding, so the system is capable of being further adjusted.

In Table 14.22, we list the active powers in transmission lines after the load shedding. It can be seen from Table 14.22 that the transmitted power in each line is below the transmission limit. This implies that the emergency load shedding actions can not only maintain the system static stability, but also reserve ample room for further adjustment.

Table 14.22 Transmission power after the load shedding

Line number	Line capability/MW	Transmission power/MW
1-2	130	40.289
1-3	130	2.105
2-4	65	49.9079
2-5	130	30.1285
4-6	90	40.5642
5-7	70	29.5902
6-7	130	−6.2188
6-8	320	26.964
6-9	65	0
6-10	32	19.6083
9-11	65	0
12-14	32	6.2581
12-15	32	−6.9269
12-16	32	−10.5299
16-17	16	−14.1153
15-18	16	−2.3067
18-19	16	−5.43
19-20	32	−14.9628
10-20	32	17.2208
10-17	32	23.5021
10-21	32	−0.001
10-22	32	−26.9124
15-23	16	−12.966
22-24	16	−1.6226
24-25	16	−10.4966
25-26	16	3.4811
25-27	16	−14.2457
28-27	65	−3.2864
27-29	16	13.4855
29-30	16	10.9365
8-28	32	−3.1585

In short, the emergency load shedding strategies can both guarantee the reliable power supply in the working areas and prevent cascading failures. It is an important tool for emergency control, and is also an indispensable method for emergency power management. Its importance also lies in that the optimal emergency load shedding strategy (obtained by solving the optimal power flow) can guarantee the secure and stable operation of the system, and at the same time reduce the load loss to the minimum.

References

[1] Zhang M. Lessons drawn from black start implementation after "9·26" blackout in Hainan. East China Electric Power, 2005, 33(11): 13 – 16 (in Chinese)
[2] Hu Y. Analysis and countermeasures discussion for large area icing accident on power grid. High Voltage Engineering, 2008, 34(2): 215 – 219 (in Chinese)
[3] Anderson G, Donalek P, Farmer R. Causes of the 2003 major grid blackouts in north America and Europe, and recommended means to improve system dynamic performance. IEEE Transactions on Power Systems, 2005, 20(4): 1922 – 1928
[4] Union for the Coordination of Electricity Transmission (UCTE). UCTE releases detailed interim report on the disturbances of 4 Nov. http://www.ucte.org. Accessed Nov. 30, 2006
[5] Bialek J W. Blackouts in the US/Canada and continental Europe in 2003: Is liberalization to blame? 2005 IEEE Power Tech(Russia), June 27 – 30, 2005
[6] Tian S M, Chen X, Zhu C Y, et al. Study on electric power emergency management platform. Power System Technology, 2008, 32(1): 26 – 30 (in Chinese)
[7] Mei S W, He F, Zhang X M, et al. An improved OPA model and blackout risk assessment. IEEE Transactions on Power Systems, 2009, 24(2): 814 – 823
[8] Hu W, Mei S W, Shen C, et al. Northeast China power grid hybrid automatic voltage control. The 4th International Conference on Power Transmission and Distribution Technology, 2003: 1215 – 1220
[9] He G Y, Sun Y Y, Ruan Q T, et al. Modern power systems control centers: From EMS to AEMS. Science in China, Series E: 2009, 52(2) 413 – 419
[10] Tian C. Studies on the forecasting, early-warning and intelligent decision-making technology for emergency management of power system blackout [D]. Beijing: Tsinghua University, 2008 (in Chinese)

Index